SEMA SIMAI Springer Series

Volume 8

More information about this series at http://www.springer.com/series/10532

Francisco Ortegón Gallego •
María Victoria Redondo Neble •
José Rafael Rodríguez Galván
Editors

Trends in Differential Equations and Applications

Springer

Editors
Francisco Ortegón Gallego
Departamento de Matemáticas
Universidad de Cádiz
Puerto Real, Spain

María Victoria Redondo Neble
Departamento de Matemáticas
Universidad de Cádiz
Puerto Real, Spain

José Rafael Rodríguez Galván
Departamento de Matemáticas
Universidad de Cádiz
Puerto Real, Spain

ISSN 2199-3041 ISSN 2199-305X (electronic)
SEMA SIMAI Springer Series
ISBN 978-3-319-32012-0 ISBN 978-3-319-32013-7 (eBook)
DOI 10.1007/978-3-319-32013-7

Library of Congress Control Number: 2016941343

Printed on acid-free paper

This Springer imprint is published by Springer Nature
The registered company is Springer International Publishing AG Switzerland

Preface

During recent years applied mathematics techniques have attained considerable dissemination within the experimental sciences and engineering. Special attention has been devoted to biomathematics and medicine, including the analysis of mathematical models for the description of tumors, blood flux in arteries, the heart and flow patterns inside an aneurysm dome. A significant element of this dissemination also derives from the applications of mathematics in industry. International meetings such as those of the European Consortium for Mathematics in Industry (ECMI) and the International Council for Industrial and Applied Mathematics (ICIAM) bear witness to these advances. In Spain, various research groups have contributed to this development; most have been based in universities across the country, sometimes acting in collaboration with nonpublic laboratories. Links and coordination with foreign groups and universities have also proved essential. The significance of the Spanish contribution is reflected in the fact that the next ECMI meeting will take place in Santiago de Compostela in June 2016, while Valencia will host the next ICIAM congress in 2019.

The XXIVth Congress on Differential Equations and Applications/XIVth Congress on Applied Mathematics was held in Cádiz (a city founded more than three millennia ago), Spain, from 8 to 12 June 2015. This biennial international conference is the most important event organized by the Spanish Society of Applied Mathematics (SEMA). Any information on the conference is available on the Society website: http://www.sema.org.es/web/index.php. The conference brought together an excellent group of international and national researchers interested in the different branches of applied mathematics. Topics ranged from tsunami prediction to modeling of epidemiological processes and encompassed mathematics in architecture, high-order long-term integration of dynamical systems, the search for exact solutions of ordinary differential equations, oceanography, numerical acoustics, mathematics in industry, numerical linear algebra, and so on. This wide variety of subject matter reflects the multidisciplinary nature of the various research projects being carried out at present by both Spanish teams and groups in other countries

The collection of articles in this book represents a selection of the contributions presented at this conference in Cádiz. Every submitted paper has undergone a standard refereeing process. The volume provides a good summary of the recent activity of the various Spanish research groups interested in the applications of mathematics to different branches of the experimental sciences and engineering.

The publication has been made possible by the contributions of a number of people. First of all, we would like to thank the authors themselves for submitting their work. Special thanks are due to the referees who agreed to participate: their comments and suggestions have resulted in improvements in most of the included contributions. Finally, we would like to express our gratitude to Francesca Bonadei from Springer for the patience, attention and support that she has shown at every stage of the editorial process.

Puerto Real, Spain — Francisco Ortegón Gallego
Puerto Real, Spain — María Victoria Redondo Neble
Puerto Real, Spain — José Rafael Rodríguez Galván
February 2016

Contents

Approximate Osher-Solomon Schemes for Hyperbolic Systems

M.J. Castro, J.M. Gallardo, and A. Marquina

Abstract The Osher-Solomon scheme is a classical Riemann solver which enjoys a number of interesting features: it is nonlinear, complete, robust, entropy-satisfying, smooth, etc. However, its practical implementation is rather cumbersome, computationally expensive, and applicable only to certain systems (compressible Euler equations for ideal gases or shallow water equations, for example). In this work, a new class of approximate Osher-Solomon schemes for the numerical approximation of general conservative and nonconservative hyperbolic systems is proposed. They are based on viscosity matrices obtained by polynomial or rational approximations to the Jacobian of the flux evaluated at some average states, and only require a bound on the maximal characteristic speeds. These methods are easy to implement and applicable to general hyperbolic systems, while at the same time they maintain the good properties of the original Osher-Solomon solver. The numerical tests indicate that the schemes are robust, running stable and accurate with a satisfactory time step restriction, and the computational cost is very advantageous with respect to schemes using a complete spectral decomposition of the Jacobians.

1 Introduction

The Osher-Solomon scheme, introduced in [12], is a nonlinear and complete Riemann solver enjoying a number of interesting features: it is robust, entropy-satisfying, smooth, and has a good behavior with slowly-moving shocks. Its main drawback is that it requires the computation of a path-dependent integral in phase space, leading to a very complex and computationally expensive Riemann solver.

M.J. Castro • J.M. Gallardo (✉)
Departamento de Análisis Matemático, Universidad de Málaga, Campus de Teatinos, 29080 Málaga, Spain
e-mail: mjcastro@uma.es; jmgallardo@uma.es

A. Marquina
Departamento de Matemática Aplicada, Universidad de Valencia, Avda. Dr. Moliner 50, 46100 Burjassot-Valencia, Spain
e-mail: marquina@uv.us

F. Ortegón Gallego et al. (eds.), *Trends in Differential Equations and Applications*, SEMA SIMAI Springer Series 8, DOI 10.1007/978-3-319-32013-7_1

Due to this difficulties, its practical application has been restricted to certain systems, e.g., the compressible Euler equations [15].

In [7], Dumbser and Toro introduced a reformulated version of the Osher-Solomon solver, denoted as DOT (Dumbser-Osher-Toro), in which the integrals in phase space are numerically approximated by means of a Gauss-Legendre quadrature formula. This leads to a scheme much simpler than the original one and applicable to general hyperbolic systems. In particular, the viscosity matrix of the numerical flux is defined as a linear combination of the absolute value matrix of the physical flux evaluated at certain quadrature points. The computation of these absolute value matrices requires the knowledge of the complete eigenstructure of the system. Thus, the scheme may be computationally expensive for systems in which the eigenstructure is not known or difficult to compute.

In this work we propose an alternative version of the DOT solver, in which the absolute value matrices are approximated using appropriate functional evaluations of the Jacobian of the flux evaluated at the quadrature points. These schemes only require a bound on the maximum speed of propagation, thus avoiding the computation of the full eigenstructure of the system. Several families of approximations have been considered. The first one is based on Chebyshev polynomials, which provide optimal uniform approximations to the absolute value function. On the other hand, it is well-known that rational functions provide more precise approximations to $|x|$ than polynomial functions. For this reason, two different families of rational approximations have also been used, based on Newman [10] and Halley [4] functions. This families of functions have also been considered in the recently introduced RVM schemes (see [6]).

The proposed approximate Osher-Solomon schemes have been applied to a number of initial value Riemann problems for ideal magnetohydrodynamics, to observe their behavior with respect to some challenging scenarios in numerical simulations. The numerical tests indicate that our schemes are robust, stable and accurate with a satisfactory time step restriction. Comparisons with the DOT solver and some other well-known schemes in the literature (e.g., Roe and HLL) have also been performed.

2 Preliminaries

Consider a hyperbolic system of conservation laws

$$\partial_t w + \partial_x F(w) = 0, \tag{1}$$

where $w(x,t)$ takes values on an open convex set $\Omega \subset \mathbb{R}^N$ and $F\colon \Omega \to \mathbb{R}^N$ is a smooth flux function. We are interested in the numerical solution of the Cauchy problem for (1) by means of finite volume methods of the form

$$w_i^{n+1} = w_i^n - \frac{\Delta t}{\Delta x}(F_{i+1/2} - F_{i-1/2}), \tag{2}$$

where w_i^n denotes the approximation to the average of the exact solution at the cell $I_i = [x_{i-1/2}, x_{i+1/2}]$ at time $t^n = n\Delta t$ (the dependence on time will be dropped unless necessary). We assume that the numerical flux is given by

$$F_{i+1/2} = \frac{F_i + F_{i+1}}{2} - \frac{1}{2}Q_{i+1/2}(w_{i+1} - w_i), \tag{3}$$

where $F_i = F(w_i)$ and $Q_{i+1/2}$ denotes the numerical *viscosity matrix*, which determines the numerical diffusion of the scheme.

The condition of hyperbolicity of system (1) states that the Jacobian matrix of the flux at each state $w \in \Omega$,

$$A(w) = \frac{\partial F}{\partial w}(w),$$

can be diagonalized as $A = PDP^{-1}$, where $D = \text{diag}(\lambda_1, \dots, \lambda_N)$, λ_i being the eigenvalues of A, and the matrix P is composed by the associated right eigenvalues of A. As it is usual, we denote the positive and negative parts of A, respectively, as $A^+ = PD^+P^{-1}$ and $A^- = PD^-P^{-1}$, where $D^\pm = \text{diag}(\lambda_1^\pm, \dots, \lambda_N^\pm)$, with $\lambda_i^+ = \max(\lambda_i, 0)$ and $\lambda_i^- = \min(\lambda_i, 0)$. It is clear that $A = A^+ + A^-$. On the other hand, the absolute value of A is defined as $|A| = A^+ - A^-$.

It is interesting to note that the well-known Roe's method [13] can be written in the form (3) with viscosity matrix $Q_{i+1/2} = |A_{i+1/2}|$, where $A_{i+1/2}$ is a Roe matrix for the system. Several numerical methods have been developed by using approximations to $|A_{i+1/2}|$ as viscosity matrices. A general approach to build such kind of approximations by means of polynomial and rational functions has recently been introduced in [5, 6]. In particular, it has been shown that a number of well-known schemes in the literature can be viewed as particular cases within this general framework: Roe, Lax-Friedrichs, Rusanov, HLL, FORCE, and many others.

3 The Osher-Solomon Scheme

The Osher-Solomon scheme [12] is a nonlinear Riemann solver that possesses a number of interesting features: it is entropy-satisfying, robust, differentiable and good behaved for slowly-moving shocks. On the contrary, its implementation is rather cumbersome, computationally expensive, and only applicable to certain systems.

Let $A(w)$ be the Jacobian of F evaluated at w, and assume the flux splitting

$$F(w) = F^{+}(w) + F^{-}(w), \tag{4}$$

where

$$A^{\pm}(w) = \frac{\partial F^{\pm}}{\partial w}(w).$$

The *classical Osher-Solomon numerical flux* is then defined as

$$F_{i+1/2} = F^{+}(w_i) + F^{-}(w_{i+1}).$$

Let now Φ be a path in the phase-space Ω linking the states w_i and w_{i+1}, i.e., $\Phi\colon [0,1] \to \Omega$ is a Lipschitz continuous function such that $\Phi(0) = w_i$ and $\Phi(1) = w_{i+1}$. Then, we can write

$$F^{-}(w_{i+1}) - F^{-}(w_i) = \int_0^1 A^{-}(\Phi(s))\Phi'(s)ds,$$

from which we deduce

$$F_{i+1/2} = F_i + \int_0^1 A^{-}(\Phi(s))\Phi'(s)ds. \tag{5}$$

Similarly, we could also write

$$F_{i+1/2} = F_{i+1} - \int_0^1 A^{+}(\Phi(s))\Phi'(s)ds. \tag{6}$$

Combining (5) and (6), the Osher-Solomon flux can be written as

$$F_{i+1/2} = \frac{F_i + F_{i+1}}{2} - \frac{1}{2}\int_0^1 \left|A(\Phi(s))\right|\Phi'(s)ds. \tag{7}$$

The expression (7) for the numerical flux depends on the path Φ in phase-space, so in general it may be difficult to compute. Osher and Solomon [12] proposed a way to build, under certain assumptions, a path which makes possible to perform the integration. Unfortunately, the resulting solver is rather complex, computationally expensive, and only applicable to certain systems.

In [7] the authors propose a way to circumvent the drawbacks of the Osher-Solomon solver, maintaining at the same time its good features. First, the path consisting in segments is chosen:

$$\Phi(s) = w_i + s(w_{i+1} - w_i), \quad s \in [0,1].$$

Thus (7) can be written in the form (3), with viscosity matrix

$$Q_{i+1/2} = \int_0^1 \left|A(w_i + s(w_{i+1} - w_i))\right| ds.$$

To avoid the analytical integration, the integral is evaluated numerically using a Gauss-Legendre quadrature formula. The resulting numerical flux, denoted as DOT (Dumbser-Osher-Toro), has the form (3) with viscosity matrix given by

$$Q_{i+1/2} = \sum_{k=1}^{q} \omega_k \left|A(w_i + s_k(w_{i+1} - w_i))\right|, \tag{8}$$

where $s_k \in [0, 1]$ and ω_k are the weights of the quadrature formula. The resulting scheme is simple to implement and applicable to general hyperbolic systems. On the other hand, it needs the full eigenstructure of the system, which must be computed numerically when it is not known or difficult to calculate.

4 Approximate Osher-Solomon Schemes

With the aim of simplifying the computation of the DOT numerical viscosity matrix (8), it would be desiderable to approximate the intermediate matrices

$$\left|A(w_i + s_k(w_{i+1} - w_i))\right|, \quad k = 1, \ldots, q,$$

in a simple and efficient way. Two approaches will be considered in this section, one based on Chebyshev polynomials and another relying on rational approximations.

Let $P(x)$ be a polynomial approximation to the absolute value function $|x|$ in the interval $[-1, 1]$, satisfying the *stability condition* [5]

$$|x| \leq P(x) \leq 1, \quad \forall\, x \in [-1, 1]. \tag{9}$$

For a given matrix A, if $\lambda_{\max}$ is the eigenvalue of A with maximum absolute value (or an upper bound of it), $|A|$ can be approximated as

$$|A| \approx |\lambda_{\max}| P(|\lambda_{\max}|^{-1} A).$$

Denote

$$A_{i+1/2}^{(k)} = A(w_i + s_k(w_{i+1} - w_i)), \quad k = 1, \ldots, q,$$

where A is the Jacobian matrix of F, and let $\lambda_{i+1/2,\max}^{(k)}$ be the eigenvalue of $A_{i+1/2}^{(k)}$ with maximum absolute value. Then, the *polynomial approximate Osher-Solomon* flux is given by (3) with viscosity matrix

$$Q_{i+1/2} = \sum_{k=1}^{q} \omega_k \widetilde{P}^{(k)}_{i+1/2}, \tag{10}$$

where

$$\widetilde{P}^{(k)}_{i+1/2} = |\lambda^{(k)}_{i+1/2,\max}| P\left(|\lambda^{(k)}_{i+1/2,\max}|^{-1} A^{(k)}_{i+1/2} \right). \tag{11}$$

Remark 1 The advantage of formula (10) with respect to (8) is that in the latter it is necessary to compute the full eigenstructure of the system, while in the former only an upper bound on the spectral radius is needed.

Notice that the closer the polynomial $P(x)$ is to $|x|$ in the uniform norm, the more similar the approximate flux (10) will be to the Osher-Solomon flux (8). This suggests to use accurate polynomial approximations to $|x|$ for building (10). In particular, Chebyshev approximations will be considered in the numerical experiments. Specifically, for a given $p \geq 1$ we take $P(x) = \tau_{2p}(x)$, where

$$\tau_{2p}(x) = \frac{2}{\pi} + \frac{4}{\pi} \sum_{j=1}^{p} \frac{(-1)^{j+1}}{(2j-1)(2j+1)} T_{2j}(x), \quad x \in [-1, 1],$$

$T_{2j}(x)$ being the Chebyshev polynomials. As it is well-known, the order of approximation of $\tau_{2p}(x)$ to $|x|$ is optimal in the $L^\infty(-1, 1)$ norm. Moreover, the recursive definition of the polynomials $T_{2k}(x)$ provides an explicit and efficient way to compute $\tau_{2p}(x)$.

As it is well-known, the order of approximation to $|x|$ can be greatly improved by using rational functions instead of polynomials. This suggests to consider *rational approximate Osher-Solomon* fluxes of the form (3) with viscosity matrix

$$Q_{i+1/2} = \sum_{k=1}^{q} \omega_k \widetilde{R}^{(k)}_{i+1/2}, \tag{12}$$

where $\widetilde{R}^{(k)}_{i+1/2}$ is defined as in (11), but taking as basis function a rational approximation $R(x)$ to $|x|$ satisfying the stability condition (9). Following [6], two different families of rational functions will be considered:

- Given a set of $r \geq 4$ distinct points $X = \{0 < x_1 < \cdots < x_r \leq 1\}$, construct the polynomial

$$p(x) = \prod_{k=1}^{r} (x + x_k).$$

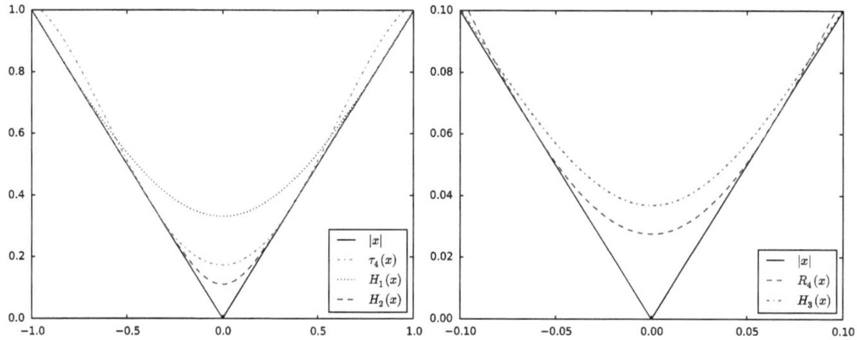

Fig. 1 *Left*: Chebyshev $\tau_4(x)$ and Halley $H_1(x)$ and $H_2(x)$ functions. *Right*: Newman $R_4(x)$ and Halley $H_3(x)$ functions. Notice the different scaling in both figures

The *Newman rational function* [10] associated to X is defined as

$$R_r(x) = x\,\frac{p(x) - p(-x)}{p(x) + p(-x)}.$$

The rate of approximation of $R_r(x)$ to $|x|$ depends on the choice of nodes X: several possibilities can be found in the literature. Here, we will take $x_k = \exp(-kr^{-1/2})$, which provides an exponential rate of approximation [10].

- The *Halley rational functions* $H_r(x)$ are recursively defined as [6]

$$H_{r+1}(x) = H_r(x)\,\frac{H_r(x)^2 + 3x^2}{3H_r(x)^2 + x^2}, \qquad H_0(x) = 1.$$

It can be proved that $\|H_r(x) - |x|\|_\infty = 3^{-r}$.

Figure 1 shows a comparison between the Chebyshev $\tau_4(x)$, Newman $R_4(x)$, and Halley $H_r(x)$ ($r = 1, 2, 3$) functions.

Both the Chebyshev polynomials $\tau_{2p}(x)$ and the Newman functions $R_r(x)$ do not satisfy the stability condition (9) strictly, although this can be easily fixed with a slight modification: see [6] for details. However, in practical computations there are no appreciable differences between both approaches. On the other hand, Halley functions $H_r(x)$ satisfy (9) by construction. As long as the functions considered do not cross the origin, no entropy-fix is needed in the presence of sonic points.

5 Application to Ideal Magnetohydrodynamics

In this section we apply the approximate Osher-Solomon schemes introduced previously to solve some challenging problems related to the ideal magnetohydrodynamics equations.

The ideal magnetohydrodynamics (MHD) equations read as

$$\begin{cases} \partial_t \rho = -\nabla \cdot (\rho \mathbf{v}), \\ \partial_t(\rho \mathbf{v}) = -\nabla \cdot \left(\rho \mathbf{v}\mathbf{v}^T + \left(P + \frac{1}{2}\mathbf{B}^2\right)I - \mathbf{B}\mathbf{B}^T\right), \\ \partial_t \mathbf{B} = \nabla \times (\mathbf{v} \times \mathbf{B}), \\ \partial_t E = -\nabla \cdot \left(\left(\frac{\gamma}{\gamma-1}P + \frac{1}{2}\rho q^2\right)\mathbf{v} - (\mathbf{v} \times \mathbf{B}) \times \mathbf{B}\right), \end{cases} \tag{13}$$

where ρ is the mass density, $\mathbf{v}$ and $\mathbf{B}$ are the velocity and magnetic fields, and E is the total energy. If q and B denote the magnitudes of the velocity and magnetic fields, the total energy can be expressed as

$$E = \frac{1}{2}\rho q^2 + \frac{1}{2}B^2 + \rho\varepsilon,$$

where the specific internal energy ε is related to the hydrostatic pressure P through the equation of state $P = (\gamma - 1)\rho\varepsilon$, γ being the adiabatic constant. The total pressure P^* is defined as $P + P_M$, where $P_M = \frac{1}{2}B^2$ is the magnetic pressure. In addition to the equations, the magnetic field satisfies the divergence-free condition

$$\nabla \cdot \mathbf{B} = 0.$$

Notice that if $\mathbf{B} = \mathbf{0}$ then the MHD system reduces to the Euler equations for ideal gases. Let us remark that the spectral structure of (13) has been widely analyzed in the literature (see, e.g., [3, 14]).

The ideal MHD equations (13) constitute a system of conservation laws. In the numerical experiments we will focus in the two-dimensional case. Then, (13) can be written as

$$\partial_t w + \partial_x F(w) + \partial_y G(w) = 0,$$

where $w = (\rho, \rho v_x, \rho v_y, \rho v_z, B_x, B_y, B_z, E)^t$,

$$F(w) = \begin{pmatrix} \rho v_x \\ \rho v_x^2 + P^* - B_x^2 \\ \rho v_x v_y - B_x B_y \\ \rho v_x v_z - B_x B_z \\ 0 \\ v_x B_y - v_y B_x \\ v_x B_z - v_z B_x \\ v_x(E + P^*) - B_x(v_x B_x + v_y B_y + v_z B_z) \end{pmatrix},$$

and

$$G(w) = \begin{pmatrix} \rho v_y \\ \rho v_x v_y - B_x B_y \\ \rho v_y^2 + P^* - B_y^2 \\ \rho v_y v_z - B_y B_z \\ v_y B_x - v_x B_y \\ 0 \\ v_y B_z - v_z B_y \\ v_y(E + P^*) - B_y(v_x B_x + v_y B_y + v_z B_z) \end{pmatrix}.$$

Let us define $(b_x, b_y, b_z) = (B_x, B_y, B_z)/\sqrt{\rho}$, $b^2 = b_x^2 + b_y^2 + b_z^2$, and the acoustic sound speed $a = \sqrt{\gamma P/\rho}$. The Alfven, fast and slow waves in the x-direction are, respectively,

$$c_a = |b_x|, \quad c_{f,s}^2 = \frac{1}{2}\left(a^2 + b^2 \pm \sqrt{(a^2 + b^2)^2 - 4a^2 b_x^2}\right)$$

(and similarly for the y-direction). The eight characteristic velocities are given by

$$\lambda_1 = v_x - c_f, \quad \lambda_2 = v_x - c_a, \quad \lambda_3 = v_x - c_s, \quad \lambda_4 = v_x,$$
$$\lambda_5 = v_x, \quad \lambda_6 = v_x + c_s, \quad \lambda_7 = v_x + c_a, \quad \lambda_8 = v_x + c_f.$$

The characteristic fields associated to $\lambda_{1,8}$, $\lambda_{3,6}$, $\lambda_{2,7}$ and $\lambda_{4,5}$ are called, respectively, the fast, slow, Alfven and entropy waves. Once the fast velocities are known, it is easy to compute $\lambda_{\max}$.

To ensure the stability and accuracy of the numerical schemes it is essential to enforce the divergence-free constraint on the magnetic field. This is done here using the technique proposed in [2], where a correction is applied at the end of every time step. Specifically, the magnetic field $\mathbf{B}$ is modified as $\mathbf{B}^c = \mathbf{B} + \nabla\phi$, where ϕ is a solution of the Poisson problem $\Delta\phi + \nabla \cdot \mathbf{B} = 0$, which is computed with a finite difference method.

For high-order schemes, the WENO-type compact third-order reconstruction operator introduced in [8] has been used. The numerical experiments have been performed using structured meshes, although they can be designed on general nonuniform quadrilateral meshes following the guidelines in [8] and the references therein.

5.1 *Smooth Isentropic Vortex*

The purpose of this test is to analyze the convergence and stability of the proposed numerical schemes. Specifically, the smooth two-dimensional convected isentropic vortex for the Euler equations proposed in [9] has been considered. The initial condition consists in a linear perturbation of an homogeneous state, of the form

$$(\rho, v_x, v_y, P) = (1 + \delta\rho, 1 + \delta v_x, 1 + \delta v_y, 1 + \delta P).$$

Denoting $r^2 = (x-5)^2 + (y-5)^2$, the perturbations of velocity, density and pressure are given by

$$\begin{pmatrix} \delta v_x \\ \delta v_y \end{pmatrix} = \frac{\varepsilon}{2\pi} e^{\frac{1-r^2}{2}} \begin{pmatrix} 5-y \\ x-5 \end{pmatrix}, \quad \delta\rho = (1+\delta T)^{\frac{1}{\gamma-1}} - 1, \quad \delta P = (1+\delta T)^{\frac{\gamma}{\gamma-1}} - 1,$$

being

$$\delta T = -\frac{(\gamma-1)\varepsilon^2}{8\gamma\pi^2} e^{1-r^2}$$

the temperature perturbation. The values $\varepsilon = 5$ and $\gamma = 1.4$ have been used.

The problem has been solved in the computational domain $[0, 10] \times [0, 10]$ with periodic boundary conditions and CFL=0.8. In Table 1 are shown the results obtained after one time period at $t = 10$ with the third-order OS-Cheb-4, OS-Newman-4, OS-Halley-2 and DOT schemes. As it can be seen, all the proposed schemes give similar results as the DOT method. We remark again that the advantage of our schemes is that the eigenstructure of the system need not to be known.

Table 1 Isentropic vortex

	OS-Cheb-4		OS-Newman-4	
N	L^1 error	L^1 order	L^1 error	L^1 order
16	1.47E+00	–	1.47E+00	–
32	7.77E–01	0.92	7.95E–01	0.89
64	1.98E–01	1.97	2.03E–01	1.97
128	1.37E–02	3.85	1.39E–02	3.87

	OS-Halley-2		DOT	
N	L^1 error	L^1 order	L^1 error	L^1 order
16	1.46E+00	–	1.45E+00	–
32	7.81E–01	0.90	7.95E–01	0.87
64	1.95E–01	2.00	1.96E–01	2.02
128	1.33E–02	3.87	1.33E–02	3.88

Third order results for the density component ρ at time $t = 10$

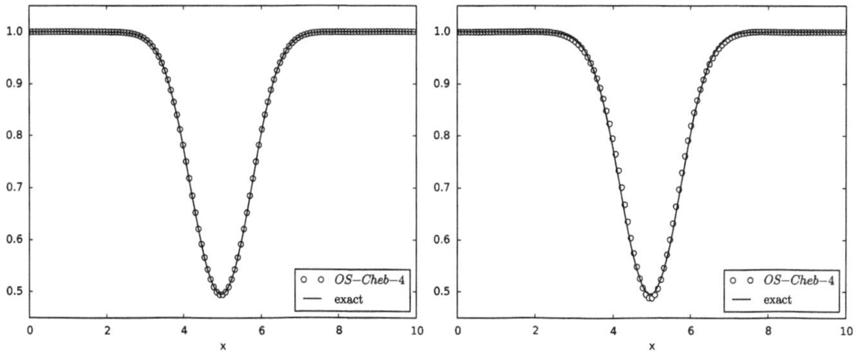

Fig. 2 Isentropic vortex. Density cut in the x-direction, computed with the third-order OS-Cheb-4 scheme. *Left*: time $t = 10$. *Right*: time $t = 100$

Furthermore, the solution has been calculated at time $t = 100$, after ten time periods. Figure 2 shows a cut through the center of the vortex in the x-direction for the density variable. The solution has been computed with the third-order OS-Cheb-4 method using 128 cells, although any of the other schemes gives a similar result. As it can be observed, the dissipation is very small in this case.

5.2 *Orszag-Tang Vortex*

The Orszag-Tang vortex system [11] has been widely analyzed in the literature, as it provides a model of complex flow containing many significant features of MHD turbulence. Starting from a smooth state, the system develops complex interactions between different shock waves generated as the system evolves in the transition to turbulence.

The initial data proposed in [17] has been considered. For $(x, y) \in [0, 2\pi] \times [0, 2\pi]$, we take

$$\rho(x, y, 0) = \gamma^2, \quad v_x(x, y, 0) = -\sin(y), \quad v_y(x, y, 0) = \sin(x),$$
$$B_x(x, y, 0) = -\sin(y), \quad B_y(x, y, 0) = \sin(2x), \quad P(x, y, 0) = \gamma,$$

with $\gamma = 5/3$. Periodic boundary conditions are imposed in the x- and y-directions. The computations have been done using a 192×192 uniform mesh and CFL=0.8.

Figure 3 shows the results obtained with the third-order OS-Cheb-4 scheme at times $t = 0.5$, $t = 2$ and $t = 3$, for the density and pressure components (analogous solutions are obtained with the third-order OS-Newman-4, OS-Halley-2,

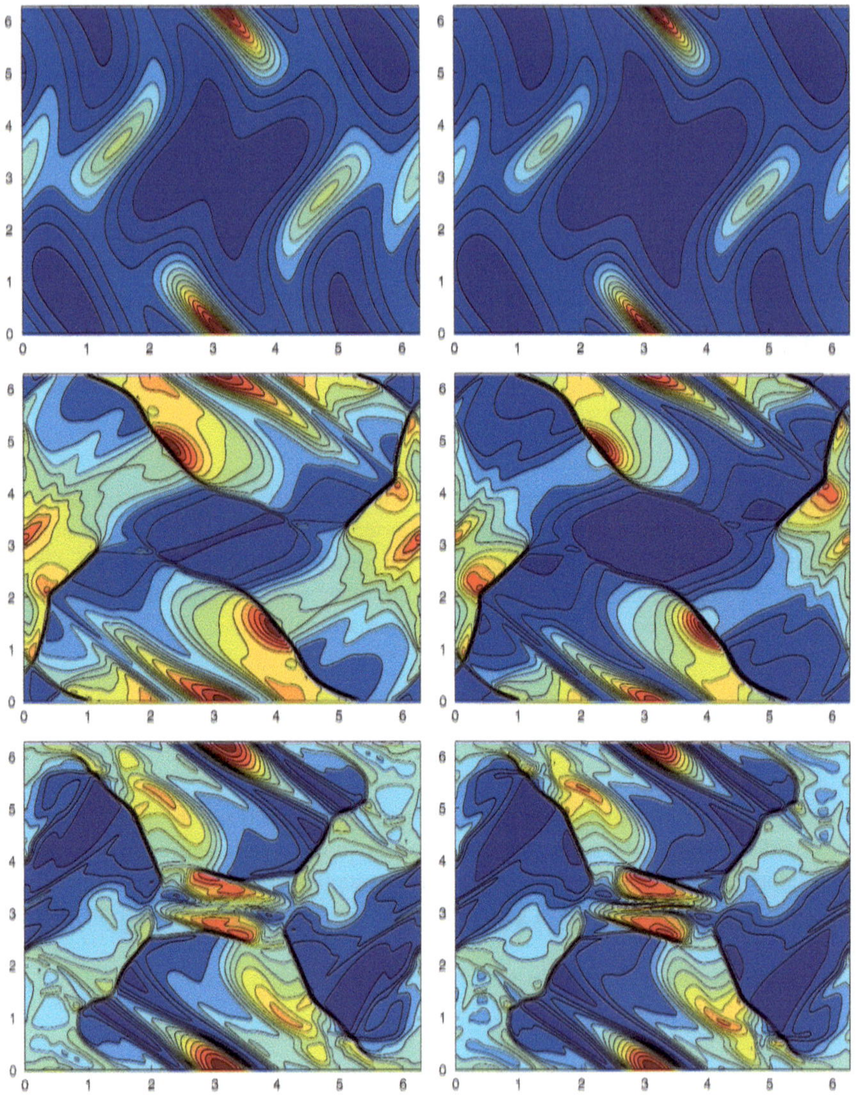

Fig. 3 Evolution of the Orszag-Tang vortex. Density (*left*) and pressure (*right*) computed at times (*top to bottom*) $t = 0.5$, $t = 2$ and $t = 3$. Results obtained with the third-order OS-Cheb-4 scheme

and DOT schemes). The results are in very good agreement with those found in the literature, which shows that our schemes are robust and accurate enough to resolve the complicated structure of this vortex system. Finally, Table 2 shows the relative CPU times with respect to the first-order DOT scheme.

Table 2 Orszag-Tang vortex

Method	CPU (first order)	CPU (third order)
DOT	1.00	5.82
OS-Cheb-4	0.16	1.04
OS-Newman-4	0.38	2.32
OS-Halley-2	0.50	2.79

Relative CPU times with respect to the first-order OS solver. Final time: $t = 0.2$

5.3 *The Rotor Problem*

In this section we consider the rotor problem proposed in [1]; see also [16]. Initially, there is a dense rotating disk at the center of the domain, while the ambient fluid remains at rest. These two areas are connected by means of a taper function, which helps to reduce the initial transient. Since the centrifugal forces are not balanced, the rotor is not in equilibrium. The rotating dense fluid will be confined into an oblate shape, due to the action of the magnetic field.

The computational domain is $[0, 1] \times [0, 1]$ with periodic boundary conditions. Define $r_0 = 0.1, r_1 = 0.115, f = (r_1 - r)/(r_1 - r_0)$ and $r = [(x-0.5)^2 + (y-0.5)^2]^{1/2}$; then, the initial conditions are given by

$$(\rho(x,y), v_x(x,y), v_y(x,y)) = \begin{cases} (10, -(y-0.5)/r_0, (x-0.5)/r_0) & \text{if } r < r_0, \\ (1+9f, -(y-0.5)f/r, (x-0.5)f/r) & \text{if } r_0 < r < r_1, \\ (1,0,0) & \text{if } r > r_1, \end{cases}$$

with $B_x = 2.5/\sqrt{4\pi}$, $B_y = 0$ and $P = 0.5$. We take $\gamma = 5/3$.

Figure 4 shows the solutions obtained with the third order OS-Cheb-4 scheme at time $t = 0.295$ on a 200×200 mesh with CFL= 0.8. The results are in good agreement with those in [1, 16]. As in the previous tests, OS-Newman-4 and OS-Halley-2 give similar results as OS-Cheb-4. On the contrary, the DOT scheme fails for this problem around time $t \approx 0.187$. Finally, Fig. 5 shows a comparison between the third-order OS-Cheb-4 and HLL methods. As it can be seen, HLL produces less precise results than OS-Cheb-4. This shows that the choice of the first order solver is important even when it is intended to be used as building block for high-order schemes.

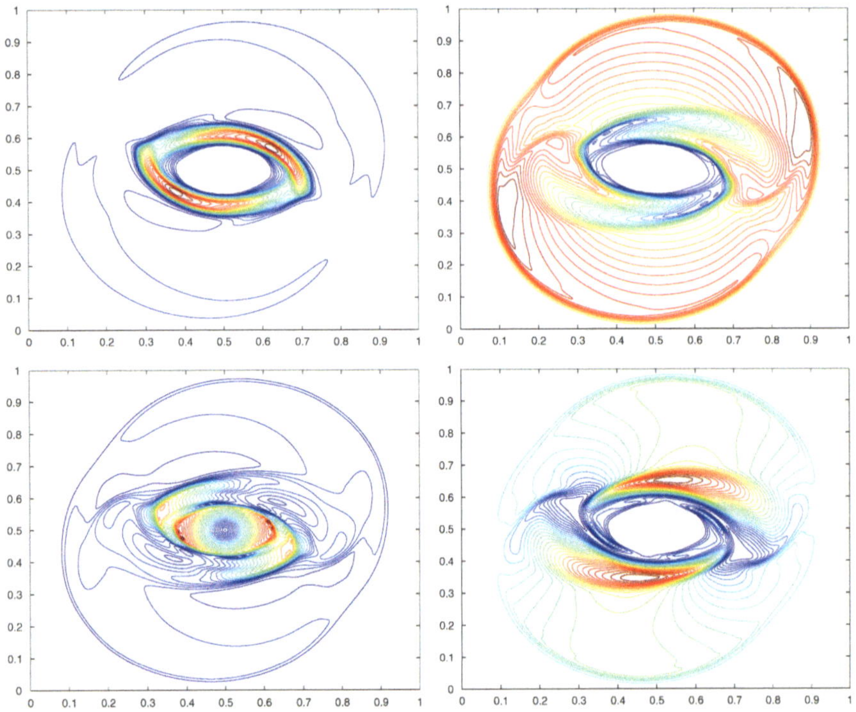

Fig. 4 Rotor problem. Density ρ (*top left*), pressure P (*top right*), Mach number $|\mathbf{v}|/a$ (*bottom left*) and magnetic pressure $|\mathbf{B}|^2/2$ (*bottom right*) computed at time $t = 0.295$. Results obtained with the third-order OS-Cheb-4 scheme with 200×200 cells

6 Conclusions

We have proposed a new kind of Riemann solvers for hyperbolic systems, which are based on a simplified version of the classical Osher-Solomon scheme. The Osher-Solomon solver relies on the evaluation of the integral of the absolute value matrix of the flux Jacobian through a path linking states in phase space. This integral can be approximated by an appropriate quadrature formula, as it is done in the DOT solver introduced in [7]. To avoid the evaluation of the absolute value matrices at the quadrature points, which require the computation of the full eigenstructure of the system, we have proposed several ways to approximate them accurately and efficiently. In particular, Chebyshev polynomial approximations and two kinds of rational approximations, based on Newman and Halley functions, have been considered. Rational functions provide much more precise approximations than polynomials; thus, for problems in which the structure of the solution is complex, rational-based methods are, in general, a more efficient choice than polynomial-based methods.

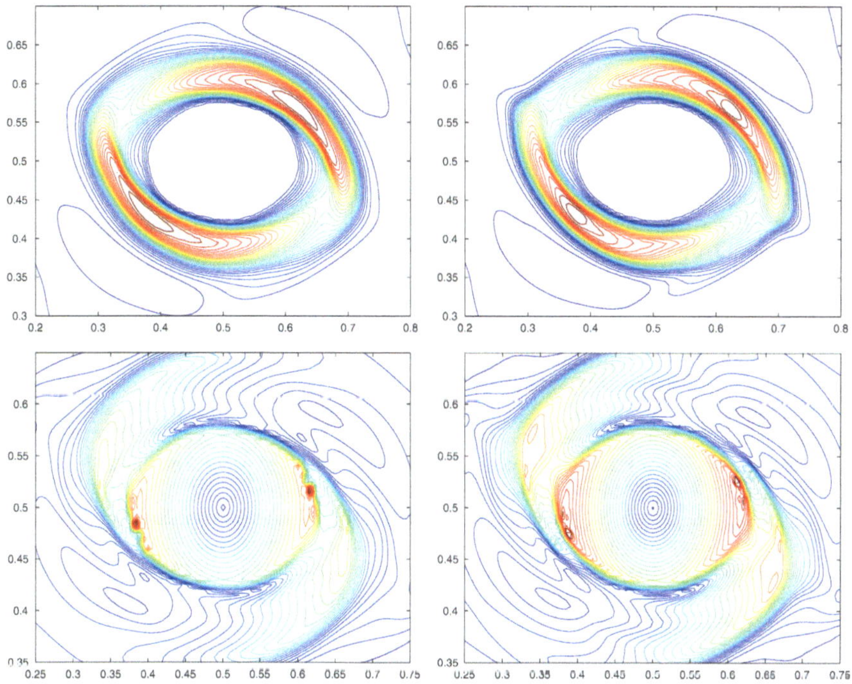

Fig. 5 Test 5.3: Comparison between the solutions obtained with the third-order HLL (*left*) and third-order OS-Cheb-4 (*right*) schemes. *Top*: density. *Bottom*: Mach number

To build the associated approximate Osher-Solomon schemes, only a bound on the spectral radius of the Jacobian is needed. The proposed schemes have been compared with the DOT, Roe, and HLL schemes. An additional feature of our schemes is that no entropy-fix is needed.

Different initial value Riemann problems for ideal magnetohydrodynamics in the two-dimensional case have been considered to test the performance of the approximate Osher-Solomon schemes. The numerical tests indicate that our schemes are robust, stable and accurate with a satisfactory time step restriction. Approximate Osher-Solomon schemes thus provide an efficient alternative when approximating time-dependent solutions in which the spectral decomposition is complex or computationally expensive.

Acknowledgements This research has been partially supported by the Spanish Government Research projects MTM2012-38383 and MTM2011-28043. The numerical computations have been performed at the Laboratory of Numerical Methods of the University of Málaga.

References

1. Balsara, D.S., Spicer, D.S.: A staggered mesh algorithm using high order Godunov fluxes to ensure solenoidal magnetic fields in magnetohydrodynamic simulations. J. Comput. Phys. **149**, 270–292 (1999)
2. Brackbill, J.U., Barnes, J.C.: The effect of nonzero $\nabla \cdot B$ on the numerical solution of the magnetohydrodynamic equations. J. Comput. Phys. **35**, 426–430 (1980)
3. Brio, M., Wu, C.C.: An upwind differencing scheme for the equations of ideal magnetohydrodynamics. J. Comput. Phys. **75**, 400–422 (1988)
4. Candela, V., Marquina, A.: Recurrence relations for rational cubic methods I: the Halley method. Computing **44**, 169–184 (1990)
5. Castro Díaz, M.J., Fernández-Nieto, E.D.: A class of computationally fast first order finite volume solvers: PVM methods. SIAM J. Sci. Comput. **34**, A2173–A2196 (2012)
6. Castro, M.J., Gallardo, J.M., Marquina, A.: A class of incomplete Riemann solvers based on uniform rational approximations to the absolute value function. J. Sci. Comput. **60**, 363–389 (2014)
7. Dumbser, M., Toro, E.F.: On universal Osher-type schemes for general nonlinear hyperbolic conservation laws. Commun. Comput. Phys. **10** 635–671 (2011)
8. Gallardo, J.M., Ortega, S., Asunción, M., Mantas, J.M.: Two-dimensional compact third-order polynomial reconstructions. Solving nonconservative hyperbolic systems using GPUs. J. Sci. Comput. **48**, 141–163 (2011)
9. Hu, C., Shu, C.-W.: Weighted essentially non-oscillatory schemes on triangular meshes. J. Comput. Phys. **150**, 97–127 (1999)
10. Newman, D.J.: Rational approximation to $|x|$. Mich. Math. J. **11**, 11–14 (1964)
11. Orszag, S.A., Tang, C.M.: Small scale structure of two-dimensional magnetohydrodynamic turbulence. J. Fluid Mech. **90**, 129–143 (1979)
12. Osher, S., Solomon, F.: Upwind difference schemes for hyperbolic conservation laws. Math. Comput. **38**, 339–374 (1982)
13. Roe, P.L.: Approximate Riemann solvers, parameter vectors, and difference schemes. J. Comput. Phys. **43**, 357–372 (1981)
14. Serna, S.: A characteristic-based nonconvex entropy-fix upwind scheme for the ideal magnetohydrodynamics equations. J. Comput. Phys. **228**, 4232–4247 (2009)
15. Toro, E.F.: Riemann Solvers and Numerical Methods for Fluid Dynamics, 3rd edn. Springer, Berlin (2009)
16. Tóth, G.: The $\nabla \cdot \mathbf{B} = 0$ constraint in shock-capturing magnetohydrodynamics codes. J. Comput. Phys. **161**, 605–652 (2000)
17. Zachary, A.L., Malagoli, A., Colella, P.: A higher-order Godunov method for multidimensional magnetohydrodynamics. SIAM J. Sci. Comput. **15**, 263–284 (1994)

Spectral Shape Analysis of the Hippocampal Structure for Alzheimer's Disease Diagnosis

G. Maicas, A.I. Muñoz, G. Galiano, A. Ben Hamza, and E. Schiavi, for the Alzheimer's Disease Neuroimaging Initiative

Abstract We present an automatic pipeline for spectral shape analysis of brain subcortical hippocampal structures with the aim to improve the Alzheimer's Disease (AD) detection rate for early diagnosis. The hippocampus is previously segmented from volumetric T1-weighted Magnetic Resonance Images (MRI) and then it is modelled as a triangle mesh (Fang and Boas, Proceedings of IEEE international symposium on biomedical imaging, pp 1142–1145, 2009) on which the spectrum of the Laplace-Beltrami (LB) operator is computed via a finite element method (Lai, Computational differential geometry and intrinsic surface processing. Doctoral dissertation. University of California, 2010). A fixed number of eigenpairs is used to compute, following (Li and Ben Hamza, Multimed Syst 20(3):253–281, 2014), three different shape descriptors at each vertex of the mesh, which are the heat kernel signature (HKS), the scale-invariant heat kernel signature (SIHKS) and the wave kernel signature (WKS). Each of these descriptors is used separately in a Bag-of-Features (BoF) framework. In this preliminary study we report on the implementation of the proposed descriptors using ADNI (adni.loni.usc.edu), and DEMCAM (T1-weighted MR images acquired on a GE Healthcare Signa HDX 3T scanner) datasets. We show that the best quality of the DEMCAM dataset images

Data used in preparation of this article were obtained from the Alzheimer's Disease Neuroimaging Initiative (ADNI) database (adni.loni.usc.edu). As such, the investigators within the ADNI contributed to the design and implementation of ADNI and/or provided data but did not participate in analysis or writing of this report. A complete listing of ADNI investigators can be found at: http://adni.loni.usc.edu/wp-content/uploads/how_to_apply/ADNI_Acknowledgement_List.pdf

G. Maicas • A.I. Muñoz (✉) • E. Schiavi
Departamento de Matemática Aplicada, Ciencia e Ingeniería de Materiales y Tecnología Electrónica, Universidad Rey Juan Carlos, ESCET, Móstoles, 28933 Madrid, Spain
e-mail: g.maicas@alumnos.urjc.es; anaisabel.munoz@urjc.es; emanuele.schiavi@urjc.es

G. Galiano
Departamento de Matematicas, Universidad de Oviedo, Oviedo 33007, Spain
e-mail: galiano@uniovi.es

A. Ben Hamza
Concordia Institute for Information Systems Engineering, Concordia University, Montreal, Canada, QC
e-mail: hamza@ciise.concordia.ca

F. Ortegón Gallego et al. (eds.), *Trends in Differential Equations and Applications*, SEMA SIMAI Springer Series 8, DOI 10.1007/978-3-319-32013-7_2

have a great impact on the AD rate of detection which can reach up to 95 %. For further development of the modelling approach, local deformation analysis is also considered through a spectral segmentation of the hippocampal structure.

1 Introduction

Alzheimer's disease (AD) is the most common form of cognitive disability in older people, and the number of affected patients is expected to considerably increase in the next future due to the population longer living. Early diagnosis of AD would greatly benefit the public health and society, resulting in patient quality of life and reduced treatment costs.

The development of magnetic resonance images (MRI) has given rise to a deeper study of the architecture of the human body. More precisely, diagnosis of Alzheimer's disease has benefited from this fact due to the possibility of studying the structure of the different components of the brain which show anatomical changes as the disease advances (see for example [18]).

The hippocampus, which is located in the medial temporal lobe of the brain, and is important for memory and spatial navigation, has been shown as one of the main components of the brain that changes in the progression of AD [1]. Its atrophy due to neurodegenerative diseases such as AD can be evaluated in terms of the global change in the volume of the hippocampus as well as through the quantification of the global and local changes in its shape. Hippocampal volumetry on MR images has been shown to be a useful tool in AD diagnosis, providing significant discrimination ability. It is, however, inadequate to fully describe the effect of the disease on the morphology of hippocampus. In addition to volumetry, hippocampal shape analysis is an emerging field enlarging the understanding of the development of the disease. Among the different methods employed to model the hippocampus and to detect the shape changes (deformation) caused by AD, shape surface processing represented by spherical harmonics [8] and statistical shape models (SSMs) have been proved to be efficient in modeling the variability in the hippocampal shapes among the population [16].

In this work, we primarily focus on spectral techniques based on the Laplace-Beltrami operator. Such techniques have been successfully applied to shape recognition of subcortical structures [10]. In [19] a heat kernel based cortical thickness estimation algorithm, which is driven by the graph spectrum and the heat kernel theory, is used to capture grey matter geometry information from in vivo brain MR. These approaches allow to compute some shape spectral descriptors such as the heat kernel signature (HKS), the scale invariant heat kernel signature (SIHKS) and wave kernel signature (WKS), which we apply to the ADNI and DEMCAM datasets. In order to assist the diagnosis of Alzheimer we merge the spectral analysis into a Bag of Features (BoF) (see [13] for details) framework proposed in [11] for shape retrieval. The diagnosis (discrimination) is then effected in the space of descriptors through the comparison of their histograms. Finally we propose a novel method for anatomical structure segmentation based on the decreasing rearrangement of the second eigenfunction of the Laplace-Beltrami (LB) operator. As an application, we consider a partition of the hippocampus into three regions exploring if just one of them mostly encapsulate the early damages caused by this dementia.

The rest of this paper is organized as follows. In Sect. 2, we consider the heat equation on a closed surface, introducing the LB operator on compact manifolds. In order to expand the solution into eigenfunctions of the LB operator we define its discretization using FEM which leads to solve a generalized eigenvalue problem. In Sect. 3, the BoF approach for shape recognition is presented and the three different shape descriptors are introduced. Local analysis is performed in Sect. 4 through a spectral segmentation algorithm which exploits the properties of the decreasing rearrangement of a function. The experiments and results obtained are described in Sect. 5. Finally we summarize the conclusions of our study which is an on going research in the framework of Project TEC2012-39095-C03-02: Mathematical Models based on Biomarkers.

2 Spectral Analysis of the LB Operator

The heat diffusion process has recently been applied successfully to shape recognition [10, 11]. In this section, we present the heat equation and eigenvalue problem on a compact manifold representing the hippocampus surface. We discretize the heat equation in a triangular mesh, which is automatically generated (see [5]) in order to find the LB spectrum using FEM.

Assume $\mathscr{M} \subset \mathscr{R}^3$, where $\mathscr{R}$ denotes the set of real numbers, to be a compact connected Riemannian manifold. Then, the heat diffusion process in the manifold is described by the following equation

$$u_t = \Delta_{\mathscr{M}} u, \qquad \forall\, (\mathbf{x}, t) \in \mathscr{M} \times [0, \infty) \tag{1}$$

$$u(\mathbf{x}, 0) = f(\mathbf{x}), \qquad \forall\, \mathbf{x} \in \mathscr{M} \tag{2}$$

where the scalar field $u : \mathscr{M} \times [0, \infty) \rightarrow \mathscr{R}$ is the amount of heat at a point on the surface (hippocampus) at time t, and $\Delta_{\mathscr{M}}$ is the LB operator defined as follows:

$$\Delta_{\mathscr{M}} f = div_{\mathscr{M}}(\nabla_{\mathscr{M}} f) = \frac{1}{\sqrt{G}} \sum_{i=1}^{2} \frac{\partial}{\partial x^i} \left(\sqrt{G} \sum_{j=1}^{2} g^{ij} \frac{\partial f}{\partial x^j} \right),$$

where $G = \det(g_{ij})$ and (g^{ij}) is the inverse of the metric matrix. Considering $u(\mathbf{x}, 0) = \delta(\mathbf{x} - \mathbf{y})$, the solution $k(\mathbf{x}, \mathbf{y}, t)$ of the Eq. (1) is called the heat kernel (HK), which is a measure of the amount of heat that moves from $\mathbf{x}$ to $\mathbf{y}$ after time t.

The HK corresponding to the solution to problem (1)–(2) can be expressed as

$$k(\mathbf{x}, \mathbf{y}, t) = \sum_{i=1}^{\infty} e^{-\lambda_i t} \phi_i(\mathbf{x}) \phi_i(\mathbf{y}). \tag{3}$$

where (λ_i, ϕ_i) are the eigenpairs (spectrum) of the LB operator. Notice that being the manifold closed, we find that $\lambda_1 = 0$ and the first eigenfunction ϕ_1 is constant. The rest of the eigenvalues satisfy $0 < \lambda_2 < \lambda_3 < \ldots$, being this sequence diverging.

In order to expand the solution in terms of the eigenpairs, we need to solve first the following eigenvalue problem:

$$\Delta_{\mathcal{M}} \phi_n = -\lambda_n \phi_n, \qquad n = 1, 2, \cdots. \tag{4}$$

Instead of solving the previous eigenvalue problem, we use a finite element method (FEM) to find numerically an approximate solution in a triangular mesh [10].

Hence, we consider the following weak formulation of the problem: Find $\phi \in H^1(\mathcal{M})$, such that for any test function $u \in H^1(\mathcal{M})$, it is satisfied

$$\int_{\mathcal{M}} (\Delta_{\mathcal{M}} \phi) u \, dV = -\lambda \int_{\mathcal{M}} \phi u \, dV. \tag{5}$$

After the weak formulation for the problem is found, its discretization is the second step according to FEM. Hence, we consider the manifold representing the hippocampus surface as a triangular mesh composed of N vertices and L triangles: $\{V = \{p_i\}_1^N, T^h = \{T_l\}_1^L\}$, where the superindex h refers to the diameter of the triangulation. Let V^h be the space generated by those functions: $V^h = \{u^h \in C(\mathcal{M}) | u_{h|_k} \in \mathcal{P}_1, k \in T^h\}$, where T^h is the set of triangles and $\mathcal{P}_1$ is the set of two-variables linear functions. Each of the elements in V^h is called a linear finite element. Following [10], the discrete version of (5) is: Find $\phi^h \in V^h$ such that

$$\sum_l \int_{T_l} \nabla_{\mathcal{M}} \phi^h \cdot \nabla_{\mathcal{M}} \psi_i^h = \lambda^h \sum_l \int_{T_l} \phi^h \psi_i^h, \quad \forall \psi_i^h \in S^h,$$

for $i = 1, \ldots, N$, where S^h is a basis of V^h consisting on the element shape functions (Fig. 1).

Considering the following matrices involving every element of the mesh: $\phi^h = \sum_1^N x_i \psi_i^h$, $\mathbf{A}^h = (a_{ij})_{N \times N}$, where $a_{ij} = \sum_l \int_{T_l} \nabla_{\mathcal{M}} \psi_i^h \nabla_{\mathcal{M}} \psi_j^h$, and $\mathbf{B}^h = (b_{ij})_{N \times N}$, with $b_{ij} = \sum_l \int_{T_l} \psi_i^h \psi_j^h$, the variational problem (5) is then equivalent to the following eigenvalue problem:

$$\mathbf{A}^h \mathbf{x} = \lambda^h \mathbf{B}^h \mathbf{x},$$

where $\mathbf{x} = (x_1, \cdots, x_N)^t$ are the unknown associated eigenfunctions (i.e. eigenvectors which can be thought of as functions on the mesh vertices). This gener-

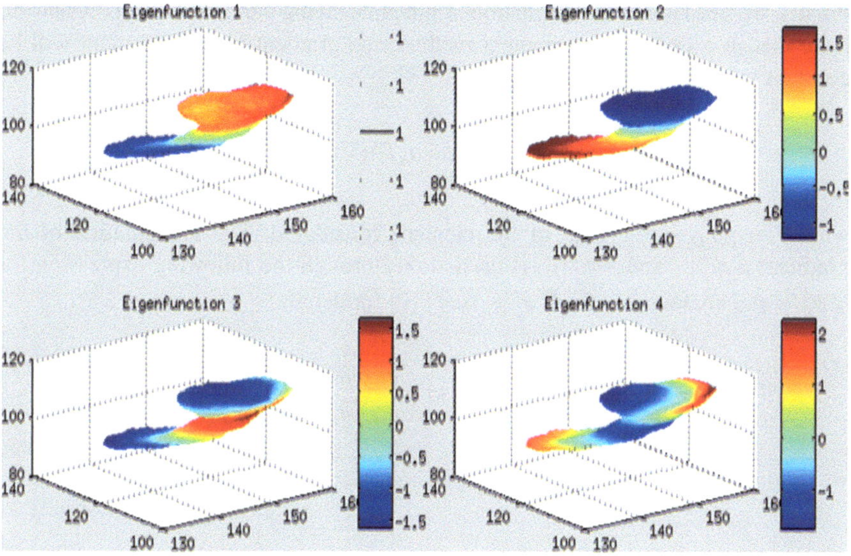

Fig. 1 Representation of the first four eigenfunctions of the LB operator. Notice that the first eigenfunction takes approximately a constant value, as expected. The second eigenfunction is known by capturing well topological features and the geometry of the shape (it corresponds to the sound we hear the best [7])

alized eigenvalue problem may be efficiently solved using the Arnoldi method of ARPACK. The computation of the local integrals $\int_{T_l} \psi_i^h \psi_j^h$ and $\int_{T_l} \nabla_{\mathcal{M}} \psi_i^h \nabla_{\mathcal{M}} \psi_j^h$, is carried out following the ideas presented in [10] based in the use of barycenter coordinates, and which we will briefly describe here. Let T_l be a triangle defined by the vertices $T_l = \{p_0, p_1, p_2\}$, $\psi_i^h \in S^h$ and $\psi_i^h = \{\psi_{i,0}, \psi_{i,1}, \psi_{i,2}\}$ the corresponding values at the vertices of the considered triangle. A point $p \subset T_l$ may be expressed in barycenter coordinates as

$$p = x^1(p_1 - p_0) + x^2(p_2 - p_0) + p_0$$

such that $0 \leq x^1, x^2, x^1 + x^2 \leq 1$. For $p \in T_l$, the values of the functions ψ_i and ψ_j might be estimated by using linear interpolation as follows:

$$\psi_i^h(p) = x^1(\psi_{i,1} - \psi_{i,0}) + x^2(\psi_{i,2} - \psi_{i,0}) + \psi_{i,0},$$

$$\psi_j^h(p) = x^1(\psi_{j,1} - \psi_{j,0}) + x^2(\psi_{j,2} - \psi_{j,0}) + \psi_{j,0}.$$

Therefore, any of the integrals needed to find matrix **B** may be approximated as

$$\int_{T_l} \psi_i^h \psi_j^h dv = \int_0^1 \int_0^{1-x^1} \psi_i^h(p) \psi_j^h(p) dx^2 dx^1.$$

Finally, we need to find an estimation of integrals taking part in the matrix **A** entries. Note that in a linear finite element method, the gradient in each element will be constant vectors. Thus, we may write

$$\int_{T_l} \nabla\psi_j^h \cdot \nabla\psi_i^h dv = area(T_l)(\nabla\psi_i^h|_{T_l} \cdot \nabla\psi_j^h|_{T_l}),$$

where $area(T_l)$ is the area of the element considered. The computation of the gradients $\nabla\psi_i^h|_{T_l}$ and $\nabla\psi_j^h|_{T_l}$, is carried out through the following expression for $\nabla\psi_i^h|_{T_l}$ and analogously for $\nabla\psi_j^h|_{T_l}$ (see [10] for details):

$$\nabla\psi_i^h|_{T_l} = (\psi_{i,1} - \psi_{i,0}, \psi_{i,2} - \psi_{i,0}) \begin{pmatrix} \partial_{x^1} \cdot \partial_{x^1} & \partial_{x^1} \cdot \partial_{x^2} \\ \partial_{x^2} \cdot \partial_{x^1} & \partial_{x^2} \cdot \partial_{x^2} \end{pmatrix}^{-1} \begin{pmatrix} p_1 - p_0 \\ p_2 - p_0 \end{pmatrix}$$

where $\partial_{x^1} = p_1 - p_0$ and $\partial_{x^2} = p_2 - p_0$.

3 Modeling Shapes

Methods for recognizing 3D shapes by their meaningful parts may be broadly divided into two categories. The first, following [11], is the *skeleton based* method (see [9]). The second one, which is the one considered in our study, is the *surface based* method. In the latter case, a shape is modelled as a frequency histogram, which is later used to compare it. The bag of features, the chosen methodology in this work, is an example of methodology that belongs to this group.

3.1 Bag of Features

The bag of features (BoF) paradigm (see [13] for details) is one of the most popular feature-based methods for shape recognition, retrieval and detection. The steps for the BoF methodology are the following: First, we detect and extract features from every shape in the training database. Second, we compute a dictionary of visual words using the training data, and allocate each feature to the closest vocabulary word. Next, we obtain the histogram of frequency for every shape. And finally, given a test shape, we model it as its histogram of frequency using the same signature, and we determine its class by majority voting of the closest training neighbors.

Local descriptors have been proven to perform well on shape recognition tasks. For every training sample, at each point of the mesh, a feature vector is computed. We build different bag of features using each of the following descriptors: the heat kernel signature (HKS), the scale-invariant heat kernel signature (SIHKS) and the wave kernel signature (WKS).

In order to quantize the feature space, the data are clustered using training samples. These data representatives are called vocabulary features. In our study, we use the k-means algorithm (see [12] for details). As each shape is modelled by a histogram, comparing shapes is tantamount to measuring histogram similarity. Two different histograms comparison metrics are used: chi-squared and Spearman distances.

3.2 Shape Descriptors

As introduced before, two different kinds of descriptors will be used: heat-diffusion and wave based descriptors. The former measures the amount of heat that remains in a point of the shape after some time t. Therefore, it is possible to capture shape information using small diffusion times and global characteristics when heat diffuses for a longer time. In addition, several times t or scales will be considered to build a feature vector for each point in the shape. The latter descriptor, which is based on the resolution of the Schrödinger equation, describes a shape by means of the probability of finding a quantum particle at a particular point of the shape.

3.2.1 HKS

At a given point of the mesh $\mathbf{p} \in \mathcal{M}$, the heat left after a time t if initially all of it was concentrated at one point, that is $u(\mathbf{p}, 0) = \delta(\mathbf{p})$, is described by $k(\mathbf{p}, \mathbf{p}, t) = K_t(\mathbf{p}, \mathbf{p})$ (see (3)), where t is the diffusion time or time scale. The heat kernel signature at each $\mathbf{p} \in \mathcal{M}$ is defined as a n-dimensional vector

$$HKS(\mathbf{p}) = (K_{t_1}(\mathbf{p}, \mathbf{p}), \cdots, K_{t_n}(\mathbf{p}, \mathbf{p})), \tag{6}$$

where $t_1, \cdots, t_n$ are different time scales.

The main advantages of the HSK are [4, 14]: it is robust to noise, it is easy to compute as it is based on the first eigenvalues and eigenfunctions, and the HKS of a shape is unique except under isometries. A major drawback of HKS is that it depends on the pixels' volume of the shape, therefore, the same hippocampus in two different scales differs on this descriptor.

3.2.2 SIHKS

In order to overcome the just mentioned dependence of HKS on the scale of the shape, Bronstein and Kokkinos [3] proposed an updated heat kernel signature which is independent on the scale-space. The scale invariant heat kernel signature, SIHKS (see [11]), which we consider in our study has been proven to improve results related to the HKS or the WKS [3, 11]. Next, we shall briefly describe the derivation of the SIHKS for reader's convenience. Given a shape $\mathscr{M}$, the heat kernel signature at a point $\mathbf{p} \in \mathscr{M}$ at time t is given by (6). Considering the same shape scaled, $\mathscr{M}' = \beta\mathscr{M}$, the relation between the eigenvalues and eigenfunctions of the Laplace-Beltrami operator of the two shapes satisfy $\lambda_i' = \beta^2\lambda_i$ and $\phi_i' = \beta\phi_i$, and the heat kernel signature for each point $\mathbf{p} \in \mathscr{M}'$ at a time t can be written as

$$K_t'(\mathbf{p}, \mathbf{p}) = \sum_{i=1}^{\infty} e^{(-\lambda_i\beta^2 t)}\phi_i\phi_i\beta^2 = \beta^2 K_{\beta^2 t}(\mathbf{p}, \mathbf{p}). \tag{7}$$

The expression (7) relates the heat kernel signature of a point in the β-scaled version $\mathscr{M}'$ at time t with the descriptor of the non-scaled version of the shape at time $\beta^2 t$. In order to accomplish the scale invariance for the HKS, we need to remove β from (7). For this purpose, we shall first write the HKS in a logarithmic time $t = \alpha^\tau$ for each point $\mathbf{p} \in \mathscr{M}$, $K_\tau = K_{\alpha^\tau}(\mathbf{p}, \mathbf{p})$. Hence, in the scaled version of the surface, $\mathscr{M}' = \beta\mathscr{M}$, the heat kernel signature can be written as follows $K_\tau' = \beta^2 K_{2log_\alpha\beta+\tau}$, and (7) is translated into

$$K_\tau' = \beta^2 K_{\tau+s} \tag{8}$$

where $s = 2log_\alpha\beta$. Now, taking logarithms in (8) and derivating with respect to τ, we obtain that

$$\frac{d}{d\tau}\log K_\tau' = \frac{d}{d\tau}\log\beta^2 + \frac{d}{d\tau}\log K_{\tau+s} = 0 + \frac{d}{d\tau}\log K_{\tau+s}, \tag{9}$$

where $\frac{d}{d\tau}\log K_\tau'$ will be computed in terms of the eigenpairs of the LB operator, through the following identity:

$$\frac{d}{d\tau}\log K_\tau' = \frac{-\sum_{i\geq 0}\lambda_i\alpha^\tau\log\alpha e^{-\lambda_i\alpha^\tau}\phi_i^2}{-\sum_{i\geq 0}e^{-\lambda_i\alpha^\tau}\phi_i^2}. \tag{10}$$

Taking the discrete Fourier transform in (9) to obtain $FK'(\omega) = FK(\omega)e^{2\pi\omega s}$, and computing the modulus of the Fourier transform, we find $|FK'(\omega)| = |FK(\omega)|$. Therefore, $|FK(\omega)|$ is scale-invariant, and we can consider the scale-invariant heat kernel signature at each $\mathbf{p} \in \mathscr{M}$ defined as a n-dimensional vector

$$SIHKS(\mathbf{p}) = (|FK(\omega_1)|, \cdots, |FK(\omega_n)|),$$

for different frequencies $\omega_1, \ldots, \omega_n$.

3.2.3 WKS

Instead of building a descriptor based on the heat diffusion on the manifold, Aubry et al. [2] proposed a signature, the wave kernel signature (WKS), based on the consideration of the Schrödinger equation

$$\frac{\partial \psi}{\partial t}(\mathbf{x}, t) = i\Delta_{\mathscr{M}} \psi(\mathbf{x}, t),$$

whose solution is a wave function which describes quantum aspects of a system. Hence, the use of WKS is in fact a quantum approach to shape analysis.

Next, we shall present a basic description of the WKS (see [2, 11] for more details). The basic idea is to characterize a point $\mathbf{p} \in \mathscr{M}$ by the average probabilities over time of quantum particles of different energy levels to be measured in $\mathbf{p}$. So, let f_E^2 be an energy probability distribution of the estimated energy E at time $t = 0$ of a quantum particle which position on the manifold is not known. Then, the wave function of the particle $\psi_E(x, t)$, if there are no repeated eigenvalues of the Laplace-Beltrami operator, can be written as

$$\psi_E(\mathbf{x}, t) = \sum_{k=0}^{\infty} e^{iE_k t} \phi_k(\mathbf{x}) f_E(E_k), \tag{11}$$

where $\{(E_k, \phi_k(\mathbf{x}))\}$ are the eigenpairs of the Laplace-Beltrami operator, which represent the energy levels of the quantum system (eigenvalues) and the corresponding wave functions (eigenfunctions) which describe the associated energy state. In fact, the probability to measure a particle at the point of the manifold $\mathbf{p} \in \mathscr{M}$, is $|\psi_E(\mathbf{p}, t)|^2$. Due to the fact that the time parameter has not clear interpretation in our analysis, it will be not taken into consideration when defining the WKS. Then, the wave kernel signature at a point $\mathbf{p}$ of the manifold $\mathscr{M}$, is the probability to measure a quantum particle overtime in an energy level

$$WKS(E, \mathbf{p}) = \lim_{T \to \infty} \frac{1}{T} \int_0^T |\psi_E(\mathbf{p}, t)|^2 dt, \tag{12}$$

which can be written as

$$WKS(E, \mathbf{p}) = \sum_{k=0}^{\infty} \phi_k^2(\mathbf{p}) f_E^2(E_k). \tag{13}$$

Regarding the energy distributions f_E^2, in [2] it is discussed that the log-normal probability distribution for f_E^2 models well the energies for our purpose. Therefore, we choose f_E^2 to be a Gaussian distribution in the logarithmic scale. Considering a logarithmic energy scale $sc = log(E)$, the wave kernel signature at $\mathbf{p} \in \mathscr{M}$ is defined as follows:

$$WKS(sc, \mathbf{p}) = C_{sc} \sum_k \phi_k^2(\mathbf{p}) e^{\frac{-(sc - \log E_k)^2}{2\sigma^2}}, \tag{14}$$

where C_{sc} is the normalizing constant $C_{sc} = \left(\sum_k e^{\frac{-(sc-\log E_k)^2}{2\sigma^2}}\right)^{-1}$. We obtain an n-dimensional vector by considering different values for sc (different energy levels) as well as σ.

Two important properties led us to include this descriptor in out study. First, it is invariant under isometries. In addition, if two shapes have the same WKS for every point of it, then both shapes are the same except for an isometry. Secondly, it is robust to noise, scale or holes in the shape.

4 Local Deformation

Recent findings suggest that the deformations on the hippocampus due to AD do not occur uniformly [19]. This leads to the necessity to develop local deformation analysis and an attempt is done here where we spectrally segment the hippocampus into different regions (classes).

We propose to apply the Neighborhood filter (NF) in terms of the decreasing rearrangement, which has recently been applied to image segmentation in [6]. In order to find a spectral segmentation of the hippocampus, we apply this technique to the quantized values of the second eigenfunction, since it is the first eigenfunction which does not take a constant value and it captures well topological features and the geometry of the shape (see [11]). In fact, the second eigenfunction of the LB operator follows the pattern of the overall shape of an object, and this geometric property is well known and used for various applications including mesh processing, feature extraction, manifold learning, data embedding, etc. (see [17]).

It is important to remark that this technique is computationally extremely efficient because the integrals involved are 1-dimensional. After applying the NF, the fixed point solution is a staircasing piecewise constant function which defines, through thresholding, a partition of the hippocampus into regions (classes) where each one of them can be understood as a segmentation of the initial shape [6].

5 Experimental Results

In our experiments we compare hippocampi using the BoF built with the three different spectral shape descriptors that we described in Sect. 3. Our aim is to achieve an acceptation rate around 80 % or above, as volume or surface area discriminate up to 80 %.

We classify hippocampi according to two disjoint classes: AD and control. To evaluate similarity between shapes, we consider two different histogram metrics, which are the chi-squared and Spearman distances. We use a total of seven eigenpairs of the LB operator to construct the descriptors. Experimentally we found that no clear improvement is achieved when using a larger number.

5.1 Database

In order to carry out our analysis, we used two datasets.

DEMCAM The DEMCAM project was a research initiative developed in Madrid for Alzheimer's dementia early detection. This dataset was collected from several hospitals of Madrid. It consists of 38 subjects, 19 control patients and 19 patients suffering from Alzheimer's disease. A total of nine subjects from each class are used as training data and the rest is test data. For each subject, a 3D high-resolution T1-weighted MR image was acquired on a GE Healthcare Signa HDX 3T scanner. All original MRIs were automatically segmented using FreeSurfer, which process included a bias field correction (N3 algorithm). Using the open source software *iso2mesh*, we obtained the left and right hippocampus represented as a triangular mesh. This mesh is described by its faces (triangles) and their vertices (Fig. 2).

ADNI The ongoing Alzheimer's disease Neuroimaging Initiative (ADNI) has been designed to provide researchers a common data framework to help in the evaluation of new methods in Alzheimer's disease detection. We considered a total of 180

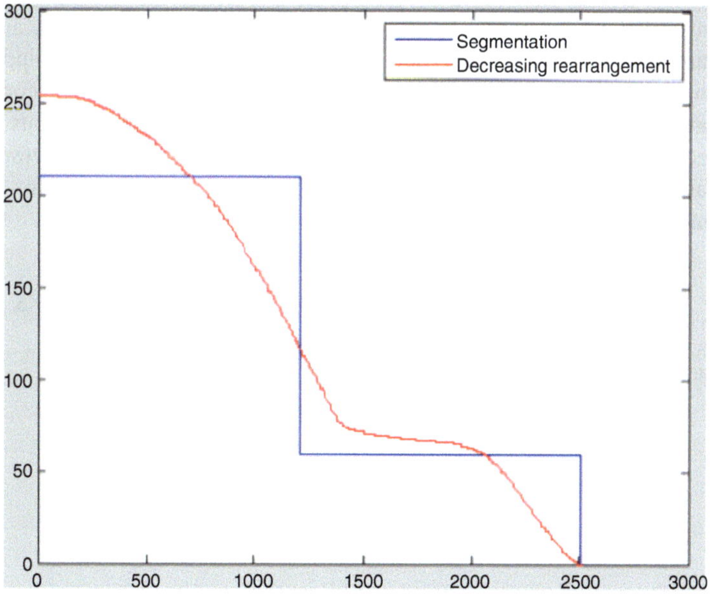

Fig. 2 NF-Decreasing rearrangement of the second eigenfunction of the Laplace-Beltrami operator quantized in 256 levels for a control hippocampus

subjects, 90 healthy patients and 90 ill subjects. We built a test data of 100 samples, including 50 of each category. Forty of the remaining patients were used as training data. Notice that the field strength (1.5T) is lower in ADNI than in DEMCAM (3T). This fact will affect the rates of AD detection.

ADNI data used in the preparation of this article were obtained from the Alzheimer's Disease Neuroimaging Initiative (ADNI) database (adni.loni.usc.edu). The ADNI (Principal Investigator Michael W. Weiner, MD) began in 2003 as a public-private partnership. The aim of ADNI has been to test whether serial magnetic resonance imaging (MRI), positron emission tomography (PET), other biological markers, and clinical and neuropsychological assessment can be combined to describe the development of mild cognitive impairment (MCI) and early Alzheimer's disease (AD).

Next, we present the experimental results obtained with the two datasets considering for classification just the left hippocampus, just the right hippocampus or the whole hippocampal structure.

5.2 DEMCAM Database

It is remarkable that SIHKS achieves the best performance with an acceptance rate of 95 % when considering both hippocampi. It outperforms WKS (90 % when considering information only of right hippocampi) and HKS (85 % when taking into account information from both hippocampi). In addition, as we expected, more information is captured by combining descriptors from both left and right hippocampi in the case of SIHKS and HKS. However, WKS uses right hippocampi to distinguish better healthier from dementia patients. This suggests that combining information leads to a better detection and also that the right hippocampus is more damaged by this disease. In fact, right hippocampus detection outperforms left hippocampus diagnosis in the maximum acceptance rates we obtained for SIHKS, WKS and HKS (see Table 1).

Table 1 Acceptance rates (%) with the standard BoF using HKS, SIHKS and WKS for DEMCAM data

DEMCAM	HKS	SIHKS	WKS
Left hippocampus	65	80	70
Right hippocampus	80	90	90
Joined	85	**95**	85

5.3 ADNI Database

Once again, SIHKS yields the best performance by correctly identifying 80 % of cases. As it was expected, information from both hippocampi (left and right) is taken into account in this outcome. HKS correctly classified 78 % of cases while WKS obtained 74 %. Also, these maxima are achieved when combining information from both left and right hippocampi (see Table 2). Regarding histogram similarity, the results show that Spearman distance leads to obtain the maximum performance for all signatures, as it occurs when considering the DEMCAM dataset.

Therefore, from this study we can argue that the scale-invariant heat kernel signature is the most suitable descriptor for detecting Alzheimer's disease. This conclusion is in agreement with [11] where it is stated that SIHKS outperformed the HKS and the WKS in most cases for shape retrieval.

5.4 Local Deformation Analysis

In order to find which zone of the hippocampus encodes more information for identifying Alzheimer's disease, we spectrally divide hippocampi into three regions, applying the Neighborhood filter (NF) in terms of the decreasing rearrangement.

We consider the zones detailed in Fig. 3 to build a BoF for each of the descriptors. In Table 3 we present the acceptance rates obtained for ADNI data for each of the just mentioned zones. The results show that SIHKS encodes most of the information for detecting Alzheimer's disease from zone 3. Region analysis using WKS as signature describes a similar behavior because encoding information from just region 2 outperforms the general WKS approach by 1 %. On the other hand, no clear information is obtained by using the HKS for local analysis. Nevertheless, 71 % of hippocampi were assigned correctly its class just by considering region 2. This descriptor needs a more global information of the shape for an accurate diagnosis. Following [11] we also model each shape by concatenating histograms corresponding to zones one, two and three, but this does not improve the descriptor performances.

Table 2 Acceptance rates (%) with the standard BoF using HKS, SIHKS and WKS for ADNI data

ADNI	HKS	SIHKS	WKS
Left hippocampus	76	71	67
Right hippocampus	68	78	73
Joined	78	**80**	74

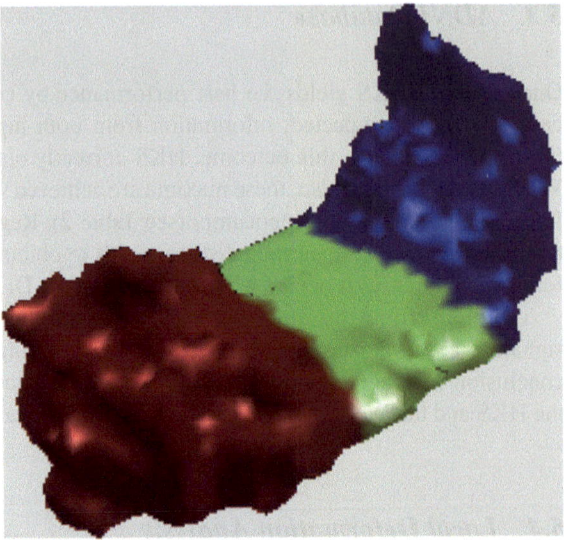

Fig. 3 Segmentation (partition) of a hippocampus into three classes using the NF in terms of the decreasing rearrangement. Zone 1: *blue-colored region*. Zone 2: *green-colored region*. Zone 3: *red-colored region*

Table 3 Acceptance rates (%) for each region of the hippocampus detailed in Fig. 3, with the standard BoF using HKS, SIHKS and WKS for ADNI data

ZONE	HKS	SIHKS	WKS
1	67	69	65
2	**71**	72	**75**
3	70	**78**	59
Concatenating	**72**	**78**	73

5.5 *Preprocessing*

In order to evaluate the effect of a preprocessing step in the hippocampus and to analyze if the noise removal involves an improvement in the detection rates of AD, we use the smoothing approach technique described in [15]. To be precise, we solve numerically the diffusion equation on the hippocampus by means of the convolution of the heat kernel, expressed as a series expansion of the eigenpairs of the LB operator, with the signal consisting of the coordinates of each of the vertices of the manifold.

According to our observations, no increase of performance is achieved by smoothing hippocampi. In fact, SIHKS and WKS best performances decreased, while HKS best acceptation rates remain constant [2]. This results suggests that the three signatures are robust to small rates of topological noise. This property is very important due to the fact that segmentation of the hippocampi from MRI may include noise. Therefore, it seems not to be necessary to apply a preprocessing step before building the BoF, which prevents from losing small details and saves computation time. In addition, the lack of precise hippocampi extraction from MRI may influence the performance of the techniques used here, as automatic segmentation of the hippocampus might not include important details to detect AD.

6 Conclusions

In this paper, we presented the use of three descriptors, namely HKS, SIHKS and WKS, in the bag-of-features framework for automatic detection of Alzheimer's disease. Our results showed that SIHKS is the best signature in detecting Alzheimer's disease in the proposed framework for both datasets. When the whole hippocampi structure is considered, the performance of our method further increases.

In an effort to study if the hippocampal structure is deformed uniformly or any of the regions is most damaged by this dementia, we proposed a spectral segmentation method of the hippocampus based on the reformulation of a NF using the decreasing rearrangement. Our preliminary results suggest that local analysis deformation usually detects a region with a greater discriminative power, but it can be different for various descriptors which makes premature any conclusion. Finally, the detection rates for 3T (DEMCAM) images are relatively greater than for 1.5T images (ADNI), which is a clear evidence that the proposed technique benefits from image quality.

Acknowledgements The second and last two authors would like to thank Ministerio de Economía y Competitividad de España for supporting Project TEC2012-39095-C03-02. Data used in preparation of this article were obtained from the Alzheimer's Disease Neuroimaging Initiative (ADNI) database and from the Hospital Fundación Reina Sofía, Madrid, Spain (DEMCAM dataset).

ADNI data: This project was funded by the Alzheimer's Disease Neuroimaging Initiative (ADNI; NIH Grant U01 AG024904; Principal Investigator: Michael Weiner). ADNI is funded by the National Institute on Aging, the National Institute of Biomedical Imaging and Bioengineering, and through generous contributions from the following: AbbVie, Alzheimer's Association; Alzheimer's Drug Discovery Foundation; Araclon Biotech; BioClinica, Inc.; Biogen; Bristol-Myers Squibb Company; CereSpir, Inc.; Eisai Inc.; Elan Pharmaceuticals, Inc.; Eli Lilly and Company; EuroImmun; F. Hoffmann-La Roche Ltd and its affiliated company Genentech, Inc.; Fujirebio; GE Healthcare; IXICO Ltd.; Janssen Alzheimer Immunotherapy Research & Development, LLC.; Johnson & Johnson Pharmaceutical Research & Development LLC.; Lumosity; Lundbeck; Merck & Co., Inc.; Meso Scale Diagnostics, LLC.; NeuroRx Research; Neurotrack Technologies; Novartis Pharmaceuticals Corporation; Pfizer Inc.; Piramal Imaging; Servier; Takeda Pharmaceutical Company; and Transition Therapeutics. The Canadian Institutes of Health Research is providing funds to support ADNI clinical sites in Canada. Industry contributions are facilitated by the Foundation for the National Institutes of Health (www.fnih.org). The grantee organization is the Northern California Institute for Research and Education, and the study is coordinated by the Alzheimer's Disease Cooperative Study at the University of California, San Diego. ADNI data are disseminated by the Laboratory for Neuro Imaging at the University of Southern California.

References

1. Aguilar, C., Muehlboeck, J.S., Mecocci, P., Velles, B., Tsolaki, M., Kloszewka, I., et al.: Application of a MRI based severity index of longitudinal atrophy change in Alzheimer's disease mild cognitive impairment and healthy older individuals in the AddNeuroMed cohort. Front. Aging Neorosci. **6**(145) (2014)

2. Aubry, M., Schlickewei, U., Cremers, D.: Pose-consistent 3D shape segmentation based on a quantum mechanical feature descriptor. Pattern Recognition, pp. 122–131. Springer, Heidelberg (2011)
3. Bronstein, M.M., Kokkinos, I.: Scale-invariant heat kernel signatures for non-rigid shape recognition. In: Proceedings of the CVPR (2010)
4. Castellani, U., Mirtuono, P., Murino, V., Bellani, M., Rambaldelli, G., Tansella, M., Brambilla, P.: A new shape diffusion descriptor for brain classification. In: Medical Image Computing and Computer-Assisted Interventional MICCAI, pp. 426–433. Springer, Heidelberg (2011)
5. Fang, Q., Boas, D.: Tetrahedral mesh generation from volumetric binary and gray-scale images. In: Proceedings of IEEE International Symposium on Biomedical Imaging, pp. 1142–1145 (2009)
6. Galiano, G., Velasco, J.: Neighborhood filters and the decreasing rearrangement. J. Math. Imaging Vis. 51(2), 279–295 (2015)
7. Gallot, S., Hulin, D., Lafontaine, J.: Riemannian Geometry. Springer, Berlin/Heidelberg (2004)
8. Gerig, G., Styner, M., Jones, D., Weinberger, D., Lieberman, J.: Shape analysis of brain ventricles using SPHARM. In: Proceedings of the IEEE Workshop on Mathematical Methods in Biomedical Image Analysis (MMBIA'01), p. 171. IEEE Computer Society (2001)
9. Kacem, A., Mohamed, W., Ben Hamza, A.: Spectral Geometric Descriptor for Deformable 3D Shape Matching and Retrieval, Image Analysis and Recognition. Lecture Notes in Computer Science, vol. 7950, pp. 181–188. Springer, Berlin (2013). http://dx.doi.org/10.1007/978-3-642-39094-4-21
10. Lai, R.: Computational differential geometry and intrinsic surface processing. Doctoral dissertation. University of California (2010)
11. Li, C., Ben Hamza, A.: Spatially aggregating spectral descriptors for nonrigid 3d shape retrieval: a comparative survey. Multimedia Syst. **20**(3), 253–281 (2014)
12. MacQueen, J.: Some methods for classification and analysis of multivariate observations. In: Proceedings of the Fifth Berkeley Symposium on Mathematical Statistics and Probability, pp. 281–297 (1967)
13. Nowak, E., Jurie, F., Triggs, B.: Sampling strategies for bag-of-features image classification. In: Computer Vision ECCV, pp. 490–503. Springer, Heidelberg (2006)
14. Raviv, D., Bronstein, M.M., Bronstein, A.M., Kimmel, R.: Volumetric heat kernel signatures. In: Proceedings of the ACM Workshop on 3D Object Retrieval, pp. 39–44. ACM, New York (2010)
15. Seo, S., Chung, M.K., Vorperian, H.K.: Heat kernel smoothing using Laplace-Beltrami eigenfunctions. In: Medical Image Computing and Computer-Assisted Intervention-MICCAI, pp. 505–512. Springer, Heidelberg (2010)
16. Shen, K., Fripp, J., Mériandeau, F., Chételat, G., Salvado, O., Bourgeaut, P., Alzheimer's Disease NeuroImaging Initiative: Detecting global and local hippocampal shape changes in Alzheimer's disease using statistical shape models. NeuroImage **59**, 2155–2166 (2012). http://dx.doi.org/10.1016/j.media.2011.10.014
17. Suzuki, K., Wang, F., Shen, D., Yan, P. (eds.): Machine learning in medical imaging. In: Second International Workshop MLMI 2011, Held in Conjunction with MICCAI, Toronto, Canada, Sep 2011 Proceedings. Lecture Notes in Computer Science, vol. 7009 (2011)
18. Teipei, S.J., Born, C., Ewers, M., Bokde, A.L., Reise, M.F., et al.: Multivariate deformation-based analysis of the brain atrophy to predict Alzheimer's disease in mild cognitive impairment. NeuroImage **38**(1), 13–24 (2007)
19. Wang, G., Zhang, X., Su, Q., Shi, J., Caselli, R.J., Wang, Y., for the Alzheimer's Disease NeuroImaging Initiative: A novel cortical thickness estimation method based on volumetric Laplace-Beltrami operator and heat kernel. Med. Image Anal. **22**, 1–20 (2015). http://dx.doi.org/10.1016/j.media.2015.01.005

Characterizations of M-Banded ASSR Matrices

P. Alonso, J.M. Peña, and M.L. Serrano

Abstract Almost strictly sign regular matrices form an important subclass of sign regular matrices, and they contain the class of almost strictly totally positive matrices. Almost strictly sign regular matrices were characterized through the Neville elimination in Alonso et al. (J Comput Appl Math 275:480–488, 2015). In this paper we present some characterizations of banded almost strictly sign regular matrices.

1 Introduction

Totally Positive (TP) matrices are matrices with all their minors nonnegative and Sign Regular (SR) matrices are matrices whose minors of the same order have the same sign. These matrices arise naturally in many areas of mathematics, statistics, mechanics, computer-aided geometric design, economics, etc. (see, for example, [3, 6]). The interest of nonsingular SR matrices comes from their characterizations as variation-diminishing linear maps: the number of changes of sign in the consecutive components of the image of a vector is bounded above by the number of changes of sign in the consecutive components of the vector.

A very important subclass of TP matrices that appears in many applications are the Almost Strictly Totally Positive (ASTP) matrices, matrices whose minors are positive if and only if all their diagonal entries are positive (see [8, 9]). Among the examples of ASTP matrices, we have Hurwitz matrices and B-spline collocation matrices. These last matrices usually present a banded structure.

In [10] the authors introduce the Almost Strictly Sign Regular (ASSR) matrices, as those whose nontrivial minors of the same order have all the same strict sign.

P. Alonso (✉) • M.L. Serrano
Departamento de Matemáticas, Universidad de Oviedo, Edificio Polivalente, Campus de Gijón, 33203 Gijón, Spain
e-mail: palonso@uniovi.es; mlserrano@uniovi.es

J.M. Peña
Departamento de Matemática Aplicada, Universidad de Zaragoza, Edificio de Matemáticas, 50009 Zaragoza, Spain
e-mail: jmpena@unizar.es

F. Ortegón Gallego et al. (eds.), *Trends in Differential Equations and Applications*, SEMA SIMAI Springer Series 8, DOI 10.1007/978-3-319-32013-7_3

Matrices that are both ASSR and TP are ASTP. On the other hand, in [1] the authors present an algorithmic characterization of ASSR matrices using Neville Elimination (NE). NE is an alternative procedure to Gaussian elimination that is especially efficient when we work with SR matrices and their subclasses or when using pivoting strategies in parallel implementations.

In this paper we present a simple characterization of banded ASSR matrices. Section 2 includes some basic results and the characterization of ASSR matrices given in [1]. Section 3 analyzes ASSR matrices whose minors of order less than or equal to a given positive integer r are nonnegative. Finally, Sect. 4 presents the characterizations of banded ASSR matrices. Results of Alonso et al. [2] for the particular case of tridiagonal matrices are also recalled.

2 Previous Results

In this work, we deal with matrices that are defined by the sign of their minors. Thus it is necessary to introduce some classical notations that will properly handle the involved submatrices.

For $k, n \in \mathbb{N}$, with $1 \leq k \leq n$, $Q_{k,n}$ denotes the set of all increasing sequences of k natural numbers not greater than n. For $\alpha = (\alpha_1, \ldots, \alpha_k)$, $\beta = (\beta_1, \ldots, \beta_k) \in Q_{k,n}$ and A an $n \times n$ real matrix, we denote by $A[\alpha|\beta]$ the $k \times k$ submatrix of A containing rows $\alpha_1, \ldots, \alpha_k$ and columns $\beta_1, \ldots, \beta_k$ of A. If $\alpha = \beta$, we denote by $A[\alpha] := A[\alpha|\alpha]$ the corresponding principal submatrix. In addition, $Q^0_{k,n}$ denotes the set of increasing sequences of k consecutive natural numbers not greater than n.

For each $\alpha \in Q_{k,n}$, we denote by the dispersion of α the number

$$d(\alpha) := \sum_{i=1}^{k-1} (\alpha_{i+1} - \alpha_i - 1) = \alpha_k - \alpha_1 - (k-1) \tag{1}$$

with the convention $d(\alpha) = 0$ for $\alpha \in Q_{1,n}$.

Note that $d(\alpha) = 0$ implies that $\alpha \in Q^0_{k,n}$.

The characterizations presented here are based on the signs of the pivots of the NE, so we will introduce briefly this procedure (see [7]).

If A is a nonsingular $n \times n$ matrix, NE consists of at most $n-1$ successive major steps, resulting in a sequence of matrices as follows:

$$A = \widetilde{A}^{(1)} \to A^{(1)} \to \cdots \to \widetilde{A}^{(n)} = A^{(n)} = U \tag{2}$$

where U is an upper triangular matrix.

For each t, $1 \leq t \leq n$, $A^{(t)} = \left(a_{ij}^{(t)}\right)_{1\leq i,j\leq n}$ has zeros in the positions $a_{ij}^{(t)}$, for $1 \leq j \leq t, j \leq i \leq n$. Besides it holds that

$$a_{it}^{(t)} = 0,\ i \geq t \Rightarrow a_{ht}^{(t)} = 0,\ \forall h \geq i. \tag{3}$$

Matrix $A^{(t)}$ is obtained from $\widetilde{A}^{(t)}$ reordering rows $t, t+1, \ldots, n$ according to a row pivoting strategy which satisfies (3).

To obtain $\widetilde{A}^{(t+1)}$ from $A^{(t)}$, zeros are introduced below the main diagonal of the tth column by subtracting a multiple of the ith row from the $(i+1)$th, for $i = n-1, \ldots, t$. The elements $\tilde{a}_{ij}^{(t+1)}$ are obtained according to the following formula

$$\begin{cases} a_{ij}^{(t)}, & 1 \leq i \leq t, \\ a_{ij}^{(t)} - \frac{a_{it}^{(t)}}{a_{i-1,t}^{(t)}} a_{i-1,j}^{(t)}, & \text{if } a_{i-1,t}^{(t)} \neq 0,\ t+1 \leq i \leq n, \\ a_{ij}^{(t)}, & \text{if } a_{i-1,t}^{(t)} = 0,\ t+1 \leq i \leq n. \end{cases} \tag{4}$$

The element

$$p_{ij} = a_{ij}^{(j)}, \quad 1 \leq i,j \leq n, \tag{5}$$

is called the (i,j) pivot of NE of A and the number

$$m_{ij} = \begin{cases} \dfrac{a_{ij}^{(j)}}{a_{i-1,j}^{(j)}} \left(= \dfrac{p_{ij}}{p_{i-1,j}}\right), & \text{if } a_{i-1,j}^{(j)} \neq 0, \\ 0, & \text{if } a_{i-1,j}^{(j)} = 0, \end{cases} \tag{6}$$

the (i,j) multiplier. Note that $m_{ij} = 0$ if and only if $p_{ij} = 0$ and by (3)

$$m_{ij} = 0 \Longrightarrow m_{hj} = 0,\ \forall h > i. \tag{7}$$

The ASSR matrices have their zero and nonzero elements grouped in certain positions (see [10]). This property inspires the following definitions:

Definition 1 A matrix $A = (a_{ij})_{1\leq i,j\leq n}$ is called type-I staircase if it satisfies simultaneously the following conditions:

- $a_{11} \neq 0,\ a_{22} \neq 0,\ \ldots,\ a_{nn} \neq 0$.
- $a_{ij} = 0, i > j \Rightarrow a_{kl} = 0, \forall l \leq j,\ i \leq k$.
- $a_{ij} = 0, i < j \Rightarrow a_{kl} = 0, \forall k \leq i,\ j \leq l$.

Definition 2 A matrix $A = (a_{ij})_{1\leq i,j\leq n}$ is called a type-II staircase matrix if it satisfies simultaneously the following conditions:

- $a_{1n} \neq 0,\ a_{2,n-1} \neq 0,\ \ldots,\ a_{n1} \neq 0.$
- $a_{ij} = 0, j > n-i+1 \Rightarrow a_{kl} = 0,\ \forall i \leq k,\ j \leq l.$
- $a_{ij} = 0, j < n-i+1 \Rightarrow a_{kl} = 0,\ \forall k \leq i,\ l \leq j.$

Observe that all entries of a matrix that is simultaneously type-I and type-II staircase are nonzero.

In order to clearly describe the zero pattern of a nonsingular matrix A type-I staircase (or type-II staircase), it is adequate to introduce the next sets of indices (see [9]):

Definition 3 Let $A = (a_{ij})_{1\leq i,j\leq n}$ be a type-I staircase matrix. We define

$$i_0 = 1, \qquad j_0 = 1, \tag{8}$$

for $k = 1, \ldots, l$:

$$i_k = \max\left\{i \,/\, a_{ij_{k-1}} \neq 0\right\} + 1\ (\leq n+1), \tag{9}$$

$$j_k = \max\left\{j \leq i_k \,/\, a_{i_k j} = 0\right\} + 1\ (\leq n+1), \tag{10}$$

where l is given in this recurrent definition by $j_l = n+1$.

Analogously we define

$$\hat{j}_0 = 1, \qquad \hat{i}_0 = 1, \tag{11}$$

for $k = 1, \ldots, r$:

$$\hat{j}_k = \max\left\{j \,/\, a_{\hat{i}_{k-1}j} \neq 0\right\} + 1\ (\leq n+1), \tag{12}$$

$$\hat{i}_k = \max\left\{i \leq \hat{j}_k \,/\, a_{i\hat{j}_k} = 0\right\} + 1\ (\leq n+1), \tag{13}$$

where $\hat{i}_r = n+1$.

In this way, we denote by $I, J, \widehat{I}$ and $\widehat{J}$ the following sets of indices

$$I = \{i_0, i_1, \ldots, i_l\},\ J = \{j_0, j_1, \ldots, j_l\},$$

$$\widehat{I} = \left\{\hat{i}_0, \hat{i}_1, \ldots, \hat{i}_r\right\},\ \widehat{J} = \left\{\hat{j}_0, \hat{j}_1, \ldots, \hat{j}_r\right\},$$

thereby defining the zero pattern in the matrix A.

In addition, for subsequent results, it is also necessary to introduce the indices j_t and $\hat{i}_t$, as well as the concept of nontrivial matrices.

Definition 4 Let A be a real $n \times n$ matrix, type-I staircase, with zero pattern $I, J, \widehat{I}$ and $\widehat{J}$. Let be $1 \leq i,j \leq n$. If $j \leq i$ we define

$$j_t = \max\{j_s \;/\; 0 \leq s \leq k-1,\; j - j_s \leq i - i_s\}, \tag{14}$$

being k the unique index satisfying that $j_{k-1} \leq j < j_k$, and if $i < j$

$$\hat{\imath}_t = \max\left\{\hat{\imath}_s \;/\; 0 \leq s \leq k' - 1,\; i - \hat{\imath}_s \leq j - \hat{\jmath}_s\right\}, \tag{15}$$

being k' the only index satisfying that $\hat{\imath}_{k'-1} \leq i < \hat{\imath}_{k'}$.

Definition 5 For a real type-I (type-II) staircase matrix $A = (a_{ij})_{1\leq i,j\leq n}$ a submatrix $A[\alpha|\beta]$, with $\alpha, \beta \in Q_{m,n}$ is nontrivial if all its main diagonal (anti-diagonal) elements are nonzero, that is, $a_{ii} \neq 0$ ($a_{i,n-i+1} \neq 0$) for all $i = 1, \ldots, n$.

The minor associated to a nontrivial submatrix ($A[\alpha|\beta]$) is called nontrivial minor ($\det A[\alpha|\beta]$).

Definition 6 An $r \times r$ matrix P_r is called backward identity matrix if the element (i,j) of matrix P_r is defined in the form

$$\begin{cases} 1, & \text{if } i + j = r + 1, \\ 0, & \text{otherwise.} \end{cases}$$

TP and ASTP matrices have been widely studied in the literature, and they are formally defined below.

Definition 7 Let $A = (a_{ij})_{1\leq i,j\leq n}$ be a real matrix. A is a TP matrix if each submatrix $A[\alpha|\beta]$, with $\alpha, \beta \in Q_{k,n}$ satisfies that

$$\det A[\alpha|\beta] \geq 0.$$

Definition 8 Let $A = (a_{ij})_{1\leq i,j\leq n}$ be a real type-I staircase matrix. Then A is an ASTP matrix, if A is TP and each nontrivial submatrix $A[\alpha|\beta]$, with $\alpha, \beta \in Q^0_{k,n}$ satisfies that

$$\det A[\alpha|\beta] > 0.$$

ASTP matrices were characterized in [8] taking into account the pivots of the NE of matrix A.

Next, we define the signature vector (± 1) that is commonly used to store the sign of the minors of order k, with $k = 1, \ldots, n$.

Definition 9 Given a vector $\varepsilon = (\varepsilon_1, \varepsilon_2, \ldots, \varepsilon_n) \in \mathbb{R}^n$, we say that ε is a signature sequence, or simply, is a signature, if $\varepsilon_i = \pm 1$ for all $i \leq n$.

Note that if a matrix is ASTP its signature sequence is $\varepsilon = (1, 1, \ldots, 1)$. Then each minor of order k has positive sign because $\varepsilon_k = +1$.

Definition 10 A real $n \times n$ matrix A is said to be SR with signature $\varepsilon = (\varepsilon_1, \varepsilon_2, \ldots, \varepsilon_n)$ if all its minors satisfy that

$$\varepsilon_m \det A[\alpha|\beta] \geq 0, \quad \alpha, \beta \in Q_{m,n}, \qquad m \leq n. \tag{16}$$

Definition 11 A real $n \times n$ matrix A is said to be ASSR with signature $\varepsilon = (\varepsilon_1, \varepsilon_2, \ldots, \varepsilon_n)$ if it is either type-I or type-II staircase and all its nontrivial minors $\det A[\alpha|\beta]$ satisfy that

$$\varepsilon_m \det A[\alpha|\beta] > 0, \quad \alpha, \beta \in Q_{m,n}, \qquad m \leq n. \tag{17}$$

Observe that an ASSR matrix is nonsingular.

In [10] the authors prove the next characterization of ASSR matrices:

Theorem 1 *Let A be a real $n \times n$ matrix and $\varepsilon = (\varepsilon_1, \varepsilon_2, \ldots, \varepsilon_n)$ be a signature. Then A is nonsingular ASSR with signature ε if and only if A is a type-I or type-II staircase matrix and all its nontrivial minors with $\alpha, \beta \in Q^0_{m,n}$, $m \leq n$, satisfy*

$$\varepsilon_m \det A[\alpha|\beta] > 0, \quad \alpha, \beta \in Q^0_{m,n}, \qquad m \leq n. \tag{18}$$

The next result (proved in [1]) establishes the relationship between the signatures of A and P_nA.

Corollary 1 *Let P_n be the $n \times n$ identity matrix. A matrix $A = (a_{ij})_{1 \leq i,j \leq n}$ is ASSR if and only if P_nA is also ASSR. Furthermore, if the signature of A is $\varepsilon = (\varepsilon_1, \varepsilon_2, \ldots, \varepsilon_n)$, then the signature of P_nA is $\varepsilon' = (\varepsilon'_1, \varepsilon'_2, \ldots, \varepsilon'_n)$, with $\varepsilon'_m = (-1)^{\frac{m(m-1)}{2}} \varepsilon_m$, for all $m = 1, \ldots, n$.*

The following result shows the relationship between ASSR and ASTP matrices.

Proposition 1 *Let A be an ASSR and TP matrix. Then A is ASTP.*

Proof If A is an ASSR matrix, then (17) is satisfied for all nontrivial minors of A. In addition, A is a TP matrix, and so $\det A[\alpha|\beta] \geq 0$ for all α, β. Then, it is immediate to conclude that $\varepsilon = (1, 1, \ldots, 1)$. Therefore A is ASTP. □

Also in [1] the authors establish the following necessary conditions for a matrix to be ASSR.

Theorem 2 *Let $A = (a_{ij})_{1\le i,j\le n}$ be a nonsingular type-I staircase matrix, with zero pattern defined by $I, J, \widehat{I}, \widehat{J}$. If A is ASSR with signature $\varepsilon = (\varepsilon_1, \varepsilon_2, \ldots, \varepsilon_n)$. Then:*

- *The NE of A can be performed without row exchanges and the pivots p_{ij} satisfy, for any $1 \le j \le i \le n$,*

$$p_{ij} = 0 \Leftrightarrow a_{ij} = 0, \tag{19}$$

$$\varepsilon_{j-j_t}\varepsilon_{j-j_t+1}p_{ij} > 0 \Leftrightarrow a_{ij} \neq 0, \tag{20}$$

  where $\varepsilon_0 = 1$ and j_t as defined in (14).
- *The NE of A^T can be performed without row exchanges and the pivots q_{ij} satisfy, for any $1 \le i \le j \le n$,*

$$q_{ij} = 0 \Leftrightarrow a_{ij} = 0, \tag{21}$$

$$\varepsilon_{i-\hat{i}_t}\varepsilon_{i-\hat{i}_t+1}q_{ij} > 0 \Leftrightarrow a_{ij} \neq 0, \tag{22}$$

  where $\varepsilon_0 = 1$ and $\hat{i}_t$ as defined in (15).

Remark 1 Let $A = \left(a_{ij}\right)_{1\le i,j\le n}$ be a type-I staircase matrix, such that the NE of A and A^T can be performed without row exchanges. If we denote by p_{ij} the pivot of NE of A when $i \ge j$ and by q_{ij} the pivot element of NE of A^T when $i < j$, then, if $a_{ij} \neq 0$, we have that:

(a) If $i \ge j$, and $j_{k-1} \le j < j_k$:

$$p_{ij} = \begin{cases} a_{ij}, & j = j_t, \\ \dfrac{\det A[i-j+j_t, \ldots, i-1, i|j_t, \ldots, j-1, j]}{\det A[i-j+j_t, \ldots, i-1|j_t, \ldots, j-1]}, & j > j_t, \end{cases} \tag{23}$$

with j_t defined in (14).

(b) If $i < j$, and $\hat{i}_{k-1} \le i < \hat{i}_k$:

$$q_{ij} = \begin{cases} a_{ij}, & i = \hat{i}_t, \\ \dfrac{\det A^T[j-i+\hat{i}_t, \ldots, j-1, j|\hat{i}_t, \ldots, i-1, i]}{\det A^T[j-i+\hat{i}_t, \ldots, j-1|\hat{i}_t, \ldots, i-1]}, & i > \hat{i}_t, \end{cases} \tag{24}$$

with $\hat{i}_t$ defined in (15).

Let $A = (a_{ij})_{1 \le i,j \le n}$ be an $n \times n$ matrix and $h = 1, \ldots, n-1$. We denote by A_h the matrix defined as

$$A_h = (a_{ij}^h)_{1 \le i,j \le n-h+1}, \quad a_{ij}^h = a_{i+h-1,j+h-1}. \tag{25}$$

Analogously, the transpose of A_h, i.e. A_h^T, is denoted as

$$A_h^T = (a_{ij}^{T,h})_{1 \le i,j \le n-h+1}, \quad a_{ij}^{T,h} = a_{j+h-1,i+h-1}. \tag{26}$$

Taking into account the previous notations, it is evident that $A_h = A[h, \ldots, n]$ and $A_h^T = A^T[h, \ldots, n]$.

Next, and using the NE of A_h and A_h^T, the characterizations given in [1] for type-I or type-II staircase ASSR are presented.

Theorem 3 *A nonsingular matrix $A = (a_{ij})_{1 \le i,j \le n}$ is ASSR with signature $\varepsilon = (\varepsilon_1, \varepsilon_2, \ldots, \varepsilon_n)$, with $\varepsilon_2 = 1$ if and only if, for every $h = 1, \ldots, n-1$, the following properties hold simultaneously:*

(i) A is type-I staircase.
(ii) The NE of the matrices A_h and A_h^T can be performed without row exchanges.
(iii) The pivots p_{ij}^h of the NE of A_h satisfy conditions corresponding to (19) and (20), and the pivots q_{ij}^h of the NE A_h^T satisfy (21) and (22).
(iv) For the positions (i^h, j^h) of matrix A_h:

- *if $i^h \ge j^h$ and $i^h - j^h = i_t^h - j_t^h$ then $\varepsilon_{j^h - j_t^h} \varepsilon_{j^h - j_t^h + 1} = \varepsilon_{j^h - 1} \varepsilon_{j^h}$,*
- *if $i^h < j^h$ and $i^h - j^h = \hat{i}_t{}^h - \hat{j}_t{}^h$ then $\varepsilon_{i^h - \hat{j}_t^h} \varepsilon_{i^h - \hat{i}_t{}^h + 1} = \varepsilon_{i^h - 1} \varepsilon_{i^h}$,*

where indices $i_t^h, j_t^h, \hat{i}_t{}^h, \hat{j}_t{}^h$ are given by conditions corresponding to (14) and (15).

Theorem 4 *Let P_r be the $r \times r$ backward identity matrix. A nonsingular matrix $A = (a_{ij})_{1 \le i,j \le n}$ is ASSR with signature $\varepsilon = (\varepsilon_1, \varepsilon_2, \ldots, \varepsilon_n)$, with $\varepsilon_2 = -1$ if and only if, for every $h = 1, \ldots, n-1$, the following properties hold simultaneously:*

(i) $B = P_n A$ is type-I staircase.
(ii) The NE of the matrices $B_h = P_{n-h+1} A_h$ and $B_h^T = P_{n-h+1} A_h^T$ can be performed without row exchanges.
(iii) The pivots p_{ij}^h of the NE of B_h satisfy conditions corresponding to (19), (20), and the pivots q_{ij}^h of the NE of B_h^T satisfy (21) and (22).
(iv) For the positions (i^h, j^h) of matrix $P_{n-h+1} A_h$:

- *if $i^h \ge j^h$ and $i^h - j^h = i_t^h - j_t^h$, then $\varepsilon_{j^h - j_t^h} \varepsilon_{j^h - j_t^h + 1} = \varepsilon_{j^h - 1} \varepsilon_{j^h}$,*
- *if $i^h < j^h$ and $i^h - j^h = \hat{i}_t{}^h - \hat{j}_t{}^h$, then $\varepsilon_{i^h - \hat{i}_t{}^h} \varepsilon_{i^h - \hat{i}_t{}^h + 1} = \varepsilon_{i^h - 1} \varepsilon_{i^h}$,*

where indices $i_t^h, j_t^h, \hat{i}_t{}^h, \hat{j}_t{}^h$ are given by conditions corresponding to (14) and (15).

Considering the previous results, in the following section certain ASSR matrices are characterized. The goal is to leverage its structure to simplify the given characterization and reduce its computational cost.

3 ASSR Matrices with Signature $(1, 1, \ldots, 1, \varepsilon_{r+1}, \ldots, \varepsilon_n)$

In this section we characterize ASSR matrices such that they (or their opposite matrices) have a signature vector whose r first positions are $+1$. In [10], the authors analyze the particular case that $r = n - 1$, that is $(1, 1, \ldots, 1, \varepsilon_n)$. For example, the tridiagonal nonnegative ASSR matrices have always this signature. In [2] the authors studied this case.

Next, we are going to present a characterization of this type of matrices, based on Theorem 3. If A is an ASSR matrix with signature $\varepsilon = (\varepsilon_1, \varepsilon_2, \ldots, \varepsilon_n)$ and $h \leq n$, then A_h (given by (25)) is ASSR matrix with signature $\varepsilon^h = (\varepsilon_1, \varepsilon_2, \ldots, \varepsilon_{n-h+1})$. In addition, when $\varepsilon_j = 1$, for $j = 1, \ldots, r$, then A_h is ASTP for $h = n - r + 1, \ldots, n$. Besides, if $h' \geq h$, then $A_{h'}$ is a submatrix of A_h, and so also $A_{h'}$ is ASTP. Then, in order to prove that A is ASSR, it is sufficient to show that A_{n-r+1} is ASTP, and checking Theorem 3 for the matrices A_h with $h = 1, \ldots, n - r$.

Theorem 5 *Let $A = (a_{ij})_{1 \leq i,j \leq n}$ be a real, nonsingular matrix. Then A is ASSR with signature $\varepsilon = (1, 1, \ldots, 1, \varepsilon_{r+1}, \ldots, \varepsilon_n)$ with $r \geq 2$ if and only if, for $h = 1, \ldots, n - r$, the following properties are simultaneously satisfied:*

(a) A is type-I staircase.
(b) The NE of A_h and A_h^T can be performed without rows exchange.
(c) The pivots p_{ij}^h of the NE of A_h satisfy conditions corresponding to (19) and (20) and the pivots q_{ij}^h of the NE of A_h^T satisfy (21) and (22).
(d) For the positions (i^h, j^h) of the matrix A_h:

- *if $i^h \geq j^h$ and $i^h - j^h = i_t^h - j_t^h$, then $\varepsilon_{j^h - j_t^h}\varepsilon_{j^h - j_t^h + 1} = \varepsilon_{j^h - 1}\varepsilon_{j^h}$,*
- *if $i^h < j^h$ and $i^h - j^h = \hat{i}_t^h - \hat{j}_t^h$, then $\varepsilon_{i^h - \hat{i}_t^h}\varepsilon_{i^h - \hat{i}_t^h + 1} = \varepsilon_{i^h - 1}\varepsilon_{i^h}$,*

where indices i_t^h, j_t^h, $\hat{i}_t^h$ and $\hat{j}_t^h$ are given by conditions corresponding to Definition 4.

(e) A_{n-r+1} is ASTP.

Proof Let us first assume that A is ASSR with signature $\varepsilon = (1, \ldots, 1, \varepsilon_{r+1}, \ldots, \varepsilon_n)$. By Theorem 3, *(a)–(d)* are satisfied. Besides, since A_{n-r+1} has signature $\varepsilon = (1, 1, \ldots, 1)$, it is ASTP and also *(e)* holds.

For the converse, let us assume that *(a)–(e)* are simultaneously hold. Condition *(a)* corresponds to the item *(i)* of Theorem 3.

Observe that the matrix $A_{n-r+1} = A[n-r+1, \ldots, n]$ contains the $n-(n-r+1)+1 = r$ last rows and columns of A. Therefore it is *ASSR* with signature $(1, 1, \ldots, 1)$. By applying *(e)*, A_{n-r+1} is ASTP, and so A_h is ASTP for each $h = n - r + 1, \ldots, n$.

Then, by the characterization given by Theorem 2.1 of [8], it is satisfied

$$p_{ij}^h \geq 0,$$
$$p_{ij}^h = 0 \Leftrightarrow a_{ij}^h = 0.$$

Taking into account that $\varepsilon_j = 1$, for all j, it is possible to write (in particular when $i \geq j$):

$$\varepsilon_{j-j_t}\varepsilon_{j-j_t+1}p_{ij}^h = p_{ij}^h > 0 \Leftrightarrow a_{ij}^h \neq 0.$$

In the cases $h = 1, \ldots, n-r+1$, by using *(c)*, (19) and (20) are satisfied, and taking into account the above arguments, for the remaining cases they are also satisfied.

Likewise, as A_h is ASTP for $h = n-r+1, \ldots, n$, then so A_h^T is, and with the same reasoning, (21) and (22) are fulfilled. Therefore *(iii)* of Theorem 3 is proven.

By Theorem 2.1 of [8], the NE of A_h and A_h^T for $h = n-r+1, \ldots, n$ can be performed without rows exchanges. This fact joint with the item *(b)* prove *(ii)*.

Finally, by applying *(d)*, if $h \leq n-r+1$ then *(iv)* is verified; if $h > n-r+1$ then $\varepsilon_s = 1$ for each $s = 1, \ldots, r$ and *(iv)* is also fulfilled, due to $\varepsilon_{j^h-j_t^h}\varepsilon_{j^h-j_t^h+1} = 1 = \varepsilon_{j^h-1}\varepsilon_{j^h}$. Then by Theorem 3, the matrix A is ASSR.

Moreover, since A_{n-r+1} is ASTP, all its signatures must be $+1$ and $\varepsilon_1 = \cdots = \varepsilon_r = 1$. □

4 Banded Matrices

In this section, ASSR matrices whose non-zero entries are confined to a diagonal band (main diagonal and parallel diagonals), are characterized. The results will be applied also to matrices with anti-diagonal band. The following definitions are introduced in [5].

Definition 12 Given a matrix $A = (a_{ij})_{1\leq i,j\leq n}$ let us consider an integer $M < n$. We say that A is M-banded matrix if $a_{ij} = 0$ when $|i-j| > M$. If, in addition $a_{ij} \neq 0$ when $|i-j| = M$, we say that A is an strictly M-banded matrix.

By Theorem 1 of [5], a nonsingular SR M-banded matrix is an strictly M-banded matrix if $a_{ij} \neq 0$ when $|i-j| \leq M$.

Definition 13 An $n \times n$ matrix A is called (strictly) anti-M-banded if P_nA is a (strictly) M-banded matrix, where P_n is the backward identity matrix given in Definition 12.

Taking into account that strictly M-banded matrices are type-I staircase, we can obtain the signature sequence for the case ASSR.

Proposition 2 *Let $A = (a_{ij})_{1\le i,j\le n}$ be a strictly M-banded ASSR matrix. If A is nonnegative, then its signature is $\varepsilon = (1, 1, \ldots, 1, \varepsilon_{n-M+1}, \ldots, \varepsilon_n)$. If A is nonpositive, then its signature is $\varepsilon = (-1, 1, -1, \ldots, (-1)^{n-M-1}, (-1)^{n-M}, \varepsilon_{n-M+1}, \ldots, \varepsilon_n)$.*

Proof Consider $k \le n - M$. Then the submatrix $A[M+1, \ldots, M+k|1, \ldots, k]$ is triangular of order k, and the associated minor is

$$\det A[M+1, \ldots, M+k|1, \ldots, k] = a_{M+1,1}a_{M+2,2}\ldots a_{M+k,k}.$$

If $a_{ij} \ge 0$, then $\det A[M+1, \ldots, M+k|1, \ldots, k] > 0$ and therefore $\varepsilon_k = 1$.

If $a_{ij} \le 0$, then the sign of $\det A[M+1, \ldots, M+k|1, \ldots, k]$ is $(-1)^k$, thus $\varepsilon_k = (-1)^k$ and the proof is complete. □

Next, we are going to characterize the nonsingular and strictly M-banded ASSR matrices. Due to the structure of these matrices, they are type-I staircase matrices, and their zero pattern is:

$$\begin{array}{rllllll}
 & (0) & (1) & (2) & & (n-M-1) & (n-M) \\
I = \{ & 1, & M+2, & M+3, & \cdots & ,n & ,n+1\}, \\
J = \{ & 1, & 2, & 3, & \cdots & ,n-M & ,n+1\}, \\
\widehat{I} = \{ & 1, & 2, & 3, & \cdots & ,n-M & ,n+1\}, \\
\widehat{J} = \{ & 1, & M+2, & M+3, & \cdots & ,n & ,n+1\}.
\end{array}$$

Look at the consequences of this zero pattern. Let us fix a position (i,j) of A such that $|i-j| \le M$, then:

- If $i \ge j$ there exists a unique k such that $j_{k-1} \le j < j_k$. When $j < n$, then $j_{k-1} = j$ and $i_{k-1} = M + j$. When $j = n$, then $j_{k-1} = n - M$ and $i_{k-1} = n$.
- If $i < j$ there exists a unique k' such that $\hat{i}_{k'-1} \le i < \hat{i}_{k'}$. Since $i < n$ then $\hat{i}_{k'-1} = i$, and $\hat{j}_{k'-1} = M + i$.

If we want to use the ASSR matrices characterization given in Theorem 3 or Theorem 5, it is important to calculate the index j_t defined in (14) and the index $\hat{i}_t$ defined in (15) for each non-zero position of A.

Lemma 1 *Let A be an $n\times n$ strictly M-banded matrix. Then, for every position (i,j) such that $|i-j| < M$ it is satisfied:*

(i) If $i = j + M$, then $j_t = j$ and $i_t = M + j$.
(ii) If $|i-j| < M$, then $j_t = 1$, $i_t = 1$ when $i \ge j$, and $\hat{i}_t = 1$, $\hat{j}_t = 1$ when $i < j$.
(iii) If $i = j - M$, then $\hat{i}_t = i$ and $\hat{j}_t = M + i$.

Proof

(i) If $i = j + M > j, j \in \{1, 2, \ldots, n - M\}$ and $i \in \{M+1, M+2, \ldots, n\}$. In this position, $j < n$ and we have $j_{k-1} = j$ and $i_{k-1} = M + j$. Therefore

$$\left.\begin{array}{l} j - j_{k-1} = 0 \\ i - i_{k-1} = (j+M) - (j+M) = 0 \end{array}\right\} \Rightarrow j - j_{k-1} = i - i_{k-1},$$

and thus $j_t = j$ and $i_t = M + j$.

(ii) Now, we are going to study the positions (i,j) such that $|i - j| < M$:

- Positions (i,j), with $i = j \in \{1, 2, \ldots, n\}$. If $i = j < n$, then $j_{k-1} = j$, $i_{k-1} = n - M$, and if $1 \leq s < k - 1$ we obtain

$$\left.\begin{array}{l} j - j_s = j - (s+1) \ \ = j - s - 1 \\ i - i_s = j - (M+s) = j - s - M \end{array}\right\} \Rightarrow j - j_s > i - i_s.$$

  If $i = j = n$ then

$$\left.\begin{array}{l} j - j_{k-1} = n - (n - M) = M \\ i - i_{k-1} = n - n \qquad\quad\ = 0 \end{array}\right\} \Rightarrow j - j_{k-1} > i - i_{k-1}.$$

  So, in both cases, we conclude $j_t = j_0 = 1$, $i_t = i_0 = 1$.

- If the position verifies $0 < i - j < M$, then $i_{k-1} = j + M$ and $j_{k-1} = j$, if $1 < s < k - 1$, it is verified:

$$\left.\begin{array}{l} j - j_s = j - (s+1) \qquad\quad = j - s - 1 \\ i - i_s = i - (M+s+1) = i - M - s - 1 \end{array}\right\} \Rightarrow j - s - 1 > j - M - s - 1 > i - s - 1,$$

  that is, $j - j_s > i - i_s$. Thus $j_t = j_0 = 1$ and $i_t = i_0 = 1$.

- Analogously, if (i,j) verifies $0 < j - i < M$, then $\hat{i}_{k-1} = i$ and $\hat{j}_{k-1} = i + 2$. We choose $1 \leq s \leq k - 1$,

$$\left.\begin{array}{l} i - \hat{i}_s = i - (s+1) \qquad\quad = i - s - 1 \\ j - \hat{j}_s = j - (M+s+1) = j - M - s - 1 \end{array}\right\} \Rightarrow i - s - 1 > j - s - 1 > j - M - s - 1,$$

  and so $i - \hat{i}_s > j - \hat{j}_s$. Therefore $\hat{i}_t = \hat{i}_0 = 1$ and $\hat{j}_t = \hat{j}_0 = 1$.

(iii) If $i = j - M$, then $i < j$ thus $i \in \{1, 2, \ldots, n - M\}$, $j \in \{M+1, \ldots, n\}$. As $i \neq n$, $\hat{i}_{k'-1} = i$ and $\hat{j}_{k'-1} = M + i$. Then:

$$\left.\begin{array}{l} i - \hat{i}_{k'-1} = i - i \qquad\qquad\quad\ = 0 \\ j - \hat{j}_{k'-1} = (i+M) - (i+M) = 0 \end{array}\right\} \Rightarrow i - \hat{i}_{k'-1} = j - \hat{j}_{k'-1},$$

and therefore $\hat{i}_t = i$ and $\hat{j}_t = M + i$. □

If a matrix A is ASSR, nonnegative and strictly M-banded, we can use Theorem 5, for $r = n - M$, because the signature of the matrix is $(1, \ldots, 1, \varepsilon_{n-M+1}, \ldots, \varepsilon_n)$, as we have seen in Proposition 2. So, we can prove the following theorem:

Theorem 6 *Let A be a strictly M-banded, real, $n \times n$ and nonsingular matrix. Then, A is ASSR with signature $\varepsilon = (1, 1, \ldots, 1, \varepsilon_{n-M+1}, \ldots, \varepsilon_n)$ if and only if, for every $h = 1, \ldots, M$, the following conditions hold simultaneously:*

(i) A is type-I staircase.
(ii) The NE of $A_h = A[h, \ldots, n]$ and A_h^T can be performed without rows exchange.
(iii) The pivots of the NE of A_h and A_h^T:

- *For $i \geq j$, if $j \leq n - M$,*

$$p_{ij}^h > 0 \Leftrightarrow a_{ij}^h \neq 0,$$

and if $j > n - M$, for $i = j, \ldots, M + j$,

$$\varepsilon_{n-M+1} p_{i,n-M+1}^h > 0,$$
$$\varepsilon_{j-1}\varepsilon_j p_{ij}^h > 0, \ \ j = n - M, \ldots, n.$$

- *For $i < j$, if $i \leq n - M$,*

$$q_{ij}^h > 0 \Leftrightarrow a_{ij^h} \neq 0,$$

and if $i > n - M$, for $j = i + 1, \ldots, M + i$,

$$\varepsilon_{n-M+1} q_{n-M+1,j}^h > 0,$$
$$\varepsilon_{i-1}\varepsilon_i q_{ij}^h > 0, \ \ i = n - M, \ldots, n.$$

(vi) $A_{n-M} = A[n - M, \ldots, n]$ is ASTP.

Proof Assume first that A is an ASSR matrix with signature sequence given by $\varepsilon = (1, 1, \ldots, 1, \varepsilon_{n-M+1}, \ldots, \varepsilon_n)$. The hypothesis of Theorem 5 are fulfilled for $r = n - M$. Items (i), (ii) and (iv) are satisfied trivially. Let see how conditions *(c)* and *(d)* of this theorem are transformed for $h = 1, \ldots, M$.

We are going to study the different possible cases:

- If $i^h = j^h + M$, we know that $j_t^h = j^h$ and $i_t^h = M + j^h$. In these positions, it is fulfilled: $j^h \in \{1, 2, \ldots, n - M\}$, $i^h \in \{M + 1, M + 2, \ldots, n\}$. In this case, $\varepsilon_{j^h - j_t^h}\varepsilon_{j^h - j_t^h + 1} = \varepsilon_0\varepsilon_1 = 1$, and therefore condition (20) is transformed into $p_{ij}^h > 0$.

  Moreover, condition *(d)* is always fulfilled, due to $i^h = j^h + M$, thus $i^h - j^h = M = i_t^h - j_t^h$, and, as $j^h - 1, j^h < n - M$, $\varepsilon_{j^h-1}\varepsilon_{j^h} = 1 = \varepsilon_{j^h - j_t^h}\varepsilon_{j^h - j_t^h + 1}$. Therefore the condition is verified.

- If $i^h = j^h$ with $j^h \in \{1, 2, \ldots, n-M\}$, then $\varepsilon_{j^h-1}\varepsilon_{j^h} = 1$, for $j^h = n-M$ $\varepsilon_{j^h-1}\varepsilon_{j^h} = \varepsilon_{n-M+1}$ and condition (20) is transformed into $p^h_{ij} > 0$, for $j^h < n-M$, $\varepsilon_{n-M+1}p^h_{n-M+1} > 0$, and for $j^h > n-M$, $\varepsilon_{j^h-1}\varepsilon_{j^h}p^h_{ij} > 0$.

  As $j^h_t = 1 = i^h_t$, $\varepsilon_{j^h-1}\varepsilon_{j^h} = \varepsilon_{j^h-j^h_t}\varepsilon_{j^h-j^h_t+1}$, condition *(d)* is fulfilled.
- If $i^h = j^h - M$, then $\hat{i}^h_t = i^h$ and $\hat{j}^h_t = M + i^h$. Moreover, $i^h \in \{M+1, \ldots, n\}$ and $j^h \in \{1, 2, \ldots, n-M\}$. In this case, $\varepsilon_{i^h-\hat{i}^h_t}\varepsilon_{i^h-\hat{i}^h_t+1} = \varepsilon_0\varepsilon_1 = 1$. Therefore condition (22) is fulfilled, resulting $q^h_{ij} > 0$.

  Condition *(d)* is also fulfilled, since being $i^h = j^h - M$, $i^h - j^h = -M = \hat{i}^h_t - \hat{j}^h_t$, and as $i^h - 1, i^h < n-M$, $\varepsilon_{i^h-1}\varepsilon_{i^h} = 1 = \varepsilon_{\hat{i}^h-\hat{i}^h_t}\varepsilon_{\hat{i}^h-\hat{i}^h_t+1}$ and the condition is fulfilled.
- Finally, for those positions for which $|i^h - j^h| < M$, either $j_t = 1$ and $i_t = 1$, or $\hat{i}_t = 1$ and $\hat{j}_t = 1$, and in this case condition *(d)* trivially holds.

  Regarding condition *(c)*, it can be rewritten as follows:

(a) For $i \geq j$:

  – If $j \leq n-M$,

$$p^h_{ij} > 0 \Leftrightarrow a^h_{ij} \neq 0.$$

  – If $i = j, \ldots, M+j$,

$$\varepsilon_{n-M+1}p^h_{i,n-M+1} > 0,$$
$$\varepsilon_{j-1}\varepsilon_j p^h_{ij} > 0, \ j = n-M, \ldots, n.$$

(b) For $i < j$,

  – If $i \leq n-M$,

$$q^h_{ij} > 0 \Leftrightarrow a_{ij^h} \neq 0.$$

  – If $j = i+1, \ldots, M+i$,

$$\varepsilon_{n-M+1}q^h_{n-M+1,j} > 0,$$
$$\varepsilon_{i-1}\varepsilon_i q^h_{ij} > 0, \ i = n-M, \ldots, n.$$

So, it has been proven that *(iii)* also holds.

To prove the converse, we use again Theorem 5. Note items *(a)*, *(b)*, *(c)* and *(e)* of this theorem are trivially fulfilled. Let's prove that *(d)* is also fulfilled, by using Lemma 1. Let (i^h, j^h) be such that $|i-j| \leq M$. Then we have the following cases:

- If $i^h = j^h + M$, then $j^h \leq N - M$, and so $\varepsilon_{j^h-1}\varepsilon_{j^h} = 1$. Moreover, $j^h_t = j^h$, then $\varepsilon_{j^h-j^h_t}\varepsilon_{j^h-j^h_t+1} = \varepsilon_0\varepsilon_1 = 1$, and therefore condition *(d)* holds.

- If $i^h = j^h - M$, then $i^h \leq N - M$, and so $\varepsilon_{\hat{i}^h-1}\varepsilon_{\hat{i}^h} = 1$. Moreover, $\hat{i}_t^h = i^h$, then $\varepsilon_{i^h-\hat{i}_t^h}\varepsilon_{i^h-\hat{i}_t^h+1} = \varepsilon_0\varepsilon_1 = 1$, and condition *(d)* holds.
- Finally, we study the case $|i^h - j^h| < M$.
  If $i^h \geq j^h$ then $j_t^h = 1$, and so $\varepsilon_{j^h-1}\varepsilon_{j^h} = \varepsilon_{j^h-j_t^h}\varepsilon_{j^h-j_t^h+1}$.
  If $i^h < j^h$ then $\hat{i}_t = 1$, and so $\varepsilon_{i^h-\hat{i}_t^h}\varepsilon_{i^h-\hat{i}_t^h+1} = \varepsilon_{\hat{i}^h-1}\varepsilon_{\hat{i}^h}$.

In any case, condition *(d)* of Theorem 5 is fulfilled, so that we can conclude that the matrix A is ASSR with signature $\varepsilon = (1, 1, \ldots, 1, \varepsilon_{n-M+1}, \ldots, \varepsilon_n)$. □

If A is a nonpositive nonsingular strictly M-banded matrix then $-A$ is a matrix satisfying hypothesis of Theorem 6. Therefore both types of matrices have been characterized, and we can deduced the following consequence of Proposition 2 and Theorem 6.

Corollary 2 *Let A be a strictly M-banded, real, $n \times n$, nonsingular matrix. Then A is ASSR if and only if either A is nonnegative and satisfies (i)–(iv) of Theorem 6 or A is nonpositive and $-A$ satisfies (i)–(iv) of Theorem 6.*

Given a strictly anti-M-banded matrix A, we can apply Corollary 2 in order to obtain the following characterization:

Corollary 3 *Let A be a strictly anti-M-banded, real, $n \times n$, nonsingular matrix. Then A is ASSR if and only if either A is nonnegative and P_nA satisfies (i)–(iv) of Theorem 6 or A is nonpositive and $-P_nA$ satisfies (i)–(iv) of Theorem 6.*

4.1 A Particular Case: 1-Banded (Tridiagonal) Matrices

Focusing on the case of tridiagonal matrices, a tridiagonal matrix of order $n \geq 2$ is a 1 banded matrix, i.e., a matrix $A = (a_{ij})_{1\leq i,j\leq n}$, with $a_{ij} = 0$ for all i, j when $|i - j| > 1$.

Using Corollary 3.4 of [4], we can see that the only signatures they can present are the following: $\varepsilon = (1, 1, \ldots, 1, \varepsilon_n)$ or $\varepsilon = (-1, 1, \ldots, (-1)^{n-1}, \varepsilon_n)$.

In [2] we have proved that a nonnegative ($A \geq 0$) tridiagonal ASSR matrix is strictly tridiagonal or it is an ASTP matrix, i.e., with signature $(1, 1, \ldots, 1)$.

Lemma 2 *Let $A = (a_{ij})_{1\leq i,j\leq n}$ be a real nonnegative tridiagonal ASSR matrix. If there exists j, $1 < j \leq n$ such that $a_{j,j-1} = 0$, or $a_{j-1,j} = 0$, then A is ASTP.*

If A is nonpositive tridiagonal matrix, its signature is $(-1, 1, -1, \ldots, (-1)^{n-1}, \varepsilon_n)$. Thus $-A$ is tridiagonal ASSR matrix with signature $(1, 1, \ldots, 1, (-1)^n\varepsilon_n)$ and we can apply the previous result to matrix $-A$. Thereby, if the matrix has an zero element in the position (i,j) with $|i - j| = 1$, then either A or $-A$ is ASTP.

In the following result (see Theorem 25 of [2]) we characterize the nonsingular and nonnegative tridiagonal ASSR matrices.

Theorem 7 *Let* $A = (a_{ij})_{1\leq i,j\leq n}$ *be a real nonnegative tridiagonal matrix and nonsingular. Then* A *is ASSR with* $\varepsilon = (1, 1, \ldots, 1, \varepsilon_n)$ *if and only if it holds that:*

(a) A *is type-I staircase.*
(b) The NE of the matrices A *and* A^T *can be performed without row changes.*
(c) The pivots p_{ij} *of the NE of* A *and the pivots* q_{ij} *of the NE of* A^T *satisfy:*

- *If* $i \geq j$,

$$p_{ij} = 0 \Leftrightarrow a_{ij} = 0, \tag{27}$$

$$\left.\begin{array}{l} j < n, \ \ p_{ij} > 0 \\ j = n, \ \varepsilon_n p_{in} > 0 \end{array}\right\} \Leftrightarrow a_{ij} \neq 0. \tag{28}$$

- *If* $i < j$,

$$q_{ij} = 0 \Leftrightarrow a_{ij} = 0, \tag{29}$$

$$q_{ij} > 0 \Leftrightarrow a_{ij} \neq 0. \tag{30}$$

(d) $A_2 = A[2, \ldots, n]$ *is ASTP.*

If $A \leq 0$ (that is, a matrix whose entries are all nonpositive), then $-A$ has signature $(1, 1, \ldots, 1, (-1)^n\varepsilon_n)$. Therefore it is possible to apply the previous result to matrix $-A$, and then all the tridiagonal ASSR matrices are characterized.

Acknowledgements This work has been partially supported by the Spanish Research Grant MTM2012-31544 and under MEC and FEDER Grant TEC2012-38142-C04-04.

References

1. Alonso, P., Peña, J.M., Serrano, M.L.: On the characterization of almost strictly sign regular matrices. J. Comput. Appl. Math. **275**, 480–488 (2015)
2. Alonso, P., Peña, J.M., Serrano, M.L.: Characterization of almost strictly sign regular matrices and some particular cases. In: Proceedings of the XVII Congress on Differential Equations and Applications and XIV Congress of Applied Mathematics, pp. 423–429 (2015)
3. Ando, T.: Total positive matrices. Linear Algebra Appl. **90**, 165–219 (1987)
4. Barreras, A., Peña, J.M.: Characterization of Jacobi sign regular matrices. Linear Algebra Appl. **436**, 381–388 (2012)
5. Barreras, A., Peña, J.M.: On tridiagonal sign regular matrices and generalizations. In: Casas, F., Martínez, V. (eds.) Advances in Differential Equations and Applications, SEMA. SIMAi Springer Series, vol. 4, pp. 239–247. Springer, Berlin (2014)
6. Fallat, S.M., Johnson, Ch.R.: Totally Nonnegative Matrices. Princeton University Press, Princeton (2011)
7. Gasca, M., Peña, J.M.: A matricial description of Neville elimination with applications to total positivity. Linear Algebra Appl. **202**, 33–54 (1994)

8. Gasca, M., Peña, J.M.: On the Characterization of almost strictly total positive matrices. Adv. Comp. Math. **3**, 239–250 (1995)
9. Gasca, M., Peña, J.M.: Characterization and decomposition of almost strictly positive matrices. SIAM J. Matrix Anal. Appl. **28**, 1–8 (2006)
10. Huang, R., Liu, J., Zhu, L.: Nonsingular almost strictly sign regular matrices. Linear Algebra Appl. **436**, 4179–4192 (2012)

A Review of Numerical Analysis for the Discretization of the Velocity Tracking Problem

Eduardo Casas and Konstantinos Chrysafinos

Abstract In this paper we are reviewing results regarding the velocity tracking problem. In particular, we focus on our work (Casas and Chrysafinos, SIAM J. Numer. Anal. 50(5):2281–2306, 2012; Casas and Chrysafinos, Numer. Math. 130:615–643, 2015; and Casas and Crysafinos, to appear in ESAIM: COCV) concerning a-priori error estimates for the velocity tracking of two-dimensional evolutionary Navier-Stokes flows. The controls are of distributed type, and subject to point-wise control constraints. The standard tracking type functional is considered, however the option of setting the penalty-regularization parameter $\lambda = 0$ in front of the $L^2(0,T;\mathbf{L}^2(\Omega))$ norm of the control in the functional is also discussed. The discretization scheme of the state and adjoint equations is based on a discontinuous time-stepping scheme combined with conforming finite elements (in space) for the velocity and pressure. Provided that the time and space discretization parameters, τ and h respectively, satisfy $\tau \leq Ch^2$, error estimates of order $\mathcal{O}(h)$, $\mathcal{O}(h^2)$ and $\mathcal{O}(h^{\frac{3}{2}-\frac{1}{p}})$ for some $p > 2$, are discussed for the difference between the locally optimal controls and their discrete approximations, when the controls are discretized by piecewise constants functions, the variational discretization approach or by using piecewise linears in space respectively for $\lambda > 0$. For the case of $\lambda = 0$, (bang-bang type controls) we also discuss various issues related to the analysis and discretization, emphasizing on the different features compared to the case $\lambda > 0$. In particular, fully-discrete estimates for the states are presented and discussed.

E. Casas
Departmento de Matemática Aplicada y Ciencias de la Computación, E.T.S.I. Industriales y de Telecomunicación, Universidad de Cantabria, Av. Los Castros s/n, 39005 Santander, Cantabria, Spain
e-mail: eduardo.casas@unican.es

K. Chrysafinos (✉)
Department of Mathematics, School of Applied Mathematics and Physical Sciences, National Technical University of Athens, Zografou Campus, Athens 15780, Greece
e-mail: chrysafinos@math.ntua.gr

F. Ortegón Gallego et al. (eds.), *Trends in Differential Equations and Applications*, SEMA SIMAI Springer Series 8, DOI 10.1007/978-3-319-32013-7_4

1 Introduction

In this paper we are reviewing various results from our works of [6–8] regarding the approximation of the velocity tracking problem. The velocity tracking problem is defined as follows: We seek velocity vector field $\mathbf{y}$, pressure p and control vector field $\mathbf{u}$ such that

$$(\mathrm{P}) \quad \begin{cases} \min J(\mathbf{u}) \\ \mathbf{u} \in \mathcal{U}_{ad} \end{cases}$$

where

$$J(\mathbf{u}) = \frac{1}{2}\int_0^T \int_\Omega |\mathbf{y_u}(t,x) - \mathbf{y}_d(t,x)|^2 \,\mathrm{d}x\mathrm{d}t + \frac{\lambda}{2}\int_0^T\int_\Omega |\mathbf{u}(t,x)|^2 \,\mathrm{d}x\mathrm{d}t.$$

Here we denote by $\mathbf{y}_d$ the given target velocity profile and $\mathbf{y_u}$ the solution of the 2d evolution Navier-Stokes equations with right hand side the control variable $\mathbf{u}$, i.e.,

$$\begin{cases} \mathbf{y}_t - \nu\Delta\mathbf{y} + (\mathbf{y}\cdot\nabla)\mathbf{y} + \nabla p = \mathbf{f} + \mathbf{u} \ \text{ in } \Omega_T = (0,T)\times\Omega, \\ \operatorname{div}\mathbf{y} = 0 \ \text{ in } \Omega_T, \ \mathbf{y}(0) = \mathbf{y}_0 \ \text{ in } \Omega, \\ \mathbf{y} = 0 \ \text{ on } \Sigma_T = (0,T)\times\Gamma. \end{cases} \tag{1}$$

The set of feasible controls is denoted by $\mathcal{U}_{ad}$ and it is defined for $-\infty < \alpha_j < \beta_j < +\infty, j = 1, 2$, by

$$\mathcal{U}_{ad} = \{\mathbf{u} \in L^2(0,T;\mathbf{L}^2(\Omega)) : \alpha_j \le u_j \le \beta_j \ \text{ a.e. in } \Omega_T, \ j = 1,2\}.$$

The physical meaning of the velocity tracking problem is to drive the velocity vector field to a given target field $\mathbf{y}_d$, by using a control function of distributed type. In our setting, the control function satisfies point-wise constraints and $\lambda \ge 0$ is a penalty parameter, which is typically small compared to the actual size of the data. There is an important distinction between the case $\lambda > 0$ and the case $\lambda = 0$. The absence of the Tikhonov regularization term from the cost functional, i.e. $\lambda = 0$, typically leads to optimal controls of bang-bang type, creating substantial difficulties for the analysis and numerical analysis, despite the fact that point-wise control constraints are being imposed. Indeed, the absence of the regularizing term leads to loss of regularity and to non-standard second order sufficient conditions, and hence to severe technical difficulties both in the analysis and in the construction of suitable numerical schemes (e.g. see for instance [5, 8, 16]). On the other hand the presence of the Tikhonov regularizing term provides the crucial relation between the control and adjoint variables facilitating the derivation of second order sufficient conditions (see for e.g. [14]) and hence the derivation of error estimates for various choices of discretization spaces for the controls when combined with piecewise constants in time, and classical finite element spaces for the velocity and the pressure (see e.g. [7]). For various related discussions, references regarding the analysis and the computational significance of such optimal control problems we refer the reader to [29]. The analysis of the above control problem is well understood, (see e.g. [1, 4, 29, 33, 46, 50] and references within), where various aspects, including first and second order necessary conditions are developed and analyzed. Our paper is organized as follows: At the remaining of the introduction, we present some related references regarding the numerical analysis of the velocity tracking problem. In Sects. 2 and 3 we formulate the optimal control problem including first and second order necessary and sufficient conditions and we stress their importance for the development of error estimates. Then, in Sect. 4, we define the discrete state and adjoint-state problems and we present the basic numerical analysis results of [6], and [7] under suitable regularity assumptions. Finally, in Sect. 5, we present the results of [7, 8] for the discretization of the optimal control problems.

1.1 Related Results

We begin with some earlier results related to the numerical analysis of the velocity tracking problem of the evolutionary Navier-Stokes equations. First, we note that in [30, 31] a gradient algorithm is analyzed for a fully discrete scheme based on the implicit Euler in time discretization combined with inf-sup stable elements for the discretization for the velocity and pressure respectively. In particular convergence of the proposed algorithm is proven, in case of distributed controls, and of bounded distributed controls respectively. Error estimates for the semi-discrete (in space) discretization are derived in [21] in case of distributed controls without control constraints by using a variational discretization approach (see [32]).

For the approximation of control problems associated to parabolic semilinear equations, error estimates are presented in [41], by using both the variational discretization and the piecewise linears for the discretization of the controls. The key feature of their analysis is a two step discretization approach. First, the state equation is discretized in time and then in space. Note that by taking advantage of the boundedness of the semi-discrete in time-space states they obtain error estimates for the controls, without imposing the assumption $\tau \leq Ch^2$. However, a strong second order necessary condition is also needed. Their approach is not easily translated to the control of Navier-Stokes systems because the non-linearity involves the gradient of the state and the boundedness of the states fails. Moreover, the discretization in time of the state equation leads to a stationary Navier-Stokes system, for which we cannot guarantee the uniqueness of a solution. Finally in [19] a convergence result for an optimal control problem related to semi-linear parabolic pdes is presented under minimal regularity assumptions on the given data.

For earlier work on these schemes within the context of optimal control problems, having states constrained to linear parabolic pdes, we refer the reader to [37, 38] for (optimal) error estimates for an optimal control problem for the heat equation, with and without control constraints respectively. Error estimates for discontinuous time-stepping schemes for distributed optimal control problems related to linear parabolic pdes with possibly time dependent coefficients, were presented in [17, 18]. An analysis of second order Petrov-Galerkin Crank-Nicolson scheme and of a Crank-Nicolson scheme, for an optimal control problem for the heat equation were analyzed in [2, 39] respectively where estimates of second-order (in time) are derived. However, the regularity assumptions on the control, state and adjoint variables are not present in the nonlinear setting of Navier-Stokes equations. Further results regarding error analysis can be found in [40, 43, 44].

We refer the reader to [23–26, 48] (see also references within) for various results related to the approximation of parabolic pdes without controls and to [20] for discontinuous time-stepping schemes of arbitrary order for the Navier-Stokes equations in 2d and 3d. Further results concerning the analysis and numerical analysis of the uncontrolled Navier-Stokes can be found in the classical works of [28, 34, 35, 47]. For several issues related to the analysis and numerics of optimal control problems we refer the reader to [49] (see also references within). Finally, we refer the reader to [9] for the analysis of control problems of 3D evolution Navier-Stokes equations.

2 Definitions and Preliminaries

Throughout this work we assume that Ω is a bounded open and convex subset in $\mathbb{R}^2$ with a C^2 boundary Γ. The outward unit normal vector to Γ at a point $x \in \Gamma$ is denoted by $\mathbf{n}(x)$. For given $0 < T < +\infty$, we denote $\Omega_T = (0,T) \times \Omega$ and $\Sigma_T = (0,T) \times \Gamma$. We use the standard notation for Sobolev spaces: $\mathbf{H}^1(\Omega) = H^1(\Omega;\mathbb{R}^2)$, $\mathbf{H}_0^1(\Omega) = H_0^1(\Omega;\mathbb{R}^2)$, $\mathbf{H}^{-1}(\Omega) = (\mathbf{H}_0^1(\Omega))'$ and $\mathbf{W}^{s,\bar{p}}(\Omega) = W^{s,\bar{p}}(\Omega;\mathbb{R}^2)$ for $1 \leq \bar{p} \leq \infty$ and $s > 0$ as well as the standard notation for the spaces of integrable functions

$$L_0^2(\Omega) = \{w \in L^2(\Omega) : \int_\Omega w(x)\,\mathrm{d}x = 0\};$$

$\mathbf{L}^{\bar{p}}(\Omega) = L^{\bar{p}}(\Omega;\mathbb{R}^2)$. Furthermore, we define (see for instance Lions and Magenes [36, Vol. 1]) the time-space space,

$$H^{2,1}(\Omega_T) = \left\{ y \in L^2(\Omega_T) : \frac{\partial y}{\partial x_i}, \frac{\partial^2 y}{\partial x_i x_j}, \frac{\partial y}{\partial t} \in L^2(\Omega_T),\ 1 \leq i,j \leq 2 \right\}$$

equipped with the standard norm. It is well known that every element of $H^{2,1}(\Omega_T)$, after a modification over a zero measure set, is a continuous function from $[0,T] \longrightarrow H^1(\Omega)$. Finally, we also denote by $\mathbf{H}^{2,1}(\Omega_T) = H^{2,1}(\Omega_T) \times H^{2,1}(\Omega_T)$.

In order to handle the case of $\lambda = 0$, we will also need to define the following generalizations of the above spaces: For given a number $1 \leq \bar{p} \leq \infty$, and we set

$$\mathbf{W}_{\bar{p}}^{2,1}(\Omega_T) = \left\{ y \in \mathbf{L}^{\bar{p}}(\Omega_T) : \frac{\partial y}{\partial x_i}, \frac{\partial^2 y}{\partial x_i x_j}, \frac{\partial y}{\partial t} \in \mathbf{L}^{\bar{p}}(\Omega_T),\ 1 \leq i,j \leq 2 \right\}$$

equipped with the standard norm. We note that $\mathbf{H}^{2,1}(\Omega_T) = \mathbf{W}_2^{2,1}(\Omega_T)$.

The usual spaces of divergence-free vector fields can be defined in a standard way:

$$\mathbf{Y}_{\bar{p}} = \{\mathbf{y} \in \mathbf{W}_0^{1,\bar{p}}(\Omega) : \operatorname{div} \mathbf{y} = 0 \text{ in } \Omega\},$$

$$\mathbf{H}_{\bar{p}} = \{\mathbf{y} \in \mathbf{L}^{\bar{p}}(\Omega) : \operatorname{div} \mathbf{y} = 0 \text{ in } \Omega \text{ and } \mathbf{y} \cdot \mathbf{n} = 0 \text{ on } \Gamma\}.$$

Finally, we define $\mathbf{W}(0,T) = \{\mathbf{y} \in L^2(0,T;\mathbf{Y}_2) : \mathbf{y}_t \in L^2(0,T;\mathbf{Y}_2^*)\}$. It is well known that $\mathbf{W}(0,T) \subset C_w([0,T],\mathbf{H}_2)$, where $C_w([0,T],\mathbf{H}_2)$ is the space of weakly continuous functions $\mathbf{y} : [0,T] \longrightarrow \mathbf{H}_2$.

We will frequently abbreviate the notation of divergence-free vector fields: $\mathbf{Y} = \mathbf{Y}_2$ and $\mathbf{H} = \mathbf{H}_2$.

Standard regularity assumptions will be imposed on the data in order to guarantee the existence of a strong solution, i.e., $\mathbf{f}, \mathbf{y}_d \in \mathbf{L}^2(\Omega_T)$ and $\mathbf{y}_0 \in \mathbf{Y}$. A weak solution of (1) will be sought in the space $\mathbf{W}(0,T) = \{\mathbf{y} \in L^2(0,T;\mathbf{Y}) : \mathbf{y}_t \in L^2(0,T;\mathbf{Y}^*)\}$. We note that $\mathbf{W}(0,T) \subset C_w([0,T],\mathbf{H})$, where $C_w([0,T],\mathbf{H})$ is the space of weakly continuous functions $\mathbf{y} : [0,T] \longrightarrow \mathbf{H}$.

To write the weak formulation of (1), we define the bilinear and trilinear forms $a : \mathbf{H}^1(\Omega) \times \mathbf{H}^1(\Omega) \longrightarrow \mathbb{R}$ and $c : \mathbf{L}^4(\Omega) \times \mathbf{H}^1(\Omega) \times \mathbf{H}^1(\Omega) \longrightarrow \mathbb{R}$ in a standard way:

$$a(\mathbf{y},\mathbf{z}) = \nu \int_\Omega (\nabla \mathbf{y} : \nabla \mathbf{z})\, \mathrm{d}x = \nu \sum_{i,j=1}^{2} \int_\Omega \partial_{x_i} y_j\, \partial_{x_i} z_j\, \mathrm{d}x$$

$$c(\mathbf{y},\mathbf{z},\mathbf{w}) = \frac{1}{2}\left[\hat{c}(\mathbf{y},\mathbf{z},\mathbf{w}) - \hat{c}(\mathbf{y},\mathbf{w},\mathbf{z})\right]$$

$$\text{with } \hat{c}(\mathbf{y},\mathbf{z},\mathbf{w}) = \sum_{i,j=1}^{2} \int_\Omega \mathbf{y}_j \left(\frac{\partial \mathbf{z}_i}{\partial x_j}\right) \mathbf{w}_i\, \mathrm{d}x.$$

Now, a weak solution of (1) is an element $\mathbf{y} \in \mathbf{W}(0,T)$ such that for a.e. $t \in (0,T)$,

$$\begin{cases} (\mathbf{y}_t, \mathbf{w}) + a(\mathbf{y},\mathbf{w}) + c(\mathbf{y},\mathbf{y},\mathbf{w}) = (\mathbf{f} + \mathbf{u}, \mathbf{w}) \quad \forall\, \mathbf{w} \in \mathbf{Y} \\ \mathbf{y}(0) = \mathbf{y}_0. \end{cases} \tag{2}$$

Equation (2) has a unique solution in $\mathbf{W}(0,T)$ and the existence of a pressure $p \in \mathcal{D}(\Omega_T)$ satisfying (1) holds in the distribution sense. Due to the regularity assumed on $\mathbf{f}$, $\mathbf{y}_0$ and Ω, then some extra regularity is proved for $(\mathbf{y}, p)$. In particular, we have that $\mathbf{y} \in \mathbf{H}^{2,1}(\Omega_T) \cap C([0,T], \mathbf{Y})$ and $p \in L^2(0,T; H^1(\Omega))$, the pressure being unique up to an additive constant; see, for instance, Ladyzhenskaya [34], Lions [35], Temam [47].

For various properties related to the trilinear term c we refer the reader to [34, 35] or [47].

The following theorem analyzes the state equation, under further non-standard regularity assumptions, and it is based on results of by Solonnikov [45, Theorem 4.2].

Theorem 1 *Suppose that the data of* (1) *satisfy:* $\nu > 0$, $\mathbf{f} \in \mathbf{L}^{\bar{p}}(\Omega_T)$ *and* $\mathbf{y}_0 \in \mathbf{W}^{2-\frac{2}{\bar{p}},\bar{p}}(\Omega) \cap \mathbf{Y}_2$, *with* $3 < \bar{p} < +\infty$. *Then, for every* $\mathbf{u} \in \mathbf{L}^{\bar{p}}(\Omega_T)$ *the state equation* (1) *has a unique solution* $\mathbf{y}_\mathbf{u} \in \mathbf{W}^{2,1}_{\bar{p}}(\Omega_T)$ *and an associate pressure* $p_\mathbf{u} \in L^{\bar{p}}(0,T; W^{1,\bar{p}}(\Omega))$, *which is unique up to the addition of a function of* $L^{\bar{p}}(0,T)$. *Moreover, the following estimate holds*

$$\|\mathbf{y}_\mathbf{u}\|_{\mathbf{W}^{2,1}_{\bar{p}}(\Omega_T)} + \|\nabla p_\mathbf{u}\|_{\mathbf{L}^{\bar{p}}(\Omega_T)} \leq C_\mathbf{u}\Big(\|\mathbf{f}+\mathbf{u}\|_{\mathbf{L}^{\bar{p}}(\Omega_T)} + \|\mathbf{y}_0\|_{\mathbf{W}^{2-\frac{2}{\bar{p}},\bar{p}}(\Omega)}\Big), \tag{3}$$

where $C_\mathbf{u}$ *depends on* $\|\mathbf{f}+\mathbf{u}\|_{\mathbf{L}^2(\Omega_T)}$ *and* $\|\mathbf{y}_0\|_{\mathbf{Y}_2}$. *Furthermore, the constant* $C_\mathbf{u}$ *in* (3) *can be chosen the same for every* $\mathbf{u} \in \mathcal{U}_{ad}$. *In addition, there exists a constant* $M_{\alpha,\beta}$ *such that* $\forall \mathbf{u} \in \mathcal{U}_{ad}$

$$\|\mathbf{y}_\mathbf{u}\|_{C([0,T];\mathbf{Y}_{\bar{p}})} + \|\mathbf{y}_\mathbf{u}\|_{\mathbf{C}^{0,1-\frac{3}{\bar{p}}}(\bar{\Omega}_T)} \leq M_{\alpha,\beta}\Big(\|\mathbf{f}+\mathbf{u}\|_{\mathbf{L}^{\bar{p}}(\Omega_T)} + \|\mathbf{y}_0\|_{\mathbf{W}^{2-\frac{2}{\bar{p}},\bar{p}}(\Omega)}\Big), \tag{4}$$

where $\mathbf{C}^{0,1-\frac{3}{\bar{p}}}(\bar{\Omega}_T)$ *is the space of Hölder functions in* $\bar{\Omega}_T$ *of order* $1-\frac{3}{\bar{p}}$.

It is well known that the mapping $G : \mathbf{L}^2(\Omega_T) \to \mathbf{H}^{2,1}(\Omega_T) \cap C([0,T]; \mathbf{Y})$, associating to each control $\mathbf{u}$ the corresponding state $G(\mathbf{u}) = \mathbf{y}_\mathbf{u}$ solution of (2), is well defined and continuous. Hence the functional $J : \mathbf{L}^2(\Omega_T) \longrightarrow \mathbb{R}$ is also well defined and continuous. The proof of the existence of at least one solution of (P) is standard. Below, we state the differentiability of G and J.

Theorem 2 ([6, 8]) *Let* $2 \leq \bar{p} < +\infty$ *and assume that* $\mathbf{f} \in \mathbf{L}^{\bar{p}}(\Omega_T)$ *and* $\mathbf{y}_0 \in \mathbf{W}^{2-\frac{2}{\bar{p}},\bar{p}}(\Omega) \cap \mathbf{Y}_2$. *Then, the mapping*

$$G : \mathbf{L}^{\bar{p}}(\Omega_T) \longrightarrow \mathbf{W}^{2,1}_{\bar{p}}(\Omega_T) \cap C([0,T]; \mathbf{Y}_{\bar{p}})$$

is of class C^∞. *Moreover, for any* $\mathbf{u}, \mathbf{v}, \mathbf{v}_i \in \mathbf{L}^{\bar{p}}(\Omega_T)$, $i = 1, 2$, *if we denote* $\mathbf{y}_\mathbf{u} = G(\mathbf{u})$, $\mathbf{z}_\mathbf{v} = G'(\mathbf{u})\mathbf{v}$, $\mathbf{z}_{\mathbf{v}_i} = G'(\mathbf{u})\mathbf{v}_i$, *and* $\mathbf{z}_{\mathbf{v}_1\mathbf{v}_2} = G''(\mathbf{u})(\mathbf{v}_1, \mathbf{v}_2)$, *then* $\mathbf{z}_\mathbf{v}$ *and* $\mathbf{z}_{\mathbf{v}_1\mathbf{v}_2}$ *are the unique solutions of the following equations*

$$\begin{cases} \dfrac{\partial \mathbf{z_v}}{\partial t} - \nu \Delta \mathbf{z_v} + (\mathbf{y_u} \cdot \nabla)\mathbf{z_v} + (\mathbf{z_v} \cdot \nabla)\mathbf{y_u} + \nabla p_\mathbf{v} = \mathbf{v} \ \ in \ \Omega_T, \\ div\, \mathbf{z_v} = 0 \ \ in \ \Omega_T, \ \mathbf{z_v}(0) = 0 \ \ in \ \Omega, \ \mathbf{z_v} = 0 \ \ on \ \Sigma_T, \end{cases} \tag{5}$$

$$\begin{cases} \dfrac{\partial \mathbf{z_{v_1 v_2}}}{\partial t} - \nu \Delta \mathbf{z_{v_1 v_2}} + (\mathbf{y_u} \cdot \nabla)\mathbf{z_{v_1 v_2}} + (\mathbf{z_{v_1 v_2}} \cdot \nabla)\mathbf{y_u} \\ \quad +(\mathbf{z_{v_2}} \cdot \nabla)\mathbf{z_{v_1}} + (\mathbf{z_{v_1}} \cdot \nabla)\mathbf{z_{v_2}} + \nabla p_{12} = 0 \ \ in \ \Omega_T, \\ div\, \mathbf{z_{v_1 v_2}} = 0 \ \ in \ \Omega_T, \ \mathbf{z_{v_1 v_2}}(0) = 0 \ \ in \ \Omega, \ \mathbf{z_{v_1 v_2}} = 0 \ \ on \ \Sigma_T, \end{cases} \tag{6}$$

for some $p_\mathbf{v}, p_{12} \in L^{\bar{p}}(0,T; W^{1,\bar{p}}(\Omega))$, *which are unique up to the addition of a function of* $L^2(0,T)$.

Theorem 3 ([6, 8]) *Under the assumptions of Theorem 2, the cost functional* J : $\mathbf{L}^{\bar{p}}(\Omega_T) \longrightarrow \mathbb{R}$ *is of class* C^∞ *and for every* $\mathbf{u}, \mathbf{v} \in \mathbf{L}^{\bar{p}}(\Omega_T)$ *we have*

$$J'(\mathbf{u})\mathbf{v} = \int_0^T \int_\Omega (\varphi_\mathbf{u} + \lambda \mathbf{u})\mathbf{v}\, \mathrm{d}x\mathrm{d}t, \tag{7}$$

$$J''(\mathbf{u})\mathbf{v}^2 = \int_0^T \int_\Omega (|\mathbf{z_v}|^2 - 2(\mathbf{z_v} \cdot \nabla)\mathbf{z_v}\varphi_\mathbf{u})\mathrm{d}x\mathrm{d}t + \lambda \int_0^T \int_\Omega |\mathbf{v}|^2 \mathrm{d}x\mathrm{d}t, \tag{8}$$

where $\mathbf{z_v} = G'(\mathbf{u})\mathbf{v}$ *is the solution of* (5) *and* $\varphi_\mathbf{u} \in \mathbf{W}^{2,1}_{\bar{p}}(\Omega_T) \cap C([0,T]; \mathbf{Y}_{\bar{p}})$ *is the unique element satisfying for every* $\mathbf{w} \in \mathbf{Y}_2$

$$\begin{cases} -(\varphi_{\mathbf{u},t}, \mathbf{w}) + a(\varphi_\mathbf{u}, \mathbf{w}) + c(\mathbf{w}, \mathbf{y_u}, \varphi_\mathbf{u}) + c(\mathbf{y_u}, \mathbf{w}, \varphi_\mathbf{u}) \\ = (\mathbf{y_u} - \mathbf{y}_d, \mathbf{w}) \quad \forall \mathbf{w} \in \mathbf{Y}, \\ \varphi_\mathbf{u}(T) = 0. \end{cases} \tag{9}$$

Before establishing the first order optimality conditions, we should observe that (P) is not convex, hence we should distinguish global and local solutions.

Definition 1 We say that a control $\bar{\mathbf{u}} \in \mathcal{U}_{ad}$ is a local minimum of (P) in the $\mathbf{L}^{\bar{p}}(\Omega_T)$ sense, $1 \leq \bar{p} \leq \infty$, if there exists $\varepsilon > 0$ such that $J(\bar{\mathbf{u}}) \leq J(\mathbf{u})$ for all $\mathbf{u} \in \mathcal{U}_{ad} \cap B_\varepsilon(\bar{\mathbf{u}})$, where $B_\varepsilon(\bar{\mathbf{u}})$ is the ball of $\mathbf{L}^{\bar{p}}(\Omega_T)$ centered at $\bar{\mathbf{u}}$ and radius ε. We say that $\bar{\mathbf{u}}$ is a strict local minimum if the previous inequality is strict for every $\mathbf{u} \neq \bar{\mathbf{u}}$.

Since $\mathcal{U}_{ad}$ is bounded in $\mathbf{L}^\infty(\Omega_T)$, it is immediate to check that $\bar{\mathbf{u}}$ is a local minimum in the $\mathbf{L}^{\bar{p}}(\Omega_T)$ sense with $\bar{p} < \infty$ if an only if it is a local minimum in the $\mathbf{L}^1(\Omega_T)$ sense. In addition, if $\bar{\mathbf{u}}$ is a local minimum in the $\mathbf{L}^\infty(\Omega_T)$ sense, then it is a local minimum in the $\mathbf{L}^{\bar{p}}(\Omega_T)$ sense for every $1 \le \bar{p} < \infty$. The contrary is not necessarily true. In the sequel, whenever we say that $\bar{\mathbf{u}}$ is a local minimum of (P), it should be understood in the $\mathbf{L}^2(\Omega_T)$ sense.

Now, the first order optimality conditions easily follow (see e.g. [6, 8, Theorem 3.3]).

Theorem 4 *Suppose that the assumptions of Theorem 2 hold. Let us assume that $\bar{\mathbf{u}}$ is a local solution of problem (P), then there exist $\bar{\mathbf{y}}, \bar{\boldsymbol{\varphi}} \in \mathbf{W}^{2,1}_{\bar{p}}(\Omega_T) \cap C([0,T], \mathbf{Y}_{\bar{p}})$ such that*

$$\begin{cases} (\bar{\mathbf{y}}_t, \mathbf{w}) + a(\bar{\mathbf{y}}, \mathbf{w}) + c(\bar{\mathbf{y}}, \bar{\mathbf{y}}, \mathbf{w}) = (\mathbf{f} + \bar{\mathbf{u}}, \mathbf{w}) \ \ \forall\, \mathbf{w} \in \mathbf{Y}, \\ \bar{\mathbf{y}}(0) = \mathbf{y}_0, \end{cases} \tag{10}$$

$$\begin{cases} -(\bar{\boldsymbol{\varphi}}_t, \mathbf{w}) + a(\bar{\boldsymbol{\varphi}}, \mathbf{w}) + c(\mathbf{w}, \bar{\mathbf{y}}, \bar{\boldsymbol{\varphi}}) + c(\bar{\mathbf{y}}, \mathbf{w}, \bar{\boldsymbol{\varphi}}) = (\bar{\mathbf{y}}_{\mathbf{u}} - \mathbf{y}_d, \mathbf{w}) \ \forall \mathbf{w} \in \mathbf{Y}, \\ \bar{\boldsymbol{\varphi}}(T) = 0, \end{cases} \tag{11}$$

$$\int_0^T \int_\Omega (\bar{\boldsymbol{\varphi}} + \lambda \bar{\mathbf{u}})(\mathbf{u} - \bar{\mathbf{u}})\, \mathrm{d}x\mathrm{d}t \ge 0 \ \ \forall \mathbf{u} \in \mathcal{U}_{ad}. \tag{12}$$

Corollary 1 *Under the assumptions of Theorem 4, the following properties hold for $j = 1, 2$:*

1- If $\lambda > 0$ then

$$\bar{u}_j(t,x) = \mathrm{Proj}_{[\alpha_j, \beta_j]}\left(-\frac{1}{\lambda}\bar{\varphi}_j(t,x)\right) \ \ \textit{for a.a. } (t,x) \in \Omega_T, \tag{13}$$

and hence $\bar{\mathbf{u}} \in \mathbf{W}^{1,\bar{p}}(\Omega_T) \cap C([0,T], \mathbf{W}^{1,\bar{p}}(\Omega_T))$ holds.

2- If $\lambda = 0$ then

$$\begin{cases} \bar{u}_j(t,x) = \alpha_j & \Rightarrow \bar{\varphi}_j(t,x) \ge 0, \\ \bar{u}_j(t,x) = \beta_j & \Rightarrow \bar{\varphi}_j(t,x) \le 0, \\ \alpha_j < \bar{u}_j(t,x) < \beta_j & \Rightarrow \bar{\varphi}_j(t,x) = 0, \end{cases} \ \textit{and} \ \begin{cases} \bar{\varphi}_j(t,x) > 0 \Rightarrow \bar{u}_j(t,x) = \alpha_j, \\ \bar{\varphi}_j(t,x) < 0 \Rightarrow \bar{u}_j(t,x) = \beta_j, \end{cases} \tag{14}$$

and hence $\bar{\mathbf{u}}$ is a bang-bang control if meas$\{(x,t) \in \Omega_T : |\bar{\varphi}(t,x)| \neq 0\} = 0$.

3 Second Order Analysis

Now, we are ready to state second order conditions. We note that it is possible to prove necessary and sufficient conditions similar to elliptic Navier–Stokes velocity tracking problem (see e.g, [15]). First, we define the cone of critical directions:

$$\mathcal{C}_{\bar{\mathbf{u}}} = \{\mathbf{v} \in L^2(0,T;\mathbf{L}^2(\Omega)) : \mathbf{v} \text{ satisfies } (16)-(17) \text{ and } J'(\bar{\mathbf{u}})\mathbf{v} = 0\}, \tag{15}$$

$$v_j(t,x) \geq 0 \text{ if } \alpha_j = \bar{u}_j(t,x), \tag{16}$$

$$v_j(t,x) \leq 0 \text{ if } \bar{u}_j(t,x) = \beta_j, \qquad j = 1,2. \tag{17}$$

Then, we have the following result; see [6].

Theorem 5 *Suppose that the assumptions of Theorem 2 hold. Let $\bar{\mathbf{u}}$ be a local solution of problem (P), then $J''(\bar{\mathbf{u}})\mathbf{v}^2 \geq 0\ \forall \mathbf{v} \in \mathcal{C}_{\bar{\mathbf{u}}}$. Conversely, if $\lambda > 0$ and $\bar{\mathbf{u}} \in \mathcal{U}_{ad}$ satisfies*

$$\begin{aligned} &J'(\bar{\mathbf{u}})(\mathbf{u}-\bar{\mathbf{u}}) \geq 0 \quad \forall \mathbf{u} \in \mathcal{U}_{ad}, \\ &J''(\bar{\mathbf{u}})\mathbf{v}^2 > 0 \quad \forall \mathbf{v} \in \mathcal{C}_{\bar{\mathbf{u}}} \setminus \{0\} \end{aligned} \tag{18}$$

then there exist $\varepsilon > 0$ and $\delta > 0$ such that

$$J(\bar{\mathbf{u}}) + \frac{\delta}{2}\|\mathbf{u}-\bar{\mathbf{u}}\|^2_{\mathbf{L}^2(\Omega_T)} \leq J(\mathbf{u}) \quad \forall \mathbf{u} \in \mathcal{U}_{ad} \cap B_\varepsilon(\bar{\mathbf{u}})$$

where $B_\varepsilon(\bar{\mathbf{u}})$ is the $\mathbf{L}^2(\Omega_T)$-ball of center $\bar{\mathbf{u}}$ and radius ε.

Now, we proceed to the case where $\lambda = 0$. We note that the absence of the Tikhonov regularizing term leads to bang-bang controls. The analysis is more delicate, and we refer the reader to [8] for the proofs. Our main focus here, is to highlight the differences between the two cases.

The condition $J''(\bar{\mathbf{u}})\mathbf{v}^2 > 0$ for all $\mathbf{v} \in \mathcal{C}_{\bar{\mathbf{u}}} \setminus \{0\}$ is not enough to deduce local optimality for $\bar{\mathbf{u}}$. This is usual in infinite dimension optimization problems. For $\lambda > 0$ the second order analysis is very similar to the finite dimensional case. For $\lambda = 0$ two differences appear. First we observe that for $\lambda > 0$ the condition $J''(\bar{\mathbf{u}})\mathbf{v}^2 > 0$ for all $\mathbf{v} \in \mathcal{C}_{\bar{\mathbf{u}}} \setminus \{0\}$ is equivalent to the existence of $\delta > 0$ such that $J''(\bar{\mathbf{u}})\mathbf{v}^2 \geq \delta\|\mathbf{v}\|^2_{\mathbf{L}^2(\Omega)}$ for all $\mathbf{v} \in \mathcal{C}_{\bar{\mathbf{u}}}$. This equivalence fails for $\lambda = 0$. Second, we have pointwise constraints for the controls, and hence we need to increase the cone of critical directions; see [22]. This extension of the cone is not necessary for $\lambda > 0$. To this end, for every $\varrho > 0$ we consider the extended cone

$$\mathcal{C}^{\varrho}_{\bar{\mathbf{u}}} = \{\mathbf{v} \in \mathbf{L}^2(\Omega_T) : \mathbf{v}\text{satisfies } (16)-(17) \text{ and } J'(\bar{\mathbf{u}})\mathbf{v} \leq \varrho\|\mathbf{z}_\mathbf{v}\|_{\mathbf{L}^2(\Omega_T)}\}, \tag{19}$$

The following theorem states the sufficient second order condition; see [8] for a detailed proof, and [3, 10, 11, 13, 14] for additional discussion on the sufficient second order conditions.

Theorem 6 *Suppose that the assumptions of Theorem 2 hold. Let us assume that* $\bar{\mathbf{u}} \in \mathcal{U}_{ad}$ *satisfies* (10)–(12) *along with the associated state and adjoint state* $(\bar{\mathbf{y}}, \bar{\boldsymbol{\varphi}}) \in (\mathbf{W}^{2,1}_{\bar{p}}(\Omega_T) \cap C([0,T]; \mathbf{Y}_{\bar{p}}))^2$. *We also suppose that*

$$\exists \varrho > 0 \ \text{and} \ \exists \delta > 0 \ : \ J''(\bar{\mathbf{u}})\mathbf{v}^2 \geq \delta \|\mathbf{z}_{\mathbf{v}}\|^2_{\mathbf{L}^2(\Omega_T)} \quad \forall \mathbf{v} \in C^{\varrho}_{\bar{\mathbf{u}}}. \tag{20}$$

Then, there exist $\varepsilon > 0$ *and* $\kappa > 0$ *such that the following inequality holds*

$$\frac{\kappa}{2} \|\mathbf{y}_{\mathbf{u}} - \bar{\mathbf{y}}\|^2_{\mathbf{L}^2(\Omega_T)} + J(\bar{\mathbf{u}}) \leq J(\mathbf{u}), \ \forall \mathbf{u} \in \mathcal{U}_{ad} \ \text{with} \ \|\mathbf{u} - \bar{\mathbf{u}}\|_{\mathbf{L}^2(\Omega_T)} < \varepsilon. \tag{21}$$

4 Approximation of the State and Adjoint-State Equations

A family of triangulations $\{\mathcal{K}_h\}_{h>0}$ of $\bar{\Omega}$, is constructed in the standard way. Two parameters h_K and ϱ_K are associated to each element $K \in \mathcal{K}_h$. Here, h_K denotes the diameter of the set K and ϱ_K is the diameter of the largest ball contained in K. The size of the mesh is denoted by $h = \max_{K \in \mathcal{K}_h} h_K$, and standard regularity assumptions on the triangulation are assumed:

1. There exist two positive constants $\varrho_{\mathcal{K}}$ and $\delta_{\mathcal{K}}$ such that $\frac{h_K}{\varrho_K} \leq \varrho_{\mathcal{K}}$ and $\frac{h}{h_K} \leq \delta_{\mathcal{K}}$ $\forall K \in \mathcal{K}_h$ and $\forall h > 0$.
2. Define $\overline{\Omega}_h = \cup_{K \in \mathcal{K}_h} K$, and let Ω_h and Γ_h denote its interior and its boundary, respectively. We assume that the vertices of $\mathcal{K}_h$ placed on the boundary Γ_h are points of Γ.

Since Ω is convex, from the last assumption we have that Ω_h is also convex. Moreover, we know that

$$|\Omega \setminus \Omega_h| \leq Ch^2; \tag{22}$$

see, for instance, [42, estimate (5.2.19)].

On the mesh $\mathcal{K}_h$ we consider two finite dimensional spaces $\mathbf{Z}_h \subset \mathbf{H}^1_0(\Omega)$ and $Q_h \subset L^2_0(\Omega)$ formed by piecewise polynomials in Ω_h, vanishing in $\Omega \setminus \Omega_h$ and satisfying the standard approximation properties of the usual finite elements considered in the discretization of Navier-Stokes equations: "Taylor-Hood", P1-Bubble finite element, and some others; see [28, Chap. 2]. In particular, we assume that:

(A1) If $\mathbf{z} \in \mathbf{H}^{1+l}(\Omega) \cap \mathbf{H}_0^1(\Omega)$, then

$$\inf_{\mathbf{z}_h \in \mathbf{Z}_h} \|\mathbf{z} - \mathbf{z}_h\|_{\mathbf{H}^s(\Omega_h)} \leq Ch^{l+1-s} \|\mathbf{z}\|_{\mathbf{H}^{1+l}(\Omega)}, \quad \text{for } 0 \leq l \leq 1 \text{ and } s = 0, 1. \tag{23}$$

(A2) If $q \in H^l(\Omega) \cap L_0^2(\Omega)$, then

$$\inf_{q_h \in Q_h} \|q - q_h\|_{L^2(\Omega_h)} \leq Ch\|q\|_{H^1(\Omega)}. \tag{24}$$

(A3) The subspaces $\mathbf{Z}_h$ and Q_h satisfy the inf-sup condition: $\exists c > 0$ such that

$$\inf_{q_h \in Q_h} \sup_{\mathbf{z}_h \in \mathbf{Z}_h} \frac{b(\mathbf{z}_h, q_h)}{\|\mathbf{z}_h\|_{\mathbf{H}^1(\Omega_h)} \|q_h\|_{L^2(\Omega_h)}} \geq c, \tag{25}$$

where $b : \mathbf{H}^1(\Omega) \times L^2(\Omega) \longrightarrow \mathbb{R}$ is defined by

$$b(\mathbf{z}, q) = \int_\Omega q(x) \operatorname{div} \mathbf{z}(x) \, dx.$$

We also consider a subspace $\mathbf{Y}_h$ of $\mathbf{Z}_h$ defined by

$$\mathbf{Y}_h = \{\mathbf{y}_h \in \mathbf{Z}_h : b(\mathbf{y}_h, q_h) = 0 \ \forall q_h \in Q_h\}.$$

The discretization in time is based on the lowest order discontinuous (in time) Galerkin approach. First, we consider a grid of points $0 = t_0 < t_1 < \ldots < t_{N_\tau} = T$, and we denote $\tau_n = t_n - t_{n-1}$. We make the following assumption,

$$\exists \varrho_0 > 0 \text{ s.t. } \tau = \max_{1 \leq n \leq N_\tau} \tau_n < \varrho_0 \tau_n \ \ \forall 1 \leq n \leq N_\tau \text{ and } \forall \tau > 0. \tag{26}$$

Given a triangulation $\mathcal{K}_h$ of Ω and a grid of points $\{t_n\}_{n=0}^{N_\tau}$ of $[0, T]$, we set $\sigma = (\tau, h)$. Finally, we consider the following spaces

$$\mathcal{Y}_\sigma = \{\mathbf{y}_\sigma \in L^2(0, T; \mathbf{Y}_h) : \mathbf{y}_\sigma|_{(t_{n-1}, t_n)} \in \mathbf{Y}_h \ \text{ for } \ 1 \leq n \leq N_\tau\},$$

$$\mathcal{Q}_\sigma = \{q_\sigma \in L^2(0, T; Q_h) : q_\sigma|_{(t_{n-1}, t_n)} \in Q_h \ \text{ for } \ 1 \leq n \leq N_\tau\}.$$

We have that the functions of $\mathcal{Y}_\sigma$ and $\mathcal{Q}_\sigma$ are piecewise constant in time. The elements of $\mathcal{Y}_\sigma$ can be written in the form

$$\mathbf{y}_\sigma = \sum_{n=1}^{N_\tau} \mathbf{y}_{n,h} \chi_n, \quad \text{with } \mathbf{y}_{n,h} \in \mathbf{Y}_h, \tag{27}$$

where χ_n is the characteristic function of (t_{n-1}, t_n). For every discrete state $\mathbf{y}_\sigma$ we will fix $\mathbf{y}_\sigma(t_n) = \mathbf{y}_{n,h}$, so that $\mathbf{y}_\sigma$ is continuous on the left. In particular, we have $\mathbf{y}_\sigma(T) = \mathbf{y}_\sigma(t_{N_\tau}) = y_{N_\tau,h}$.

4.1 The Discrete State Equation

Our first goal is to discretize the state equation (1) or equivalently (2). We employ the lowest order discontinuous time-stepping Galerkin method in time, i.e., piecewise constants in time while for the spatial discretization we use conforming finite element spaces. For any $\mathbf{u} \in L^2(0,T;\mathbf{L}^2(\Omega))$ the discrete state equation is given by

$$\begin{cases} \text{For } n = 1, \ldots, N_\tau, \\ \left(\dfrac{\mathbf{y}_{n,h} - \mathbf{y}_{n-1,h}}{\tau_n}, \mathbf{w}_h\right) + a(\mathbf{y}_{n,h}, \mathbf{w}_h) + c(\mathbf{y}_{n,h}, \mathbf{y}_{n,h}, \mathbf{w}_h) \\ = (\mathbf{f}_n + \mathbf{u}_n, \mathbf{w}_h) \quad \forall \mathbf{w}_h \in \mathbf{Y}_h, \\ \mathbf{y}_{0,h} = \mathbf{y}_{0h}, \end{cases} \tag{28}$$

where

$$(\mathbf{f}_n, \mathbf{w}_h) = \frac{1}{\tau_n}\int_{t_{n-1}}^{t_n} (\mathbf{f}(t), \mathbf{w}_h)\mathrm{d}t,\ (\mathbf{u}_n, \mathbf{w}_h) = \frac{1}{\tau_n}\int_{t_{n-1}}^{t_n} (\mathbf{u}(t), \mathbf{w}_h)\mathrm{d}t, \tag{29}$$

$$\mathbf{y}_{0h} \in \mathbf{Y}_h \text{ with } \|\mathbf{y}_0 - \mathbf{y}_{0h}\|_{\mathbf{L}^2(\Omega_h)} \le Ch, \text{ and } \|\mathbf{y}_{0h}\|_{\mathbf{H}^1(\Omega_h)} \le C. \tag{30}$$

It well known that the discrete equation (28) has at least one solution. Concerning uniqueness and error estimates under the prescribed regularity assumptions, the following results were proven in [6, Theorem 4.7], and [7, Theorem 12]

Theorem 7 *Given* $\mathbf{u} \in \mathbf{L}^2(\Omega_T)$, *let us denote the solution of* (2) *by* $\mathbf{y} \in \mathbf{H}^{2,1}(\Omega_T) \cap C([0,T];\mathbf{Y})$, *and let* $\mathbf{y}_\sigma \in \mathcal{Y}_\sigma$ *be any solution of* (28)-(29)-(30). *Then, there exists a constant* $C > 0$ *independent of* $\mathbf{u}$, $\mathbf{y}$ *and* σ *such that*

$$\max_{1\le n\le N_\tau} \|\mathbf{y}(t_n) - \mathbf{y}_\sigma(t_n)\| + \|\mathbf{y} - \mathbf{y}_\sigma\|_{L^2(0,T;\mathbf{H}^1(\Omega_h))}$$
$$\le C\left\{\frac{\tau}{h}\|\mathbf{y}'\|_{L^2(0,T;\mathbf{L}^2(\Omega))} + h\|\mathbf{y}\|_{L^2(0,T;\mathbf{H}^2(\Omega))} + h\|\mathbf{y}_0\|_{\mathbf{H}^1(\Omega)}\right\}. \tag{31}$$

$$\|\mathbf{y} - \mathbf{y}_\sigma\|_{L^\infty(0,T;\mathbf{L}^2(\Omega_h))} \le C\left\{\left(\frac{\tau}{h} + \sqrt{\tau}\right)\|\mathbf{y}'\|_{L^2(0,T;\mathbf{L}^2(\Omega))}\right.$$
$$\left. + h\|\mathbf{y}\|_{L^2(0,T;\mathbf{H}^2(\Omega))} + h\|\mathbf{y}_0\|_{\mathbf{H}^1(\Omega)}\right\}. \tag{32}$$

Moreover, if there exists a constant $C_0 > 0$ *such that* $\tau \le C_0 h^2$ *for every* $\sigma = (\tau, h)$, *then* $\{\mathbf{y}_\sigma\}_\sigma$ *is bounded in* $L^\infty(0,T;\mathbf{H}^1(\Omega_h))$ *and* (28) *has a unique solution.*

Moreover, the following estimate holds:

$$\|\mathbf{y} - \mathbf{y}_\sigma\|_{L^2(0,T;\mathbf{L}^2(\Omega))} \leq Ch^2 \quad \forall \mathbf{u} \in \mathcal{U}_{ad}, \tag{33}$$

where C is independent of σ.

A few remarks are under way:

1. Standard techniques developed for the numerical analysis of the uncontrolled Navier-Stokes equations can not be directly applied in the optimal control setting due to the limited regularity in presence of control constraints. We note that standard bootstrap arguments fail to increase regularity for the state, adjoint and control variables. The case of $\lambda = 0$ is even more restrictive in terms of the available regularity. Hence, we view the results of the above theorem (in terms of the available regularity) optimal.
2. The proposed numerical scheme, based on the discontinuous time-stepping Galerkin dG(0) scheme, is the implicit Euler scheme. However the analysis of the scheme is performed in a totally discontinuous (in time) fashion, in order to avoid any additional regularity assumption. In particular, we note that the proof is based the construction on locally (in time) $L^2(t^{n-1}, t^n; \mathbf{L}^2(\Omega))$ projections, as well as on suitable duality arguments in a way to avoid the use of global (in time) interpolants.
3. The two parameters τ and h need to satisfy the assumption $\tau \leq Ch^2$ in order to prove that the discrete equation has a unique solution, and our estimate is optimal in $L^2(0, T; \mathbf{H}^1(\Omega))$ norms for the state and adjoint. We emphasize that if we discretize the state equation only in time, not in space, then we cannot prove uniqueness of a solution for the resulting elliptic system. Indeed, this discrete elliptic system is very close to the stationary Navier-Stokes system, for which there is no a uniqueness result. Therefore, it is not surprising that the discretization parameter τ is needed to be small compared with h if we want to prove the uniqueness of a solution for the full discrete system. The key idea of [6] was to utilize ideas from [15] developed for the stationary Navier-Stokes, together with a detailed error analysis of the uncontrolled state and adjoint equations of the underlying scheme.

4.2 The Discrete Adjoint-State Equation

Associated to the discrete state equation (28), the cost functional J is approximated by $J_\sigma : \mathbf{L}^2(\Omega_T) \longrightarrow \mathbb{R}$

$$J_\sigma(\mathbf{u}) = \frac{1}{2}\int_0^T \int_{\Omega_h} |\mathbf{y}_\sigma - \mathbf{y}_d|^2 \, \mathrm{d}x \, \mathrm{d}t + \frac{\lambda}{2}\int_0^T \int_{\Omega_h} |\mathbf{u}|^2 \, \mathrm{d}x \, \mathrm{d}t$$

and we have a first expression of its derivative as follows

$$J'_\sigma(\mathbf{u})\mathbf{v} = \int_0^T \int_{\Omega_h} (\mathbf{y}_\sigma - \mathbf{y}_d)\mathbf{z}_\sigma \,\mathrm{d}x\mathrm{d}t + \lambda \int_0^T \int_{\Omega_h} \mathbf{u}\mathbf{v} \,\mathrm{d}x\mathrm{d}t,$$

where $\mathbf{y}_\sigma = \mathbf{y}_\sigma(\mathbf{u})$ is the discrete state corresponding to the control $\mathbf{u}$ and $\mathbf{z}_\sigma$ is the solution of the linearized equation

$$\begin{cases} \text{For } n = 1, \ldots, N_\tau, \\ \left(\dfrac{\mathbf{z}_{n,h} - \mathbf{z}_{n-1,h}}{\tau_n}, \mathbf{w}_h\right) + a(\mathbf{z}_{n,h}, \mathbf{w}_h) + c(\mathbf{z}_{n,h}, \mathbf{y}_{n,h}, \mathbf{w}_h) \\ +c(\mathbf{y}_{n,h}, \mathbf{z}_{n,h}, \mathbf{w}_h) = \frac{1}{\tau_n} \int_{t_{n-1}}^{t_n} (\mathbf{v}(t), \mathbf{w}_h) \,\mathrm{d}t \ \forall \mathbf{w}_h \in \mathbf{Y}_h, \\ \mathbf{z}_{0,h} = 0; \end{cases} \tag{34}$$

see [6, Theorem 4.12]. By using the adjoint state equation

$$\begin{cases} \text{for } n = N_\tau, \ldots, 1, \ \text{ and } \forall \mathbf{w}_h \in \mathbf{Y}_h, \\ \left(\dfrac{\boldsymbol{\varphi}_{n,h} - \boldsymbol{\varphi}_{n+1,h}}{\tau_n}, \mathbf{w}_h\right) + a(\boldsymbol{\varphi}_{n,h}, \mathbf{w}_h) + c(\mathbf{w}_h, \mathbf{y}_{n,h}, \boldsymbol{\varphi}_{n,h}) \\ +c(\mathbf{y}_{n,h}, \mathbf{w}_h, \boldsymbol{\varphi}_{n,h}) = \frac{1}{\tau_n} \int_{t_{n-1}}^{t_n} (\mathbf{y}_{n,h} - \mathbf{y}_d(t), \mathbf{w}_h) \,\mathrm{d}t, \\ \boldsymbol{\varphi}_{N_\tau+1,h} = 0, \end{cases} \tag{35}$$

the derivative of J_σ can be expressed as

$$J'_\sigma(\mathbf{u})\mathbf{v} = \int_0^T \int_{\Omega_h} (\boldsymbol{\varphi}_\sigma + \lambda \mathbf{u})\mathbf{v} \,\mathrm{d}x\mathrm{d}t. \tag{36}$$

Observe that in the above system (35), first we compute $\boldsymbol{\varphi}_{N_\tau,h}$ from $\boldsymbol{\varphi}_{N_\tau+1,h} = 0$ and then we descend in n until $n = 1$. Unlike to the discrete states $\mathbf{y}_\sigma$ where we fix $\mathbf{y}_\sigma(t_n) = \mathbf{y}_{n,h}$, we will set for the discrete adjoint states $\boldsymbol{\varphi}_\sigma(t_{n-1}) = \boldsymbol{\varphi}_{n,h}$ for every $1 \le n \le N_\tau$. Analogously to Theorem 7, we have the following result.

Theorem 8 *Let* $\mathbf{u} \in \mathbf{L}^2(\Omega_T)$, $\boldsymbol{\varphi}_u \in \mathbf{H}^{2,1}(\Omega_T) \cap C([0,T];\mathbf{Y})$ *the solution of* (9) *and* $\boldsymbol{\varphi}_\sigma \in \mathcal{Y}_\sigma$ *the solution of the discrete equation* (35). *Suppose that there exists a constant* $C_0 > 0$ *such that* $\tau \le C_0 h^2$ *for every* $\sigma = (\tau, h)$. *Then, there exists a constant* $C > 0$ *independent of* σ *such that for all* $\mathbf{u} \in \mathcal{U}_{ad}$

$$\begin{aligned} &\|\boldsymbol{\varphi}_\mathbf{u} - \boldsymbol{\varphi}_\sigma\|_{L^2(0,T;\mathbf{L}^2(\Omega_h))} \\ &+h(\|\boldsymbol{\varphi}_\mathbf{u} - \boldsymbol{\varphi}_\sigma\|_{L^\infty(0,T;\mathbf{L}^2(\Omega_h))} + \|\boldsymbol{\varphi}_\mathbf{u} - \boldsymbol{\varphi}_\sigma\|_{L^2(0,T;\mathbf{H}^1(\Omega_h))}) \le Ch^2. \end{aligned} \tag{37}$$

5 Error Estimates of the Discrete Optimal Control Problem

We define the discrete control problem as follows

$$(\mathrm{P}_\sigma) \quad \begin{cases} \min J_\sigma(\mathbf{u}_\sigma) \\ \mathbf{u}_\sigma \in \mathcal{U}_{\sigma,ad} \end{cases}$$

where different possibilities will be analyzed for $\mathcal{U}_{\sigma,ad} \equiv \mathcal{U}_\sigma \cap \mathcal{U}_{ad}$.

In this final section we present the main results of [7, 8]. We also comment on the fine relation between regularity of optimal solutions and the corresponding rates of convergence.

5.1 *The Case* $\lambda > 0$

In this case three different choices for $\mathcal{U}_{\sigma,ad}$ are considered.

1. *Piecewise constant controls:*

$$\mathbf{U}_h = \mathbf{U}_{h,0} = \{\mathbf{u}_h \in \mathbf{L}^2(\Omega_h) : \mathbf{u}_{h|K} \equiv \mathbf{u}_K \in \mathbb{R}^2 \ \ \forall K \in \mathcal{K}_h\}$$

and

$$\mathcal{U}_\sigma = \mathcal{U}_{\sigma,0} = \{\mathbf{u}_\sigma \in L^2(0,T;\mathbf{U}_h) : \mathbf{u}_\sigma|_{(t_{n-1},t_n]} \in \mathbf{U}_h, \ \text{ for } 1 \le n \le N_\tau\}.$$

2. *Piecewise linear controls:*

$$\mathbf{U}_h = \mathbf{U}_{h,1} = \{\mathbf{u}_h \in \mathbf{C}(\bar{\Omega}_h) : \mathbf{u}_{h|K} \in \mathcal{P}_1(K)^2 \ \ \forall K \in \mathcal{K}_h\}$$

and

$$\mathcal{U}_\sigma = \mathcal{U}_{\sigma,1} = \{\mathbf{u}_\sigma \in L^2(0,T;\mathbf{U}_h) : \mathbf{u}_\sigma|_{(t_{n-1},t_n)} \in \mathbf{U}_h \ \text{ for } \ 1 \le n \le N_\tau\}.$$

3. *Variational discretization:*

$$\mathbf{U}_h = \mathbf{U}_{h,2} = \mathbf{L}^2(\Omega_h) \text{ and } \mathcal{U}_\sigma = \mathcal{U}_{\sigma,2} = L^2(0,T;\mathbf{L}^2(\Omega_h)).$$

For any of the above choices, the discrete problem (P_σ) has at least one solution, and the family of problems (P_σ) realizes a good approximation of problem (P). We refer the reader to [6, Theorems 4.13 and 4.15] for a detailed proof.

Theorem 9 *For every* $\sigma = (\tau, h)$ *let* $\bar{\mathbf{u}}_\sigma$ *be a global solution of problem* (P_σ)*. Then the sequence* $\{\bar{\mathbf{u}}_\sigma\}_\sigma$ *is bounded in* $\mathbf{L}^2(\Omega_T)$ *and there exist subsequences, denoted in* *the same way, converging to a point* $\bar{\mathbf{u}}$ *weakly in* $\mathbf{L}^2(\Omega_T)$. *Any of these limit points is a solution of problem (P). Moreover, we have*

$$\lim_{\sigma\to 0}\|\bar{\mathbf{u}}-\bar{\mathbf{u}}_\sigma\|_{L^2(0,T;\mathbf{L}^2(\Omega_h))}=0 \quad \text{and} \quad \lim_{\sigma\to 0}J_\sigma(\bar{\mathbf{u}}_\sigma)=J(\bar{\mathbf{u}}). \tag{38}$$

In addition, let $\bar{\mathbf{u}}$ *be a strict local minimum of (P), then there exists a sequence* $\{\bar{\mathbf{u}}_\sigma\}_\sigma$ *of local minima of problems* (P_σ) *such that* (38) *holds.*

Now, we are ready to proceed to main results regarding convergence rates. In the remaining of this section, $\bar{\mathbf{u}}$ denotes a local solution of (P) and for every σ, $\bar{\mathbf{u}}_\sigma$ denotes a local solution of (P_σ) such that $\|\bar{\mathbf{u}}-\bar{\mathbf{u}}_\sigma\|_{L^2(0,T;\mathbf{L}^2(\Omega_h))}\to 0$; see Theorem 9. We also denote by $\bar{\mathbf{y}}$ and $\bar{\boldsymbol{\varphi}}$ the state and adjoint state associated to $\bar{\mathbf{u}}$, and $\bar{\mathbf{y}}_\sigma$ and $\bar{\boldsymbol{\varphi}}_\sigma$ will denote the discrete state and adjoint state corresponding to $\bar{\mathbf{u}}_\sigma$. The goal is to estimate the rate of the convergence $(\bar{\mathbf{u}}_\sigma,\bar{\mathbf{y}}_\sigma,\bar{\boldsymbol{\varphi}}_\sigma)\to(\bar{\mathbf{u}},\bar{\mathbf{y}},\bar{\boldsymbol{\varphi}})$.

As in [6, 12, Sect. 4], all the elements $\mathbf{u}_\sigma\in\mathcal{U}_\sigma$, for $\mathcal{U}_\sigma=\mathcal{U}_{\sigma,0}$ and $\mathcal{U}_\sigma=\mathcal{U}_{\sigma,1}$, are extended to $(0,T)\times\Omega$ by setting $\mathbf{u}_\sigma(x,t)=\bar{\mathbf{u}}(x,t)$ for $(x,t)\in(0,T)\times(\Omega\setminus\Omega_h)$.

Let us write $\bar{\mathbf{u}}=(\bar{u}_1,\bar{u}_2)$. Associated to the components $\bar{u}_j$, $j=1,2$, for every $t\in(0,T)$, we split the elements (I_n,K), with $I_n=(t_{n-1},t_n]$ and $K\in\mathcal{K}_h$, as follows: $\mathcal{T}_\sigma=\mathcal{T}^j_{\sigma,1}\cup\mathcal{T}^j_{\sigma,2}\cup\mathcal{T}^j_{\sigma,3}$, $j=1,2$, where

$$\mathcal{T}_\sigma=\{K\times I_n : 1\le n\le N_\tau \ \text{ and } \ K\in\mathcal{K}_h\},$$

$$\mathcal{T}^j_{\sigma,1}=\{K\times I_n\in\mathcal{T}_\sigma : \bar{\varphi}_j(x,t)+\lambda\bar{u}_j(x,t)\neq 0\ \forall(x,t)\in K\times I_n\},$$

$$\mathcal{T}^j_{\sigma,2}=\{K\times I_n\in\mathcal{T}_\sigma : \bar{\varphi}_j(x,t)+\lambda\bar{u}_j(x,t)= 0\ \forall(x,t)\in K\times I_n\},$$

$$\mathcal{T}^j_{\sigma,3}=\mathcal{T}_\sigma\setminus(\mathcal{T}^j_{\sigma,1}\cup\mathcal{T}^j_{\sigma,2}).$$

Finally, let us denote

$$\begin{aligned}E_\sigma &= \|\bar{\mathbf{u}}-\bar{\mathbf{u}}_\sigma\|_{L^2(0,T;\mathbf{L}^2(\Omega_h))}\\ &\quad+\|\bar{\mathbf{y}}-\bar{\mathbf{y}}_\sigma\|_{L^2(0,T;\mathbf{L}^2(\Omega_h))}+\|\bar{\boldsymbol{\varphi}}-\bar{\boldsymbol{\varphi}}_\sigma\|_{L^2(0,T;\mathbf{L}^2(\Omega_h))},\end{aligned} \tag{39}$$

$$\begin{aligned}\mathcal{E}_\sigma &= \|\bar{\mathbf{y}}-\bar{\mathbf{y}}_\sigma\|_{L^\infty(0,T;\mathbf{L}^2(\Omega_h))}+\|\bar{\mathbf{y}}-\bar{\mathbf{y}}_\sigma\|_{L^2(0,T;\mathbf{H}^1(\Omega_h))}\\ &\quad+\|\bar{\boldsymbol{\varphi}}-\bar{\boldsymbol{\varphi}}_\sigma\|_{L^\infty(0,T;\mathbf{L}^2(\Omega_h))}+\|\bar{\boldsymbol{\varphi}}-\bar{\boldsymbol{\varphi}}_\sigma\|_{L^2(0,T;\mathbf{H}^1(\Omega_h))}.\end{aligned} \tag{40}$$

Then we have the following error estimates.

Theorem 10 *Suppose that* (18) *holds, and there exists a constant* $C_0>0$ *such that* $\tau\le C_0h^2$ *for every* $\sigma=(\tau,h)$. *Moreover, if* $\mathcal{U}_\sigma=\mathcal{U}_{\sigma,1}$ *we also assume that* $\mathbf{y}_d\in\mathbf{L}^p(\Omega_T))$ *with* $3<p<+\infty$ *and for some constant* $M>0$

$$\sum_{j=1}^{2}\sum_{K\times I_n\in\mathcal{T}^j_{\sigma,3}}|K|\tau_n\le Mh. \tag{41}$$

Then, then we have the following estimates

$$E_\sigma \leq \begin{cases} Ch & if\, \mathcal{U}_\sigma = \mathcal{U}_{\sigma,0}, \\ Ch^{\frac{3}{2}-\frac{2}{p}} & if\, \mathcal{U}_\sigma = \mathcal{U}_{\sigma,1}, \\ Ch^2 & if\, \mathcal{U}_\sigma = \mathcal{U}_{\sigma,2}, \end{cases} \tag{42}$$

$$\mathcal{E}_\sigma \leq Ch \text{ in all cases.} \tag{43}$$

A few comments follow:

- For the estimates of (43), we point out that they are optimal in the natural energy norm under the given regularity. In particular, given the regularity of strong solutions for the evolutionary Stokes and Navier-Stokes equations, it is not possible to improve the estimate. Hence, it seems that if we are interested in the energy norm, then the choice of piecewise constants controls is the most effective in the computational point of view.
- The splitting of the general elements $\mathcal{T}_\sigma = \mathcal{T}^j_{\sigma,1} \cup \mathcal{T}^j_{\sigma,2} \cup \mathcal{T}^j_{\sigma,3}, j = 1, 2$ is performed in way to distinguish the role of active and inactive space - time elements. The related assumption (41) is similar to one of [40, 44], and it is valid in many cases (see related discussions in [40, 44]). Let us observe that the set of points $\{(x,t) : \bar{\varphi}_j(x,t) + \lambda \bar{u}_j(x,t) = 0\}$ is usually formed by isolated points (x,t), curves or surfaces in Ω_T. The amount of cells $K \times I_n$ intersecting such surfaces is typically at most of order $\frac{1}{h\tau}$. This number is smaller for points or curves. This justifies the assumption (41).

5.2 The Case $\lambda = 0$

When the Tikhonov regularization term is absent the situation is much more complicated. First the absence of the projection formula (13) severely complicates the numerical analysis since there is no possibility to recover additional regularity for the controls through the adjoint via classical bootstrap arguments. As a consequence it is not clear if there is any possibility to recover the improved rate in $\mathbf{L}^2(\Omega_T)$ norm when piecewise linears are being used for the approximation of the controls. To this end we restrict our results to the cases $\mathcal{U}_\sigma = \mathcal{U}_{\sigma,0}$. Hence, the discrete control problem can be formulated as follows:

$$(\mathrm{P}_\sigma) \quad \begin{cases} \min J_\sigma(\mathbf{u}_\sigma) \\ \mathbf{u}_\sigma \in \mathcal{U}_{\sigma,ad} \end{cases}$$

where $\mathcal{U}_{\sigma,ad} = \mathcal{U}_{ad} \cap \mathcal{U}_\sigma$.

It is easy to prove that for every σ, the control problem (P_σ) has at least one solution. In the next theorem we state the results of [8, Theorems 4.4 and 4.6] regarding the convergence of the solutions of (P_σ) towards solutions of (P).

Theorem 11 *1. Let $\{\bar{\mathbf{u}}_\sigma\}_\sigma$ be a sequence of solutions of problems (P_σ) and let $\{\bar{\mathbf{y}}_\sigma\}_\sigma$ be the associated discrete states. Then, if $\bar{\mathbf{u}}$ is the weak limit in $\mathbf{L}^2(\Omega_T)$ of $\{\bar{\mathbf{u}}_\sigma\}_\sigma$ as $\sigma \to 0$, then $\bar{\mathbf{u}}$ is a solution of (P). Moreover, $\{\bar{\mathbf{y}}_\sigma\}_\sigma$ converges strongly to $\bar{\mathbf{y}}$ in $\mathbf{L}^2(\Omega_T)$, where $\bar{\mathbf{y}}$ is the continuous state associated with $\bar{\mathbf{u}}$. In addition, if $\bar{\mathbf{u}}$ is a bang-bang control, then $\bar{\mathbf{u}}_\sigma \to \bar{\mathbf{u}}$ as $\sigma \to 0$ strongly in $\mathbf{L}^p(\Omega_T)$ for every $1 \leq p < +\infty$.*

2. *Let $\bar{\mathbf{u}}$ be a strict local minimum of (P). Let $\bar{\mathbf{y}}$ and $\bar{\varphi}$ be the state and adjoint state, respectively. Let us assume that $\bar{\mathbf{u}}$ is bang-bang control. Then, there exist $\varepsilon > 0$, $\sigma_0 > 0$ and a sequence $\{\bar{\mathbf{u}}_\sigma\}_{|\sigma| \leq \sigma_0}$, such that each $\bar{\mathbf{u}}_\sigma$ is a local solution of (P_σ) satisfying*

   (a) $J_\sigma(\bar{\mathbf{u}}_\sigma) \leq J_\sigma(\mathbf{u}_\sigma)\ \forall \mathbf{u}_\sigma \in \mathcal{U}_{\sigma,ad} \cap \bar{B}_\varepsilon(\bar{\mathbf{u}})$, $\bar{B}_\varepsilon(\bar{\mathbf{u}})$ denoting the $\mathbf{L}^2(\Omega_T)$ ball.
   (b) $\bar{\mathbf{u}}_\sigma \to \bar{\mathbf{u}}$ strongly in $\mathbf{L}^p(\Omega_T)$ for every $1 \leq p < +\infty$.

Finally, we state the result regarding convergence rates:

Theorem 12 *Let $\bar{\mathbf{u}}$ be a local solution of (P) with associated state $\bar{\mathbf{y}}$ and adjoint state $\bar{\varphi}$. Suppose that* (20) *holds, and there exists a constant $C_0 > 0$ such that $\tau \leq C_0 h^2$ for every $\sigma = (\tau, h)$. Let $\{\bar{\mathbf{u}}_\sigma\}_\sigma$ be a sequence of local minima of problems (P_σ) such that $J_\sigma(\bar{\mathbf{u}}_\sigma) \leq J_\sigma(\mathbf{u}_\sigma)\ \forall \mathbf{u}_\sigma \in \mathcal{U}_{\sigma,ad} \cap \bar{B}_\varepsilon(\bar{\mathbf{u}})$, where $\bar{B}_\varepsilon(\bar{\mathbf{u}})$ denotes a $\mathbf{L}^2(\Omega_T)$ ball, and $\bar{\mathbf{u}}_\sigma \to \bar{\mathbf{u}}$ in $\mathbf{L}^2(\Omega_T)$. Let $\{\bar{\mathbf{y}}_\sigma\}_\sigma$ be the corresponding discrete states. Then, there exists a constant $C > 0$ independent of σ such that*

$$\lim_{|\sigma| \to 0} \frac{1}{\sqrt{|\sigma|}} \|\bar{\mathbf{y}}_\sigma - \bar{\mathbf{y}}\|_{\mathbf{L}^2(\Omega_T)} = 0. \tag{44}$$

Remark 1 We note that when $\lambda = 0$, the discrete optimality condition takes the form

$$\int_0^T \int_{\Omega_h} \bar{\varphi}_\sigma(x,t)(\mathbf{u}_\sigma(x,t) - \bar{\mathbf{u}}_\sigma(x,t))\, dx\, dt \geq 0 \ \ \forall \mathbf{u}_\sigma \in \mathcal{U}_{\sigma,ad}.$$

Similar to the continuous case, this is a fundamental difference between the case $\lambda > 0$ where the discrete optimality condition takes the form,

$$\int_0^T \int_{\Omega_h} (\bar{\varphi}_\sigma(x,t) + \lambda \bar{\mathbf{u}}_\sigma(x,t))(\mathbf{u}_\sigma(x,t) - \bar{\mathbf{u}}_\sigma(x,t))\, dx\, dt \geq 0 \ \ \forall \mathbf{u}_\sigma \in \mathcal{U}_{\sigma,ad}.$$

In this case the standard projection relation between the control and adjoint is available, and the estimates of Theorems 7 and 8 combined with the second order analysis of Theorem 5, imply estimates for the control, state and adjoint variables (see e.g. [6, 7]). On the other hand when $\lambda = 0$, in addition to the lack of regularity for the controls, the classical 'bootstrap' argument fails to imply estimates for the control and the adjoint, when combined with the second order analysis of Theorem 6 (see e.g. [8]). However, it is still possible to obtain an estimate for the difference between the state, and its discrete approximation. This is due to the fact that using Theorem 6, only the difference between the state and its discrete version is involved when deriving error estimates for the states. We also note that that despite the lack of regularity on the controls, the estimates presented in Theorem 7 are still applicable.

Acknowledgements The author "Eduardo Casas" was partially supported by the Spanish Ministerio de Economía y Competitividad under project MTM2014-57531-P.

References

1. Abergel, F., Temam, R.: On some control problems in fluid mechanics. Theor. Comput. Fluid Dyn. **1**, 303–325 (1990)
2. Apel, T., Flaig, T.: Crank-Nicolson schemes for optimal control problems with evolution equations. SIAM J. Numer. Anal. **50**(3), 1484–1512 (2012)
3. Bonnans, J.F., Zidani, H.: Optimal control problems with partially polyhedric constraints. SIAM J. Control Optim. **37**(6), 1726–1741 (1999)
4. Casas, E.: An optimal control problem governed by the evolution Navier-Stokes equations. In: Sritharan, S.S. (ed.) Optimal Control of Viscous Flows. Frontiers in Applied Mathematics. SIAM, Philadelphia (1998)
5. Casas, E.: Second order analysis for bang-bang control problems of PDEs. SIAM J. Control Optim. **50**(4), 2356–2372 (2012)
6. Casas, E., Chrysafinos, K.: A discontinuous Galerkin time-stepping scheme for the velocity tracking problem. SIAM J. Numer. Anal. **50**(5), 2281–2306 (2012)
7. Casas, E., Chrysafinos, K.: Error estimates for the discretization of the velocity tracking problem. Numer. Math. **130**, 615–643 (2015)
8. Casas, E., Chrysafinos, K.: Error estimates for the approximation of the velocity tracking problem with bang-bang controls. ESAIM: COCV (to appear)
9. Casas, E., Chrysafinos, K.: Analysis of the velocity tracking control problem for the 3D evolutionary Navier-Stokes equations. SIAM J. Control Optim. **54**(1), 99–128 (2016)
10. Casas, E., Mateos, M.: Second order optimality conditions for semilinear elliptic control problems with finitely many state constraints. SIAM J. Control Optim. **40**(5), 1431–1454 (2002)
11. Casas, E., Raymond, J.-P.: Error estimates for the numerical approximation of Dirichlet boundary control for semilinear elliptic equations. SIAM J. Control Optim. **45**(5), 1586–1611 (2006)
12. Casas, E., Tröltzsch, F.: A general theorem on error estimates with application to elliptic optimal control problems. Comput. Optim. Appl. (2012) doi:10.1007/s10589-011-9453-8
13. Casas, E., Troltzsch, F.: Second order analysis for optimal control problems: improving results expected from abstract theory. SIAM J. Optim. **22**(1), 261–279 (2012)
14. Casas, E., Troltzsch, F.: Second order optimality conditions and their role in pde control. Jahresbericht der Deutschen Mathematiker-Vereinigung **117**, 3–44 (2015)
15. Casas, E., Mateos, M., Raymond, J.-P.: Error estimates for the numerical approximation of a distributed control problem for the steady-state navier-stokes equations. SIAM J. Control Optim. **46**(3), 952–982 (2007)
16. Casas, E., Ryll, C., Tröltzsch, F.: second order and stability analysis for optimal sparse control of the FitzHugh-Nagumo equation. SIAM J. Control Optim. **53**(4), 2168–2202 (2015)
17. Chrysafinos, K.: Discontinuous Galerkin approximations for distributed optimal control problems constrained by parabolic pde's. Int. J. Numer. Anal. Model. **4**(3–4), 690–712 (2008)
18. Chrysafinos, K.: Analysis and finite element approximations for distributed optimal control problems for implicit parabolic PDE's. J. Comput. Appl. Math. **231**, 327–348 (2009)

19. Chrysafinos, K.: Convergence of discontinuous Galerkin approximations of an optimal control problem associated to semilinear parabolic pde's. ESAIM: M^2AN **44**(1), 189–206 (2010)
20. Chrysafinos, K., Walkington, N.J.: Discontinuous Galerkin approximations of the Stokes and Navier-Stokes equations. Math. Comput. **79**(272), 2135–2167 (2010)
21. Deckelnick, K., Hinze, M.: Semidiscretization and error estimates for distributed control of the instationary Navier-Stokes equations. Numer. Math. **97**, 297–320 (200)
22. Dunn, J.C.: On second order sufficient optimality conditions for structured nonlinear programs in infinite-dimensional function spaces. In: Fiacco, A. (ed.) Mathematical Programming with Data Perturbations, pp. 83–107. Dekker, New York (1998)
23. Eriksson, K., Johnson, C.: Adaptive finite element methods for parabolic problems. I. A linear model problem. SIAM J. Numer. Anal. **28**(1), 43–77 (1991)
24. Eriksson, K., Johnson, C.: Adaptive finite element methods for parabolic problems. II. Optimal error estimates in $l_\infty(l^2)$ and $l_\infty(l_\infty)$. SIAM J. Numer. Anal. **32**(3), 706–740 (1995)
25. Eriksson, K., Johnson, C.: Adaptive finite element methods for parabolic problems IV: nonlinear problems. SIAM J. Numer. Anal. **32**(6), 1729–1749 (1995)
26. Estep, D., Larsson, S.: The discontinuous Galerkin method for semilinear parabolic equations. RAIRO Modél. Math. Anal. Numér. **27**, 35–54 (1993)
27. Geissert, M., Hess, M., Hieber, M., Schwarz, C., Stavrakidis, K.: Maximal L^p-L^q-estimates for the Stokes equation: a short proof of Solonnikov's theorem. J. Math. Fluid Mech. **12**, 47–60 (2010)
28. Girault, P., Raviart, P.A.: Finite Element Methods for Navier-Stokes Equations. Theory and Algorithms. Springer, Berlin, Heidelberg, New York, Tokyo (1986)
29. Gunzburger, M.D.: Perspectives in Flow Control and Optimization. Advances in Design and Control. SIAM, Philadelphia (2003)
30. Gunzburger, M.D., Manservisi, S.: The velocity tracking problem for navier-stokes flows with bounded distributed control. SIAM J. Control Optim. **37**(6), 1913–1945 (1999)
31. Gunzburger, M.D., Manservisi, S.: Analysis and approximation of the velocity tracking problem for Navier-Stokes flows with distributed control. SIAM J. Numer. Anal. **37**, 1481–1512 (2000)
32. Hinze, M.: A variational discretization concept in control constrained optimization: the linear quadratic case. Comput. Optim. and Appl. **30**, 45–61 (2005)
33. Hinze, M., Kunisch, K.: Second order methods for optimal control of time-dependent fluid flow. SIAM J. Control Optim. **40**(3), 925–946 (2001)
34. Ladyzhenskaya, O.A.: The Mathematical Theory of Viscous Incompressible Flow, 2nd edn. Gordon and Breach, New York (1969). English translation
35. Lions, J.L.: Quelques Méthodes de Résolution des Problèmes aux Limites non Linéaires. Dunod, Paris (1969)
36. Lions, J.L., Magenes, E.: Problèmes aux Limites non Homogènes. Dunod, Paris (1968)
37. Meidner, D., Vexler, B.: A priori error estimates for the space-time finite element discretization of parabolic optimal control problems. Part I: Problems without control constraints. SIAM J. Control Optim. **47**(3), 1150–1177 (2008)
38. Meidner, D., Vexler, B.: A priori error estimates for the space-time finite element discretization of parabolic optimal control problems. Part II: problems with control constraints. SIAM J. Control Optim. **47**(3), 1301–1329 (2008)
39. Meidner, D., Vexler, B.: A-priori error analysis of the Petrov-Galerkin Crank-Nicolson scheme for parabolic optimal control problems. SIAM J. Control Optim. **49**(5), 2183–2211 (2011)
40. Meyer, C., Rösch, A.: Superconvergence properties of optimal control problems. SIAM J. Control Optim. **43**, 970–985 (2004)
41. Neitzel, I., Vexler, B.: A priori error estimates for space-time finite element discretization of semilinear parabolic optimal control problems. Numer. Math. **120**, 345–386 (2012)
42. Raviart, P.A., Thomas, J.M.: Introduction à L'analyse Numérique des Equations aux Dérivées Partielles. Masson, Paris (1983)
43. Rösch, A.: Error estimates for parabolic optimal control problems with control constraints. Z. Anal. Anwendungen **23**, 353–376 (2004)

44. Rösch, A., Vexler, B.: Optimal control of the Stokes equations: a priori error analysis for finite element discretization with postprocessing. SIAM J. Numer. Anal. **44**, 1903–1920 (2006)
45. Solonnikov, V.A.: Estimates for solutions of nonstationary Navier-Stokes equations. J. Sov. Math. **8**, 213–317 (1977)
46. Sritharan, S.S.: Optimal Control of Viscous Flow. SIAM, Philadelphia (1998)
47. Temam, R.: Navier-Stokes Equations. North-Holland, Amsterdam (1979)
48. Thomée, V.: Galerkin Finite Element Methods for Parabolic Problems. Springer, Berlin (1997)
49. Tröltzsch, F.: Optimal control of partial differential equations. In: Graduate Studies in Mathematics, vol. 112. American Mathematical Society, Philadelphia (2010)
50. Tröltzsch, F., Wachsmuth, D.: Second-order sufficient optimality conditions for the optimal control of Navier-Stokes equations. ESAIM: COCV **12**, 93–119 (2006)

Asymptotic Analysis of a Viscous Flow in a Curved Pipe with Elastic Walls

Gonzalo Castiñeira and José M. Rodríguez

Abstract This communication is devoted to the presentation of our recent results regarding the asymptotic analysis of a viscous flow in a tube with elastic walls. This study can be applied, for example, to the blood flow in an artery. With this aim, we consider the dynamic problem of the incompressible flow of a viscous fluid through a curved pipe with a smooth central curve. Our analysis leads to the obtention of an one dimensional model via singular perturbation of the Navier-Stokes system as ε, a non dimensional parameter related to the radius of cross-section of the tube, tends to zero. We allow the radius depend on tangential direction and time, so a coupling with an elastic or viscoelastic law on the wall of the pipe is possible. To perform the asymptotic analysis, we take a change of variable to a reference domain where we assume the existence of asymptotic expansions on ε for both velocity and pressure which, upon substitution on Navier-Stokes equations, leads to the characterization of various terms of the expansion. This allows us to obtain an approximation of the solution of the Navier-Stokes equations.

1 Introduction

Last decades, applied mathematics have been involved in some new fields where they had not been applied before. One of these fields is biomedicine, from which new methods to improve the diagnosis and treatment of different diseases are demanded. In particular, in the case of cardiovascular problems, modeling the blood flow in veins and arteries is a difficult problem.

G. Castiñeira (✉)
Facultad de Matemáticas, Departamento de Matemática Aplicada, Univ. de Santiago de Compostela, 15782 Santiago de Compostela, Spain
e-mail: gonzalo.castineira@usc.es

J.M. Rodríguez
Departamento de Métodos Matemáticos y de Representación, E.T.S. Arquitectura, Universidade da Coruña, 15071 A Coruña, Spain
e-mail: jose.rodriguez.seijo@udc.es

F. Ortegón Gallego et al. (eds.), *Trends in Differential Equations and Applications*, SEMA SIMAI Springer Series 8, DOI 10.1007/978-3-319-32013-7_5

A large number of articles have studied the flow of a viscous fluid through a pipe. For example, in [3, 6, 12] the flow behavior inside the pipe is related with the curvature and torsion of its middle line. In [3] the main term of the asymptotic expansion of the solution is compared with a Poiseuille flow inside a pipe with rigid walls. In [9], the same problem but with visco-elastic walls is considered, leading to a fluid-structure problem. In [4] the secondary flow is studied, the boundary layer in [11], both depending on values of Dean number. More recently, the non-steady case in tube structures, has been considered in [7, 8], where estimates of the error between exact solution and the asymptotic approximation are proved.

There are also articles where the flow in blood vessels is modeled. An one dimensional model is presented in [2], where clinical procedures where this model can be useful are highlighted. Another model for blood flow in arteries is developed in [10], relating blood pulse and flow patterns, and remarking how this kind of models can help with the design of treatments for particular diseases.

In this article, we shall follow the spirits of [5], where asymptotic analysis is used to find a model for a steady flow through a curved pipe with rigid walls. We shall consider, instead, an unsteady flow and elastic walls. The structure of this article is the following: in Sect. 2 we shall describe the problem in a reference domain, in Sect. 3 we shall suppose the existence of an asymptotic expansion of the solution and we shall identify the first terms of this expansion, in Sect. 4 we shall show some examples of the tangential and transversal velocity, and finally, we shall present some conclusions in Sect. 5.

2 Setting the Problem in a Reference Domain

Let us suppose that central curve of the pipe is parametrized by $\mathbf{c}(s)$, where $s \in [0, L]$ is the arc-length parameter, and the interior points of the pipe are given by

$$(x, y, z) = \mathbf{c}(s) + \varepsilon\, r\, R(t, s)\, [(\cos\theta)\mathbf{N}(s) + (\sin\theta)\mathbf{B}(s)]\,,$$

where $r \in [0, 1]$, $\theta \in [0, 2\pi]$, $\{\mathbf{T} = \mathbf{c}', \mathbf{N}, \mathbf{B}\}$ is the Frenet-Serret frame of $\mathbf{c}$, and $\varepsilon R(t, s)$ is the radius of the cross-section of the pipe at point $\mathbf{c}(s)$ and time t (see Fig. 1). The non dimensional parameter ε represents the different scale of magnitude between the pipe diameter and its length, so we shall assume that $\varepsilon << 1$.

Fig. 1 Domain of the original problem. Note that εR denotes the radius of the pipe

Fig. 2 Reference domain after the change of variable, we obtain a cylinder of radius one

Let us introduce the following notation, $s_1 := s, s_2 := \theta, s_3 := r$ for the variables, and $\{\mathbf{v}_1 := \mathbf{T}, \mathbf{v}_2 := \mathbf{N}, \mathbf{v}_3 := \mathbf{B}\}$, for the Frenet-Serret frame of $\mathbf{c}$. This new notation will allow us to use Einstein summation convention in what follows.

Let be the subsets of $\mathbb{R}^3$ defined by $\Omega^\varepsilon = [0, L] \times [0, 2\pi] \times [0, \varepsilon]$ and $\Omega = [0, L] \times [0, 2\pi] \times [0, 1]$. We define the maps

$$\phi_1^\varepsilon : \Omega \to \Omega^\varepsilon,$$
$$\phi_2^\varepsilon : \Omega^\varepsilon \to \hat{\Omega}_t^\varepsilon,$$

given by the expressions

$$\begin{aligned} \phi_1^\varepsilon(s_1, s_2, s_3) &= (s_1, s_2, \varepsilon s_3) =: (s_1^\varepsilon, s_2^\varepsilon, s_3^\varepsilon), \\ \phi_2^\varepsilon(s_1^\varepsilon, s_2^\varepsilon, s_3^\varepsilon) &= \mathbf{c}(s_1^\varepsilon) + s_3^\varepsilon R(t, s_1^\varepsilon)[(\cos s_2^\varepsilon)\mathbf{v}_2(s_1^\varepsilon) + (\sin s_2^\varepsilon)\mathbf{v}_3(s_1^\varepsilon)]. \end{aligned} \tag{1}$$

We can then introduce the change of variable from the reference domain Ω (see Fig. 2),

$$\phi^\varepsilon = (\phi_2^\varepsilon \circ \phi_1^\varepsilon) : \Omega \to \hat{\Omega}_t^\varepsilon,$$

$$\begin{aligned} \phi^\varepsilon(s_1, s_2, s_3) = \mathbf{c}(s_1) + \varepsilon s_3 R(t, s_1)[(\cos s_2)\mathbf{v}_2(s_1) \\ +(\sin s_2)\mathbf{v}_3(s_1)] =: (x_1^\varepsilon, x_2^\varepsilon, x_3^\varepsilon). \end{aligned} \tag{2}$$

Let us consider the incompressible Navier-Stokes equations in the domain $\hat{\Omega}_t^\varepsilon$ given by,

$$\frac{\partial \mathbf{u}^\varepsilon}{\partial t} + (\nabla \mathbf{u}^\varepsilon)\mathbf{u}^\varepsilon = \frac{1}{\rho_0}\operatorname{div} \mathbf{T}^\varepsilon + \mathbf{b}_0^\varepsilon, \tag{3}$$

$$\operatorname{div} \mathbf{u}^\varepsilon = 0, \tag{4}$$

where $\mathbf{u}^\varepsilon$ stands for the velocity field, $\mathbf{b}_0^\varepsilon$ is the density of body forces and $\mathbf{T}^\varepsilon$ is the stress tensor given by

$$\mathbf{T}^\varepsilon = -p^\varepsilon \mathbf{I} + 2\mu \boldsymbol{\Sigma}^\varepsilon,$$

where p^ε is the pressure field, μ the dynamic viscosity and where

$$\boldsymbol{\Sigma}^\varepsilon = \frac{1}{2}\left(\nabla \mathbf{u}^\varepsilon + (\nabla \mathbf{u}^\varepsilon)^T\right).$$

Let $\nu = \mu/\rho_0$ be the kinematic viscosity, so we can write these equations,

$$\frac{\partial \mathbf{u}^\varepsilon}{\partial t} + (\nabla \mathbf{u}^\varepsilon)\mathbf{u}^\varepsilon + \frac{1}{\rho_0}\nabla p^\varepsilon - \nu \Delta \mathbf{u}^\varepsilon = \mathbf{b}_0^\varepsilon, \tag{5}$$

$$\operatorname{div} \mathbf{u}^\varepsilon = 0. \tag{6}$$

We shall consider continuity between the fluid and the wall of the pipe displacements. Let us suppose that only radial displacements of the wall are allowed. Then the boundary condition at the interface of the fluid and the wall of the pipe can be expressed as

$$\mathbf{u}^\varepsilon = \left(\varepsilon \frac{\partial R}{\partial t}\right) \mathbf{n}^\varepsilon \text{ at } s_3^\varepsilon = \varepsilon, \tag{7}$$

where $\mathbf{n}^\varepsilon$ is the outward unitary normal at $s_3^\varepsilon = \varepsilon$.

Our next step is to write the equations of the problem in the reference domain Ω. Taking into account the change of variable (2), we can associate to each vector field $\mathbf{w}^\varepsilon$ in $\hat{\Omega}_t^\varepsilon$, a new vector field $\mathbf{w}(\varepsilon)$ defined in Ω, as follows

$$w_i^\varepsilon = \mathbf{w}^\varepsilon \cdot \mathbf{e}_i = (w_k^\varepsilon \mathbf{e}_k) \cdot \mathbf{e}_i = (w_k(\varepsilon)\mathbf{v}_k) \cdot \mathbf{e}_i =: w_k(\varepsilon) v_{ki},$$

where $\{\mathbf{e}_1, \mathbf{e}_2, \mathbf{e}_3\}$ is an orthonormal basis, we are using the Einstein summation convention (where latin indices indicate sum from 1 to 3), and we denote $v_{ki} := \mathbf{v}_k \cdot \mathbf{e}_i$.

In the case of a scalar field p^ε in $\hat{\Omega}_t^\varepsilon$, we associate a new scalar field $p(\varepsilon)$ defined in Ω, as

$$p^\varepsilon(t^\varepsilon, x_1^\varepsilon, x_2^\varepsilon, x_3^\varepsilon) = p(\varepsilon)(t, s_1, s_2, s_3).$$

With these considerations, the incompressible Navier-Stokes equations in the reference domain can be written

$$D_t(u_k(\varepsilon)v_{ki}) + \left(\frac{\partial(u_k(\varepsilon)v_{ki})}{\partial s_q}\frac{\partial s_q}{\partial x_j^\varepsilon}\right)(u_m(\varepsilon)v_{mj})$$

$$-\nu\frac{\partial}{\partial s_m}\left(\frac{\partial(u_k(\varepsilon)v_{ki})}{\partial s_q}\frac{\partial s_q}{\partial x_j^\varepsilon}\right)\frac{\partial s_m}{\partial x_j^\varepsilon} = -\frac{1}{\rho_0}\frac{\partial p(\varepsilon)}{\partial s_q}\frac{\partial s_q}{\partial x_i^\varepsilon} + b_{0k}(\varepsilon)v_{ki}, \tag{8}$$

$$\frac{\partial}{\partial s_q}(u_k(\varepsilon)v_{kj})\frac{\partial s_q}{\partial x_j^\varepsilon} = 0, \tag{9}$$

where we have used the operator defined by

$$D_t := \frac{\partial}{\partial t} - \frac{s_3}{R}\frac{\partial R}{\partial t}\frac{\partial}{\partial s_3}.$$

Finally, from the boundary condition (7) at $s_3^\varepsilon = \varepsilon$, we obtain

$$\begin{cases} u_1(\varepsilon) = 0 & \text{at } s_3 = 1, \\ u_2(\varepsilon) = \varepsilon \dfrac{\partial R}{\partial t}\cos s_2 & \text{at } s_3 = 1, \\ u_3(\varepsilon) = \varepsilon \dfrac{\partial R}{\partial t}\sin s_2 & \text{at } s_3 = 1. \end{cases} \tag{10}$$

3 Asymptotic Expansion of the Solution

Following [5], we assume that there exists a formal expansion on powers of ε for the components of velocity and pressure fields of the form,

$$u_k(\varepsilon) = u_k^0 + \varepsilon u_k^1 + \varepsilon^2 u_k^2 + \dots \tag{11}$$

$$p(\varepsilon) = \frac{1}{\varepsilon^2}p^0 + \frac{1}{\varepsilon}p^1 + p^2 + \dots \tag{12}$$

If we substitute (11) and (12) into (8)–(9), and group the terms multiplying the same powers of ε, we are able to identify the first terms of expansion (11)–(12). Identifying these terms is a very hard and long work, and we refer the interested reader to our future work [1], currently under development. However, we shall present as an example, how the zeroth-order terms have been obtained.

Upon substitution of (11) and (12) in (8), we group the terms multiplied by ε^{-3} in the continuity equation, obtaining the following equations related with the zeroth-order term of pressure,

$$-\frac{\sin s_2}{s_3}\frac{\partial p^0}{\partial s_2} + \cos s_2 \frac{\partial p^0}{\partial s_3} = 0, \tag{13}$$

$$\frac{\cos s_2}{s_3}\frac{\partial p^0}{\partial s_2} + \sin s_2 \frac{\partial p^0}{\partial s_3} = 0. \tag{14}$$

Therefore, it is clear that

$$p^0 = p^0(t, s_1). \tag{15}$$

This is, the zeroth-order term of pressure does not depend on the cross-sectional variables and only depends on time and on the point s_1 of the middle line of the curved pipe. We group terms multiplied by ε^{-2} on the continuity equation, obtaining the equations

$$\frac{1}{(rs_3)^2}\frac{\partial^2 u_1^0}{\partial s_2^2} + \frac{1}{r^2 s_3}\frac{\partial u_1^0}{\partial s_2} + \frac{1}{r^2}\frac{\partial^2 u_1^0}{\partial s_3^2} = \frac{1}{\nu\rho_0}\frac{\partial p^0}{\partial s_1}, \tag{16}$$

$$\frac{1}{(rs_3)^2}\frac{\partial^2 u_2^0}{\partial s_2^2} + \frac{1}{r^2 s_3}\frac{\partial u_2^0}{\partial s_2} + \frac{1}{r^2}\frac{\partial^2 u_2^0}{\partial s_3^2} = \frac{1}{\nu\rho_0}\left(-\frac{\sin s_2}{rs_3}\frac{\partial p^1}{\partial s_2} + \frac{\cos s_2}{r}\frac{\partial p^1}{\partial s_3}\right), \tag{17}$$

$$\frac{1}{(rs_3)^2}\frac{\partial^2 u_3^0}{\partial s_2^2} + \frac{1}{r^2 s_3}\frac{\partial u_3^0}{\partial s_2} + \frac{1}{r^2}\frac{\partial^2 u_3^0}{\partial s_3^2} = \frac{1}{\nu\rho_0}\left(\frac{\cos s_2}{rs_3}\frac{\partial p^1}{\partial s_2} + \frac{\sin s_2}{r}\frac{\partial p^1}{\partial s_3}\right). \tag{18}$$

Now, firstly, let us introduce the local cartesian coordinates at cross section of the pipe at s_1,

$$z = (z_2, z_3) = (s_3 \cos s_2, s_3 \sin s_2). \tag{19}$$

Using the change of variable (19) in (16) , we obtain the following problem for the axial component of the zeroth-order term of velocity,

$$\begin{cases} \Delta_{\mathbf{z}} u_1^0 = \dfrac{r^2}{\nu\rho_0}\dfrac{\partial p^0}{\partial s_1}, & \text{in } \omega, \\ u_1^0 = 0 & \text{at } s_3 = 1, \end{cases} \tag{20}$$

where $\omega = \{(z_2, z_3)/z_2^2 + z_3^2 \leq 1\}$. The problem (20) has a unique solution which expression is

$$u_1^0 = \frac{R^2}{4\rho_0\nu}\frac{\partial p^0}{\partial s_1}(s_3^2 - 1). \tag{21}$$

Now, grouping terms multiplied by ε^{-1} in the incompressibility equation, we find that

$$-\frac{\sin u_2^0}{s_3}\frac{\partial u_2^0}{\partial s_2} + \cos s_2 \frac{\partial u_2^0}{\partial s_3} + \frac{\cos s_2}{s_3}\frac{\partial u_3^0}{\partial s_2} + \sin s_2 \frac{\partial u_3^0}{\partial s_3} = 0. \tag{22}$$

Now, we use (19) in this equation and in (17)–(18). Then, the cross-sectional components of the zeroth-order term of velocity denoted by $\mathbf{U}^0 = (u_2^0, u_3^0)$ and the first order term of pressure, are solution of the problem,

$$\begin{cases} \Delta_{\mathbf{z}} \mathbf{U}^{\mathbf{0}} = \dfrac{r}{\nu \rho_0} \nabla_{\mathbf{z}} p^1 \\ \mathrm{div}_{\mathbf{z}} \mathbf{U}^{\mathbf{0}} = 0, \\ \mathbf{U}^{\mathbf{0}} = \mathbf{0} \quad \text{at } s_3 = 1. \end{cases} \tag{23}$$

By the Theorem 2.4 in [13], this problem has uniqueness of solution up to a constant depending on s_1 for the pressure term. This solution is

$$u_2^0 = u_3^0 - 0, \quad p^1 = p^1(t, s_1). \tag{24}$$

We sum up the terms identified in this work in what follows. We shall identify here $\mathbf{u}^0, \mathbf{u}^1, \mathbf{u}^2, p^0, p^1$ and p^2.

As we presented, the term of order zero of velocity, $\mathbf{u}^0$, verifies

$$u_1^0 = \frac{R^2}{4\rho_0 \nu} \frac{\partial p^0}{\partial s_1} (s_3^2 - 1), \tag{25}$$

$$u_2^0 - u_3^0 = 0. \tag{26}$$

Moreover, the zero order term of pressure, p^0, is the solution of the problem,

$$\frac{\partial}{\partial s_1} \left(R^4 \frac{\partial p^0}{\partial s_1} \right) = 16 \nu \rho_0 R \frac{\partial R}{\partial t}, \tag{27}$$

plus suitable boundary conditions.

The components of the next order term of velocity, $\mathbf{u}^1$, are

$$u_1^1 = \left[\frac{3R^3 \kappa s_3 \cos s_2}{16 \nu \rho_0} \frac{\partial p^0}{\partial s_1} + \frac{R^2}{4 \nu \rho_0} \frac{\partial p^1}{\partial s_1} \right] (s_3^2 - 1), \tag{28}$$

$$u_2^1 = \frac{s_3 R}{16 \rho_0 \nu} \left[2 \frac{\partial}{\partial s_1} (R^2 \frac{\partial p^0}{\partial s_1}) - R^2 s_3^2 \frac{\partial^2 p^0}{\partial s_1^2} \right] \cos s_2, \tag{29}$$

$$u_3^1 = \frac{s_3 R}{16 \rho_0 \nu} \left[2 \frac{\partial}{\partial s_1} (R^2 \frac{\partial p^0}{\partial s_1}) - R^2 s_3^2 \frac{\partial^2 p^0}{\partial s_1^2} \right] \sin s_2, \tag{30}$$

where $\kappa = \kappa(s_1)$ is the curvature of the middle line of the pipe at $\mathbf{c}(s_1)$, and first order term of pressure, p^1, is the solution of the problem,

$$\frac{\partial}{\partial s_1}\left(R^4\frac{\partial p^1}{\partial s_1}\right) = 0, \tag{31}$$

where we also have to consider the appropriate boundary conditions.

The first component of the second order term of velocity, $\mathbf{u}^2$, is

$$\begin{aligned}
u_1^2 &= \frac{R^2}{16}\left[\frac{R^2}{4\rho_0\nu^2}\frac{\partial^2 p^0}{\partial t\partial s_1} - \frac{R^4}{16\rho_0^2\nu^3}\frac{\partial p^0}{\partial s_1}\frac{\partial^2 p^0}{\partial s_1^2} - \frac{R^2}{2\rho_0\nu}\frac{\partial^3 p^0}{\partial s_1^3}\right.\\
&\left.+\frac{11\kappa^2R^2}{8\rho_0\nu}\frac{\partial p^0}{\partial s_1}\right](s_3^4-1) + \frac{R^2}{4}\left[-\frac{1}{4\rho_0\nu^2}\frac{\partial}{\partial t}\left(R^2\frac{\partial p^0}{\partial s_1}\right)\right.\\
&+\frac{R^2}{16\rho_0^2\nu^3}\frac{\partial p^0}{\partial s_1}\frac{\partial}{\partial s_1}\left(R^2\frac{\partial p^0}{\partial s_1}\right) + \frac{1}{4\rho_0\nu}\frac{\partial^2}{\partial s_1^2}\left(R^2\frac{\partial p^0}{\partial s_1}\right)\\
&\left.-\frac{7\kappa^2R^2}{16\rho_0\nu}\frac{\partial p^0}{\partial s_1} + \frac{1}{\rho_0\nu}\frac{\partial p_0^2}{\partial s_1} - \frac{b_{01}}{\nu}\right](s_3^2-1)\\
&+\frac{R^6}{1152\rho_0^2\nu^3}\frac{\partial p^0}{\partial s_1}\frac{\partial^2 p^0}{\partial s_1^2}(s_3^6-1) + \frac{3\kappa R^3}{16\rho_0\nu}\frac{\partial p^1}{\partial s_1}(s_3^3-s_3)\cos s_2\\
&+\frac{5\kappa R^4}{64\rho_0\nu}\frac{\partial p^0}{\partial s_1}(s_3^4-s_3^2)\cos(2s_2),
\end{aligned} \tag{32}$$

and the second order term of pressure, p^2, is

$$p^2 = -\frac{R^2}{4}\frac{\partial^2 p^0}{\partial s_1^2}s_3^2 + p_0^2(t,s_1), \tag{33}$$

where $p_0^2(t,s_1)$ is the solution, with the adequate boundary conditions, of the problem

$$\begin{aligned}
\frac{\partial}{\partial s_1}\left(R^4\frac{\partial p_0^2}{\partial s_1}\right) &= \frac{\partial}{\partial s_1}\left[-\frac{3R^8}{64\rho_0\nu^2}\frac{\partial p^0}{\partial s_1}\frac{\partial^2 p^0}{\partial s_1^2} - \frac{R^6}{12}\frac{\partial^3 p^0}{\partial s_1^3} - \frac{\kappa^2R^6}{48}\frac{\partial p^0}{\partial s_1}\right.\\
&-\frac{R^7}{8\rho_0\nu^2}\frac{\partial R}{\partial s_1}\left(\frac{\partial p^0}{\partial s_1}\right)^2 - \frac{R^4}{2}\left(\frac{\partial R}{\partial s_1}\right)^2\frac{\partial p^0}{\partial s_1}\\
&-\frac{R^5}{2}\frac{\partial^2 R}{\partial s_1^2}\frac{\partial p^0}{\partial s_1} - R^5\frac{\partial R}{\partial s_1}\frac{\partial^2 p^0}{\partial s_1^2} + \frac{R^5}{2\nu}\frac{\partial R}{\partial t}\frac{\partial p^0}{\partial s_1} + \frac{R^6}{6\nu}\frac{\partial^2 p^0}{\partial t\partial s_1}\\
&\left.+R^4\rho_0 b_{01}\right].
\end{aligned} \tag{34}$$

Let $\mathbf{U}^2 = (u_2^2, u_3^2)$. Therefore $(\mathbf{U}^2, p^3)$ solves the following problem

$$\begin{cases} \Delta_z \mathbf{U}^2 = \dfrac{R}{\rho_0 \nu} \nabla_z p^3 + \mathbf{F} \text{ in } \omega, \\ \operatorname{div} \mathbf{U}^2 = g \qquad \text{in } \omega, \\ \mathbf{U}^2 = \mathbf{0} \qquad \text{in } \partial\omega. \end{cases} \tag{35}$$

The scalar field g and the vectorial field $\mathbf{F}$, derived from the asymptotic procedure (see [1] for details), are defined by the following expressions respectively

$$\begin{aligned} g &:= -\kappa s_3 R^2 \cos s_2 \left(\frac{1}{4\rho_0 \nu} \frac{\partial}{\partial s_1} \left(R^2 \frac{\partial p^0}{\partial s_1} \right) (s_3^2 - 1) - \frac{\partial R}{\partial s_1} \frac{R}{2\rho_0 \nu} s_3^2 \right) \\ &- \frac{3R}{16\nu\rho_0} \frac{\partial}{\partial s_1} \left(R^3 \kappa \frac{\partial p^0}{\partial s_1} \right) s_3 \cos s_2 (s_3^2 - 1) - \frac{R}{4\nu\rho_0} \frac{\partial}{\partial s_1} \left(R^2 \frac{\partial p^1}{\partial s_1} \right) (s_3^2 - 1) \\ &+ \kappa \frac{s_3 R^2}{16\rho_0 \nu} \cos s_2 \left[2 \frac{\partial}{\partial s_1} \left(R^2 \frac{\partial p^0}{\partial s_1} \right) - R^2 s_3^2 \frac{\partial^2 p^0}{\partial s_1^2} \right] \\ &- \frac{3R^4 \kappa \tau \sin s_2}{16\rho_0 \nu} \frac{\partial p^0}{\partial s_1} (s_3^2 - 1) + s_3 \frac{\partial R}{\partial s_1} \left[\frac{3R^3 \kappa \cos s_2}{16\rho_0 \nu} \frac{\partial p^0}{\partial s_1} (3s_3^2 - 1) \right. \\ &\left. + \frac{R^2}{2\rho_0 \nu} \frac{\partial p^1}{\partial s_1} s_3 \right], \end{aligned} \tag{36}$$

$$\begin{aligned} \mathbf{F} &:= \left(\frac{\kappa R^6}{16\rho^2 \nu^3} \left(\frac{\partial p^0}{\partial s_1} \right)^2 (s_3^4 + 1) \right. \\ &+ \left(-\frac{\kappa R^6}{8\rho_0^2 \nu^3} \left(\frac{\partial p^0}{\partial s_1} \right)^2 - \frac{9R^4 \kappa}{16\rho_0 \nu} \frac{\partial^2 p^0}{\partial s_1^2} - \frac{R^4}{4\rho_0 \nu} \frac{\partial \kappa}{\partial s_1} \frac{\partial p^0}{\partial s_1} \right) s_3^2 \\ &- \frac{R^4 \kappa}{8\rho_0 \nu} \frac{\partial^2 p^0}{\partial s_1^2} s_3^2 \cos^2 s_2 + \frac{5R^2 \kappa}{8\rho_0 \nu} \frac{\partial}{\partial s_1} \left(R^2 \frac{\partial p^0}{\partial s_1} \right) \\ &+ \frac{R^4}{4\rho_0 \nu} \frac{\partial \kappa}{\partial s_1} \frac{\partial p^0}{\partial s_1} - \frac{R^2}{\nu} b_{02}, \end{aligned} \tag{37}$$

$$\left. - \frac{\kappa \tau R^4}{4\rho_0 \nu} \frac{\partial p^0}{\partial s_1} (s_3^2 - 1) - \frac{2R^4 \kappa}{16\rho_0 \nu} \frac{\partial^2 p^0}{\partial s_1^2} s_3^2 \cos s_2 \sin s_2 - \frac{R^2}{\nu} b_{03} \right). \tag{38}$$

Problem (35) has an unique solution $(\mathbf{U}^2, p^3)$ (where $\mathbf{U}^2$ is unique, but p^3 is unique except for an arbitrary function depending only on s_1), if a compatibility condition ($\int_\omega g = 0$) is fulfilled (see [13]). That is, multiplying by a test function and integrating over ω the divergence equation of (35), we have that

$$\int_\omega \mathrm{div}\mathbf{U}^2 v = \int_\omega g v,$$

using the Green formula we find that

$$-\int_\omega \mathbf{U}^2 \cdot \nabla_z v + \int_{\delta\omega} \left(\mathbf{U}^2 \cdot n\right) v = \int_\omega g v.$$

Hence, taking into account that $\mathbf{U}^2 = 0$ on $\delta\omega$ and considering a constant test function, we conclude

$$\int_\omega g = 0. \tag{39}$$

Computing this condition using the expression in (36), and together with Eq. (31), we can see that (39) is verified, so we can ensure existence of a unique solution of the problem (35).

As we have observed just above, the expressions of g and $\mathbf{F}$ are polynomial on s_3, so $\mathbf{U}^2$ can be explicitly computed and is also polynomial on s_3.

To finish, we need to close the equations with a law on the wall of the pipe. There are different possibilities (for example, elastic or viscoelastic laws). In the simplest case (see [2]), we can consider an algebraic elastic law:

$$p^0 - p_e = \frac{Eh_0}{R_0^2}(R - R_0) \tag{40}$$

where E is the Young modulus of the wall, h_0 its thickness, R_0 the radius of the cross-section at rest, and p_e is the external pressure.

4 Some Examples

In this section we shall present some examples in order to illustrate the behavior of the approximated solution obtained in the previous section.

We start plotting the main tangential velocity u_1^0 and its corrections u_1^1 and u_1^2. We observe in Fig. 3 that u_1^0 is a Poiseuille flow (other works as [3, 9] have also shown this behavior). In Fig. 4 we can see that u_1^1 is a correction of u_1^0 that takes into

Fig. 3 Plot of u_1^0 field

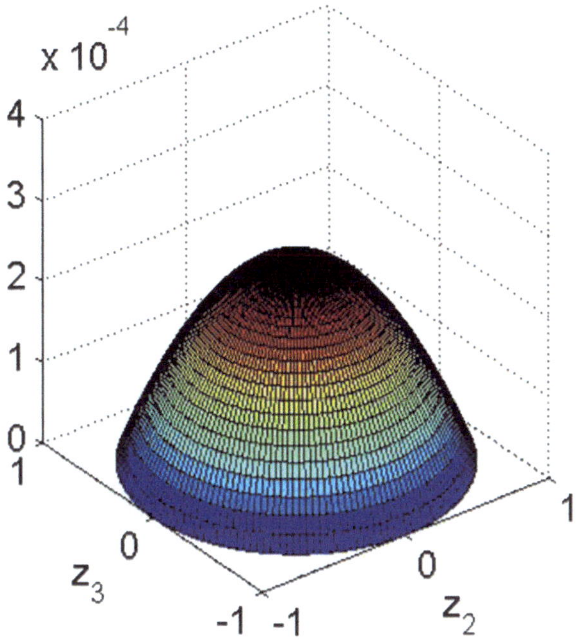

Fig. 4 Plot of u_1^1 field

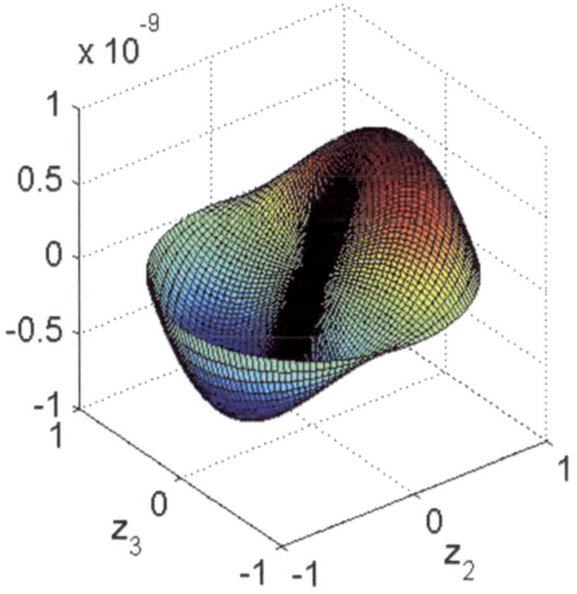

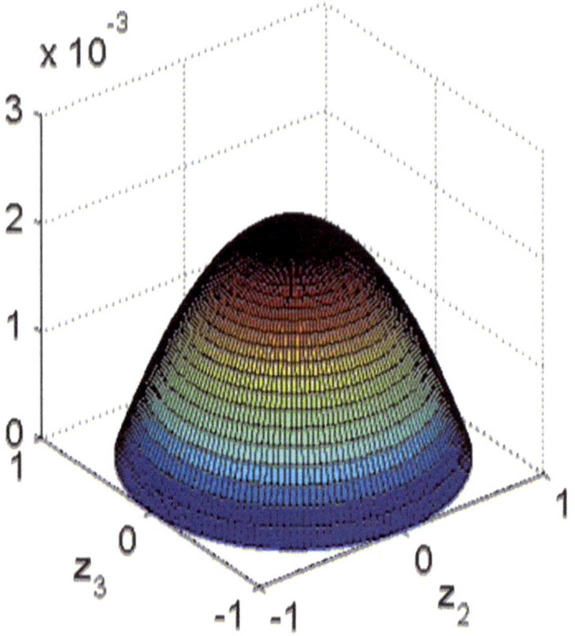

Fig. 5 Plot of u_1^2 field

account the curvature of the middle line (the fluid is faster in the side of the cross section of the pipe pointing to **N**). The correction of order two u_1^2, has a complex dependence on various terms (32), but we get also a Poiseuille flow (Fig. 5).

We have seen at (26) that, at order zero, the transversal velocity is zero, so the tangential velocity is dominant. The first order correction, $\mathbf{U}^1 = (u_2^1, u_3^1)$, is related with the expansion and contraction of the pipe wall in radial direction. We can see in Fig. 6 different cases depending on the value of $\frac{\partial p^0}{\partial s_1}$ (dp1), $\frac{\partial^2 p^0}{\partial s_1^2}$ (dp2) and $\frac{\partial r}{\partial s_1}$ (dr).

The second order correction of transversal velocity, $\mathbf{U}^2 = (u_2^2, u_3^2)$, is related with the recirculation of the fluid in the cross section of the pipe, as we can see in Fig. 7, where we show different cases depending on the curvature (k), its derivative (dk) and the torsion (tau) of the middle line of the pipe.

5 Conclusions

A transient model for a newtonian fluid through a curved pipe with elastic walls has been obtained. The asymptotic expansions have allowed us to find out the main components of velocity and their corrections. Furthermore, we have verified that our model reduces to the obtained in [5], when steady case and rigid walls are

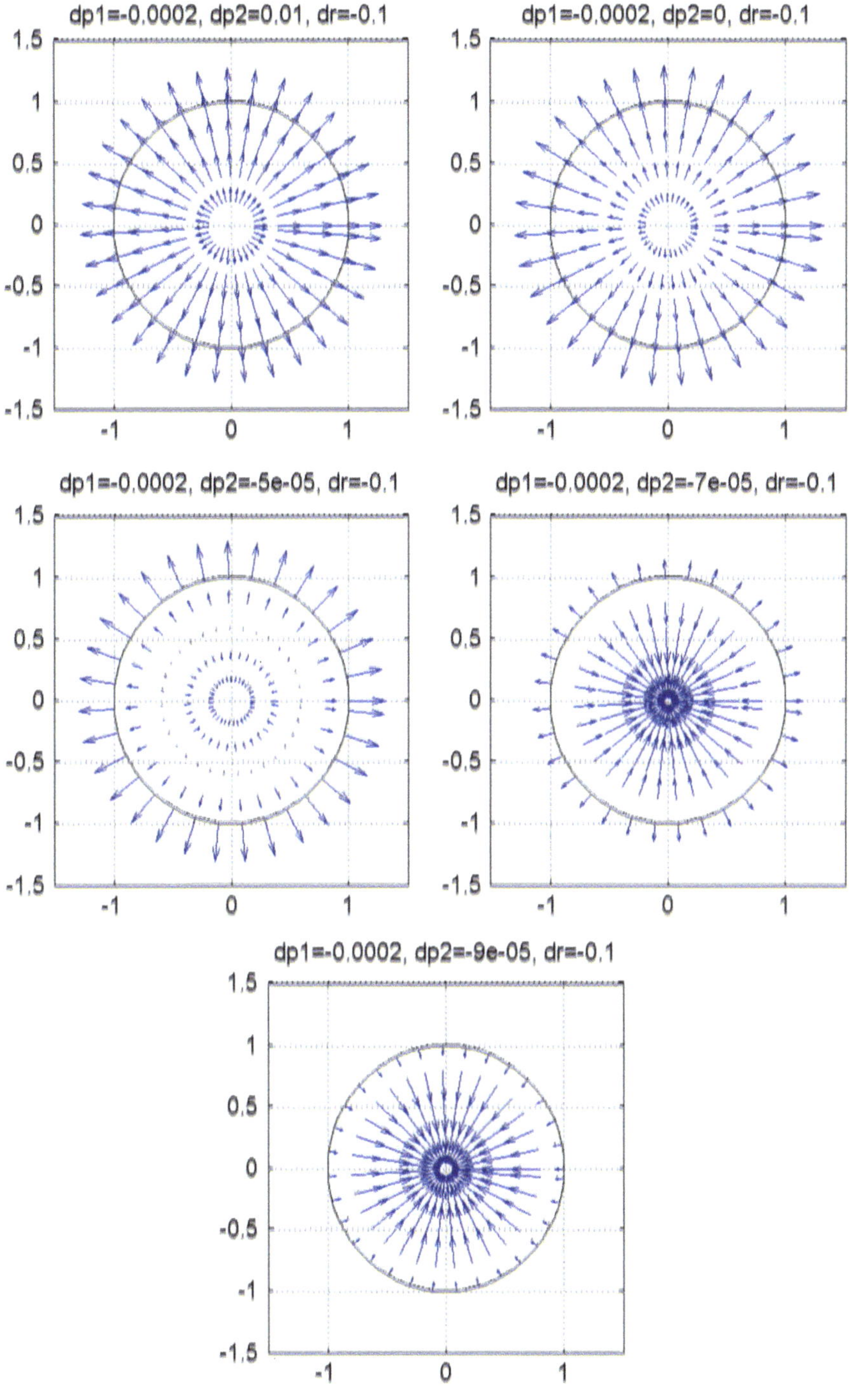

Fig. 6 Plot of (u_2^1, u_3^1) field

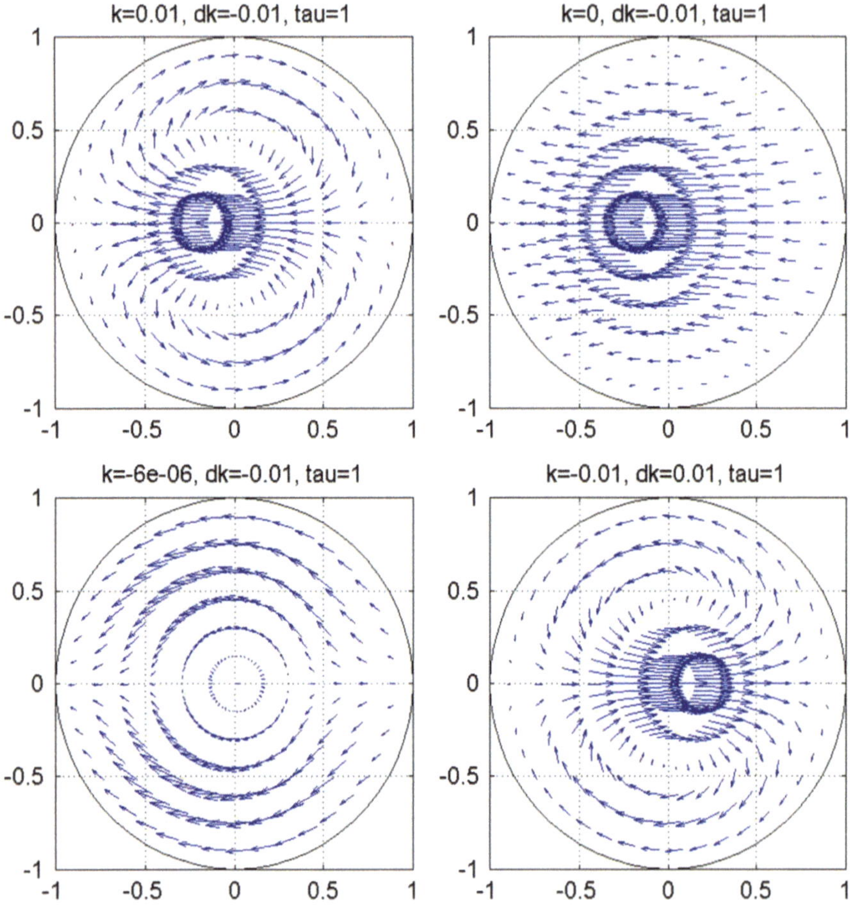

Fig. 7 Plot of (u_2^2, u_3^2) field

considered, that is, by replacing (40) by the equality $R = R_0$. Plots presented here (see Figs. 3–7) compare very well with real patterns of blood flow and agree with the data available in the literature. A simple algebraic elastic law for the pipe wall has been considered in (40), but other more general laws can be used.

Acknowledgements This research was partially supported by Ministerio de Economía y Competitividad under grant MTM2012-36452-C02-01 with the participation of FEDER.

References

1. Castiñeira, G., Rodríguez, J.M.: Asymptotic analysis of a viscous flow in a curved pipe with moving walls. arXiv:1602.06121 (http://arxiv.org/abs/1602.06121)
2. Formaggia, L., Lamponi, D., Quarteroni, A.: One-dimensional models for blood flow in arteries. J. Eng. Math. **47**, 251–276 (2003)
3. Gammack, D., Hydon, P.E.: Flow in pipes with non-uniform curvature and torsion. J. Fluid Mech. **433**, 357–382 (2001)
4. Lyne, W.H.: Unsteady viscous flow in a curved pipe. J. Fluid. Mech. **45**, 13–31 (1970)
5. Marušić-Paloka, E.: The effects of flexion and torsion on a fluid flow through a curved pipe. Appl. Math. Optim. **44**, 245–272 (2001)
6. Marušić-Paloka, E., Pažanin, I.: Fluid flow through a helical pipe. Z. Angew. Math. Phys. **58**, 81–89 (2007)
7. Panasenko, G.P., Pileckas, K.: Asymptotic analysis of the non-steady Navier-Stokes equations in a tube structure. I. The case without boundary-layer-in-time. Nonlinear Anal. **122**, 125–168 (2015)
8. Panasenko, G.P., Pileckas, K.: Asymptotic analysis of the non-steady Navier-Stokes equations in a tube structure. II. General case. Nonlinear Anal. **125**, 582–607 (2015)
9. Panasenko, G.P., Stavre, R.: Asymptotic analysis of a periodic flow in a thin channel with visco-elastic wall. J. Math. Pures Appl. **85**, 558–579 (2006)
10. Pedley, T.J.: Mathematical modelling of arterial fluid dynamics. J. Eng. Math. **47**, 419–444 (2003)
11. Riley, N.: Unsteady fully-developed flow in a curved pipe. J. Eng. Math. **34**, 131–141, (1998)
12. Smith, F.T.: Fluid flow into a curved pipe. Proc. R. Soc. Lond. A **351**, 71–87 (1976)
13. Temam, R.: Navier-Stokes Equations: Theory and Numerical Analysis. AMS Chelsea Publishing, New York (2000)

A Two-Scale Homogenization Approach for the Estimation of Porosity in Elastic Media

Joaquín Mura and Alfonso Caiazzo

Abstract We propose a novel method for estimating the porosity of an elastic medium starting from inner displacement measurement, such as the ones that can be obtained from seismogram data for the study of soils or from magnetic resonance elastography for the diagnosis of tissue diseases. The approach is based on a two-scale homogenization, which relates geometrical characteristics of the void-elastic solid mixture at the small (mesoscopic) scale of the pore with an effective elasticity tensor at the large (macroscopic) scale of the effective material. Through semi-analytical approximations of the homogenized equations, the idea can be further extended considering slight variations in the shape of the pore. This procedure leads eventually to an inverse problem formulation that enable us to recover approximately the porosity field by means of the finite element formulation of the effective macroscale problem only. We validate the multiscale approximation and the two-scale porosity estimation method with numerical examples.

1 Introduction

The behavior of elastic materials characterized by the presence of small cavities has been largely studied since it is of utmost importance for a vast amount of applications in material sciences, biomechanics and engineering. When describing mathematically this class of materials, the interaction between the mesoscopic scale (the cavities) and the macroscopic one (the matrix) is not negligible and lead to corrective terms in classical Navier equations. These coefficients can be explicitly derived under certain conditions, using two-scale or periodic homogenization, assuming geometrical periodicity, or via small amplitude homogenization, if the

J. Mura
School of Civil Engineering, Pontificia Universidad Católica de Valparaíso, Av. Brasil 2147, Valparaíso, Chile
e-mail: joaquin.mura@ucv.cl

A. Caiazzo (✉)
Weierstrass Institute for Applied Analysis and Stochastics (WIAS), Leibniz Institute im Forschungsverbund Berlin e.V., Mohrenstrasse 39, 10117, Berlin, German
e-mail: caiazzo@wias-berlin.de

F. Ortegón Gallego et al. (eds.), *Trends in Differential Equations and Applications*, SEMA SIMAI Springer Series 8, DOI 10.1007/978-3-319-32013-7_6

contrast between material coefficients is small. These two approaches can be considered as special applications of a more general Homogenization theory (see, e.g., [14]), providing mathematical results that lately have set the basis for a new class of numerical methods, aiming at accurately solving multiscale problems without excessive use of computational resources [2, 7, 11].

In this paper we focus on porous soils, and, in particular, on detecting the regions characterized by different porosities starting from seismic data. This problems corresponds to characterize the porosity of a soil sample starting from the inner displacement response to a harmonic excitation. To this aim, we perform the following steps: (1) we obtain an upscaled problem splitting the dynamics on the micro- and the macroscale, via a homogenized model for a solid matrix with an array of small cavities; (2) we derive semi-analytical approximations of the solution of the mesoscopic cell problems, which allow to write the effective elasticity tensor as a function of the porosity and of the material constants; (3) we construct an optimization framework to recover the porosity (mesoscopic parameter) solving a homogenized inverse problem at the macroscale. To solve the inverse problem, we adopt an iterative variational approach based on the solution of an adjoint problem (see e.g. [8, 9]).

The algorithm is validated using synthetic displacement data, obtained simulating the full problem.

The rest of the paper is organized as follows. In Sect. 2 we describe the multiscale elasticity problem and the corresponding two-scale homogenization. In Sect. 3 we derive semi-analytical expressions of the effective tensor coefficients, which are then used in Sect. 4 to define a multiscale variational inverse problem for the estimation of porosity starting from measurement of the internal displacement field. Numerical results are presented in Sect. 5, while Sect. 6 draws the conclusion.

2 Two-Scale Modeling of Elastic Media with Void Inclusions

Let us assume to deal with a biphasic material, composed of an elastic matrix, with shear modulus μ, compression modulus (Lamé first parameter) λ, and small void inclusions (empty pores) of different sizes. Furthermore, we assume that the following conditions are satisfied: (1) the inclusions are very small with respect to the matrix and isolated from each other; (2) they can be organized into a periodic array; (3) they have spherical shape.

In this setting, the properties of the effective material can be obtained via a two-scale homogenization, in order to characterize the effect of the mesoscale geometry (i.e. the pores) only from a macroscopic point of view [12, 13]. In particular, we adopt the two scales homogenization presented in [4] (and extended in [5] to the time harmonic regime), which is summarized in this section. For other recently proposed approaches concerning homogenization in the context of elasticity problems, we refer the reader to, e.g., [3, 6].

Let a bounded domain $\Omega \subset \mathbb{R}^d$ ($d = 2, 3$) represent the space occupied by the material and let the small parameter $0 < \epsilon \ll 1$, denote the aspect ratio between the inclusions and the container matrix. Under the above assumptions, we consider the composite material as the combination of a large number of equal *mesoscopic* cells, i.e.

$$\Omega^\epsilon = \bigcup_i \Omega_{\epsilon,i} ,$$

where $\Omega_{\epsilon,i} := \epsilon Y + \mathbf{x}_i$ contain a single inclusion, $Y = [0, 1]^N$ is the unitary cell in $\mathbb{R}^N$ ($N = 2, 3$), and $\mathbf{x}_i$ denotes the center of the cell (see Fig. 1).

In this configuration, the unitary cell Y can be decomposed as

$$Y = Y_S \cup \Gamma \cup Y_F$$

where Y_S and Y_F stand for the domains occupied by the matrix and the void, respectively, and Γ denotes the interface between them (as shown in Fig. 1).

Remark 1 The *porosity* of the material, defined as the fraction of void volume, is given by

$$\phi = |Y_F| .$$

Remark 2 (Effective Density) Let $\chi^{k,\epsilon}(y)$ the characteristic function of the solid domain in the cell k. If the solid matrix has a constant density ρ_S, then the effective

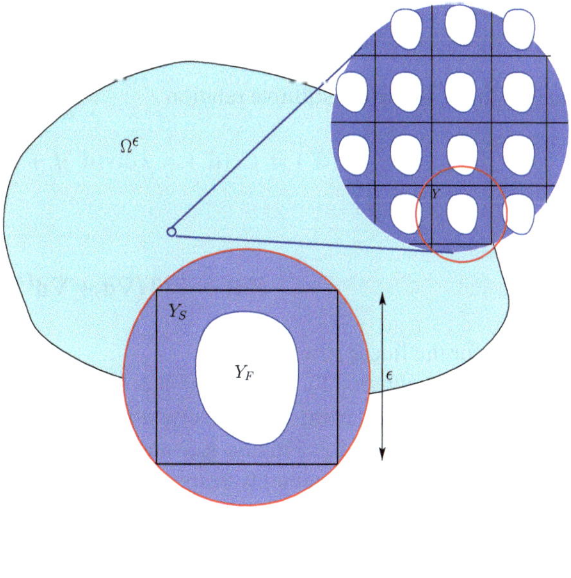

Fig. 1 Sketch of the mesoscopic and macroscopic scales in the composite material

density in the microscopic cell is given by

$$\rho_\epsilon(y) = \sum_k \rho_S \chi_S^{k,\epsilon}(y)$$

(summing up all the contribution of the cells), while the cell-averaged density is

$$\overline{\rho} = \frac{1}{|Y|}\int_Y \rho_\epsilon(y)\,dy = \frac{|Y_S|}{|Y|}\rho_S = (1-\phi)\rho_S\,. \tag{1}$$

2.1 Linear Elasticity in Harmonic Regime

Since our main motivation is to address the estimation of soil properties starting from seismogram data, we focus on a time harmonic regime, with the pulsation driven by a known frequency $\omega > 0$. In this configuration, the displacement of the material can be described by a vector field $\mathbf{d}^\epsilon : \Omega \to \mathbb{R}^d$ ($d = 2, 3$) obeying the following equation:

$$-\rho_\epsilon\omega^2\mathbf{d}^\epsilon - \operatorname{div}\sigma(\mathbf{d}^\epsilon) = \mathbf{f} \text{ in } \Omega^\epsilon, \tag{2}$$

completed with the boundary conditions

$$\begin{cases} \sigma(\mathbf{d}^\epsilon)\mathbf{n}^\epsilon = 0 \text{ on } \Gamma^{\epsilon,k},\ k = 1,\dots \\ \sigma(\mathbf{d}^\epsilon)\mathbf{n}^\epsilon = \mathbf{g}_{\text{ext}} \text{ on } \Gamma_N, \\ \mathbf{d}^\epsilon = \mathbf{d}_{\text{ext}} \text{ on } \Gamma_D, \end{cases} \tag{3}$$

and with the linear constitutive relation

$$\sigma(\mathbf{d}^\epsilon) = Ce(\mathbf{d}^\epsilon) = \lambda\operatorname{div}(\mathbf{d}^\epsilon)I + 2\mu e(\mathbf{d}^\epsilon),$$

where

$$e(\mathbf{d}) = 1/2\left(\nabla\mathbf{d} + \nabla\mathbf{d}^T\right)$$

denotes for the linear strain tensor.

In (3), the subsets Γ_N and Γ_D of the $\partial\Omega^\epsilon$ denote the external Neumann and the Dirichlet boundaries, respectively, while $\mathbf{n}^\epsilon$ stands for the outgoing normal vector on the boundary. Finally, $\mathbf{g}_{\text{ext}}$ and $\mathbf{d}_{\text{ext}}$ denote the external forces and the external imposed displacements, respectively, which are assumed to be known from experimental data.

2.2 Asymptotic Expansion

In order to obtain the effective equations in the limit of small inclusions ($\epsilon \to 0$), we use the following multiscale ansatz for the displacement field

$$\mathbf{d}^{\epsilon}(x) = \sum_{k=0}^{2} \epsilon^k \mathbf{d}^k(x, x/\epsilon) + O(\epsilon^3). \tag{4}$$

Furthermore, let us denote with $y = x/\epsilon$ the so-called *fast variable* (defined in the unit cell Y), and let us introduce the splitting of the spatial derivative

$$\partial = \partial_x + (1/\epsilon)\partial_y. \tag{5}$$

Inserting (4)–(5) in (2) and collecting the terms of the same orders in ϵ, one obtains a system of equations for the variables x and y, describing the macroscopic dynamics (leading order) and the dynamics at the pore scale at different orders in ϵ. For the zeroth order it holds

$$\begin{cases} -\mathrm{div}_y \sigma_y(\mathbf{d}^0) = 0 & \text{in } \Omega \times Y_S, \\ \sigma_y(\mathbf{d}^0)\mathbf{n} = 0 & \text{on } \Omega \times \Gamma, \\ \mathbf{d}^0 \quad Y - \text{periodic}, & \end{cases} \tag{6}$$

whose solution is given by a displacement field $\mathbf{d}^0$ constant in y (but not in x). Collecting the terms of first order in ϵ yields an equation for $\mathbf{d}^1$:

$$\begin{cases} -\mathrm{div}_y \sigma_y(\mathbf{d}^1) = 0 \text{ in } \Omega \times Y_S, \\ \sigma_y(\mathbf{d}^1)\mathbf{n} = -\sigma_x(\mathbf{d}^0)\mathbf{n} \text{ on } \Omega \times \Gamma. \\ \mathbf{d}^1 \quad Y - \text{periodic} \end{cases} \tag{7}$$

while the second order field $\mathbf{d}^2$ is described by

$$\begin{cases} -\mathrm{div}_y \sigma_y(\mathbf{d}^2) = \overline{\rho}\omega^2 \mathbf{d}^0 + \mathrm{div}_x(\sigma_y(\mathbf{d}^1) + \sigma_x(\mathbf{d}^0)) \\ \quad + \mathrm{div}_y \sigma_x(\mathbf{d}^1) \;\text{ in } \Omega \times Y_S, \\ \sigma_y(\mathbf{d}^2)\mathbf{n} = -\sigma_x(\mathbf{d}^1)\mathbf{n} \;\text{ on } \Omega \times \Gamma, \\ \mathbf{d}^2 \text{ Y} - \text{periodic}. \end{cases} \tag{8}$$

Notice that, since the frequency is independent from ϵ, Eqs. (6)–(8) do not depend on the harmonic waves, when solving the dependence on y of $\mathbf{d}^0$, $\mathbf{d}^1$ and $\mathbf{d}^2$, respectively. In other words, at the spatial scale of the pores, the length-wave of the excitation is, at first order in ϵ, too large to be perceived with the fast variable y.

The only contribution of the term associated with the frequency just appear in the right hand side of (8). This equation will be used later to obtain the formula of the effective tensor.

2.3 Homogenized Elasticity Tensor

In what follows, let us denote with

$$\langle f\rangle_{Y_S}(x) = \frac{1}{Y_S}\int_{Y_S} f(x,y)\,dy$$

the average of a function f over Y_S with respect to the fast variable. The variable $\mathbf{d}^2$ can be eliminated by direct integration of (8), thanks to the so-called compatibility condition in Y_S with respect to y [12]. Hence, one gets the following differential problem for the macroscopic variable $\mathbf{d}^0$:

$$\begin{cases} -\overline{\rho}\omega^2\mathbf{d}^0 - \mathrm{div}_x\left(\sigma_x(\mathbf{d}^0) + \langle\sigma_y(\mathbf{d}^1)\rangle_{Y_S}\right) = \mathbf{f} \text{ in } \Omega, \\ \qquad\qquad\qquad\qquad \sigma_x(\mathbf{d}^0)\mathbf{n} = \mathbf{g}_{\mathrm{ext}} \text{ on } \Gamma_N, \\ \qquad\qquad\qquad\qquad\qquad \mathbf{d}^0 = \mathbf{d}_{ext} \text{ on } \Gamma_D, \end{cases} \tag{9}$$

For the study problem (9) it is convenient to introduce the change of variable [4]

$$\mathbf{d}^1(x,y) = \sum_{k,l=1}^{N} [e_x(\mathbf{d}^0)(x)]_{kl}\,\boldsymbol{\chi}^{kl}(y). \tag{10}$$

Hence, by inserting (10) into (9), one can conclude that (9) describes the dynamics of a compressible material, with the effective elasticity tensor given by

$$C^{eff}_{ijkl} = \lambda\delta_{ij}\delta_{kl} + 2\mu\delta_{ijkl} + \langle[\sigma_y(\boldsymbol{\chi}^{kl})]_{ij}\rangle_{Y_S}, \tag{11}$$

where $\delta_{ijkl} := \frac{1}{2}(\delta_{li}\delta_{kj} + \delta_{ki}\delta_{lj})$, and the variable $\boldsymbol{\chi}^{kl}$ can be obtained solving the following problem on the cell Y_S [4]:

$$\begin{cases} -\mathrm{div}_y\sigma_y(\boldsymbol{\chi}^{kl}) = 0 \text{ in } Y_S, \\ \quad \sigma_y(\boldsymbol{\chi}^{kl})\mathbf{n} = -T^{kl}\mathbf{n} \text{ on } \Gamma, \\ \qquad\quad \boldsymbol{\chi}^{kl} \quad \text{is } Y-\text{periodic}, \end{cases} \tag{12}$$

with

$$T^{kl}_{ij} = -\lambda\delta_{ij}\delta_{kl} + 2\mu\delta_{ijkl}\,. \tag{13}$$

In the two-dimensional case, the following expressions for T^{kl} hold:

$$T^{11} = \begin{bmatrix} \lambda + 2\mu & 0 \\ 0 & \lambda \end{bmatrix}, T^{22} = \begin{bmatrix} \lambda & 0 \\ 0 & \lambda + 2\mu \end{bmatrix}, T^{12} = T^{21} = \begin{bmatrix} 0 & \mu \\ \mu & 0 \end{bmatrix}, \tag{14}$$

while for a three-dimensional problem one obtains

$$T^{11} = \begin{bmatrix} \lambda + 2\mu & 0 & 0 \\ 0 & \lambda & 0 \\ 0 & 0 & \lambda \end{bmatrix}, \quad T^{12} = T^{21} = \begin{bmatrix} 0 & \mu & 0 \\ \mu & 0 & 0 \\ 0 & 0 & 0 \end{bmatrix},$$

$$T^{22} = \begin{bmatrix} \lambda & 0 & 0 \\ 0 & \lambda + 2\mu & 0 \\ 0 & 0 & \lambda \end{bmatrix}, \quad T^{13} = T^{31} = \begin{bmatrix} 0 & 0 & \mu \\ 0 & 0 & 0 \\ \mu & 0 & 0 \end{bmatrix},$$

$$T^{23} = T^{32} = \begin{bmatrix} 0 & 0 & 0 \\ 0 & 0 & \mu \\ 0 & \mu & 0 \end{bmatrix}, \quad T^{33} = \begin{bmatrix} \lambda & 0 & 0 \\ 0 & \lambda & 0 \\ 0 & 0 & \lambda + 2\mu \end{bmatrix}.$$

In practice, in order to solve for the macroscopic dynamics for given λ, μ and inclusion radius (i.e. for given porosity), the effective tensor coefficients shall be every time evaluated solving the corresponding cell problems (12) for the variable χ^{kl}. In order to obtain a faster procedure, which does not require the solution of the cell problems for any configuration of parameters, in the following Section we derive semi-analytic approximations of the effective tensor coefficients (similarly as in [5]), exploiting the symmetry and linearity properties of (12) and using numerical interpolation.

3 Approximation of the Two-Dimensional Effective Tensor

For the sake of simplicity, in what follows we will restrict to the two-dimensional case. Moreover, for the application of interest (estimation of porosity in soil), we will focus on small porosities and $\lambda = O(\mu)$.

According to (11), the element C_{ijkl} of the effective tensor depends on the microscale through the integrals

$$a_{ijkl} = \int_{Y_S} [\sigma_y(\chi_{kl})]_{ij} \, dy \,,$$

which, for symmetry reasons (see, e.g., [4]), satisfy

$$\begin{aligned} a_{1111} &= a_{2222}\,, \\ a_{1122} &= a_{2211}\,, \\ a_{1212} &= a_{1221} = a_{2121} = a_{2112}\,, \\ a_{1112} &= a_{1121} = a_{1211} = a_{2212} = a_{1222} = a_{2111} = 0\,. \end{aligned} \tag{15}$$

Hence, the effective elasticity tensor is completely defined by just three coefficients

$$\begin{aligned} C^{eff}_{1111} &= C^{eff}_{2222} = \lambda + 2\mu + \langle [\sigma_y(\chi^{11})]_{11} \rangle_{Y_S}, \\ C^{eff}_{1122} &= C^{eff}_{2211} = \lambda + \langle [\sigma_y(\chi^{11})]_{22} \rangle_{Y_S}, \\ C^{eff}_{1212} &= C^{eff}_{1221} = C^{eff}_{2112} = C^{eff}_{2121} = \mu + \langle [\sigma_y(\chi^{12})]_{12} \rangle_{Y_S}. \end{aligned} \tag{16}$$

Observing that $|Y_S| = (1 - \phi)$, we introduce the notations

$$\begin{aligned} a_1 &= a_{1111} = (1-\phi)\langle [\sigma_y(\chi^{11})]_{11} \rangle_{Y_S} \\ a_2 &= a_{1212} = (1-\phi)\langle [\sigma_y(\chi^{12})]_{12} \rangle_{Y_S} \\ a_3 &= a_{1122} = (1-\phi)\langle [\sigma_y(\chi^{11})]_{22} \rangle_{Y_S} = (1-\phi)\langle [\sigma_y(\chi^{22})]_{11} \rangle_{Y_S}. \end{aligned} \tag{17}$$

3.1 *The Coefficients a_1 and a_3*

First, let us introduce the quantity

$$a_+ = a_1 + a_3 = (1-\phi)\langle [\sigma_y(\chi^{11} + \chi^{22})]_{11} \rangle_{Y_S} \tag{18}$$

Hence, a_+ can be computed from the solution to the differential problem obtained adding up the cell problems for $(k, l) = (1, 1)$ and $(k, l) = (2, 2)$:

$$\begin{cases} -\mathrm{div}_y \sigma_y(\chi^+) = 0 & \text{in } Y_S, \\ \sigma_y(\chi^+)\mathbf{n} = -\begin{pmatrix} 2(\lambda+\mu) & 0 \\ 0 & 2(\lambda+\mu) \end{pmatrix} \mathbf{n} & \text{on } \Gamma, \\ \chi^+ \quad \text{is } Y-\text{periodic}. \end{cases} \tag{19}$$

Problem (19) is now linear in $(\lambda + \mu)$, due to the symmetric boundary condition $(19)_2$. Hence, one can consider the variable $\hat{\chi}^+ := \frac{\chi^+}{\lambda+\mu}$ which satisfies

$$\begin{cases} -\mathrm{div}_y \sigma_y(\hat{\chi}^+) = 0 & \text{in } Y_S, \\ \sigma_y(\hat{\chi}^+)\mathbf{n} = -\begin{pmatrix} 2 & 0 \\ 0 & 2 \end{pmatrix} \mathbf{n} \text{ on } \Gamma, \\ \chi^+ \quad \text{is } Y - \text{periodic}. \end{cases} \tag{20}$$

and rewrite (18) as

$$a_+ = (1-\phi)(\lambda+\mu)\int_{Y_S} \sigma_y(\hat{\chi}^+)_{11}.$$

Furthermore, we observe that, multiplying both λ and μ by the same constant, the solution to (20) does not change. Hence, the integral $\int_{Y_S} \sigma_y(\hat{\chi}^+)_{11}$ shall depend on the Lamé coefficient only through their ratio.

Let $\delta = \frac{\lambda}{\mu}$. Based on the above considerations, we search for an approximation of a_+ using the ansatz

$$a_+ = (\lambda+\mu)\frac{A(\delta)\phi}{B(\delta)+C(\delta)\phi}.$$

The coefficients A, B and C have been fitted from several numerical simulations, which resulted in the approximation

$$a_1 + a_3 = a_+ = (\lambda+\mu)\left(-\frac{5(1+\delta)\phi}{2+4(2+\delta)\phi}\right). \tag{21}$$

As next, we obtained by numerical interpolation the following approximation

$$a_1 - a_3 = a_+\left(\frac{\delta^2}{\delta^4+2\delta^2-1} - \frac{\phi}{\pi\delta}\right), \tag{22}$$

which, combined with (21) yields

$$\begin{aligned} a_1 &= a_+\left(\frac{1}{2} + \frac{\delta^2}{2(\delta^4+2\delta^2-1)} - \frac{\phi}{2\pi\delta}\right), \\ a_3 &= a_+\left(\frac{1}{2} - \frac{\delta^2}{2(\delta^4+2\delta^2-1)} + \frac{\phi}{2\pi\delta}\right). \end{aligned} \tag{23}$$

3.2 The Coefficient a_2

The remaining integral a_2 is obtained from the solution of the cell problem

$$\begin{cases} -\mathrm{div}_y \sigma_y(\chi^{12}) = 0 & \text{in } Y_S, \\ \sigma_y(\chi^{12})\mathbf{n} = -\begin{pmatrix} 0 & -\mu \\ -\mu & 0 \end{pmatrix} \mathbf{n} \text{ on } \Gamma, & \\ \chi^{+} \quad \text{is } Y-\text{periodic}, & \end{cases} \tag{24}$$

whose solution is linear in μ. Once more, considering the differential problem satisfied by the function $\frac{\chi^{12}}{\mu}$, one can conclude that $\frac{a_2}{\mu}$ depends on the Lamé coefficients only through δ. Moreover, numerical evidence showed that, for small porosity and in the regime $\lambda = O(\mu)$, a_2 behaves almost linearly in ϕ, Hence, in order to determine the coefficient, we considered a numerical interpolation for the variable δ, which resulted in the following expression

$$a_2 = -\mu\phi\left(\frac{\pi}{2} + \frac{4(5-\delta^2)(\delta^2-2)}{15\delta^2(\delta^2+1)}\right). \tag{25}$$

3.3 Semi-Analytical Effective Tensor

In conclusion, the non-zero entries of the effective elasticity tensor are approximated as

$$\begin{aligned} &C^{eff}_{1111} = C^{eff}_{2222} = \\ &\quad \lambda + 2\mu + (\lambda+\mu)\left(-\frac{5(1+\delta)\phi}{2+4(2+\delta)\phi}\right)\left(\frac{1}{2} + \frac{\delta^2}{2(\delta^4+2\delta^2-1)} - \frac{\phi}{2\pi\delta}\right), \\ &C^{eff}_{1122} = C^{eff}_{2211} = \lambda + (\lambda+\mu)\left(-\frac{5(1+\delta)\phi}{2+4(2+\delta)\phi}\right)\left(\frac{1}{2} - \frac{\delta^2}{2(\delta^4+2\delta^2-1)} + \frac{\phi}{2\pi\delta}\right), \\ &C^{eff}_{1212} = C^{eff}_{1221} = C^{eff}_{2112} = C^{eff}_{2121} = \mu\left(1 - \phi\left(\frac{\pi}{2} + \frac{4(5-\delta^2)(\delta^2-2)}{15\delta^2(\delta^2+1)}\right)\right). \end{aligned} \tag{26}$$

Even if based on heuristic considerations and a coarse numerical interpolation, the results showed that the formulas (23) and (25) provide satisfactory approximations in the range of small porosity and for $\lambda = O(\mu)$. A detailed comparison of the coefficients for small porosities, and for two particular choices of λ and μ, is shown in Fig. 2.

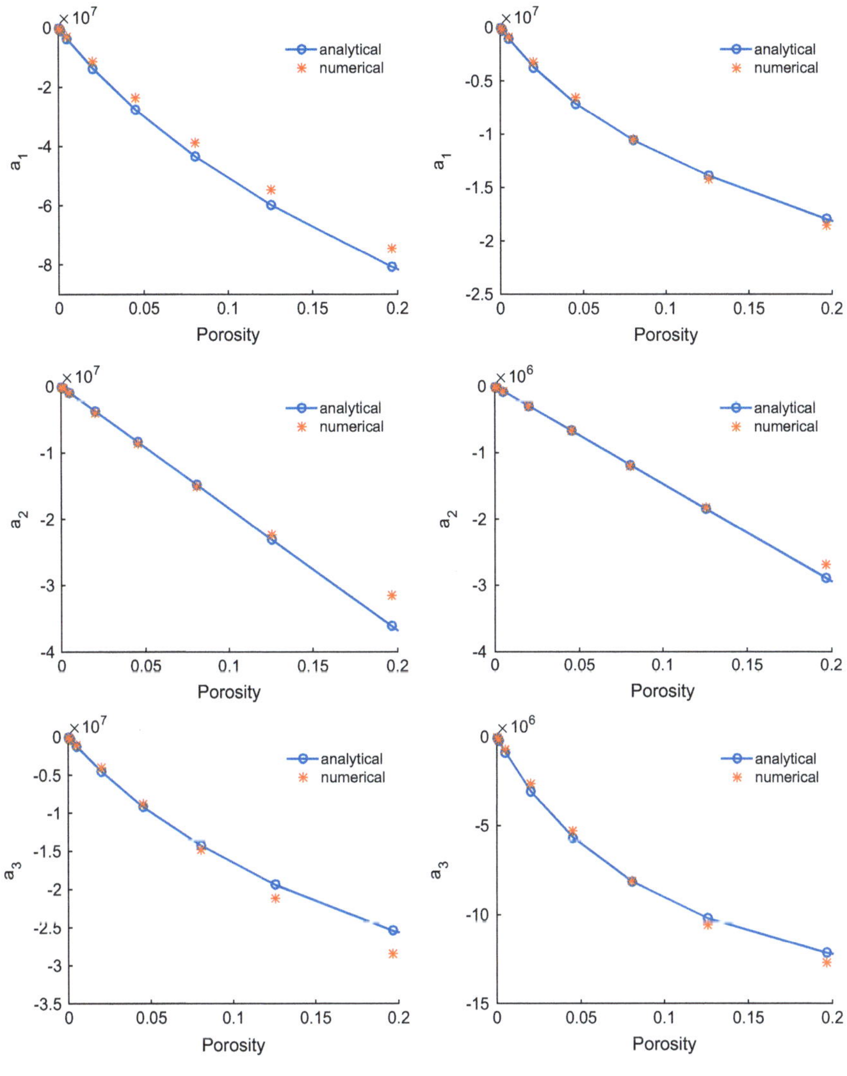

Fig. 2 Comparisons of the coefficients of the effective elasticity tensor computed numerically (*red*) and with the approximations (23) and (25) (*blue*). *Left*: $\lambda = \mu = 100\,\text{MPa}$. *Right*: $\lambda = 30\,\text{MPa}$, $\mu = 10\,\text{MPa}$

4 Estimation of Porosity Through the Homogenized Model

Finally, we exploit the results of Sects. 2 and 3 to define a two-scale variational-based estimation algorithm for the identification of material porosity using harmonic wave analysis. In particular, the analytical approximations (26) of the effective tensor coefficients allows us to solve the inverse problem without resolving numerically the dynamics at the fine scale (of the pores).

For the set up of the problem, we assume that the porosity field $\phi(\mathbf{x}) : \Omega \to [0, 1]$ is an unknown function of space, and that a set of displacement measurements (for a given frequency ω) at different location inside Ω is available, described by a field

$$\mathbf{d}_{\text{exp}}^{\omega} : \mathcal{M} \to \mathbb{R}^d,$$

with ($\mathcal{M} \subset \Omega$). The inverse problem is formulated as a minimization problem for the objective functional

$$J(\phi) = \frac{1}{2} \int_{\mathcal{M}} \left| \mathbf{d}_{\phi} - \mathbf{d}_{\text{exp}}^{\omega} \right|^2, \tag{27}$$

where $\mathbf{d}_{\phi}$ denotes the numerical solution for the macroscopic displacement corresponding to a particular porosity field ϕ:

$$\begin{cases} -\overline{\rho}\omega^2 \mathbf{d}_{\phi} - \operatorname{div}\left(C^{\mathit{eff}} e_x(\mathbf{d}_{\phi})\right) = 0, \text{ in } \Omega, \\ C^{\mathit{eff}} e_x(\mathbf{d}_{\phi})\mathbf{n} = \mathbf{0} \text{ on } \Gamma_N, \\ \mathbf{d}_{\phi} \cdot \mathbf{n} = d_{\text{bd}} \text{ on } \Gamma_D. \end{cases} \tag{28}$$

The minimization problem for $J(\phi)$ is solved with a variational procedure similar to the one described in [8, 9], which is shortly summarized below. First, we compute the derivative of $J(\phi)$ with respect to a given increment θ of the porosity field by perturbing the system in the variable ϕ, yielding a new problem for the sensitivity of $\mathbf{d}_{\phi}$, which will be denoted as $\partial_{\phi}\mathbf{d}_{\phi}$. Then, the gradient of the functional $J(\phi)$ can be obtained as the Fréchet derivative of (27):

$$\left\langle \frac{\partial J}{\partial \phi}, \theta \right\rangle = \int_{\mathcal{M}} \left((\mathbf{d}_{\phi} - \mathbf{d}_{\text{exp}}^{\omega}) \cdot \partial_{\phi}\mathbf{d}_{\phi} \right) \theta \, dx \tag{29}$$

along any increment direction θ. Introducing the adjoint problem

$$\begin{cases} -\overline{\rho}\omega^2 \mathbf{z}_{\phi} - \operatorname{div}\left(C^{\mathit{eff}} e_x(\mathbf{z}_{\phi})\right) = \mathbf{d}_{\phi} - \mathbf{d}_{\text{exp}}^{\omega} & \text{in } \Omega, \\ C^{\mathit{eff}} e_x(\mathbf{z}_{\phi})\mathbf{n} = 0 & \text{on } \Gamma_N, \\ \mathbf{z}_{\phi} \cdot \mathbf{n} = 0 & \text{on } \Gamma_D, \end{cases} \tag{30}$$

one can rewrite Eq. (29) as

$$\left\langle \frac{\partial J}{\partial \phi}, \theta \right\rangle = -\int_{\mathcal{M}} \left(\omega^2 \rho_S \mathbf{d}_{\phi} \cdot \mathbf{z}_{\phi} + \left[\partial_{\phi} C^{\mathit{eff}}\right] e(\mathbf{d}_{\phi}) : e(\mathbf{z}_{\phi}) \right) \theta \, dx, \tag{31}$$

which is obtained testing the variational formulation of (28) with $\mathbf{v} = \mathbf{z}_{\phi}$ and testing the variational formulation associated with (30) with $\partial_{\phi}\mathbf{d}_{\phi}$.

Using (31), a descent direction for J can be obtained defining the increment as $\theta = -\alpha S$, where $\alpha > 0$ is a free parameter, controlling the length of the step along the θ-direction and

$$S = -\omega^2 \rho_S \mathbf{d}_\phi \cdot \mathbf{z}_\phi - \left[\partial_\phi C^{eff}\right] e(\mathbf{d}_\phi) : e(\mathbf{z}_\phi). \tag{32}$$

In particular, the tensor $\partial_\phi C^{eff}$ in (32), i.e. the derivative of C^{eff} with respect to ϕ, can be computed using (26), without the need of multiple numerical solution of the microscopic cell problem (12) for different porosity (i.e. pore size in the cell problem).

4.1 The Algorithm

Our two-scale estimation algorithm can be summarized as follows. As initial conditions, let be given a porosity field $\phi^{(0)}$ (e.g. equal to a constant ϕ_0), let $\alpha^{(0)} = 1$, and let $S^{(0)} = S(\phi^{(0)})$ [as in (32)].

Until a convergence criterium is satisfied, do:

1 Compute $\phi^{(k)} = \phi^{(k-1)} - \alpha^{(k)} S^{(k)}$ (possibly restrict $\phi^{(k)}$ between 0 and 1).
2 Evaluate the homogenized tensor $C^{eff}(\phi^{(k)})$ using (26).
3 Solve (28) and compute the macroscopic solution $\mathbf{d}^{(k)}$,
4 Evaluate $J^{(k)} = J(\phi^{(k)})$:
  if $J^{(k)} \geq J^{(k-1)} \Rightarrow$ set $\phi^{(k+1)} = \phi^{(k)}$, $\alpha^{(k+1)} = \frac{\alpha^{(k)}}{2}$ and go back to step 1
  else (if $J^{(k)} < J^{(k-1)}$).
5 Evaluate the derivatives of $C^{eff}(\phi^{(k)})$ through (26).
6 Solve the adjoint problem (30) for $\mathbf{z}^{(k)}$.
7 Compute $S^{(k)}$ using (32).

In the numerical results presented below we used, as indicators of convergence, either a lower bound on the magnitude of the increment in porosity or the relative decrease of the objective functional between successive iterations.

5 Numerical Results

This section is devoted to the numerical results. In all cases, the finite element formulations have been implemented and solved with FreeFem++ [1, 10]. The programs used for the numerical tests are available for download at http://wias-berlin.de/people/caiazzo/FreeFem/cedya2015.zip.

In order to validate the two-scale homogenization model, we consider a square computational domain $\Omega = \left[0, \frac{2}{3}\right]^2$ consisting of an elastic matrix and several small void inclusions of circular shape, with radii such that the resulting porosity is equal to 0.07 in the outer part and to 0.15 in the inner part (Fig. 3, left). The synthetic

Fig. 3 *Left*: The domain with two sub-regions of different porosity (*outer*: $\phi = 0.07$, *inner*: $\phi = 0.15$) used for testing the two-scale porosity estimation method. *Right*: The coarse mesh used for solving the homogenized problems within the estimation algorithm

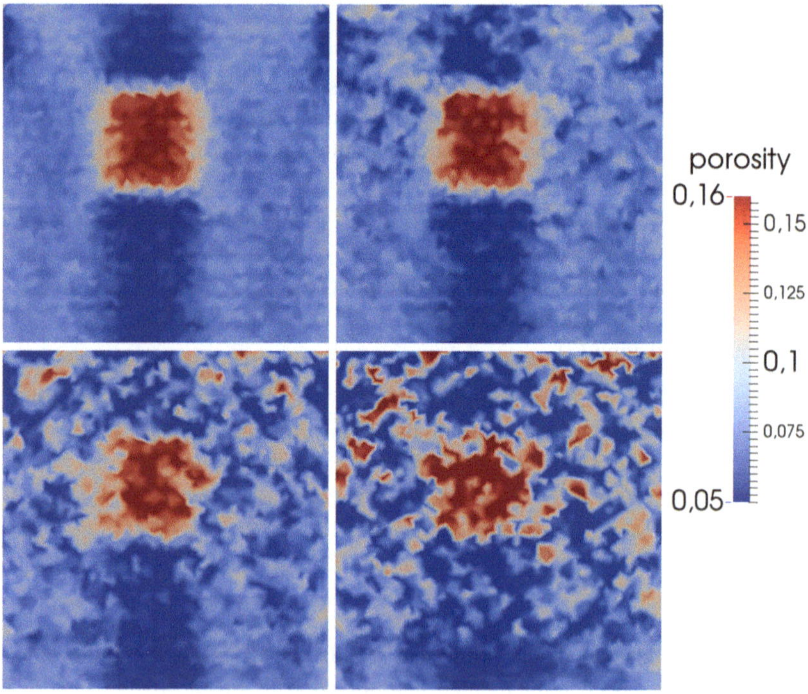

Fig. 4 Solution obtained with the estimation algorithm in the case $\lambda = \mu = 100$ MPa. *Top-Left*: result without adding noise to the mesoscopic solution. *Top-Right*: result with 1 % of noise. *Bottom-Left*: 2.5 % of noise. *Bottom-Right*: 5 % of noise. All the simulations have been performed using the homogenized formulation (on the coarse mesh shown in Fig. 3, *right*)

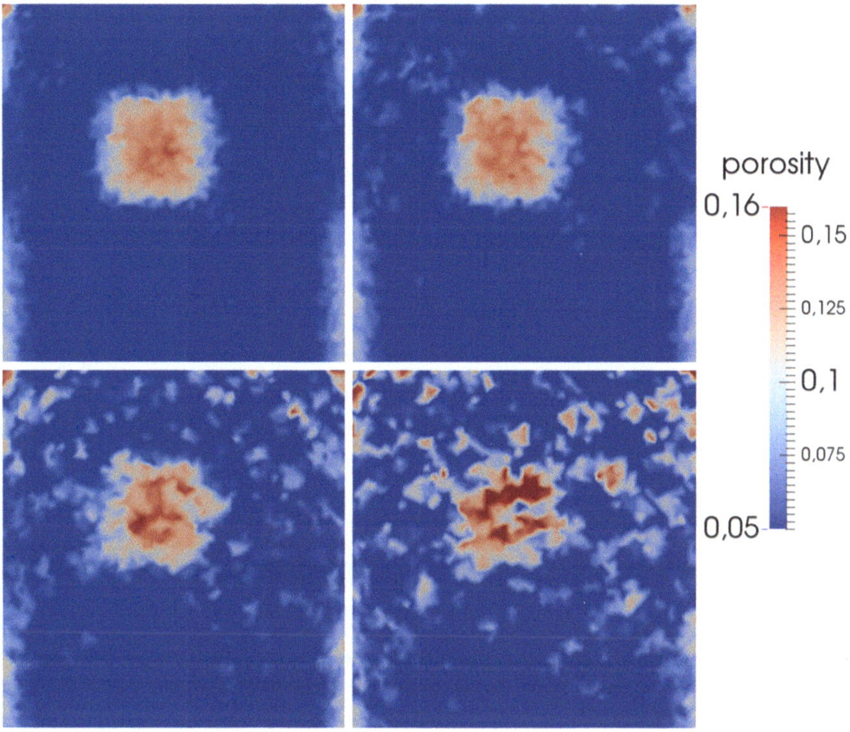

Fig. 5 Solution obtained with the estimation algorithm in the case $\lambda = 30\,\text{MPa}$, $\mu = 10\,\text{MPa}$. *Top-Left*: result without adding noise to the mesoscopic solution. *Top-Right*: result with 1 % of noise. *Bottom-Left*: 2.5 % of noise. *Bottom-Right*: 5 % of noise. All the simulations have been performed using the homogenized formulation (on the coarse mesh shown in Fig. 3, *right*)

measurements to feed the minimization algorithm have been constructed solving the full scale problem (2)–(3) on a very fine mesh (around 130K nodes and 256K triangles) then interpolating the displacement on a much coarser mesh (3K nodes and 6K triangles), that does not resolve the geometry of the inclusions (Fig. 3, right). Finally, the obtained displacement field has been perturbed with Gaussian noise. The frequency has been fixed to 50 Hz.

Figures 4 and 5 show the estimated porosity for $\lambda = \mu = 100\,\text{MPa}$ and $\lambda = 30\,\text{MPa}$, $\mu = 10\,\text{MPa}$, respectively. In both cases, the estimation algorithm is able to detect the larger porosity, also perturbing the interpolated measurements with random noise (up to an intensity of 5 %).

The detailed behavior of the objective functional $J(\phi)$ for $\lambda = \mu = 100\,\text{MPa}$ is shown in Fig. 6. A similar descent behavior has been obtained in the case $\lambda = 30\,\text{MPa}$, $\mu = 10\,\text{MPa}$ (not shown). In particular, we observe that, also in the noisy cases, the variational approach is still able to find a descent direction of the functional.

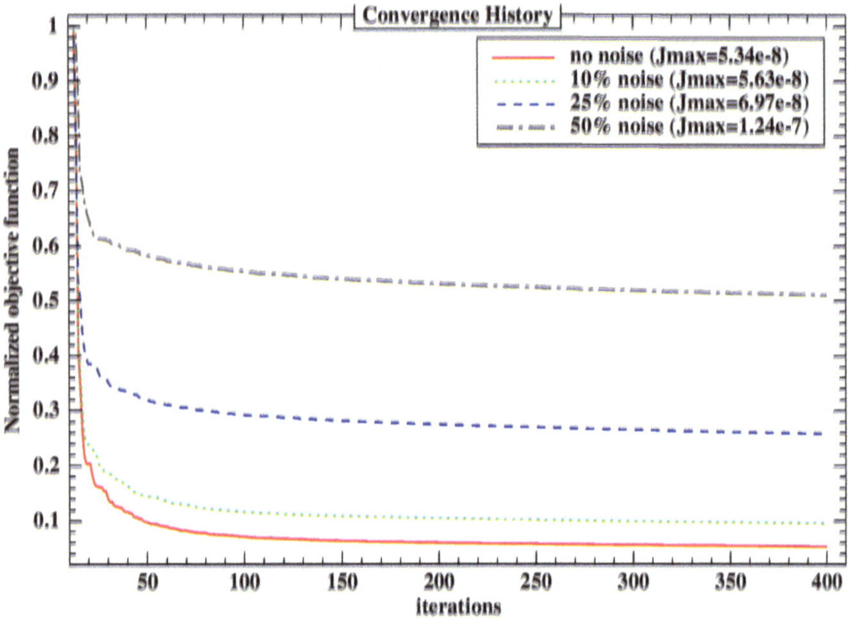

Fig. 6 Convergence history of the objective functional for $\lambda = \mu = 100$ MPa, for different level of noise. For the sake of visualization, the curves have been renormalized with respect to the initial value (indicated in the legend)

Finally, it is worth noticing that the estimation algorithm, although seeking the minimum with respect to a displacement field of the small scale problem (2)–(3), is purely based on the solutions of the homogenized problems (28)–(30). The numerical results, besides drastically improving the efficiency of the computation, provide an intrinsic validation of the two-scale approximation of the original elasticity problem in terms of the local cell problems (12) and the semi-analytical approximations (23) and (25).

6 Conclusions

We presented a novel algorithm for the detection of porosity in elastic media based on a variational optimization procedure, extending the two-scale approach firstly presented in [5]. The main advantage of the proposed methodology is that it allows to formulate the inverse problem directly on the homogenized version of the original equation, parametrizing semi-analytically the upscaled elasticity tensor in terms of mesoscopic geometrical properties (e.g. the shape of the pores). We showed that the semi-analytical formulas provide a good approximation for the detection of void pores in selected cases of interests (small porosity and $\lambda = O(\mu)$) and in moderate harmonic regime. However, further developments are needed in order to obtain more robust approximations for a wider range of Lamé coefficients and to extend the procedure to the high frequency regime.

Acknowledgements The research of J. Mura has been supported by the *Fondecyt-Initiation to Research* project no. 11121606 (Conicyt/Chile).

References

1. Auliac, S., Le Hyaric, A., Morice, J., Hecht, F., Ohtsuka, K., Pironneau, O.: FreeFem++, 3rd edn. Version 3.31–2 (2014). http://www.freefem.org/ff++/ftp/freefem++doc.pdf
2. Aurialt, J.L., Boutin, C., Geindreau, C.: Homogenization of Coupled Phenomena in Heterogeneous Media. Wiley, New York (2009)
3. Ávila, A., Griso, G., Miara, B., Rohan, E.: Multiscale modeling of elastic waves: theoretical justification and numerical simulation of band gaps. SIAM Multiscale Model. Simul. **7**, 1–21 (2008)
4. Baffico, L., Grandmont, C., Maday, Y., Osses, A.: Homogenization of elastic media with gaseous inclusions. SIAM Multiscale Model. Simul. **7**, 432–465 (2008)
5. Caiazzo, A., Mura, J.: Multiscale modeling of weakly compressible elastic materials in the harmonic regime and applications to microscale structure estimation. SIAM J. Multiscale Model. Simul. **12**(2), 514–537 (2014)
6. Cioranescu, D., Piatnitski, A.: Homogenization in perforated domains with rapidly pulsing perforations. ESAIM Control Optim. Calc. Var. **9**, 461–483 (2003)
7. Efendiev, Y., Hou, T.: Multiscale Finite Element method, Theory and Applications. Surveys and Tutorials in the Applied Mathematical Sciences, vol. 4. Springer, New York (2009)
8. Gutiérrez, S., Mura, J.: Small amplitude homogenization applied to inverse problems. Comp. Mech. **41**, 699–706 (2008)
9. Gutiérrez, S., Mura, J.: An adaptive procedure for inverse problems in elasticity. Comptes Rendus Mécanique **338**, 402–411 (2010)
10. Hecht, F.: New development in FreeFem++. J. Numer. Math. **20**(3–4), 251–265, 65Y15 (2012)
11. Hornung, U.: Homogenization and Porous Media. Springer, New York, (1997)
12. Sanchez-Palencia, E.: Non-homogeneous Media and Vibration Theory. Springer, Berlin (1980)
13. Sanchez-Palencia, E., Zaoui, A.: Homogenization Techniques for Composite Media. Lecture Notes in Physics, vol. 272. Springer, Berlin (1987)
14. Tartar, L.: H-measures, a new approach for studying homogenisation, oscillations and concentration effects in partial differential equations. Proc. Roy. Soc. Edinb. **115**(3–4), 193–230 (1990)

A Matrix Approach to the Newton Formula and Divided Differences

J.M. Carnicer, Y. Khiar, and J.M. Peña

Abstract The Crout factorization of a Vandermonde matrix is related with the Newton polynomial interpolation formula expressed in terms of divided differences. Another triangular factorization, which can be related with the Newton formula in terms of finite differences, is provided by the Doolittle factorization. The influence of the order of the nodes on the conditioning of the corresponding linear system is analyzed, considering the three cases of increasing order, Leja order and increasing distances to the origin. The lower triangular systems for the computation of divided and finite differences are analyzed and the conditioning of the corresponding lower triangular matrices is studied. Numerical examples are included.

1 Introduction

In this paper we analyze the problem of estimating the coefficients of the polynomial interpolant with respect to different bases. First we consider the monomial basis, which gives rise to a linear system whose coefficient matrix is the Vandermonde matrix. Then we use the Newton basis. The problem of computing the divided differences leads to a linear system whose coefficient matrix is lower triangular.

The Vandermonde matrix can be decomposed into triangular matrices by means of the Newton formula for the Lagrange interpolation polynomial problem. The resulting triangular factorization coincides with the Crout factorization because the upper triangular matrix has unit diagonal, as recalled in Sect. 2. In Sect. 2, we also recall the results of [3] showing that the computation of both triangular factors and their inverses can be performed with high relative accuracy. We present in Sect. 3 the results for the Doolittle factorization of the Vandermonde matrix, where the lower triangular factor has unit diagonal. We also present in this section three different orderings of the nodes: natural, Leja and a new ordering that will be called central ordering. We discuss their influence on the conditioning (cf. [7]) of the Vandermonde linear system. The numerical experiments of Sect. 3 show the nice

J.M. Carnicer • Y. Khiar (✉) • J.M. Peña
Departamento de Matemática Aplicada, Universidad de Zaragoza, C/Pedro Cerbuna 12, 50009 Zaragoza, Spain
e-mail: carnicer@unizar.es; yasmina@unizar.es; jmpena@unizar.es

F. Ortegón Gallego et al. (eds.), *Trends in Differential Equations and Applications*, SEMA SIMAI Springer Series 8, DOI 10.1007/978-3-319-32013-7_7

properties of this new ordering when the interval is centered at the origin. Finally, in Sects. 4 and 5 we analyze the systems $L\mathbf{d} = \mathbf{f}$ and $\tilde{L}\Delta = \mathbf{f}$, where $\mathbf{f}$ is the vector of the initial data, $\mathbf{d}$ and Δ are the divided and finite differences, and L and $\tilde{L}$ are the lower triangular matrices of the Crout and the Doolittle factorizations, respectively. The conditioning of $\tilde{L}$ under several orderings is studied. For equidistant nodes, it is shown that $\tilde{L}$ is better conditioned with the Leja ordering than with the natural ordering. Numerical experiments are also included.

2 Lagrange and Newton Formulas

Let us pose the Lagrange interpolation problem. Given an $(n+1)$-dimensional function space U, distinct nodes $x_0, \ldots, x_n$, and values $f_0, \ldots, f_n$, we want to find $u \in U$ such that $u(x_i) = f_i$, $i = 0, \ldots, n$.

Let $(u_0, \ldots, u_n)$ be a basis of U. Then we can write the solution u, which is called the interpolant, respect to this basis with coefficients $c_0, \ldots, c_n$, $u = \sum_{i=0}^{n} c_i u_i$, and the interpolation problem is reduced to the linear system

$$M\begin{pmatrix} u_0, \ldots, u_n \\ x_0, \ldots, x_n \end{pmatrix} \mathbf{c} = \mathbf{f},$$

where $\mathbf{c} = (c_0, \ldots, c_n)^T, \mathbf{f} = (f_0, \ldots, f_n)^T$ in $\mathbf{R}^{n+1}$ and the matrix

$$M\begin{pmatrix} u_0, \ldots, u_n \\ x_0, \ldots, x_n \end{pmatrix} = (u_j(x_i))_{i,j=0,\ldots,n}$$

in $\mathbf{R}^{(n+1)\times(n+1)}$ is called the *collocation matrix* of the basis at the nodes $x_0, \ldots, x_n$.

We consider the particular case $U = P_n$, the space of the polynomials of degree less than or equal to n, and the monomial basis $\mathbf{m} = (m_0, \ldots, m_n)^T$, $m_j(x) := x^j$, $j = 0, \ldots, n$. Then the problem of finding the coefficients $\mathbf{c}$ of the interpolant is reduced to solve the system

$$V\mathbf{c} = \mathbf{f}, \tag{1}$$

where

$$V = V(x_0, \ldots, x_n) := M\begin{pmatrix} m_0, m_1, \ldots, m_n \\ x_0, x_1, \ldots, x_n \end{pmatrix} \tag{2}$$

is the *Vandermonde matrix* with nodes $x_0, \ldots, x_n$, whose (i,j) entry is $v_{ij} = x_i{}^j$, $i,j = 0, \ldots, n$.

We denote the interpolation polynomial by $p(x)$. The solution of the Lagrange interpolation problem can be expressed using Lagrange formula

$$p(x) = \sum_{j=0}^{n} f(x_j) l_j(x), \qquad l_j(x) = \prod_{\substack{k=0 \\ k \neq j}}^{n} \frac{x - x_k}{x_j - x_k}, \quad j = 0, \ldots, n,$$

where the functions l_j are the Lagrange polynomials. Let us denote by $\mathbf{l} = (l_0, \ldots, l_n)^T$ the Lagrange basis.

If we compare the expressions of the interpolant with respect to both bases $\mathbf{l}$ and $\mathbf{m}$, we have

$$\mathbf{l}^T \mathbf{f} = \mathbf{m}^T \mathbf{c}.$$

Using (1), we have $\mathbf{l}^T \mathbf{f} = \mathbf{m}^T V^{-1} \mathbf{f}$ for all $\mathbf{f}$, and so we deduce that

$$\mathbf{l}^T = \mathbf{m}^T V^{-1}, \tag{3}$$

i.e., the matrix of change of basis between the Lagrange basis and the monomial basis is the inverse of the Vandermonde matrix. Expanding the elements of the Lagrange basis in terms of the monomial basis, we obtain that the entry (i, j) of V^{-1} is

$$v_{ij}^{(-1)} = \frac{(-1)^{n-i} \sum_{k_1 < \cdots < k_{n-i} \in \{0, \ldots, n\} \setminus \{j\}} x_{k_1} \cdots x_{k_{n-i}}}{\prod_{k \neq j} (x_j - x_k)}. \tag{4}$$

The *Newton formula*

$$p(x) = \sum_{j=0}^{n} [x_0, \ldots, x_j] f \omega_j(x)$$

expresses the polynomial interpolant in terms of the Newton basis, $\boldsymbol{\omega} = (\omega_0, \ldots, \omega_n)$, and the divided differences $d_j = [x_0, \ldots, x_j] f$, $j = 0, \ldots, n$. The elements of the Newton basis are

$$\omega_j(x) = (x - x_0) \cdots (x - x_{j-1}), \quad j = 0, \ldots, n. \tag{5}$$

Each element ω_j of the Newton basis is a monic polynomial of degree j such that $\omega_j(x_i) = 0$ for $i < j$ and for this reason the collocation matrix of the Newton basis

$M\binom{\omega_0,\dots,\omega_n}{x_0,x_1,\dots,x_n}$ is a lower triangular matrix. Let us denote by L this matrix

$$L = \begin{pmatrix} 1 & 0 & \cdots & 0 \\ 1 & x_1 - x_0 & \cdots & 0 \\ 1 & x_2 - x_0 & \ddots & \vdots \\ \vdots & \vdots & \ddots & 0 \\ 1 & x_n - x_0 & \cdots & (x_n - x_0)\cdots(x_n - x_{n-1}) \end{pmatrix}, \tag{6}$$

with

$$l_{ij} = \omega_j(x_i). \tag{7}$$

Observe that L is the matrix of change of basis between the Newton basis and the Lagrange basis

$$\boldsymbol{\omega}^T = \mathbf{l}^T L, \tag{8}$$

and so

$$L\mathbf{d} = \mathbf{f}, \tag{9}$$

where $\mathbf{d} = (d_0, \dots, d_n)^T$ is the vector of divided differences.

Let

$$U := \begin{pmatrix} 1 & x_0 & x_0^2 & \cdots & x_0^n \\ 0 & 1 & [x_0, x_1]x^2 & \cdots & [x_0, x_1]x^n \\ 0 & 0 & 1 & \ddots & \vdots \\ \vdots & \vdots & \ddots & \ddots & [x_0, \dots, x_{n-1}]x^n \\ 0 & 0 & \cdots & 0 & 1 \end{pmatrix}. \tag{10}$$

From the Newton formula applied to the monomials, $m_0, \dots, m_n$, we deduce

$$\mathbf{m}^T = \boldsymbol{\omega}^T U. \tag{11}$$

Relating the formulas (3), (8) and (11) we have

$$V = LU, \tag{12}$$

that is, the matrices L and U form the (unique) Crout factorization of the Vandermonde matrix. The Crout factorization is characterised by the fact that L is a lower triangular matrix whose diagonal entries are the pivots of the Gaussian elimination and U is an upper triangular matrix with unit diagonal.

The LU factorization is used frequently to solve linear systems. In order to solve $V\mathbf{c} = \mathbf{f}$, with $V = LU$, we consider the following two triangular systems

$$L\mathbf{d} = \mathbf{f}, \quad U\mathbf{c} = \mathbf{d}. \tag{13}$$

The solution of the system $V\mathbf{c} = \mathbf{f}$ is reduced to the successive solution of these triangular systems with matrices L and U. These systems link the solution with the intermediate vector $\mathbf{d}$, the vector of the divided differences, and, therefore, they are directly related with the Newton formula. We will study the lower triangular system in Sect. 4.

In [3] we proposed the following algorithms to compute L and U. The first algorithm computes the entries of the matrix L

$$\begin{aligned} &l_{i0} = 1, \quad i = 0, \ldots, n, \\ &l_{ij} = l_{i,j-1}(x_i - x_{j-1}), \quad i = j, \ldots, n, \quad j = 1, \ldots, n. \end{aligned} \tag{14}$$

The second method computes U as follows

$$\begin{aligned} &u_{00} = 1, \quad u_{0j} = x_0 u_{0,j-1}, \quad j = 1, \ldots, n, \\ &u_{ii} = 1, \quad u_{ij} = u_{i-1\,j-1} + x_i u_{i\,j-1}, \quad j = i+1, \ldots, n, \quad i = 1, \ldots, n. \end{aligned} \tag{15}$$

The following algorithms compute the entries of L^{-1} and U^{-1}, respectively

$$\begin{aligned} l_{ij}^{(-1)} &= -\frac{l_{i-1,j}^{(-1)}}{x_i - x_j}, \quad j = 0, \ldots, i-1, \\ l_{ii}^{(-1)} &= \frac{1}{\prod_{j=0}^{i-1}(x_l - x_j)}, \quad i = 0, \ldots, n, \end{aligned} \tag{16}$$

$$\begin{aligned} &u_{00}^{(-1)} = 1, \quad u_{0j}^{(-1)} = -x_{j-1} u_{0,j-1}^{(-1)}, \quad j = 1, \ldots, n, \\ &u_{ii}^{(-1)} = 1, \quad u_{ij}^{(-1)} = u_{i-1,j-1}^{(-1)} - x_{j-1} u_{i,j-1}^{(-1)}, \quad j = i+1, \ldots, n, \quad i = 1, \ldots, n. \end{aligned} \tag{17}$$

Let us recall that a value $\mathbf{X}$ can be obtained with *high relative accuracy* (HRA) if the relative error of the computed value $\widehat{X}$ can be bounded as follows:

$$\frac{||\mathbf{X} - \widehat{X}||}{||\mathbf{X}||} \leq Cu,$$

where C is a positive constant independent of the arithmetic precision and u is the unit roundoff.

In [5], it was shown we can compute with high relative accuracy the products, quotients and true additions (addition of numbers with the same sign) of expressions that can be computed with high relative accuracy. The subtractions (addition of numbers with the opposite sign) are permitted only with initial data of the problem.

It is well-known that the inverse of a Vandermonde matrix with nonnegative increasing nodes can be computed with HRA because the matrix is totally nonnegative and its bidiagonal factorization can be obtained with high relative accuracy (see [4] and [8]). The following result (corresponding to Theorem 1 of [3]) shows that such computation is possible whenever all distinct nodes have the same nonstrict sign and for all possible order configurations of the nodes.

Theorem 1 *Let V be the Vandermonde matrix in (2) corresponding to a sequence of distinct nodes of the same nonstrict sign. Then V^{-1} can be computed with HRA through the formula (4).*

We also have the following results, corresponding to Theorems 2 and 3 of [3] respectively, about the matrices L and U and their inverses. For the matrices L and L^{-1} there are no restrictions on the nodes, that is, HRA can be ensured for all possible signs and order configurations of the nodes.

Theorem 2 *Let L be the lower triangular matrix in (6). Then L and L^{-1} can be computed with HRA for any distinct nodes, x_i, $i = 0, \ldots, n$, using algorithms (14) and (16), respectively.*

In contrast to Theorem 2, the following result requires the same restrictions on the nodes as in Theorem 1 to ensure that U and U^{-1} can be computed with HRA.

Theorem 3 *Let U be the upper triangular matrix in (10) corresponding to a sequence of nodes with the same nonstrict sign. Then U and U^{-1} can be computed with HRA, using algorithms (15) and (17), respectively.*

3 Triangular Factorizations and the Influence of the Order of the Nodes

In the previous section we have used the Crout factorization LU of a Vandermonde matrix V. However, the most common triangular factorization used with Gaussian elimination is $V = \tilde{L}\tilde{U}$ where $\tilde{L}$ is a lower triangular matrix with unit diagonal and $\tilde{U}$ an upper triangular with the pivots of the Gaussian elimination on the main diagonal. This factorization is also called the Doolittle factorization. Let D be the diagonal matrix with the pivots of the Gaussian elimination on the main diagonal. Then we have

$$\tilde{L} = LD^{-1}, \quad \tilde{U} = DU. \tag{18}$$

Since L, U and their inverses can be computed with HRA by the results of the previous section, it easily follows that $\tilde{L}$ and $\tilde{U}$ and their inverses can be computed with HRA under certain conditions on the nodes.

We have seen that the Crout factorization $V = LU$ is related with the Newton formula in terms of divided differences. The Doolittle factorization $V = \tilde{L}\tilde{U}$ can be related with the Newton formula in terms of finite differences

$$p(x) = \sum_{j=0}^{n} \Delta(x_0, \ldots, x_j) f \tilde{\omega}_j(x),$$

where

$$\tilde{\omega}_j(x) = \frac{\omega_j(x)}{\omega_j(x_j)}, \quad j = 0, \ldots, n, \tag{19}$$

and

$$\Delta(x_0, \ldots, x_j) f = (x_j - x_0) \cdots (x_j - x_{j-1})[x_0, \ldots, x_j] f, \quad j = 0, \ldots, n,$$

are a different normalization of the divided differences.

Different orderings of the nodes are related with different row pivoting strategy of Gaussian elimination and this leads to different conditioning of the triangular factors. A common ordering consists of setting the nodes in increasing order, that is, $x_0 < \cdots < x_n$. This ordering will be called *natural ordering*.

If the nodes are positive and increasing $0 < x_0 < \cdots < x_n$, then the Vandermonde matrix has all its minors nonnegative, i.e., it is a totally positive matrix. Some properties of the totally positive matrices suggest that Gaussian elimination without reordering of rows gives good stability results (see [1]), in particular for the conditioning of $\tilde{U}$ (see [9]). This fact provides a motivation to work with the nodes in increasing order.

However, other orderings of the nodes may also lead to stability in the computations. In the diagonal of the matrix L we have the pivots of Gaussian elimination. Through a strategy of partial pivoting, we try to maximize the multipliers in each step. Note that partial pivoting is equivalent to reorder the nodes. In fact, this way of arranging the nodes is equivalent to Leja ordering (see [2] and [6]), leading to nice properties of the lower triangular factors. The *Leja ordering* is achieved using the following strategy (see [10]).

1. We can choose as the first node x_0 any node in the set. However, to maximize $|x_1 - x_0|$ in the second step, we may choose one extreme point, either the minimum or the maximum.

2. In the second step, x_1 is chosen such that

$$|x_1 - x_0| = \max_{j=1,\dots,n} |x_j - x_0|.$$

So x_1 is the other extreme (minimum or maximum).
3. In the i-th step, we select x_i such that

$$\prod_{k=0}^{i-1} |x_i - x_k| = \max_{j=i,\dots,n} \prod_{k=0}^{i-1} |x_j - x_k|.$$

We consider another ordering of the nodes that we call the *central ordering*, where the nodes are ordered increasingly according to a center c. If $x_0, \dots, x_n$ are ordered following the central ordering then

$$|x_0 - c| \leq |x_1 - c| \leq \cdots \leq |x_n - c|.$$

Since the Vandermonde matrix computes the coefficients with respect to the powers centered at the origin, we shall only consider in our problem the center $c = 0$.

In order to test the central ordering, we shall perform our experiments with intervals centered at the origin. We also consider intervals starting at zero, due to the total positivity of the Vandermonde matrix when the nodes are ordered increasingly. Observe that the natural order is the central order for an interval starting at zero.

Remark 1 Since L can be expressed in terms of differences of nodes, the translation of the nodes does not have any have influence on the behaviour of L, in contrast to the behaviour of U. Then, for any order, the properties of the matrix L will be analyzed in intervals of different lengths.

The condition number of a nonsingular matrix is given by

$$\kappa_\infty(A) := ||A||_\infty ||A^{-1}||_\infty$$

and it is a measure of the sensitivity of the solution of the linear systems respect to the perturbations of the initial data. The high conditioning of the Vandermonde matrix explains the difficulty to estimate the coefficients of the interpolant with respect to the monomial basis. If we have a triangular factorization of the Vandermonde matrix, then we can find the solution by solving two triangular linear systems. Then the conditioning of each triangular matrix will influence in each step. So, the product of both condition numbers will provide an upper bound of the sensitivity of the solution of the linear systems with respect to data perturbations when using triangular factorization algorithm. We call this product, $\kappa_\infty(L)\kappa_\infty(U)$ or $\kappa_\infty(\tilde{L})\kappa_\infty(\tilde{U})$, *joint condition number*.

For our tests, we consider two types of intervals for the nodes: centered at the origin and starting at zero. We take the intervals $[-1, 1]$ and $[-1.5, 1.5]$ as examples of the first type and $[0, 1]$ and $[0, 3]$ as examples of the second type.

We will take equidistant nodes in $[a, b]$

$$x_i := a + \sigma(i)\frac{b-a}{n}, \quad i = 0, \ldots, n,$$

where σ is a permutation of the set $\{0, \ldots, n\}$ corresponding to the associated order.

We consider three orderings for these nodes:

- Natural: the nodes are ordered increasingly.
- Leja: the nodes are ordered according to the strategy (1), (2) and (3) described previously in this section.
- Central: the nodes are ordered increasingly according to their distance to the origin.

As mentioned in the previous section we shall consider in our numerical experiments two triangular factorizations of the Vandermonde matrix V, the Crout factorization $V = LU$ and the Doolittle factorization $V = \tilde{L}\tilde{U}$.

The following tables collect information about the condition number of the Vandermonde matrix and the product of the condition numbers of the triangular factors for different dimensions.

Let us compare the joint condition number of both factorizations. Let us start with nodes in $[0, 1]$.

In Table 1 we can see that the best order of the nodes for the LU factorization is the natural ordering. However, Table 2 shows that the best joint condition number for the Doolittle factorization corresponds to the Leja ordering but there are no

Table 1 $\kappa_\infty(L)\kappa_\infty(U)$ in $[0, 1]$

		Natural	Leja
n	$\kappa_\infty(V)$	$\kappa_\infty(L)\kappa_\infty(U)$	
3	216	416	672
4	1.7067×10^3	3.4333×10^3	5.7600×10^3
5	1.2500×10^4	3.3700×10^4	5.6386×10^4
9	4.8184×10^7	3.3789×10^8	7.6455×10^8
19	5.0877×10^{16}	5.3085×10^{18}	2.7287×10^{19}

Table 2 $\kappa_\infty(\tilde{L})\kappa_\infty(\tilde{U})$ in $[0, 1]$

		Natural	Leja
n	$\kappa_\infty(V)$	$\kappa_\infty(\tilde{L})\kappa_\infty(\tilde{U})$	
3	216	576	800
4	1.7067×10^3	6.8267×10^3	7.6800×10^3
5	1.2500×10^4	9.0667×10^4	6.0450×10^4
9	4.8184×10^7	3.6280×10^9	4.1542×10^8
19	5.0877×10^{16}	1.9664×10^{21}	6.3140×10^{17}

Table 3 $\kappa_\infty(L)\kappa_\infty(U)$ in $[0,3]$

		Natural	Leja
n	$\kappa_\infty(V)$	$\kappa_\infty(L)\kappa_\infty(U)$	
3	266.6667	512	2240
4	2.8681×10^3	6.8661×10^3	2.2511×10^4
5	2.6963×10^4	7.2298×10^4	3.2855×10^5
9	4.0049×10^8	1.4074×10^9	5.5271×10^9
19	2.0837×10^{19}	1.6470×10^{20}	3.2911×10^{20}

Table 4 $\kappa_\infty(\tilde{L})\kappa_\infty(\tilde{U})$ in $[0,3]$

		Natural	Leja
n	$\kappa_\infty(V)$	$\kappa_\infty(\tilde{L})\kappa_\infty(\tilde{U})$	
3	266.6667	938.6667	1.1852×10^3
4	2.8681×10^3	9.9000×10^3	1.5775×10^4
5	2.6963×10^4	1.1612×10^5	1.6111×10^5
9	4.0049×10^8	2.2396×10^9	3.9113×10^9
19	2.0837×10^{19}	2.3145×10^{20}	2.6917×10^{20}

Table 5 $\kappa_\infty(L)\kappa_\infty(U)$ in $[-1,1]$

		Natural	Leja	Central
n	$\kappa_\infty(V)$	$\kappa_\infty(L)\kappa_\infty(U)$		
3	18	327.5556	193.3333	78
4	53.3333	2.2295×10^3	580	245.3333
5	187.5000	1.2696×10^4	2.0157×10^3	915.488
9	2.0562×10^4	1.5169×10^7	2.1156×10^5	1.2454×10^5
19	1.7511×10^9	8.6376×10^{14}	5.8935×10^{10}	2.122×10^{10}

significative differences between the $\tilde{L}\tilde{U}$ factorization with the Leja ordering and the LU factorization with natural ordering.

The system $V\mathbf{c} = \mathbf{f}$ is worse conditioned in the interval $[0,3]$ than in the interval $[0,1]$, as can be seen in Tables 3 and 4. We can also see in Tables 3 and 4 that the differences between the joint condition number and the conditioning of V are much smaller for the interval $[0,3]$ than for the previous interval $[0,1]$. Besides, the differences among the considered factorizations and orderings are less significative for the interval $[0,3]$. Even if the interpolant takes the same values in each interval, their coefficients are scaled in a different way and this fact affects the joint conditions.

Tables 5 and 6 correspond to the interval $[-1,1]$. We see that $\kappa_\infty(L)\kappa_\infty(U)$ is lower with the central order. Table 6 shows that the Doolittle factorization behaves better with the Leja order. If we compare the two factorizations for high values of n, we see that the lowest value is achieved for the LU factorization with the central order.

Table 7 collects the joint condition number of the LU factorization with the three orderings in the interval $[-1.5,1.5]$. We can see that the best ordering is the central

Table 6 $\kappa_\infty(\tilde{L})\kappa_\infty(\tilde{U})$ in $[-1, 1]$

		Natural	Leja	Central
n	$\kappa_\infty(V)$	$\kappa_\infty(\tilde{L})\kappa_\infty(\tilde{U})$		
3	18	784	113.3333	96
4	53.3333	7.4667×10^3	360	672
5	187.5000	6.1512×10^4	1.2788×10^3	2.9451×10^3
9	2.0562×10^4	3.0868×10^8	1.6768×10^5	1.7532×10^6
19	1.7511×10^9	4.5246×10^{17}	2.2233×10^{10}	2.4734×10^{13}

Table 7 $\kappa_\infty(L)\kappa_\infty(U)$ in $[-1.5, 1.5]$

		Natural	Leja	Central
n	$\kappa_\infty(V)$	$\kappa_\infty(L)\kappa_\infty(U)$		
3	18.9583	942.5000	381.8750	137.5000
4	62.5185	7.1397×10^3	905.6909	490.9722
5	192.4190	5.1245×10^4	3.4541×10^3	1.6736×10^3
9	1.8257×10^4	1.3326×10^8	4.4569×10^5	2.0001×10^5
19	2.3696×10^9	7.0152×10^{16}	8.4964×10^{10}	6.5344×10^{10}

Table 8 $\kappa_\infty(\tilde{L})\kappa_\infty(\tilde{U})$ in $[-1.5, 1.5]$

		Natural	Leja	Central
n	$\kappa_\infty(V)$	$\kappa_\infty(\tilde{L})\kappa_\infty(\tilde{U})$		
3	18.9583	1.5275×10^3	276.2500	405
4	62.5185	1.3504×10^4	633	2268
5	192.4190	1.4203×10^5	1.6537×10^3	9.0202×10^3
9	1.8257×10^4	1.4643×10^9	2.7830×10^5	6.2784×10^6
19	2.3696×10^9	2.0495×10^{19}	4.1203×10^{10}	8.2891×10^{13}

ordering followed closely by the Leja ordering. In Table 8 we consider the $\tilde{L}\tilde{U}$ factorization for the same interval. In this case, the Leja ordering gives better results for the joint condition number $\kappa_\infty(\tilde{L})\kappa_\infty(\tilde{U})$.

The conclusions of the numerical experiments can be summarized as follows. For the Crout factorization the best choices are central ordering, for intervals centered at the origin, and natural ordering, for intervals starting at zero. For the Doolittle factorization the best choice corresponds to the Leja ordering. Finally, for intervals centered at the origin the natural order gives the worst results.

4 Computation of Divided Differences

In the previous section, we have focused in the joint condition number because we were interested in estimating the coefficients of the interpolant with respect to the monomial basis. In this section we are going to analyze how the intervals and the different orderings of the nodes affect to the lower triangular matrices L and $\tilde{L}$. The solution of the triangular system $L\mathbf{d} = \mathbf{f}$ is the vector $\mathbf{d}$ of divided differences, which are the coefficients of the interpolant respect to the basis $\boldsymbol{\omega} = (\omega_0, \ldots, \omega_n)^T$ given by (5), is reduced to solve one of the triangular systems of (13)

$$L\mathbf{d} = \mathbf{f}.$$

Therefore, the condition number $\kappa_\infty(L)$ is related with the stability of the computation of the divided differences.

Tables 9, 10 and 11 show the condition number of the matrix L with nodes in the different intervals that we have proposed for the numerical experiments in the previous section.

By analyzing these tables we conclude that the best choice for all proposed intervals is the Leja order. The fact that the condition number of L is lower with the Leja order is not surprising since we have already mentioned in Sect. 3 that the Leja ordering controls the size of the entries of the lower triangular matrix.

Table 9 $\kappa_\infty(L)$ in $[0, 1]$

	Natural	Leja
n	$\kappa_\infty(L)$	
3	104	72
4	549.3333	341.3333
5	2.9253×10^3	1.6760×10^3
9	2.4370×10^6	1.1165×10^6
19	5.2459×10^{13}	1.7479×10^{13}

Table 10 $\kappa_\infty(L)$ in $[-1, 1]$

	Natural	Leja	Central
n	$\kappa_\infty(L)$		
3	33.5000	14.5000	22.5000
4	112	38.6667	61.3333
5	373.4548	99.0417	170.2917
9	4.5301×10^4	4.3480×10^3	9.6775×10^3
19	6.9906×10^9	6.7112×10^7	2.2788×10^8

Table 11 $\kappa_\infty(L)$ in $[-1.5, 1.5]$

	Natural	Leja	Central
n	$\kappa_\infty(L)$		
3	32	8	22
4	101.2222	13.9095	56.8889
5	302.7358	25.2631	139.4173
9	2.3969×10^4	260.5714	4.8004×10^3
19	1.4329×10^9	7.5274×10^4	3.3873×10^7

5 Computation of Finite Differences

Let us considerer the Newton formula in terms of the finite differences

$$p(x) = \sum_{j=0}^{n} \Delta(x_0, \ldots, x_j)\tilde{\omega}_j(x).$$

We denote by $\Delta_j = \Delta(x_0, \ldots, x_j)$ and by $\boldsymbol{\Delta} = (\Delta_0, \ldots, \Delta_n)^T$. Hence to find the coefficients of the interpolant respect to the basis $\tilde{\boldsymbol{\omega}} = (\tilde{\omega}_0, \ldots, \tilde{\omega}_n)^T$ given by (19) we have to solve the system

$$\tilde{L}\boldsymbol{\Delta} = \mathbf{f}.$$

We denote by $\tilde{l}_{ij}$ and $\tilde{l}_{ij}^{(-1)}$ the (i,j) entries of $\tilde{L}$ y $\tilde{L}^{-1}$, respectively. By (7) and (18) we have

$$\tilde{l}_{ij} = \frac{l_{ij}}{\omega_j(x_j)} = \frac{\omega_j(x_i)}{\omega_j(x_j)}, \quad i,j = 0, \ldots, n. \tag{20}$$

We use the formula

$$[x_0, \ldots, x_i]f = \sum_{j=0}^{i} \frac{f(x_j)}{\prod_{k \in \{0,\ldots,i\} \setminus \{j\}} (x_j - x_k)} = \sum_{j=0}^{i} \frac{f(x_j)}{\omega'_{i+1}(x_j)}$$

which allows us to establish the following relation between $\mathbf{d}$ and $\mathbf{f}$

$$d_i = \sum_{j=0}^{i} \frac{f_j}{\omega'_{i+1}(x_j)}, \quad i = 0, \ldots, n.$$

By (9), $\mathbf{d} = L^{-1}\mathbf{f}$, and we deduce that the (i,j) entry of the matrix L^{-1} is

$$l_{ij}^{(-1)} = \frac{1}{\omega'_{i+1}(x_j)}$$

when $j \leq i$ and 0 otherwise. So, by (18), the (i,j) entry of $\tilde{L}^{-1}$ is

$$\tilde{l}_{ij}^{(-1)} = \omega_i(x_i) l_{ij}^{(-1)} = \frac{\omega_i(x_i)}{\omega'_{i+1}(x_j)}. \tag{21}$$

Remark 2 Taking into account (20) and (21) we can ensure that $\tilde{L}$ and its inverse are invariant under afine transformations of the nodes because their elements are products of quotients of differences of the nodes.

We now present a result for the condition number of $\tilde{L}$ when using the Leja ordering for not necessarily equidistant nodes.

Proposition 1 *Let $x_0, \ldots, x_n$ nodes following the Leja ordering. Let $\tilde{L}$ be the matrix associated with the representation of the Newton formula with finite differences. Then*

$$\kappa_\infty(\tilde{L}) \leq n2^n.$$

Proof For each vector s whose entries satisfy $s_i \in \{-1, 0, 1\}$, $i = 0, \ldots, n$, let c be the solution of the system $\tilde{L}c = s$.

Let us see by induction on k that $|c_k| \leq 2^k$, $k = 0, \ldots, n$. For $k = 0$ it is trivial because $\tilde{L}^{-1}$ has unit diagonal and therefore

$$c_0 = s_0 \Longrightarrow |c_0| \leq 1.$$

Assume that the results holds for $k - 1$. Since the nodes follow the Leja order, we have

$$\omega_j(x_j) \geq \omega_j(x_i), \quad \forall i \geq j.$$

Then, taking into account (20),

$$|\tilde{l}_{ij}| \leq 1, \quad j \leq i.$$

and so

$$||\tilde{L}||_\infty \leq n.$$

Due to the following equality

$$c_k = s_k + \sum_{j=0}^{k-1} \tilde{l}_{kj} c_j,$$

we have

$$|c_k| \leq |s_k| + \sum_{j=0}^{k-1} |\tilde{l}_{kj}|\, |c_j| \leq 1 + (1 + 2 + \cdots + 2^{k-1}) = 2^k.$$

For each k, we take $s = (\text{sign}(\tilde{l}_{kj}^{(-1)}))_{j=0,\ldots,n}$. Then the k-th entry of the vector $c := \tilde{L}^{-1}s$ is

$$c_k = \sum_{j=0}^{k} \tilde{l}_{kj}^{(-1)} \text{sign}(\tilde{l}_{kj}^{(-1)}) = \sum_{j=0}^{k} |\tilde{l}_{kj}^{(-1)}| \leq 2^k,$$

and therefore

$$||\tilde{L}^{-1}||_\infty = \max_{k=0,\dots,n} \sum_{j=0}^{k} |\tilde{l}_{kj}^{(-1)}| \le 2^n.$$

Finally, we have

$$\kappa_\infty(\tilde{L}) = ||\tilde{L}||_\infty ||\tilde{L}^{-1}||_\infty \le n2^n. \qquad \square$$

We have seen in the numerical experiments in Tables 9, 10 and 11 of Sect. 4 that the Leja ordering gives better conditioning of L than the natural ordering for equidistant nodes. We want to prove that an analogous property for $\tilde{L}$ always holds. We first obtain the conditioning for $\tilde{L}$ for equidistant nodes and natural order.

Proposition 2 *Let $x_i = a + \frac{i}{n}(b-a)$, $i = 0,\dots,n$, be equidistant nodes in the interval $[a,b]$ and let $\tilde{L}$ be the lower triangular matrix in (18). Then, $\tilde{L}$ is the lower triangular Pascal matrix and*

$$\kappa_\infty(\tilde{L}) = 4^n.$$

Proof We evaluate the basis functions ω_j given by (5) at equidistant nodes

$$\begin{aligned}\omega_j(x_i) &= \prod_{k=0}^{j-1}(x_i - x_k) = \prod_{k=0}^{j-1}\frac{b-a}{n}(i-k)\\ &= \left(\frac{b-a}{n}\right)^j i(i-1)\cdots(i+1-j) = \left(\frac{b-a}{n}\right)^j \frac{i!}{(i-j)!}.\end{aligned} \tag{22}$$

Analogously, we compute $\omega'_{i+1}(x_j) = \prod_{k\in\{0,\dots,i\}\setminus\{j\}}(x_j - x_k)$ for equidistant nodes

$$\begin{aligned}\omega'_{i+1}(x_j) &= \prod_{k\in\{0,\dots,i\}\setminus\{j\}}(x_j - x_k) = \left(\frac{b-a}{n}\right)^i \prod_{k\in\{0,\dots,i\}\setminus\{j\}}(j-k)\\ &= (-1)^{i-j}\left(\frac{b-a}{n}\right)^i j!(i-j)!.\end{aligned} \tag{23}$$

By (20) and (22) we obtain

$$\tilde{l}_{ij} = \frac{i!}{(i-j)!j!} = \binom{i}{j} \quad j = 0,\dots,i, \quad i = 0,\dots n.$$

So, $\tilde{L}$ is the lower triangular Pascal matrix. Now we can compute the infinity norm of $\tilde{L}$.

$$||\tilde{L}||_\infty = \max_{i=0,\dots,n} \sum_{j=0}^{i} \binom{i}{j} = \max_{i=0,\dots,n} 2^i = 2^n.$$

By (21)–(23), the (i,j) entry of $\tilde{L}^{-1}$ is

$$\tilde{l}_{ij}^{(-1)} = \frac{\omega_i(x_i)}{\omega'_{i+1}(x_j)} = (-1)^{i+j}\binom{i}{j}.$$

Then

$$||\tilde{L}^{-1}||_\infty = \max_{i=0,\dots,n} \sum_{j=0}^{i} \binom{i}{j} = \max_{i=0,\dots,n} 2^i = 2^n.$$

And finally, we have

$$\kappa_\infty(\tilde{L}) = ||\tilde{L}||_\infty ||\tilde{L}^{-1}||_\infty = 4^n.$$

□

In the following result we compare the natural and the Leja ordering using the previous results.

Corollary 1 *Let $\tilde{L}$ be the lower triangular matrix in (18). The condition number of $\tilde{L}$ with nodes following the Leja ordering is lower than the condition number of $\tilde{L}$ with equidistant nodes following the natural ordering.*

Proof By Propositions 1 and 2 it is sufficient to prove that

$$n2^n \leq 4^n, \quad n = 1, 2, 3, \dots$$

And it is equivalent to prove

$$n \leq 2^n,$$

and this last inequality holds for all n. □

In Table 12 we can see the numerical experiments of these previous results. Due to Remark 2, $\kappa_\infty(\tilde{L})$ does not depend on the interval but only on the ordering. We can also observe that the conditioning of $\tilde{L}$ for Leja ordering is considerably lower than the upper bound $n2^n$. In this sense, we recall the observation by Reichel in [10] on the subexponential growth of a condition number associated to the Newton formula with Leja nodes.

Table 12 $\kappa_\infty(\tilde{L})$ for natural and Leja ordering

	Natural	Leja
n	$\kappa_\infty(\tilde{L})$	
3	64	8.8889
4	256	16
5	1.0240×10^3	14.8800
9	2.6214×10^5	46.1569
19	2.7488×10^{11}	91.9666

Finally we summarize the results of the two last sections. We have seen experimentally in Tables 9, 10 and 11 that the best choice for the L matrix is the Leja ordering. Moreover, we have proved, for equidistant nodes, that the condition number of $\tilde{L}$ is lower with the Leja ordering than with the natural ordering.

Acknowledgements This work has been partially supported by the Spanish Research Grant MTM2015-65433, by Gobierno the Aragón and Fondo Social Europeo.

References

1. de Boor, C., Pinkus, A.: Backward error analysis for totally positive linear systems. Numer. Math. **27**, 485–490 (1976/1977)
2. Bos, L., de Marchi, S., Sommariva, A., Vianello, M.: Computing multivariate Fekete and Leja points by numerical linear algebra. SIAM J. Numer. Anal. **48**(5), 1984–1999 (2010)
3. Carnicer, J., Khiar, Y., Peña, J.M.: Factorization of Vandermonde matrix and the Newton formula. In: Thirteenth International Conference Zaragoza-Pau on Mathematics and its Applications, Ahusborde, É., Amrouche, C., et al. (eds.) Monografías Matemáticas "García Galdeano", vol. 40, pp. 53–60. Universidad de Zaragoza, Zaragoza (2015)
4. Demmel, J., Koev, P.: The accurate and efficient solution of a totally positive generalized Vandermonde linear system. SIAM J. Matrix Anal. Appl. **27**, 142–152 (2005)
5. Demmel, J., Gu, M., Eisenstat, S., Slapniňar, I., Veselić, K., Drmač, Z.: Computing the singular value decomposition with high relative accuracy. Lineal Algebra Appl. **299**, 21–80 (1999)
6. Higham, N.J.: Stability analysis of algorithms for solving confluent Vandermonde-like systems. SIAM J. Matrix Anal. Appl. **1**, 23–41 (1990)
7. Higham, N.J.: Accuracy and Stability of Numerical Algorithms. Society for Industrial and Applied Mathematics, Philadelphia (1996)
8. Koev, P.: Accurate computations with totally nonnegative matrices. SIAM J. Matrix Anal. Appl. **29**, 731–751 (2007)
9. Peña, J.M.: Pivoting strategies leading to small bounds of the errors for certain linear systems. IMA J. Numer. Anal. **16**, 141–153 (1996)
10. Reichel, L.: Newton interpolation at Leja points. BIT **30**(2), 332–346 (1990)

Long-Time Behavior of a Cahn-Hilliard-Navier-Stokes Vesicle-Fluid Interaction Model

Blanca Climent-Ezquerra and Francisco Guillén-González

Abstract A model about the dynamic of vesicle membranes in incompressible viscous fluids is introduced. The system consists of the Navier-Stokes equations with an extra stress depending on the membrane, coupled with a Cahn-Hilliard phase-field equation in 3D domains. This problem has a time dissipative energy which leads, in particular, to the existence of global in time weak solutions. By using some extra regular estimates, we prove that every weak solution is strong and unique for sufficiently large times. Moreover, the asymptotic behavior of these solutions is analyzed. We prove that the w-limit set is a subset of the set of equilibrium points. By using a Lojasiewic-Simon type inequality and a continuity result with respect to the initial values, we demonstrate the convergence of the whole trajectory to a single equilibrium.

1 Introduction

In this paper, we consider a model for the dynamic of vesicle membranes within incompressible viscous fluids. This type of models was introduced by Helfrich [7]. The model that we will consider in this paper consists of the Navier-Stokes equations with an extra stress depending on the membrane, coupled with a Cahn-Hilliard phase-field equation transported by the fluid.

Membranes are formed by lipid bilayers. Under appropriate conditions, they withdraw into itself forming a sort of bag, named vesicle. The equilibrium configurations of vesicle membranes can be obtained minimizing the Helfrich bending elastic energy [7], fixing its surface area and volume.

A phase function can be used to model the vesicle membrane as a diffuse interface. In the literature, a coupled Allen-Cahn and Navier-Stokes problem is studied approaching both constrains, area and volume, in a penalized manner. The existence of global in time weak solutions of this model is proven in [5]. In [8] authors prove the existence and uniqueness of local in time solution. Under periodic

B. Climent-Ezquerra (✉) • F. Guillén-González
Dpto. Ecuaciones Diferenciales y Análisis Numérico, Universidad de Sevilla, Aptdo. 1160, 41080 Sevilla, Spain
e-mail: bcliment@us.es, guillen@us.es

F. Ortegón Gallego et al. (eds.), *Trends in Differential Equations and Applications*, SEMA SIMAI Springer Series 8, DOI 10.1007/978-3-319-32013-7_8

boundary conditions, the stability near of local minimizers of the elastic bending energy is investigated in [9].

On the other hand, a Cahn-Hilliard phase-field model is introduced in [1], without taking into account the vesicle-fluid interaction. In [6], the long-time behavior for a 2D Cahn-Hilliard-Navier-Stokes model, without membranes, is studied.

Now a Cahn-Hilliard-Navier-Stokes model with Neumann boundary conditions will be considered. In particular, by using the volume conservation of the Cahn-Hilliard equation, the volume constraint will be implicitly imposed and therefore, only the surface area constraint must be approximated via penalization.

We obtain existence of global weak solutions for arbitrary initial data and prove that any global weak solution becomes a bounded strong solution after a sufficiently large time. Then, its long-time behavior is studied identifying a unique critical point as the limit of the whole trajectory as the time goes to infinity, by means of an appropriate Lojasiewicz-Simon's Lemma and a local in time continuous dependence result with respect to regular initial conditions.

2 The Model

We will analyze the case where the bending energy E_b is given by a simplified elastic Willmore energy modified to penalize the surface area constraint [4]:

$$E_b(\phi) = \frac{1}{2\varepsilon}\int_\Omega (-\varepsilon\Delta\phi + \frac{1}{\varepsilon}F'(\phi))^2\,dx + \frac{1}{2}M(A(\phi)-\alpha)^2 \tag{1}$$

where $F'(\phi) = (|\phi|^2-1)\phi$ for each $\phi \in \mathbf{R}$, being $F(\phi) = \frac{1}{4}(\phi^2-1)^2$ the Ginzburg-Landau potential, $M > 0$ is the penalization constant, ε is related to the interface width, and

$$A(\phi) = \int_\Omega \left(\frac{\varepsilon}{2}|\nabla\phi|^2 + \frac{1}{\varepsilon}F(\phi)\right)\,dx,$$

approaches the surface area.

Remark 1 The results of this paper can be extended to the case of replacing the surface area $A(\phi)$ only by $A(\phi) = \int_\Omega \frac{\varepsilon}{2}|\nabla\phi|^2\,dx$ considered in [1].

We are going to consider a Cahn-Hilliard phase-field model which is conservative for the phase function. Therefore, the volume constraint

$$V(\phi) = \int_\Omega \phi(x,t)\,dx = V_0\left(= \int_\Omega \phi_0(x)\,dx\right)$$

is satisfied implicitly. By denoting

$$w := \frac{\delta E_b(\phi)}{\delta \phi} \tag{2}$$

(depending on $\mu := \dfrac{\delta A(\phi)}{\delta \phi} = -\varepsilon \Delta \phi + \dfrac{1}{\varepsilon} F'(\phi)$), we will analyze the following Navier-Stokes-Cahn-Hilliard equations in $\Omega \times (0, +\infty)$:

$$\partial_t \boldsymbol{u} + (\boldsymbol{u} \cdot \nabla)\boldsymbol{u} - \nu \Delta \boldsymbol{u} - \lambda w \nabla \phi + \nabla q = 0, \tag{3}$$

$$\nabla \cdot \boldsymbol{u} = 0, \tag{4}$$

$$\partial_t \phi + \boldsymbol{u} \cdot \nabla \phi - \gamma \Delta w = 0. \tag{5}$$

Remark 2 In both cases, the variational derivative $w := \dfrac{\delta E_b(\phi)}{\delta \phi}$ and $\mu := \dfrac{\delta A(\phi)}{\delta \phi}$ are identified as a $L^2(\Omega)$-function via the $L^2(\Omega)$ scalar product.

The constants $\nu > 0$, $\lambda > 0$ and $\gamma > 0$ are coefficients depending on viscosity, elasticity and mobility, respectively. The system (2)–(5) is completed with the boundary conditions

$$\boldsymbol{u}|_{\partial\Omega} = 0, \quad \partial_n \phi|_{\partial\Omega} = 0, \quad \partial_n \Delta \phi|_{\partial\Omega} = 0, \quad \partial_n w|_{\partial\Omega} = 0, \tag{6}$$

and the initial conditions

$$\boldsymbol{u}(0) = \boldsymbol{u}_0, \qquad \phi(0) = \phi_0 \quad \text{in } \Omega. \tag{7}$$

For compatibility, we assume $\boldsymbol{u}_0|_{\partial\Omega} = 0$ with $\nabla \cdot \boldsymbol{u}_0 = 0$ and $\partial_n \phi_0|_{\partial\Omega} = 0$.

By using in the w-equation (5) the free divergence constraint $\nabla \cdot \boldsymbol{u} = 0$, the non-slip condition $\boldsymbol{u}|_{\partial\Omega} = 0$, and the last boundary condition $\partial_n w|_{\partial\Omega} = 0$, it is easy to deduce that $\dfrac{d}{dt} \displaystyle\int_\Omega \phi(x,t)\, dx = 0$, that is, the conservative character of ϕ in Ω. Therefore, the total volume is conserved:

$$\int_\Omega \phi(x,t)\, dx = \int_\Omega \phi_0(x)\, dx := V_0 \in \mathbf{R}.$$

On the other hand, for all $\bar{\phi} \in H^2$,

$$\begin{aligned} \left\langle \frac{\delta E_b(\phi)}{\delta \phi}, \bar{\phi} \right\rangle &= \frac{1}{\varepsilon} \int_\Omega \mu(-\varepsilon \Delta \bar{\phi} + \frac{1}{\varepsilon} F'(\phi) \bar{\phi}) \\ &\quad + M(A(\phi) - \alpha)(\int_\Omega \varepsilon \nabla \phi \cdot \nabla \bar{\phi} + \frac{1}{\varepsilon} F'(\phi) \bar{\phi}) \end{aligned}$$

Observe that from the boundary conditions given in (6) for ϕ, we also have $\nabla\mu\cdot\boldsymbol{n}|_{\partial\Omega}=0$. This enables us, after some integrations by parts, using $\nabla\mu\cdot\boldsymbol{n}|_{\partial\Omega}=0$ and $\partial_n\bar{\phi}|_{\partial\Omega}=0$, to identify $\dfrac{\delta E_b(\phi)}{\delta\phi}$ with w via the $L^2(\Omega)$-scalar product, where

$$w=-\Delta\mu+\frac{1}{\varepsilon^2}\mu F'(\phi)+M(A(\phi)-\alpha)\mu=\varepsilon\Delta^2\phi+G(\phi)$$

with

$$G(\phi)=-\frac{1}{\varepsilon}\Delta F'(\phi)+\frac{1}{\varepsilon^2}F'(\phi)\mu+M(A(\phi)-\alpha)\mu. \tag{8}$$

Since $\partial_n\phi|_{\partial\Omega}=0$, in particular $\partial_nF'(\phi)|_{\partial\Omega}=0$, hence

$$\int_\Omega -\Delta F'(\phi)\,dx=\int_{\partial\Omega}-F''(\phi)(\nabla\phi\cdot n)\,dx=0.$$

Integrating (8),

$$\int_\Omega w\,dx=\int_\Omega G(\phi)\,dx=\frac{1}{\varepsilon^2}\int_\Omega \mu F'(\phi)\,dx+M(A(\phi)-\alpha)\int_\Omega \mu\,dx. \tag{9}$$

By using $w=\varepsilon\Delta^2\phi+G(\phi)$ as auxiliary variable, we rewrite the problem (2)–(5) as

$$\partial_t\boldsymbol{u}+(\boldsymbol{u}\cdot\nabla)\boldsymbol{u}-\nu\Delta\boldsymbol{u}-\lambda w\nabla\phi+\nabla q=0, \tag{10}$$

$$\nabla\cdot\boldsymbol{u}=0, \tag{11}$$

$$\partial_t\phi+\boldsymbol{u}\cdot\nabla\phi-\gamma\Delta w=0, \tag{12}$$

$$\varepsilon\Delta^2\phi+G(\phi)-w=0. \tag{13}$$

By denoting $m_0=\langle\phi_0\rangle=\dfrac{1}{|\Omega|}\displaystyle\int_\Omega\phi_0(x)\,dx$, we define the following mean-value variables

$$\psi(x,t)=\phi(x,t)-m_0 \quad\text{and}\quad z=w-\langle G(\phi)\rangle.$$

Observe that $\langle G(\phi)\rangle=\dfrac{1}{\varepsilon^2}\langle F'(\phi)\mu\rangle+M(A(\phi)-\alpha)\langle\mu\rangle$.

Let us consider the following mean-value spaces:

$$L^2_* = \left\{ w \in L^2(\Omega);\ \int_\Omega w = 0 \right\},$$
$$H^k_* = \left\{ w \in H^k(\Omega);\ \int_\Omega w = 0 \right\} \qquad k \geq 1,$$
$$H^2_1 = \left\{ w \in H^2_*(\Omega);\ \partial_n w = 0 \text{ on } \partial\Omega \right\}$$
$$H^k_2 = \left\{ w \in H^k_*;\ \partial_n w|_{\partial\Omega} = 0,\ \partial_n \Delta w|_{\partial\Omega} = 0 \right\} \qquad k \geq 4.$$

We will consider Ω regular enough to use the regularity of the two following elliptic Laplacian-Neuman, Bilaplacian-Dirichlet-Neumann problems, respectively:

$$\begin{cases} -\Delta z = f & \text{in } \Omega \\ \partial_n z|_{\partial\Omega} = 0, & \displaystyle\int_\Omega z\, dx = 0, \end{cases} \qquad \begin{cases} \Delta^2 \psi = f & \text{in } \Omega \\ \partial_n \psi|_{\partial\Omega} = 0, & \partial_n \Delta\psi|_{\partial\Omega} = 0, \\ \displaystyle\int_\Omega \psi\, dx = 0 \end{cases}$$

where $f : \Omega \mapsto \mathbf{R}, f \in L^2(\Omega), \int_\Omega f\, dx = 0$. From the H^2-regularity of the first problem, we have the following equivalents norms:

$$\|z\|_2 \approx |\Delta z|_2 \quad \text{in } H^2_1, \tag{14}$$

and from the H^4, H^5 and H^6-regularity of the second problem,

$$\|\psi\|_4 \approx |\Delta^2\psi|_2 \text{ in } H^4_2, \quad \|\psi\|_5 \approx \|\Delta^2\psi\|_1 \text{ in } H^5_2, \quad \|\psi\|_6 \approx \|\Delta^2\psi\|_2 \text{ in } H^6_2. \tag{15}$$

By rewriting the Eqs. (10)–(13) in the variables

$$\psi(x,t) = \phi(x,t) - m_0 \quad \text{and} \quad z = w - \langle G(\phi)\rangle$$

we obtain

$$\partial_t \boldsymbol{u} + (\boldsymbol{u}\cdot\nabla)\boldsymbol{u} - \nu\Delta\boldsymbol{u} - \lambda z\nabla\psi + \nabla\tilde{q} = 0, \tag{16}$$
$$\nabla\cdot\boldsymbol{u} = 0, \tag{17}$$
$$\partial_t\psi + \boldsymbol{u}\cdot\nabla\psi - \gamma\Delta z = 0, \tag{18}$$
$$\varepsilon\Delta^2\psi + \overline{G}(\psi) - z = 0, \tag{19}$$

where $\tilde{q} = q - \lambda \langle G(\phi) \rangle \psi$ and

$$\overline{G}(\psi) = G(\psi + m_0) - \langle G(\psi + m_0) \rangle$$
$$= -\frac{1}{\varepsilon} \Delta F'(\psi + m_0) + \frac{1}{\varepsilon^2} F'(\psi + m_0)(-\varepsilon \Delta \psi + \frac{1}{\varepsilon} F'(\psi + m_0))$$
$$+ M(A(\psi + m_0) - \alpha)(-\varepsilon \Delta \psi + \frac{1}{\varepsilon} F'(\psi + m_0)) - \langle G(\psi + m_0) \rangle.$$

Observe that, $\int_\Omega \psi \, dx = 0$ and $\int_\Omega z \, dx = 0$. The system (16)–(19) is completed with the boundary conditions

$$\boldsymbol{u}|_{\partial\Omega} = 0, \quad \partial_n \psi|_{\partial\Omega} = 0, \quad \partial_n \Delta\psi|_{\partial\Omega} = 0, \quad \partial_n z|_{\partial\Omega} = 0, \tag{20}$$

and the initial conditions

$$\boldsymbol{u}(0) = \boldsymbol{u}_0, \qquad \psi(0) = \psi_0 := \phi_0 - \langle \phi_0 \rangle \quad \text{in } \Omega. \tag{21}$$

Finally, by denoting

$$\overline{E}_b(\psi) = E_b(\psi + m_0) \tag{22}$$

then,

$$\frac{\delta \overline{E}_b(\psi)}{\delta \psi} = \varepsilon \Delta^2 \psi + \overline{G}(\psi) = z. \tag{23}$$

3 Some Preliminary Results

We are going to consider the following notations:

- In general, the notation will be abridged. We set $L^p = L^p(\Omega)$, $p \geq 1$, $H_0^1 = H_0^1(\Omega)$, etc. If $X = X(\Omega)$ is a space of functions defined in the open set Ω, we denote by $L^p(0, T; X)$ the Banach space $L^p(0, T; X(\Omega))$. Also, boldface letters will be used for vectorial spaces, for instance $\mathbf{L}^2 = L^2(\Omega)^N$.
- The L^p-norm is denoted by $|\cdot|_p$, $1 \leq p \leq \infty$, the H^m-norm by $\|\cdot\|_m$ (in particular $|\cdot|_2 = \|\cdot\|_0$). The inner product of $L^2(\Omega)$ is denoted by $(\cdot,\cdot)$. The boundary $H^s(\partial\Omega)$-norm is denoted by $\|\cdot\|_{s;\partial\Omega}$.
- We set $\mathcal{V}$ the space formed by all fields $\boldsymbol{u} \in C_0^\infty(\Omega)^N$ satisfying $\nabla \cdot \boldsymbol{u} = 0$. We denote $\boldsymbol{H}$ (respectively $\boldsymbol{V}$) the closure of $\mathcal{V}$ in $\boldsymbol{L}^2$ (respectively $\boldsymbol{H}^1$). $\boldsymbol{H}$ and $\boldsymbol{V}$ are Hilbert spaces for the norms $|\cdot|_2$ and $\|\cdot\|_1$, respectively. Furthermore,

$$\boldsymbol{H} = \{\boldsymbol{u} \in \boldsymbol{L}^2; \nabla\cdot\boldsymbol{u} = 0, \boldsymbol{u}\cdot\mathbf{n} = 0 \text{ on } \partial\Omega\}, \qquad \boldsymbol{V} = \{\boldsymbol{u} \in \boldsymbol{H}^1; \nabla\cdot\boldsymbol{u} = 0, \boldsymbol{u} = 0 \text{ on } \partial\Omega\}.$$

- From now on, $C > 0$ will denote different constants, depending only on the fixed data of the problem.

Lemma 1 *Given $\mu \in L^2(\Omega)$, we consider ψ as the solution of the elliptic problem*

$$\begin{cases} -\varepsilon \Delta\psi + \dfrac{1}{\varepsilon} F'(\psi + m_0) = \mu & \textit{in } \Omega, \\ \partial_n \psi|_{\partial\Omega} = 0, \quad \displaystyle\int_\Omega \psi \, dx = 0. \end{cases}$$

Then, the following inequalities hold

$$\|\psi\|_1 \le C(1 + |\mu|_2), \tag{24}$$

$$\|\psi\|_2 \le C(1 + |\mu|_2 + \|\psi\|_1^3). \tag{25}$$

Proof Firstly, by taking $\psi + m_0$ as test function, we obtain

$$\varepsilon|\nabla\psi|_2^2 + \frac{1}{\varepsilon}|\psi + m_0|_4^4 = (\mu, \psi + m_0) + \frac{1}{\varepsilon}|\psi + m_0|_2^2.$$

By using Young and Holder inequalities, and the Poincaré inequality for mean-value functions, we obtain:

$$C\varepsilon\|\psi\|_1^2 + \frac{1}{\varepsilon}|\psi + m_0|_4^4 \le \frac{1}{2}|\mu|_2^2 + (\frac{1}{2} + \frac{1}{\varepsilon})|\psi + m_0|_2^2 \le \frac{1}{2}|\mu|_2^2 + \frac{1}{2\varepsilon}|\psi + m_0|_4^4 + C,$$

hence,

$$\|\psi\|_1^2 \le C|\mu|_2^2 + C$$

and (24) holds. Secondly, from the regularity of the problem (bootstrap's argument)

$$\begin{cases} -\varepsilon \Delta\psi = \mu - \dfrac{1}{\varepsilon} F'(\psi + m_0) & \text{in } \Omega \\ \partial_n \psi|_{\partial\Omega} = 0, \quad \displaystyle\int_\Omega \psi \, dx = 0, \end{cases}$$

we obtain that $\|\psi\|_2 \le C(|\mu|_2 + \frac{1}{\varepsilon}|F'(\psi + m_0)|_2)$. From the definition of F',

$$|F'(\psi + m_0)|_2 \le C(|\psi + m_0|_6^3 + |\psi + m_0|_2) \le C(\|\psi + m_0\|_1^3 + |\psi + m_0|_2) \le C(1 + \|\psi\|_1^3).$$

Therefore, $\|\psi\|_2 \le C(1 + |\mu|_2 + \|\psi\|_1^3)$ and (25) holds. □

Assume the following starting point:

Let $E, \Phi \in L^1_{loc}(0, +\infty)$ be two positive functions with $E \in H^1(0, T)$ $\forall T > 0$, satisfying

$$E'(t) + \Phi(t) \leq 0, \quad \text{a.e. } t \in (0, +\infty). \tag{26}$$

Therefore, E is a decreasing function with $E \in L^\infty(0, +\infty)$ and

$$\exists \lim_{t \to +\infty} E(t) = E_\infty \geq 0. \tag{27}$$

Moreover, by integrating (26), one has $\Phi \in L^1(0, +\infty)$.

The following results are proved in [3] and [2].

Lemma 2 *Let $\Phi, B \in L^1(0, +\infty)$ be two positive functions such that $\Phi \in H^1(0, T)$ $\forall T > 0$, which satisfies*

$$\Phi'(t) \leq C(\Phi(t)^3 + B(t)). \tag{28}$$

Then, there exists a sufficiently large $T^* \geq 0$ such that $\Phi \in L^\infty(T^*, +\infty)$ and

$$\exists \lim_{t \to +\infty} \Phi(t) = 0.$$

In order to apply the previous result to a sequence of approximate solutions (furnished for instance by the Galerkin method), an extension of Lemma 2 for sequences of functions will be necessary in order to get uniform bounds with respect to the index of sequence. Specifically,

Lemma 3 *Let $\Phi^m \in L^1(0, +\infty)$, $E^m \in L^\infty(0, +\infty)$, be two positive sequences of functions satisfying (26) and (28) for some constant $C > 0$ independent of m. Let $E(t) = \lim_{m \to +\infty} E^m(t)$ a.e. $t \in (0, +\infty)$ (assuming that the limit exists). Therefore, for each $\delta \in (0, 1)$, there exists a sufficiently large time $T^* = T^*(\delta) \geq 0$, independent of m, such that $\Phi^m \in L^\infty(T^*, +\infty)$ and $\|\Phi^m\|_{L^\infty(T^*, +\infty)} \leq \delta$.*

The proof of the following Lojasiewicz-Simon inequality is like the one that appears in Lemma 5.2 of [9] changing periodic by Neumann boundary conditions and periodic by zero mean spaces.

Lemma 4 (Lojasiewicz-Simon Inequality) *Let $\mathcal{S}$ be the following set of equilibrium points related to the bending energy (1)*

$$\mathcal{S} = \{\psi \in H^4_2(\Omega) : \varepsilon \Delta^2 \psi + \overline{G}(\psi) = 0 \text{ a.e in } \Omega\}. \tag{29}$$

If $\psi_\infty \in \mathcal{S}$, there are three positive constants C, α, and $\theta \in (0,1/2)$ (depending on ψ_∞), such that for all $\psi \in H_2^4$ and $\|\psi - \psi_\infty\|_2 \le \beta$, then

$$|\overline{E}_b(\psi) - \overline{E}_b(\psi_\infty)|^{1-\theta} \le C\,|z|_2 \tag{30}$$

where $z = z(\psi) := \varepsilon\Delta^2\psi + \overline{G}(\psi)$.

4 Weak Solutions

Definition 1 Let $u_0 \in \boldsymbol{H}$ and $\psi_0 = \phi_0 - m_0 \in H_1^2$, we say that $(\boldsymbol{u}, \psi, z)$ is a global weak solution of (16)–(21) in $(0, +\infty)$, if

$$\begin{gathered} \boldsymbol{u} \in L^2(0,+\infty; \boldsymbol{V}) \cap L^\infty(0,+\infty; \boldsymbol{H}), \quad z \in L^2(0,+\infty; H_*^1) \\ \psi \in L^\infty(0,+\infty; H_1^2), \end{gathered} \tag{31}$$

satisfying

$$\langle \partial_t \boldsymbol{u}, \overline{\boldsymbol{u}}\rangle + ((\boldsymbol{u}\cdot\nabla)\boldsymbol{u}, \overline{\boldsymbol{u}}) + \nu(\nabla \boldsymbol{u}, \nabla\overline{\boldsymbol{u}}) - \lambda(z\nabla\psi, \overline{\boldsymbol{u}}) = 0 \quad \forall\, \overline{\boldsymbol{u}} \in \boldsymbol{V}, \tag{32}$$

$$\langle \partial_t\psi, \overline{z}\rangle + (\boldsymbol{u}\cdot\nabla\psi, \overline{z}) + \gamma(\nabla z, \nabla\overline{z}) = 0, \quad \forall\, \overline{z} \in H_*^1 \tag{33}$$

$$\varepsilon(\Delta\psi, \Delta\overline{\psi}) + (\overline{G}(\psi), \overline{\psi}) - (z, \overline{\psi}) = 0, \quad \forall\, \overline{\psi} \in H_1^2, \tag{34}$$

and the initial conditions (21).

Observe that the initial conditions, (21), have sense from (31)–(33). Moreover, $\partial_t \boldsymbol{u} \in L^{4/5}(0,+\infty; \boldsymbol{V}')$ and $\partial_t\psi \in L^2(0,+\infty; (H_*^1)')$.

4.1 Energy Equality and Large Time Estimates

In a formal manner, we assume that $(\boldsymbol{u}, \psi, z)$ is a regular enough solution of (16)–(21). By taking $\bar{\boldsymbol{u}} = \boldsymbol{u}$, $\bar{z} = z$ and $\bar{\psi} = \partial_t\psi$ as test function in (32), (33) and (34) respectively, we have

$$\frac{1}{2}\frac{d}{dt}|\boldsymbol{u}|_2^2 + \nu|\nabla\boldsymbol{u}|_2^2 - \lambda(z\,\nabla\psi, \boldsymbol{u}) = 0,$$

$$(\partial_t\psi, z) + (\boldsymbol{u}\cdot\nabla\psi, z) + \gamma|\nabla z|_2^2 = 0,$$

$$\varepsilon\frac{d}{dt}\frac{1}{2}|\Delta\psi|_2^2 + (\overline{G}(\psi), \partial_t\psi) - (z, \partial_t\psi) = 0.$$

Adding the first equality plus the second and third ones multiplied by λ, the term $(z, \partial_t \psi)$ cancels, as well as the nonlinear convective term $(\boldsymbol{u} \cdot \nabla \psi, z)$ with the elastic term $-(z \nabla \psi, \boldsymbol{u})$, arriving at the following equality

$$\frac{1}{2}\frac{d}{dt}(|\boldsymbol{u}|_2^2 + \lambda\varepsilon|\Delta\psi|_2^2) + \lambda(\overline{G}(\psi), \partial_t\psi) + \nu|\nabla\boldsymbol{u}(t)|_2^2 + \lambda\gamma|\nabla z(t)|_2^2 = 0. \tag{35}$$

Since $\dfrac{\delta \overline{E}_b(\psi)}{\delta\psi} = z,$

$$\frac{d}{dt}\overline{E}_b(\psi(t)) = \left\langle \frac{\delta \overline{E}_b(\psi)}{\delta\psi}, \partial_t\psi \right\rangle = (z, \partial_t\psi)$$
$$= \varepsilon(\Delta^2\psi, \partial_t\psi) + (\overline{G}(\psi), \partial_t\psi) = \varepsilon\frac{1}{2}\frac{d}{dt}|\Delta\psi|_2^2 + (\overline{G}(\psi), \partial_t\psi).$$

We define the total free energy $\overline{E}(\boldsymbol{u}, \psi) = E_k(\boldsymbol{u}) + \lambda\overline{E}_b(\psi)$, being $\overline{E}_b(\psi)$ the bending energy defined in (22) and $E_k(\boldsymbol{u}) = \frac{1}{2}\int_\Omega |\boldsymbol{u}|^2$ the kinetic energy. Then, equality (35) can be rewritten as the following *energy equality*:

$$\frac{d}{dt}\overline{E}(\boldsymbol{u}(t), \psi(t)) + \nu|\nabla\boldsymbol{u}(t)|_2^2 + \lambda\gamma|\nabla z(t)|_2^2 = 0, \tag{36}$$

which shows the dissipative character of the model with respect to the total free energy $\overline{E}(\boldsymbol{u}(t), \psi(t))$. Moreover, assuming the initial estimates $(\boldsymbol{u}_0, \psi_0)$ in $\boldsymbol{H} \times H^1_*$, the following uniform "weak" bounds in the infinite time interval $(0, +\infty)$ hold:

$$\boldsymbol{u} \text{ in } L^\infty(0, +\infty; \boldsymbol{H}) \cap L^2(0, +\infty; \boldsymbol{V}), \quad \mu \text{ in } L^\infty(0, +\infty; L^2), \quad z \text{ in } L^2(0, +\infty; H^1). \tag{37}$$

From the bound of μ and Lemma 1, we have also:

$$\psi \text{ in } L^\infty(0, +\infty; H^2_1). \tag{38}$$

4.2 Additional Estimates for ψ in H^5_2

By using previous weak estimate (38), we can deduce

$$|\boldsymbol{u} \cdot \nabla\psi|_2^2 \le |\boldsymbol{u}|_6^2 |\nabla\psi|_3^2 \le C\|\boldsymbol{u}\|_1^2. \tag{39}$$

Since $\psi \in L^\infty(0,+\infty;H^2)$, in particular $\psi \in L^\infty(0,+\infty;L^\infty)$, then $F'(\psi)$, $F''(\psi)$ and $F'''(\psi)$ are also bounded in $L^\infty(0,+\infty;L^\infty)$. Therefore, we have

$$|\overline{G}(\psi)|_2 \le C, \tag{40}$$

$$|\nabla\overline{G}(\psi)|_2 \le C(1+\|\psi\|_3), \tag{41}$$

and

$$|\Delta\overline{G}(\psi)|_2 \le C(1+\|\psi\|_4). \tag{42}$$

From the ψ-equation (19), by using (15), (40)–(42), we obtain

$$\|\psi\|_4 \le C(1+|\overline{G}(\psi)|_2+|z|_2) \le C(1+|z|_2).$$

In particular, from (42) and Poincaré inequality for the mean-value function z,

$$|\Delta\overline{G}(\psi)|_2 \le C(1+|z|_2) \le C(1+|\nabla z|_2). \tag{43}$$

On the other hand, again from (19), by using (41) and the interpolation inequality $\|\psi\|_3 \le \|\psi\|_2^{1/2}\|\psi\|_4^{1/2}$, we deduce $\|\psi\|_5 \le C(\|z\|_1+\|\psi\|_2^{1/2}\|\psi\|_4^{1/2})$ and, therefore, since $\|\psi\|_2 \le C$ and $\|\psi\|_4 \le C(1+|z|_2)$, then $\|\psi\|_5 \le C(\|z\|_1+1)$. In particular,

$$\psi \in L^2_{loc}(0,+\infty;H^5). \tag{44}$$

For instance, weak solutions furnished by a limit of Galerkin approximate solutions satisfy the corresponding energy inequality (changing in (36) the equality $=0$ by ≤ 0) and this inequality energy suffices to prove rigorously all previous estimates (37) and (44) for instance for the Galerking approximations.

5 Strong Solution

Definition 2 Let $u_0 \in \boldsymbol{V}$ and $\psi_0 = \phi_0 - m_0 \in H^3_2$, we say that $(\boldsymbol{u},\psi,z)$ is a global strong solution of (16)–(19), (21) in $(0,+\infty)$, if

$$\begin{gathered} \boldsymbol{u} \in L^\infty([T^*_{reg},+\infty);\boldsymbol{H}^1) \cap L^2([T^*_{reg},+\infty);\boldsymbol{H}^2), \qquad z \in L^2_{loc}(0,+\infty;H^2_1) \\ \psi \in L^\infty(0,+\infty;H^3_2), \qquad \psi \in L^\infty_{loc}(0,+\infty;H^6_2), \end{gathered} \tag{45}$$

satisfies the system (16)–(19), almost everywhere in $(0, +\infty) \times \Omega$ and the initial conditions (21)

Moreover, $\partial_t \boldsymbol{u} \in L^2([T^*_{reg}, +\infty); \boldsymbol{L}^2)$ and $\partial_t \psi \in L^2_{loc}(0, +\infty; L^2)$.

5.1 Strong Estimates for Velocity for Large Times

By taking into account that $(z\nabla\psi, \bar{\boldsymbol{u}}) = -(\psi\ \nabla z, \bar{\boldsymbol{u}})$, $\bar{\boldsymbol{u}} \in \boldsymbol{V}$, (32) can be rewrite as:

$$\langle \partial_t \boldsymbol{u}, \bar{\boldsymbol{u}}\rangle + ((\boldsymbol{u}\cdot\nabla)\boldsymbol{u}, \bar{\boldsymbol{u}}) + \nu(\nabla\boldsymbol{u}, \nabla\bar{\boldsymbol{u}}) - \lambda(\nabla z\ \psi, \bar{\boldsymbol{u}}) = 0 \quad \forall\, \bar{\boldsymbol{u}} \in \boldsymbol{V},$$

by means of taking $-A\boldsymbol{u} + \partial_t \boldsymbol{u}$ as a test function (A being the Stokes operator), by applying interpolation, Hölder and Young's inequalities, we attain:

$$\begin{aligned}
&\frac{d}{dt}|\nabla\boldsymbol{u}|_2^2 + \nu|A\boldsymbol{u}|_2^2 + |\partial_t\boldsymbol{u}|_2^2 \le C\left(|(\boldsymbol{u}\cdot\nabla)\boldsymbol{u}|_2^2 + |(\nabla z)\psi|_2^2\right) \\
&\le C\left(|\boldsymbol{u}|_6^2|\nabla\boldsymbol{u}|_3^2 + |\nabla z|_2^2|\psi|_\infty^2\right) \le C\left(\|\boldsymbol{u}\|_1^3\|\boldsymbol{u}\|_2 + |\nabla z|_2^2\|\psi\|_1\|\psi\|_2\right) \\
&\le \frac{\nu}{2}\|\boldsymbol{u}\|_2^2 + C\left(\|\boldsymbol{u}\|_1^6 + |\nabla z|_2^2\right).
\end{aligned}$$

Therefore, we obtain

$$\frac{d}{dt}\|\boldsymbol{u}\|_1^2 + \frac{\nu}{2}\|\boldsymbol{u}\|_2^2 + |\partial_t\boldsymbol{u}|_2^2 \le C\left(\|\boldsymbol{u}\|_1^6 + |\nabla z|_2^2\right). \tag{46}$$

By denoting

$$\Phi_1(t) := \|\boldsymbol{u}\|_1^2, \qquad \Psi_1(t) := \frac{\nu}{2}\|\boldsymbol{u}\|_2^2 + |\partial_t\boldsymbol{u}|_2^2, \qquad B_1(t) := |\nabla z|_2^2,$$

Eq. (46) can be rewritten as

$$\Phi_1' + \Psi_1 \le C(\Phi_1^3 + B_1). \tag{47}$$

Notice that, owing to (37), $B_1(t) \in L^1(0, +\infty)$.

From (47), we can deduce two different results:

- There exists $\widehat{T} = \widehat{T}(\|\boldsymbol{u}(0)\|_1^2)$ such that

$$\boldsymbol{u} \in L^\infty([0, \widehat{T}]; \boldsymbol{H}^1) \cap L^2([0, \widehat{T}]; \boldsymbol{H}^2), \quad \partial_t\boldsymbol{u} \in L^2([0, \widehat{T}]; \boldsymbol{L}^2).$$

  This fact has been proved in [9] or [8] for a Navier-Stokes Allen-Cahn model with different boundary conditions that we are considering in this paper.

- Since the hypothesis of Lemma 2 holds, there exists $T^*_{reg} \geq 0$ (sufficiently large) such that

$$u \in L^\infty([T^*_{reg}, +\infty); H^1), \tag{48}$$

and $\|u(t)\|_1 \to 0$ as $t \uparrow +\infty$. Moreover, integrating (47) in $[0, t]$ for all $t > 0$, we obtain the following regularity, also in $[T^*_{reg}, +\infty)$:

$$u \in L^2([T^*_{reg}, +\infty); H^2), \quad \partial_t u \in L^2([T^*_{reg}, +\infty); L^2).$$

5.2 Global in Time Strong Estimates for ψ

By taking $\overline{z} = \partial_t \psi \in H^1_*$ in the z-equation (33), we obtain:

$$|\partial_t \psi|^2 + (u \cdot \nabla \psi, \partial_t \psi) + \gamma(\nabla z, \nabla \partial_t \psi) = 0. \tag{49}$$

By taking $\overline{\psi} = \Delta \partial_t \psi \in H^2_1$ (see (20)) in the ψ-equation (34) multiplied by γ and integrating respectively, once, twice and once by parts the first, second and third term, taking into account that $\nabla \partial_t \psi \cdot n|_{\partial\Omega} = 0$, $\nabla \overline{G}(\psi) \cdot n|_{\partial\Omega} = 0$ and $\nabla z \cdot n|_{\partial\Omega} = 0$, then, we obtain:

$$\varepsilon \frac{\gamma}{2} \frac{d}{dt} |\nabla \Delta \psi|_2^2 + \gamma(\Delta \overline{G}(\psi), \partial_t \psi) - \gamma(\nabla z, \nabla \partial_t \psi) = 0. \tag{50}$$

Adding (49) and (50), the term $\gamma(\nabla z, \nabla \partial_t \psi)$ cancels, and it remains:

$$\varepsilon \frac{\gamma}{2} \frac{d}{dt} |\nabla \Delta \psi|_2^2 + |\partial_t \psi|_2^2 = -(u \cdot \nabla \psi, \partial_t \psi) + \gamma(\Delta \overline{G}(\psi), \partial_t \psi).$$

In particular,

$$\varepsilon \frac{d}{dt} |\nabla \Delta \psi|_2^2 + |\partial_t \psi|_2^2 \leq C(|u \cdot \nabla \psi|_2^2 + |\Delta \overline{G}(\psi)|_2^2). \tag{51}$$

We can bound the convective term as

$$|u \cdot \nabla \psi|_2^2 \leq |u|_6^2 \, |\nabla \psi|_3^2 \leq C \, \|u\|_1^2$$

From (15), (19), (39) and (42), we have that $\|\psi\|_6 \leq C(1 + \|\psi\|_4 + |\partial_t \psi|_2 + \|u\|_1)$. By interpolation, $\|\psi\|_4 \leq C \|\psi\|_2^{1/4} \|\psi\|_4^{1/2} \|\psi\|_6^{1/4}$, hence, $\|\psi\|_4 \leq C \|\psi\|_6^{1/2}$. Therefore,

$$\|\psi\|_6^2 \leq C(1 + |\partial_t \psi|_2^2 + \|u\|_1^2). \tag{52}$$

By using (52) in (51), we obtain

$$\frac{d}{dt}|\nabla\Delta\psi|_2^2 + C_0\|\psi\|_6^2 \le C\left(1 + \|\boldsymbol{u}\|_1^2 + |\boldsymbol{u}\cdot\nabla\psi|_2^2 + |\Delta\overline{G}(\psi)|_2^2\right)$$

and owing to (39) and (43), we have

$$\frac{d}{dt}|\nabla\Delta\psi|_2^2 + C_0\|\psi\|_6^2 \le C\left(1 + \|\boldsymbol{u}\|_1^2 + |\nabla z|_2^2\right). \tag{53}$$

Since $\|\psi\|_3$ is equivalent to $(|\Delta\psi|_2 + |\nabla\Delta\psi|_2)$ and, taking into account the weak estimate of ψ, equivalent to $(1 + |\nabla\Delta\psi|_2)$, from (53), we obtain

$$\frac{d}{dt}\|\psi\|_3^2 + C_0\|\psi\|_6^2 \le C\left(1 + \|\boldsymbol{u}\|_1^2 + |\nabla z|_2^2\right). \tag{54}$$

By denoting

$$\Phi_2(t) := \|\psi\|_3^2, \qquad B_2(t) := \|\boldsymbol{u}\|_1^2 + |\nabla z|_2^2,$$

from (54), in particular,

$$\Phi_2' + C_0\Phi_2 \le C(1 + B_2). \tag{55}$$

Multiplying (55) by e^t and integrating in time, we obtain

$$\Phi_2(t) \le \Phi_2(0)e^{-C_0 t} + Ce^{-C_0 t}\int_0^t e^{C_0 s}(1 + B_2(s))\,ds.$$

Therefore, $\Phi_2(t) \le \Phi_2(0) + C(1 - e^{-C_0 t}) + C\int_0^t B_2(s)\,ds$. Since $B_2(t) \in L^1(0, +\infty)$, we have that $\Phi_2 \in L^\infty([0, +\infty))$. Moreover, integrating (51) and (54) in $[0, t]$, we obtain

$$\psi \in L^\infty(0, +\infty; H_2^3), \qquad \psi \in L^2_{loc}(0, +\infty; H_2^6), \qquad \partial_t\psi \in L^2_{loc}(0, +\infty; L_*^2). \tag{56}$$

Moreover, from (19),

$$z \in L^2_{loc}(0, +\infty; H_1^2).$$

Observe that in this model, it has been possible to obtain the estimates for the velocity and for the phase separately of each other.

Remark 3 The phase equation is satisfied any everywhere, globally in time, if the data are sufficiently regular.

Consequently, fixed the initial datum $(\boldsymbol{u}_0, \psi_0) \in \boldsymbol{H}\times H^2_1$, by using a Galerkin Method and proceeding in analogous way to Sect. 3.3 of [2], one can prove existence of weak solutions of (16)–(21), in $(0, +\infty)$, and also, existence (and uniqueness) of strong solution in $(T^*_{reg}, +\infty)$ for a large sufficiently time $T^*_{reg} \geq 0$.

6 Convergence at Infinite Time

From the energy-inequality, (36), we have for each $t \geq 0$

$$\overline{E}(\boldsymbol{u}(t), \psi(t)) - \overline{E}(\boldsymbol{u}_0, \psi_0) + \int_0^t (\nu|\nabla \boldsymbol{u}|_2^2 + \lambda\gamma|\nabla z|_2^2)\, d\tau \leq 0. \tag{57}$$

In particular, there exists a number $E_\infty \geq 0$ such that the total energy satisfies

$$\overline{E}(\boldsymbol{u}(t), \psi(t)) \searrow E_\infty \text{ in } \mathbf{R} \quad \text{as } t \uparrow +\infty. \tag{58}$$

The ω-limit set of a fixed global weak solution, $(\boldsymbol{u}, \psi)$, associated to the initial data, $(\boldsymbol{u}_0, \psi_0) \in \boldsymbol{V} \times H^3_2$, can be defined as follows:

$$\omega(\boldsymbol{u}, \psi) = \{(\boldsymbol{u}_\infty, \psi_\infty) \in \boldsymbol{V} \times H^3_2 : \exists\{t_n\} \uparrow +\infty \text{ s.t. } (\boldsymbol{u}(t_n), \psi(t_n)) \to (\boldsymbol{u}_\infty, \psi_\infty) \text{ weakly in } \boldsymbol{V} \times H^3_2\}.$$

Let $\mathscr{S}$ be the set of equilibrium points of (16)–(19) (see also (29)):

$$\mathscr{S} = \{(0, \psi) : \psi \in H^4_2(\Omega) : \varepsilon\Delta^2\psi + \overline{G}(\psi) = 0 \text{ a.e in } \Omega\}.$$

Lemma 5 *If $(\boldsymbol{u}^\varepsilon, \psi^\varepsilon, z^\varepsilon)$, for some $\varepsilon > 0$, and $(\boldsymbol{u}^0, \psi^0, z^0)$ are two regular solutions in $(0, T^*)$ of (16)–(21); associated to the different initial conditions, $(\boldsymbol{u}^\varepsilon_0, \psi^\varepsilon_0) \in \boldsymbol{H}^1 \times H^3_2$ and $(\boldsymbol{u}_0, \psi_0) \in \boldsymbol{H}^1 \times H^3_2$, respectively, then $\boldsymbol{u}^\varepsilon - \boldsymbol{u}^0$, $\psi^\varepsilon - \psi^0$ and $z^\varepsilon - z^0$ depend continuously of the initial values in the following sense: If $\boldsymbol{u}^\varepsilon_0 \to \boldsymbol{u}_0$ weakly in $\boldsymbol{H}^1$ (and strongly in $\boldsymbol{L}^2$) and $\psi^\varepsilon_0 \to \psi_0$ weakly in H^3_2 (and strongly in H^2_1), then,*

$$\begin{aligned} \boldsymbol{u}^\varepsilon - \boldsymbol{u}^0 \to 0 \quad & \text{in } L^\infty(0, T^*; \boldsymbol{L}^2) \cap L^2(0, T^*; \boldsymbol{H}^1), \\ \psi^\varepsilon - \psi^0 \to 0 \quad & \text{in } L^\infty(0, T^*; H^2) \cap L^2(0, T^*; H^5). \end{aligned}$$

Proof We denote $\boldsymbol{u} = \boldsymbol{u}^\varepsilon - \boldsymbol{u}^0$, $\psi = \psi^\varepsilon - \psi^0$ and $z = z^\varepsilon - z^0$ By means of taking $\boldsymbol{u}$, z and $\partial_t\psi$, respectively, as test functions in the difference between the equations for $(\boldsymbol{u}^\varepsilon, \psi^\varepsilon, z^\varepsilon)$ and $(\boldsymbol{u}, \psi, z)$, the term $(z, \partial_t\psi)$ cancels, as well as the term $(\boldsymbol{u} \cdot \nabla\psi, z)$

with $-(z\nabla\psi, \boldsymbol{u})$ then, the following equality is attained:

$$\frac{1}{2}\frac{d}{dt}(|\boldsymbol{u}|_2^2 + \varepsilon^2\lambda|\Delta\psi|_2^2) + \nu|\nabla\boldsymbol{u}|_2^2 + \lambda\gamma|\nabla z|_2^2 = -((\boldsymbol{u}\cdot\nabla)\boldsymbol{u}^0, \boldsymbol{u})$$
$$+\lambda(z^\varepsilon\nabla\psi, \boldsymbol{u}) - \lambda(\boldsymbol{u}^\varepsilon\nabla\psi, z) + \lambda(\overline{G}(\psi^\varepsilon) - \overline{G}(\psi^0), \partial_t\psi) := \sum_{i=1}^4 I_i. \tag{59}$$

Observe that

$$\begin{aligned}\overline{G}(\psi^\varepsilon) - \overline{G}(\psi^0) &= -[F'''(\psi^\varepsilon + m_0)\nabla(\psi^\varepsilon + \psi^0)\nabla\psi + 6\psi(\nabla\psi^0)^2] \\ &\quad -2[F'(\psi^\varepsilon + m_0)\Delta\psi + 3(\psi^\varepsilon + \psi^0 + 2m_0)\psi\Delta\psi^0] \\ &\quad +\frac{1}{\varepsilon^2}[F'(\psi^\varepsilon + m_0)\psi((\psi^\varepsilon + m_0)^2 + (\psi^\varepsilon + m_0)(\psi^0 + m_0) + (\psi^0 + m_0)^2 - 1) \\ &\quad +3\psi(\psi^\varepsilon + \psi^0 + 2m_0)F'(\psi^0 + m_0)] \\ &\quad +\frac{M\varepsilon^2}{2}(-\Delta\psi^\varepsilon + \frac{1}{\varepsilon}F'(\psi^\varepsilon + m_0))\int_\Omega \nabla(\psi^\varepsilon + \psi^0)\nabla\psi\, dx \\ &\quad +M\varepsilon(A(\psi^0) - \alpha)(-\Delta\psi + \frac{1}{\varepsilon}(F'(\psi^\varepsilon + m_0) - F'(\psi^0 + m_0))).\end{aligned}$$

By applying the regularity already obtained, we can infer that

$$|\overline{G}(\psi^\varepsilon) - \overline{G}(\psi^0)|_2 \le C\|\psi\|_2 \qquad |\nabla\overline{G}(\psi^\varepsilon) - \nabla\overline{G}(\psi^0)|_2 \le C\|\psi\|_3, \tag{60}$$

hence the terms on the right hand side of (59) can be to bounded as follows (here $\delta > 0$ is a sufficiently small constant):

$$I_1 \le |\boldsymbol{u}|_2|\nabla\boldsymbol{u}^0|_3|\boldsymbol{u}|_6 \le \delta\|\boldsymbol{u}\|_1^2 + C\|\boldsymbol{u}^0\|_1\|\boldsymbol{u}^0\|_2|\boldsymbol{u}|_2^2,$$

$$I_2 \le |z^\varepsilon|_3|\nabla\psi|_6|\boldsymbol{u}|_2 \le C|\boldsymbol{u}|_2^2 + C\|z^\varepsilon\|_1^2\|\psi\|_2^2,$$

$$I_3 \le |\boldsymbol{u}^\varepsilon|_2|\nabla\psi|_3|z|_6 \le \delta|\nabla z|_2^2 + C\|\psi\|_2^2,$$

$$I_4 = \lambda(\overline{G}(\psi^\varepsilon) - \overline{G}(\psi^0), -\boldsymbol{u}\nabla\psi^0 - \boldsymbol{u}^\varepsilon\nabla\psi + \gamma\Delta z) \le C(\|\psi\|_2|\boldsymbol{u}|_2 + \|\psi\|_2^2 + I_{41}).$$

where

$$\begin{aligned}I_{41} &= |(\nabla\overline{G}(\psi^\varepsilon) - \nabla\overline{G}(\psi^0), \nabla z)| \le C\|\psi\|_3\|z\|_1 \le C\|\psi\|_2^{1/2}\|\psi\|_4^{1/2}\|z\|_1 \\ &\le C(\|\psi\|_2^{1/2}(|\overline{G}(\psi^\varepsilon) - \overline{G}(\psi^0)|_2^{1/2} + |z|_2^{1/2})\|z\|_1 \le C(\|\psi\|_2\|z\|_1 + \|\psi\|_2^{1/2}\|z\|_1^{3/2}) \\ &\le C(\|\psi\|_2(|\nabla z|_2 + \|\psi\|_2) + \|\psi\|_2^{1/2}(|\nabla z|_2 + \|\psi\|_2)^{3/2} \le \delta|\nabla z|_2^2 + C\|\psi\|_2^2.\end{aligned}$$

Therefore, taking into account the equivalence of the norms $\|\psi\|_2$ and $|\Delta\psi|_2$, we arrive at

$$\frac{d}{dt}\left(|\boldsymbol{u}|_2^2 + \varepsilon^2\lambda|\Delta\psi|_2^2\right) \le a(t)(|\boldsymbol{u}|_2^2 + |\Delta\psi|_2^2)$$

where $a(t)$ is bounded in $L^1(0,T)$ for all $T > 0$. Applying Gronwall's Lemma and taking into account that $|\boldsymbol{u}(0)|_2^2 = |\boldsymbol{u}_0^\varepsilon - \boldsymbol{u}_0|_2^2$ and $|\Delta\psi(0)|_2^2 \le \|\psi_0^\varepsilon - \psi_0\|_2^2$, one has

$$|\boldsymbol{u}(t)|_2^2 + \varepsilon^2\lambda|\Delta\psi(t)|_2^2 \le (|\boldsymbol{u}_0^\varepsilon - \boldsymbol{u}_0|_2^2 + \|\psi_0^\varepsilon - \psi_0\|_2^2)\exp\left(\int_0^t a(s)\,ds\right)$$

and the convergence of $\boldsymbol{u}^\varepsilon$ in $L^\infty(0,T^*;\boldsymbol{L}^2)$ and ψ^ε in $L^\infty(0,T^*;H_1^2)$ is obtained. Coming back to the inequality

$$\frac{d}{dt}(|\boldsymbol{u}|_2^2 + \varepsilon^2\lambda|\Delta\psi|_2^2) + C(\nu|\nabla\boldsymbol{u}|_2^2 + \lambda\gamma|\nabla z|_2^2) \le a(t)(|\boldsymbol{u}|_2^2 + |\Delta\psi|_2^2),$$

we obtain the convergence of $\boldsymbol{u}^\varepsilon$ in $L^2(0,T^*;\boldsymbol{H}^1)$ and of z in $L^2(0,T^*;H_*^1)$, and in particular, the convergence of ψ in $L^2(0,T^*;H_2^5)$. □

Theorem 1 *The set $\omega(\boldsymbol{u},\psi)$ is nonempty and $\omega(\boldsymbol{u},\psi) \subset \mathcal{S}$. Moreover, for any $(0,\psi_\infty) \in \omega(\boldsymbol{u},\psi)$, then $\overline{E}_b(\psi_\infty) = E_\infty$ holds.*

Proof By applying Lemma 2 in (47), we obtain that $\boldsymbol{u}(t) \to 0$ in $\boldsymbol{H}_0^1$, therefore, $\boldsymbol{u}_\infty = 0$. Let $(\boldsymbol{u},\psi)$ be a weak solution of problem of (16)–(20), associated to the initial conditions, $(\boldsymbol{u}(0),\psi(0)) = (\boldsymbol{u}_0,\psi_0)$ and let $(0,\psi_\infty)$ be an element of the ω-limit set $\omega(\boldsymbol{u},\psi)$, that is,

$$\exists\{t_n\} \uparrow +\infty \text{ s.t. } (\boldsymbol{u}(t_n),\psi(t_n)) \to (0,\psi_\infty) \text{ weakly in } \boldsymbol{V}\times H_2^3.$$

Let $t_n \ge T_{reg}^*$ be and $(\boldsymbol{v},\xi)$ the unique regular solution in $(0,\widehat{T})$ of (16)–(20), associated to the initial condition $(0,\psi_\infty)$. Since $\overline{E}(\boldsymbol{u}(t),\psi(t)) \searrow E_\infty$ in $\mathbf{R}$ as $t\uparrow +\infty$, we also have

$$\overline{E}(\boldsymbol{u}(t_n+\bar{t}),\psi(t_n+\bar{t})) \searrow E_\infty \text{ in } \mathbf{R} \quad \text{as } n\uparrow +\infty$$

for $\bar{t} \in [0,\widehat{T}]$. By applying Lemma 5 with $u_0 = 0$, $\psi_0 = \psi_\infty$, $u_0^\varepsilon = u(t_n)$ and $\psi_0^\varepsilon = \psi(t_n)$, and denoting

$$\boldsymbol{u}_n(\bar{t}) := \boldsymbol{u}(t_n+\bar{t}) \qquad \psi_n(t) := \psi(t_n+\bar{t}),$$

then

$$\boldsymbol{u}_n \to \boldsymbol{v} \quad \text{in } L^\infty(0,\widehat{T};\boldsymbol{L}^2)\cap L^2(0,\widehat{T};\boldsymbol{H}^1),$$
$$\psi_n \to \xi \quad \text{in } L^\infty(0,\widehat{T};H^2)\cap L^2(0,\widehat{T};H^5)$$

as $n\uparrow+\infty$. In particular, $\overline{E}(\boldsymbol{u}_n(\bar t),\psi_n(\bar t))\to\overline{E}(\boldsymbol{v}(\bar t),\xi(\bar t))$ in $\mathbf{R}$, for all $\bar t\in[0,\widehat{T}]$. Therefore,

$$\overline{E}(\boldsymbol{v}(\bar t),\xi(\bar t)) = E_\infty \qquad \forall\,\bar t\in[0,\widehat{T}].$$

Since $\frac{d}{dt}\overline{E}(\boldsymbol{v}(\bar t),\xi(\bar t))=0$, from the energy equality for $(\boldsymbol{v},\xi)$, we obtain

$$\nu|\nabla\boldsymbol{v}(\bar t)|_2^2+\lambda\gamma|\nabla\tilde z(\bar t)|_2^2=0 \qquad \forall\,\bar t\in[0,\widehat{T}],$$

where $\tilde z=\varepsilon^2\Delta^2\xi+\overline{G}(\xi)$. Taking into account that $\boldsymbol{v}(0)=0$, then for each $\bar t\in[0,\widehat{T}]$, $\boldsymbol{v}(\bar t)\equiv 0$ and also, $\tilde z(\bar t)$ is constant, hence in particular $\Delta\tilde z=0$. Therefore, from the $\tilde z$-equation, $\partial_t\xi+\boldsymbol{v}\cdot\nabla\xi=0$ and hence, $\partial_t\xi=0$. Consequently, $v(\bar t)=0$ and $\xi(\bar t)=\psi_\infty$ for all $\bar t\in[0,\widehat{T}]$. □

Theorem 2 *Under the hypotheses of Theorem 1, there exists $\psi_\infty\in H_2^4$ such that $\psi(t)\to\psi_\infty$ in H_2^3 weakly as $t\uparrow+\infty$, i.e. $\omega(\boldsymbol{u},\psi)=\{(0,\psi_\infty)\}$.*

Proof Let $(0,\psi_\infty)\in\omega(\boldsymbol{u},\psi)\subset\mathscr{S}$, i.e, there exists $t_n\uparrow+\infty$ such that $\boldsymbol{u}(t_n)\to 0$ in $\boldsymbol{V}$ and $\psi(t_n)\to\psi_\infty$ in H_2^3 weakly.

Without any loss of generality, it can be assumed that $\overline{E}(\boldsymbol{u}(t),\psi(t))>\overline{E}(0,\psi_\infty)(=E_\infty)$ for all $t>0$, because otherwise, if it exists some $\tilde t>0$ such that $\overline{E}(\boldsymbol{u}(\tilde t),\psi(\tilde t))=\overline{E}(0,\psi_\infty)$, then, from the energy equality (36),

$$\overline{E}(\boldsymbol{u}(t),\psi(t))=\overline{E}(0,\psi_\infty),\quad |\nabla\boldsymbol{u}(t)|_2^2=0 \quad\text{and}\quad |\nabla z(t)|_2^2=0 \text{ for each } t\ge\tilde t.$$

Therefore, $\boldsymbol{u}(t)=0$ and $z(t)$ is constant for each $t\ge\tilde t$. In particular, by using the z-equation (18), then $\partial_t\psi(t)=0$, hence $\psi(t)=\psi_\infty$ for each $t\ge\tilde t$. In this situation the convergence of the ψ-trajectory is trivial.

Assuming $\overline{E}(\boldsymbol{u}(t),\psi(t))>\overline{E}(0,\psi_\infty)(=E_\infty)$ for all $t>0$, the proof is now divided into three steps.

Step 1: *Assuming there exists $t_\star>T^*_{reg}$ such that*

$$\|\psi(t)-\psi_\infty\|_2\le\beta \quad\text{and}\quad |\boldsymbol{u}(t)|_2\le 1 \quad \forall\, t\ge t_\star$$

where the solution is strong in $(T^*_{reg}, +\infty)$ *and* $\alpha > 0$ *is the constant appearing in Lemma 4 (of Lojasiewicz-Simon's type), then the following inequalities hold:*

$$\frac{d}{dt}\Big((\overline{E}(\boldsymbol{u}(t), \psi(t)) - \overline{E}(0, \psi_\infty))^\theta\Big) + C\theta\,(|\nabla \boldsymbol{u}(t)|_2 + |\nabla z(t)|_2) \le 0, \quad \forall\, t \ge t_\star \tag{61}$$

$$\int_{t_1}^{t_2} \|\partial_t \psi\|_{(H^1_*)'} \le \frac{C}{\theta}(\overline{E}(\boldsymbol{u}(t_1), \psi(t_1)) - E(0, \psi_\infty)))^\theta, \qquad \forall\, t_2 > t_1 \ge t_\star, \tag{62}$$

where $\theta \in (0, 1/2]$ *is the constant appearing in Lemma 4.*
Indeed, the energy equality (36) can be written as

$$\frac{d}{dt}(\overline{E}(\boldsymbol{u}(t), \psi(t)) - E_\infty) + C\left(|\nabla \boldsymbol{u}(t)|_2^2 + |\nabla z(t)|_2^2\right) = 0.$$

Hence, in particular, from Poincaré inequality:

$$\frac{d}{dt}(\overline{E}(\boldsymbol{u}(t), \psi(t)) - E_\infty) + C(|\boldsymbol{u}(t)|_2 + |z(t)|_2)\,(|\nabla \boldsymbol{u}(t)|_2 + |\nabla z(t)|_2) \le 0, \quad \forall t \ge 0.$$

Therefore, by taking the time derivative of the (strictly positive) function

$$H(t) := (\overline{E}(\boldsymbol{u}(t), \psi(t)) - E_\infty)^\theta > 0,$$

and the Poincaré inequality, we obtain

$$\begin{aligned} &\frac{dH(t)}{dt} + \theta(\overline{E}(\boldsymbol{u}(t), \psi(t)) - E_\infty)^{\theta-1} \\ &\cdot C(|\boldsymbol{u}(t)|_2 + |z(t)|_2)\,(|\nabla \boldsymbol{u}(t)|_2 + |\nabla z(t)|_2) \le 0,\ \forall t \ge 0. \end{aligned} \tag{63}$$

On the other hand, by recalling that the unique critical point of the kinetic energy is $\boldsymbol{u} = 0$, by taking into account that $|E_k(\boldsymbol{u}) - E_k(0)| = \frac{1}{2}|\boldsymbol{u}|_2^2$ and since $2(1-\theta) > 1$ and $|\boldsymbol{u}(t)|_2 \le 1$, then

$$|E_k(\boldsymbol{u}(t)) - E_k(0)|^{1-\theta} = \frac{1}{2^{1-\theta}}|\boldsymbol{u}(t)|_2^{2(1-\theta)} \le C|\boldsymbol{u}(t)|_2 \quad \forall\, t \ge t_\star.$$

Therefore, by using the Lojasiewicz-Simon inequality (given in Lemma 4):

$$\begin{aligned} &(\overline{E}(\boldsymbol{u}(t), \psi(t)) - E_\infty)^{1-\theta} \le |E_k(\boldsymbol{u}(t)) - E_k(0)|^{1-\theta} \\ &\quad + |\overline{E}_b(\psi(t)) - \overline{E}_b(\psi_\infty)|^{1-\theta} \le C(|\boldsymbol{u}(t)|_2 + |z(t)|_2), \end{aligned} \tag{64}$$

From (63) and (64), we obtain

$$\frac{dH(t)}{dt} + \theta\, C(|\nabla \boldsymbol{u}(t)|_2 + |\nabla z(t)|_2) \leq 0, \quad \forall\, t \geq t_\star$$

and (61) is proved.
Integrating (61) into $[t_1, t_2]$ for any $t_2 > t_1 \geq t_\star$, we have

$$\begin{aligned}(\overline{E}(\boldsymbol{u}(t_2), \psi(t_2)) - E_\infty)^\theta + \theta\, C \int_{t_1}^{t_2} (|\nabla \boldsymbol{u}(t)|_2 + |\nabla z(t)|_2) dt \\ \leq (\overline{E}(\boldsymbol{u}(t_1), \psi(t_1)) - E_\infty)^\theta .\end{aligned} \tag{65}$$

By using the weak estimate $\|\psi(t)\|_2 \leq C$ in the z-equation $\partial_t \psi = -\boldsymbol{u}\cdot\nabla\psi + \Delta z$, one has

$$\|\partial_t \psi\|_{(H^1_*)'} \leq C(|\nabla \boldsymbol{u}|_2 + |\nabla z|_2).$$

By using this inequality in (65), (62) is attained.
Step 2: *There exists a sufficiently large n_0 such that $t_{n_0} \geq T^*_{reg}$ and $\|\psi(t) - \psi_\infty\|_2 \leq \beta$ and $|\boldsymbol{u}(t)|_2 \leq 1$ for all $t \geq t_{n_0}$, where β is the constant appearing in Lemma 4.*
The bound $|\boldsymbol{u}(t)|_2 \leq 1$ is based on $\boldsymbol{u}(t) \to 0$ in $\boldsymbol{H}^1_0$. We now focus on the bound for $\|\psi(t) - \psi_\infty\|_2$. Since $\psi(t_n) \to \psi_\infty$ in H^2 and $\overline{E}(\boldsymbol{u}(t_n), \psi(t_n)) \to E_\infty = \overline{E}_b(\psi_\infty)$, then for any $\varepsilon \in (0, \alpha)$, there exists an integer $N(\varepsilon)$ such that, for all $n \geq N(\varepsilon)$,

$$\|\psi(t_n) - \psi_\infty\|_2 \leq \varepsilon \quad \text{and} \quad \frac{1}{\theta}(\overline{E}_b(\boldsymbol{u}(t_n), \psi(t_n)) - E_\infty)^\theta \leq \varepsilon. \tag{66}$$

For each $n \geq N(\varepsilon)$, we define

$$\overline{t}_n := \sup\{t : t > t_n,\ \|\psi(s) - \psi_\infty\|_2 < \beta \quad \forall s \in [t_n, t)\}.$$

It suffices to prove that $\overline{t}_{n_0} = +\infty$ for some n_0. Assume by contradiction that $t_n < \overline{t}_n < +\infty$ for all n, hence in this way $\|\psi(\overline{t}_n) - \psi_\infty\|_2 = \beta$ and $\|\psi(t) - \psi_\infty\|_2 < \beta$ for all $t \in [t_n, \overline{t}_n)$. By applying step 1 for all $t \in [t_n, \overline{t}_n]$, from (62) and (66) we obtain,

$$\int_{t_n}^{\overline{t}_n} \|\partial_t \psi\|_{(H^1_*)'} \leq C\varepsilon, \quad \forall\, n \geq N(\varepsilon).$$

Therefore,

$$\|\psi(\overline{t}_n) - \psi_\infty\|_{(H^1_*)'} \leq \|\psi(t_n) - \psi_\infty\|_{(H^1_*)'} + \int_{t_n}^{\overline{t}_n} \|\partial_t \psi\|_{(H^1_*)'} \leq (1 + C)\varepsilon,$$

which implies that $\lim_{n\to+\infty}\|\psi(\bar{t}_n)-\psi_\infty\|_{(H^1_*)'}=0$. Since ψ is bounded in $L^\infty(t^*,+\infty;H^3_2)$, $(\psi(t))_{t\geq t^*}$ is relatively compact in H^2. Therefore, there exists a subsequence of $\psi(\bar{t}_n)$, which is still denoted as $\psi(\bar{t}_n)$, that converges to ψ_∞ in H^2. Hence, $\|\psi(\bar{t}_n)-\psi_\infty\|_2<\beta$ for a sufficiently large n, which contradicts the definition of $\bar{t}_n$.

Step 3: *There exists a unique ψ_∞ such that $\psi(t)\to\psi_\infty$ weakly in H^3_2 as $t\uparrow+\infty$.*

By using Steps 1 and 2, (62) can be applied, for all $t_2>t_1\geq t_{n_0}$, hence

$$\|\psi(t_2)-\psi(t_1)\|_{(H^1_*)'}\leq\int_{t_1}^{t_2}\|\partial_t\psi\|_{(H^1_*)'}\to 0,\quad \text{as } t_1,t_2\to+\infty.$$

Therefore, $(\psi(t))_{t\geq t_{n_0}}$ is a Cauchy sequence in $(H^1_*)'$ as $t\uparrow+\infty$, hence the $(H^1_*)'$-convergence of the whole trajectory is deduced, i.e. there exists a unique $\overline{\psi}\in(H^1_*)'$ such that $\psi(t)\to\psi_\infty$ in $(H^1_*)'$ as $t\uparrow+\infty$. Finally, the weak H^4_2-convergence by sequences of $\psi(t)$ proved in Theorem 1, yields $\psi(t)\to\psi_\infty$ in H^4_2 weakly. □

Acknowledgements This research was partially supported by MINECO grant MTM2012-32325 with the participation of FEDER.

References

1. Campelo, F., Hernández-Machado, A.: Model for curvature-driven pearling instability in membranes. Phys. Rev. Lett. **99**(8), 1–4 (2007)
2. Climent-Ezquerra, B., Guillén-González, F.: Convergence to equilibrium for smectic-A liquid crystals in 3D domains without constraints for the viscosity. Nonlinear Anal. **102**, 208–219 (2014)
3. Climent-Ezquerra, B., Guillén-González, F., Rodríguez-Bellido, M.A.: Stability for Nematic liquid crystals with stretching terms. Int. J. Bifurcations Chaos **20**, 2937–2942 (2010)
4. Du, Q., Liu, C., Wang, X.: A phase field approach in the numerical study of the elastic bending energy for vesicle membranes. J. Comput. Phys. **198**, 450–468 (2004)
5. Du, Q., Li, M., Liu, C.: Analysis of a phase field Navier-Stokes vesicle-fluid interaction model. Discret. Contin. Dyn. Syst. B **8**(3), 539–556 (2007)
6. Gal, C.G., Grasselli, M.: Asymptotic behavior of a Cahn-Hilliard-Navier-Stokes system in 2D. Ann. I. H. Poincaré (C) Non Linear Anal. **27**, 401–436 (2010)
7. Helfrich, W.: Elastic properties of lipid bilayers-theory and possible experiments. Z. Naturforsch. C **28**, 693–703 (1973)
8. Liu, C., Takahashi, T., Tucsnak, M.: Strong solutions for a phase-filed Navier-Stokes Vesicle-Fluid interaction model. J. Math. Fluid Mech. **14**, 177–195 (2012)
9. Wu, H., Xu, X.: Strong solutions, global regularity and stability of a hydrodynamic system modeling vesicle and fluid interactions. SIAM J. Math. Anal. **45**(1), 181–214 (2013)

Explicit Blow-Up Time for Two Porous Medium Problems with Different Reaction Terms

Giuseppe Viglialoro

Abstract This paper deals with the blow-up phenomena of classical solutions to porous medium problems, defined in a bounded domain of $\mathbb{R}^n$, with $n \geq 1$. We distinguish two situations: in the first case, no gradient nonlinearity is present in the reaction term contrarily to the other case. Specifically, some theoretical and general results concerning the mathematical model, existence analysis and estimates of the blow-up time t^* of unbounded solutions to these problems are summarized and discussed. More exactly, for both problems, explicit lower bounds of t^* if blow-up occurs are derived in the case $n = 3$ and in terms of an auxiliary function. On the other hand, in order to compute the real blow-up times of such blowing-up solutions and discuss their properties, a general resolution method is proposed and used in some two-dimensional examples.

1 Introduction

Partial differential equations (PDEs) represent one of the most powerful and efficacious mathematical techniques used to model several real world phenomena. Subsequently, it is very important to control the solutions of the corresponding problems; herein we present theoretical and numerical approaches capable to infer explicit and accurate estimates of the solution of two specific reaction-diffusion problems.

In line with all the above, different reaction-diffusion phenomena, naturally appearing in various physical, chemical and biological applications, are exactly formulated through nonlinear parabolic PDEs. In this sense, the solutions of these time-dependent equations can be commonly characterized by global boundedness in time or, contrarily, by unboundedness in finite or infinite time. For results dealing with global existence and nonexistence, blow-up, blow-up rates and lower and upper bounds of blow-up time of solutions to different and general parabolic problems and systems we refer the reader to [3, 5–7, 11, 14, 15, 18, 24, 25].

G. Viglialoro (✉)
Department of Mathematics and Computer Science, University of Cagliari, Cagliari, Italy
e-mail: giuseppe.viglialoro@unica.it

F. Ortegón Gallego et al. (eds.), *Trends in Differential Equations and Applications*, SEMA SIMAI Springer Series 8, DOI 10.1007/978-3-319-32013-7_9

As far as the blowing-up solutions are concerned, due to its importance in real applications, many authors pay attention to estimates of the blow-up time to those problems whose source (reaction) term depends, generally not linearly, on the solution and/or also on its gradient (see, for instance, [12, 13, 23, 28, 29] and the references therein).

In this sense, if $u = u(\mathbf{x}, t)$ represents the real value of the unknown at the spatial point $\mathbf{x}$ and at the instant time t, here we are concerned with this reaction-diffusion problem, the *complete Porous Medium Equation* (see [27]),

$$u_t = \Delta(u^m) + f(u, |\nabla u|),$$

where m belongs to a suitable subset of $\mathbb{R}^+$ and where the reaction f may also contain the convection term associated to $|\nabla u|$.

Precisely, the main goal of this work is to control the blow-up time t^* of unbounded solutions to the following two problems

$$\begin{cases} u_t = \Delta(u^m) + k_1 u^p, & \mathbf{x} \in \Omega, t \in (0, t^*), \\ u = 0, & \mathbf{x} \in \partial\Omega, t \in (0, t^*), \\ u = u_0(\mathbf{x}) > 0, & \mathbf{x} \in \Omega, \end{cases} \qquad (P_1)$$

and

$$\begin{cases} u_t = \Delta(u^m) + k_1 u^p - k_2|\nabla u|^q, & \mathbf{x} \in \Omega, t \in (0, t^*), \\ u = 0, & \mathbf{x} \in \partial\Omega, t \in (0, t^*), \\ u = u_0(\mathbf{x}) > 0, & \mathbf{x} \in \Omega, \end{cases} \qquad (P_2)$$

where Ω is a bounded and smooth domain of $\mathbb{R}^n$ (with $n \geq 1$), $k_1 > 0$ and $k_2 > 0$, for some constants $p \geq 2$ and $p > q > \frac{3}{2}$, and where $u_0(\mathbf{x})$ is positive in Ω, satisfying the compatibility condition $u_0(\mathbf{x}) = 0$ on $\partial\Omega$. We primarily dedicate our discussion to nonnegative classical solutions of (P_1) and (P_2) which exist for a certain period of time but that eventually may present a delta function at some finite time t^*.

We also precise that in (P_2) we set $p > q$, since for $p \leq q$ the negative convection gradient term, which has a damping effect, may contrast the power source term and the solution does not blow up in finite time (see [20]).

The rest of this paper is organized as follows. A reduced mathematical model of the motion of a gas in a porous medium is derived in Sect. 2; moreover an interpretation of the concept of porosity is also here given.

In Sect. 3 we present the main theoretical results, summarized in Theorems 2 and 3. Specifically, once an appropriate time-dependent energy function is defined (the so called E-energy), it is possible to derive, in the three-dimensional setting, lower bounds for t^* to solutions to problems (P_1) and (P_2) that are unbounded in such an energy.

We deal with the resolution method of the two problems in Sect. 4; starting from a common weak formulation, we propose an algorithm based on a mixed Finite Element Method in space and Euler Method in time capable to numerically solve them. This resolution approach is implemented in the 2D case; hence, we analyze the behaviors of the solutions and, subsequently, compute the blow-up time t^*, precisely in terms of the aforementioned E-energy.

Finally, the paper is complemented with some conclusions (Sect. 5).

2 Mathematical Model of the Porous Medium Equation

Very often, studying the evolution of physical, biological and chemical phenomena, the spatial diffusion of the quantity in consideration is modeled through the heat diffusion law: the flow is proportional to the gradient of the temperature and has the opposite direction, so that there is a motility towards zones of reduced density. Such a model is based on the not so realistic assumption that the quantity spreads with infinity velocity and reaches all the points instantly.

In this sense, a more suitable model is the one in which the diffusion arises with finite velocity, physically appropriate to describe processes involving flow of an isentropic gas through porous media like, for instance, sand or gravel. Anyway, an exhaustive analysis of the general microscopic phenomenon is out of the scope of this notes (we refer the interested reader to [10, 16]); therefore, in the sequel we only derive a simplified mathematical model.

Essentially, if $\Omega \subset \mathbb{R}^n$ ($n \geq 1$) identifies the domain occupied by the porous medium, for any time $t > 0$ a macroscopic approach of the diffusion of a gas through such a medium leads to take into account these equations in $\Omega \times (0, \infty)$:

$$\begin{cases} p = p_0 \rho^\alpha, & \text{equation of state,} \\ \chi \rho_t + \operatorname{div}(\rho \vec{v}) = 0, & \text{conservation of mass,} \\ \vec{v} = -\frac{\mu}{\nu} \nabla p, & \text{Darcy's Law,} \end{cases}$$

where $p = p(\mathbf{x}, t)$, $\rho = \rho(\mathbf{x}, t)$ and $\vec{v} = \vec{v}(\mathbf{x}, t)$ represent the pressure, the density and the velocity at the point $\mathbf{x}$ and instant t of the gas, respectively, and with p_0, χ, μ, ν and α positive constants, where in addition $\alpha \geq 1$.

By arranging the previous three equations we obtain

$$\rho_t = \frac{\mu p_0}{\chi \nu} \operatorname{div}(\rho \nabla \rho^\alpha) = \frac{\mu p_0 \alpha}{\chi \nu (\alpha + 1)} \Delta \rho^{\alpha+1},$$

so that if $\alpha = m - 1$, after an appropriate linear transformation of the temporal variable t, we infer the *Porous Medium Equations* (also known as P.M.E.)

$$u_t = \Delta(u^m), \quad \mathbf{x} \in \Omega, t \in (0, \infty), \tag{1}$$

with $m \in [2, \infty)$ and where $u = u(\mathbf{x}, t)$ represents the density of the gas in the new variables.

Remark 1 Let us point out that:

- the parameter χ depends on the porosity of the medium (and defines the portion of the medium that can be crossed by the gas), μ on its permeability and ν on the viscosity of the gas.
- Although Eq. (1) was derived for $m \geq 2$, herein we will consider $m \in (1, \infty)$, since $m \in (1, 2)$ is also meaningful and concerns the fusion of the ionized gases.
- For $m = 1$ the heat diffusion model is recovered while that $m \in (0, 1)$ represents the mathematical problem of the so called plasma diffusion phenomena; we will not dedicate to this case because it presents strong differences with the case $m > 1$.

2.1 Interpretation of Porosity

As explained, from the mathematical point of view, it is well known that the diffusion term affecting u for the heat transfer phenomena (infinity velocity) is described by Δu; for the flow in porous media (corresponding to finite velocity) this term is replaced by $\Delta(u^m)$, m representing precisely the porosity degree.

Heuristically and intuitively, it is possible to connect the spread velocity $|\overrightarrow{v}|$ of the gas and the porosity coefficient m of the medium: the larger m is, the smaller $|\overrightarrow{v}|$, that is $|\overrightarrow{v}| = |\overrightarrow{v(m)}|$ is a decreasing function of m:

$$|\overrightarrow{v(m_1)}| < |\overrightarrow{v(m_2)}|, \quad \text{for any} \quad m_1 > m_2. \tag{2}$$

In particular, the limit case $\lim_{m \to 1^+} |\overrightarrow{v(m)}| = \infty$ holds.

2.2 Some Properties of the Solutions to the P.M.E.

Although the efficiency of the computers allows us to achieve very precise numerical approximations to a number of difficult problems, it is always very useful rely on analytical solutions or some of their qualitative properties. In this sense, we want to briefly summarize some general aspects connected to the solutions of the P.M.E.

Specifically, under symmetry assumptions on the domain $\Omega \subset \mathbb{R}^n$, Barenblatt and Pattle found an explicit formula for solutions to Eq. (1) in terms of a delta

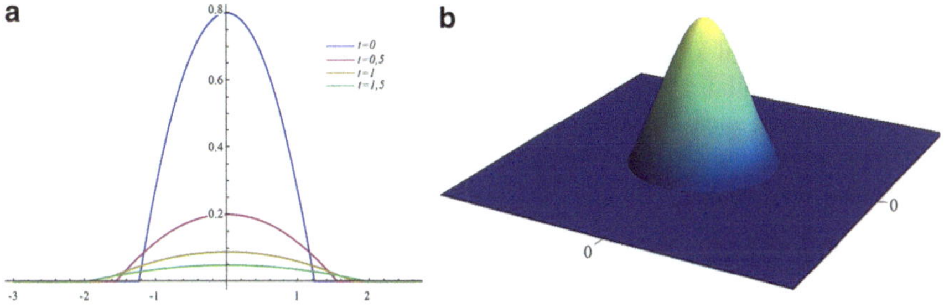

Fig. 1 Representations of solutions to P.M.E. (**a**) Barenblatt-Pattle solution, for $n = 1$ with $C = 0.2$ and $m = 2$. (**b**) Barenblatt-Pattle solution, for $n = 2$ with $C = 0.2$ and $m = 2$, at $t = 2$

function-type initial condition with integral C (see [4, 17]), whose expression is

$$u(|\mathbf{x}|, t) = \max \left\{0, t^{-a}\left[C - \frac{a(m-1)}{2nm}\frac{|\mathbf{x}|^2}{t^{\frac{2a}{n}}}\right]^{\frac{1}{m-1}}\right\}, \tag{3}$$

where $a = (m - 1 + 2/n)^{-1}$.

The solution u presented in (3) has some interesting features; we mention, for instance:

1. the support of u is a ball with radius increasing in time;
2. the maximum of u decreases in time, for any $\mathbf{x}$ given;
3. in the absence of flow through the boundary, $\int_{\mathbb{R}^n} u(|\mathbf{x}|, t)d\mathbf{x} = C$, constant in time (mass conservation);
4. $\lim_{t \to 0^+} u(|\mathbf{x}|, t) = C\delta(\mathbf{x})$, in the sense of distributions.

In line with this, for further completeness, Fig. 1a, b provide graphical representations of the analytical expression (3) in one and two dimensions, respectively.

2.3 *Existence Results and Blow-Up Phenomena*

In the previous section, we focused on analytical and classical solutions of Eq. (1); now we are interested in a more general analysis regarding the following Cauchy problem:

$$\begin{cases} u_t = \Delta(u^m) + f(u), & \mathbf{x} \in \Omega, t \in (0, \infty), \\ u = 0, & \mathbf{x} \in \partial\Omega, t \in (0, \infty), \\ u = u_0(\mathbf{x}) \geq 0, & \mathbf{x} \in \Omega. \end{cases} \tag{4}$$

Since $\Delta(u^m) = \operatorname{div}(mu^{m-1}\nabla u)$, the porous medium operator does not obey the *uniform parabolicity condition* when the initial datum vanishes in an open subset of

Ω (observe that, on the contrary, we fix strictly positive $u_0(\mathbf{x})$ in (P_1) and (P_2)); hence, no classical solution of the equation exists. Subsequently, an appropriate variational formulation of the previous Cauchy problem has to be introduced to ensure existence and uniqueness of weak solutions (see, for instance, [2, 27]). Here, we only provide the forthcoming theorem.

Firstly, we give this

Definition 1 We say that u is a weak solution of problem (4) in $\Omega \times (0, \infty)$ if for any $T > 0$,

- $u \in L^1(\Omega \times (0, T))$.
- $u^m \in L^1((0, T); W_0^{1,1}(\Omega))$.
- The identity

$$\int_0^T \int_\Omega (\nabla u^m \cdot \nabla \phi - u\phi_t) d\mathbf{x}dt = \int_\Omega u_0(\mathbf{x})\phi(\mathbf{x}, 0)d\mathbf{x} + \int_0^T \int_\Omega f\phi d\mathbf{x}dt,$$

holds for any $\phi \in C^1(\Omega \times (0, T))$ which vanishes on $\partial\Omega \times [0, T)$ and for $t = T$.

Theorem 1 *Let f be a global Lipschitz continuous function such that $f(0) = 0$ and $m > 1$. Then, for any $u_0(\mathbf{x}) \geq 0$, continuous and bounded in Ω, there exists $\tau > 0$, with $\tau \leq \infty$, such that problem* (4) *admits a unique weak solution in* $[0, \tau)$.

Proof See [21]. ■

When f is superlinear (or more generally not globally Lipschitz) the situation is quite different and the weak solution exists and is bounded at least for a small time interval $0 < t < t_1$. However, it may become unbounded in a finite time t^*:

$$\lim_{t \to t^*} \sup ||u||_{L^\infty(\Omega)} = \infty. \tag{5}$$

In this case we say that u blows up at $t = t^*$ in the L^∞-norm.

3 Lower Bound for Blow-Up Time

In this section we establish results concerning lower bounds for t^* of unbounded, classical and nonnegative solutions to both problems (P_1) and (P_2).

We, previously, need this

Definition 2 For any nonnegative solution u of (P_1), or (P_2), let us introduce the E-energy

$$E(t) = \int_\Omega u^{m(p-1)} d\mathbf{x}, \tag{6}$$

with $E(0) = \int_\Omega u_0^{m(p-1)} d\mathbf{x} > 0$.

We say that u blows up (or is unbounded) in E-energy (6) at finite time t^* if

$$\lim_{t\to t^*} E(t) = \infty.$$

Hence, we can present these two fundamental results:

Theorem 2 *Let Ω be a bounded domain of $\mathbb{R}^3$ with Lipschitz boundary. Assume $p \geq 2$ and $d = (m-1)/(p-1)$, with $2 - 1/p < m < p$. If u is a nonnegative classical solution of (P_1) becoming unbounded in E-energy (6) at time $t = t^*_{P_1}$, then*

$$t^*_{P_1} \geq b_1 \int_{E(0)}^{\infty} \frac{d\eta}{\eta^{\gamma}} = T_{P_1}, \tag{7}$$

where

$$\gamma = \frac{2m+3d-1}{2m+3d-3} > 1,$$

b_1 being a positive computable constant depending on the data.

Proof See [22]. ∎

Theorem 3 *Let Ω be a bounded domain of $\mathbb{R}^3$ with Lipschitz boundary. Assume $p \geq 2$ and δ a real number verifying*

$$1 < \delta < \frac{2}{3}(m+d)\Big(\frac{2m+3d-3}{2m+3d-1}\Big),$$

where $d = (m-1)/(p-1)$, with $2 - 1/p < m < p$ and $p > q > \frac{3}{2}$. If u is a nonnegative classical solution of (P_2) becoming unbounded in E-energy (6) at time $t^*_{P_2}$, then

$$t^*_{P_2} \geq \int_{E(0)}^{\infty} \frac{d\eta}{c_7\eta^{\alpha} + c_8\eta^{\alpha\beta}} = T_{P_2}, \tag{8}$$

where

$$\begin{cases} \alpha = \frac{2(m+d)-\delta}{2(m+d)-3\delta} > 1, \\ \beta = \frac{2m+3d-1}{2m+3d-3\alpha} > 1, \end{cases}$$

c_7 and c_8 being two positive computable constants depending on the data.

Proof For any nonnegative solution u of (P_2), let us set $s = p - 1$. Due to the divergence theorem and the boundary condition, we lead to

$$\begin{aligned} E'(t) &= ms \int_\Omega u^{ms-1}[\Delta(u^m) + k_1 u^p - k_2|\nabla u|^q]d\mathbf{x} \\ &= -ms \int_\Omega \nabla u^{ms-1} \cdot \nabla(u^m)dx \\ &\quad + msk_1 \int_\Omega u^{s(m+1)}d\mathbf{x} - msk_2 \int_\Omega u^{ms-1}|\nabla u|^q d\mathbf{x} \\ &= -m^2 s(ms-1) \int_\Omega u^{ms-3+m}|\nabla u|^2 d\mathbf{x} \\ &\quad + msk_1 \int_\Omega u^{s(m+1)}d\mathbf{x} - msk_2 \int_\Omega u^{ms-1}|\nabla u|^q d\mathbf{x}. \end{aligned} \tag{9}$$

Now, using inequality (2.10) in [19], we achieve

$$msk_2 \int_\Omega u^{ms-1}|\nabla u|^q d\mathbf{x} = msk_2\Big(\frac{q}{ms+q-1}\Big)^q \int_\Omega |\nabla u^{\frac{ms+q-1}{q}}|^q d\mathbf{x} \geq msk \int_\Omega u^{ms+q-1}d\mathbf{x}, \tag{10}$$

where $k = k_2\Big(\frac{2\sqrt{\lambda_1}}{ms+q-1}\Big)^q$, λ_1 being the first positive eigenvalue of the fixed membrane problem

$$\Delta w + \lambda w = 0 \text{ in } \Omega, \;\; w > 0 \text{ in } \Omega, \;\; w = 0 \text{ on } \partial\Omega.$$

Let us observe that, as specified for instance in [1], λ_1 represents the optimal constant of the classical Poincaré inequality.

For simplicity, we indicate

$$u^s = V, \; \mu = \frac{q-1}{s} < 1, \; d = \frac{m-1}{s} < 1.$$

Furthermore, let us also note that

$$|\nabla V|^2 = s^2 u^{2(s-1)}|\nabla u|^2. \tag{11}$$

As a consequence, since $m > 2 - \frac{1}{p}$, using (10) and (11), relation (9) becomes

$$\begin{aligned} E'(t) &\leq -c_1 \int_\Omega V^{(m-2)+d}|\nabla V|^2 d\mathbf{x} + c_2 \int_\Omega V^{m+1}d\mathbf{x} - kms \int_\Omega V^{m+\mu}d\mathbf{x} \\ &= -c_3 \int_\Omega |\nabla V^{\frac{m+d}{2}}|^2 d\mathbf{x} + c_2 \int_\Omega V^{m+1}d\mathbf{x} - kms \int_\Omega V^{m+\mu}d\mathbf{x}, \end{aligned} \tag{12}$$

where

$$\begin{cases} c_1 = \frac{m^2(ms-1)}{s}, \\ c_2 = msk_1, \\ c_3 = \frac{4}{(m+d)^2} c_1. \end{cases}$$

On the other hand, the Hölder inequality yields

$$\int_\Omega V^{m+1} d\mathbf{x} \le \Big(\int_\Omega V^{m+\mu} d\mathbf{x}\Big)^{\frac{\gamma-1}{\gamma-\mu}} \Big(\int_\Omega V^{m+\gamma} d\mathbf{x}\Big)^{\frac{1-\mu}{\gamma-\mu}},$$

for some positive constant $\gamma > 1$. Therefore, by means of

$$a^r b^{1-r} \le ra + (1-r)b, \tag{13}$$

valid for $a, b > 0$ and $0 < r < 1$, we have

$$\int_\Omega V^{m+1} d\mathbf{x} \le \frac{\gamma-1}{\gamma-\mu} \varepsilon_1 \int_\Omega V^{m+\mu} d\mathbf{x} + \frac{1-\mu}{\gamma-\mu} \varepsilon_1^{-\frac{\gamma-1}{1-\mu}} \int_\Omega V^{m+\gamma} d\mathbf{x}, \tag{14}$$

where ε_1 is a positive constant to be chosen. Specifically, if $\varepsilon_1 = k\frac{\gamma-\mu}{k_1(\gamma-1)}$, replacing (14) into (12), we get

$$E'(t) \le -c_3 \int_\Omega |\nabla V^{\frac{m+d}{2}}|^2 d\mathbf{x} + c_4 \int_\Omega V^{m+\gamma} d\mathbf{x}, \tag{15}$$

where

$$c_4 = c_2 \frac{1-\mu}{\gamma-\mu} \varepsilon_1^{-\frac{\gamma-1}{1-\mu}}.$$

From now on, let δ be such that

$$1 < \delta < \frac{2}{3}(m+d)\Big(\frac{2m+3d-3}{2m+3d-1}\Big). \tag{16}$$

For $\gamma = d + \delta > 1$, the Hölder inequality leads to

$$\int_\Omega V^{m+\gamma} d\mathbf{x} = \int_\Omega V^{(m+d)+\delta} d\mathbf{x} \le \Big(\int_\Omega V^{m+d} d\mathbf{x}\Big)^{\frac{2(m+d)-\delta}{2(m+d)}} \Big(\int_\Omega (V^{\frac{m+d}{2}})^6 d\mathbf{x}\Big)^{\frac{\delta}{2(m+d)}}. \tag{17}$$

In addition, since $u = 0$ on $\partial\Omega$, the Sobolev embedding in $\mathbb{R}^3$, $W_0^{1,2} \hookrightarrow L^6$, provides

$$\int_\Omega \left(V^{\frac{m+d}{2}}\right)^6 d\mathbf{x} \le \Gamma^6 \Big(\int_\Omega |\nabla V^{\frac{m+d}{2}}|^2 d\mathbf{x}\Big)^3, \tag{18}$$

$\Gamma = 4^{\frac{1}{2}} 3^{-\frac{1}{2}} \pi^{-\frac{2}{3}}$ being the best Sobolev constant (see [26]).

Replacing (18) into (17), we obtain

$$\int_\Omega V^{(m+d)+\delta} d\mathbf{x} \le \Gamma^{\frac{3\delta}{m+d}} \Big(\int_\Omega V^{m+d} d\mathbf{x}\Big)^{\frac{2(m+d)-\delta}{2(m+d)}} \Big(\int_\Omega |\nabla V^{\frac{m+d}{2}}|^2 d\mathbf{x}\Big)^{\frac{3\delta}{2(m+d)}},$$

and, introducing a positive constant ε_2, through (13) we get (recall (16))

$$\begin{aligned}
&\int_\Omega V^{(m+d)+\delta} d\mathbf{x} \\
&\le \Gamma^{\frac{3\delta}{m+d}} \left(\varepsilon_2 \Big(\int_\Omega V^{m+d} d\mathbf{x}\Big)^{\frac{2(m+d)-\delta}{2(m+d)-3\delta}}\right)^{\frac{2(m+d)-3\delta}{2(m+d)}} \left(\varepsilon_2^{1-\frac{2(m+d)}{3\delta}} \int_\Omega |\nabla V^{\frac{m+d}{2}}|^2 d\mathbf{x}\right)^{\frac{3\delta}{2(m+d)}} \\
&\le \Gamma^{\frac{3\delta}{m+d}} \varepsilon_2 \frac{2(m+d)-3\delta}{2(m+d)} \Big(\int_\Omega V^{m+d} d\mathbf{x}\Big)^{\frac{2(m+d)-\delta}{2(m+d)-3\delta}} \\
&\quad + \Gamma^{\frac{3\delta}{m+d}} \varepsilon_2^{1-\frac{2(m+d)}{3\delta}} \frac{3\delta}{2(m+d)} \int_\Omega |\nabla V^{\frac{m+d}{2}}|^2 d\mathbf{x}.
\end{aligned} \tag{19}$$

To bound the term $\left(\int_\Omega V^{m+d} d\mathbf{x}\right)^{\frac{2(m+d)-\delta}{2(m+d)-3\delta}}$, let us observe that the Hölder and the Schwarz inequalities give, respectively,

$$\int_\Omega V^{m+1} d\mathbf{x} \le \Big(\int_\Omega V^{2(m+d)} d\mathbf{x}\Big)^{\frac{1}{m+2d}} \Big(\int_\Omega V^m d\mathbf{x}\Big)^{\frac{m+2d-1}{m+2d}}, \tag{20}$$

and

$$\int_\Omega V^{2(m+d)} d\mathbf{x} \le \Big[\int_\Omega \left(V^{\frac{m+d}{2}}\right)^6 d\mathbf{x} \int_\Omega V^{m+d} d\mathbf{x}\Big]^{\frac{1}{2}}. \tag{21}$$

Now, using in (21) relation (18), we have

$$\int_\Omega V^{2(m+d)} d\mathbf{x} \le \Gamma^3 \Big(\int_\Omega |\nabla V^{\frac{m+d}{2}}|^2 d\mathbf{x}\Big)^{\frac{3}{2}} \Big(\int_\Omega V^{m+d} d\mathbf{x}\Big)^{\frac{1}{2}}.$$

Subsequently, (20) becomes

$$\int_{\Omega} V^{m+1} d\mathbf{x} \leq \Gamma^{\frac{3}{m+2d}} \Big(\int_{\Omega} |\nabla V^{\frac{m+d}{2}}|^2 d\mathbf{x} \Big)^{\frac{3}{2(m+2d)}} \Big(\int_{\Omega} V^{m+d} d\mathbf{x} \Big)^{\frac{1}{2(m+2d)}} \Big(\int_{\Omega} V^{m} d\mathbf{x} \Big)^{\frac{m+2d-1}{m+2d}}. \tag{22}$$

In addition, we first use the Hölder inequality to lead to

$$\int_{\Omega} V^{m+d} d\mathbf{x} \leq \Big(\int_{\Omega} V^{m+1} d\mathbf{x} \Big)^{d} \Big(\int_{\Omega} V^{m} d\mathbf{x} \Big)^{1-d}, \tag{23}$$

and then insert this estimate in (22); combining terms, applying (13) and setting

$$\frac{2(m+d)-\delta}{2(m+d)-3\delta} = \alpha > 1,$$

we have

$$\begin{aligned}
\Big(\int_{\Omega} V^{m+1} d\mathbf{x} \Big)^{\alpha} &\leq \Gamma^{\frac{6\alpha}{2m+3d}} \Big(\int_{\Omega} |\nabla V^{\frac{m+d}{2}}|^2 d\mathbf{x} \Big)^{\frac{3\alpha}{2m+3d}} \Big(\int_{\Omega} V^{m} d\mathbf{x} \Big)^{\alpha \frac{2m+3d-1}{2m+3d}} \\
&\leq \Gamma^{\frac{6\alpha}{2m+3d}} \frac{3\alpha}{2m+3d} \int_{\Omega} |\nabla V^{\frac{m+d}{2}}|^2 d\mathbf{x} \\
&+ \Gamma^{\frac{6\alpha}{2m+3d}} \frac{2m+3d-3\alpha}{2m+3d} \Big(\int_{\Omega} V^{m} d\mathbf{x} \Big)^{\alpha \frac{2m+3d-1}{2m+3d-3\alpha}},
\end{aligned} \tag{24}$$

where we have also taken into account assumption (16).

Hence, rearranging again (23) with (13) we attain

$$\begin{aligned}
\Big(\int_{\Omega} V^{m+d} d\mathbf{x} \Big)^{\alpha} &\leq \Big[\Big(\int_{\Omega} V^{m+1} d\mathbf{x} \Big)^{d} \Big(\int_{\Omega} V^{m} d\mathbf{x} \Big)^{1-d} \Big]^{\alpha} \\
&\leq d \Big(\int_{\Omega} V^{m+1} d\mathbf{x} \Big)^{\alpha} + (1-d) \Big(\int_{\Omega} V^{m} d\mathbf{x} \Big)^{\alpha},
\end{aligned}$$

so that in view of (24) expression (19) becomes (recall $\gamma = d + \delta$)

$$\begin{aligned}
\int_{\Omega} V^{m+\gamma} d\mathbf{x} \leq & \Gamma^{\frac{3\delta}{m+d} + \frac{6\alpha}{2m+3d}} 3\alpha d \frac{\sigma}{2m+3d} \varepsilon_2 \int_{\Omega} |\nabla V^{\frac{m+d}{2}}|^2 d\mathbf{x} \\
&+ \Gamma^{\frac{3\delta}{m+d} + \frac{6\alpha}{2m+3d}} d\sigma \frac{2m+3d-3\alpha}{2m+3d} \varepsilon_2 \Big(\int_{\Omega} V^{m} d\mathbf{x} \Big)^{\alpha\beta} \\
&+ (1-d)\varepsilon_2 \Gamma^{\frac{3\delta}{m+d}} \sigma \Big(\int_{\Omega} V^{m} d\mathbf{x} \Big)^{\alpha} \\
&+ \Gamma^{\frac{3\delta}{m+d}} \frac{3\delta}{2(m+d)} \varepsilon_2^{1-\frac{2(m+d)}{3\delta}} \int_{\Omega} |\nabla V^{\frac{m+d}{2}}|^2 d\mathbf{x},
\end{aligned} \tag{25}$$

where $\frac{2(m+d)-3\delta}{2(m+d)} = \sigma$ and

$$\beta = \frac{2m+3d-1}{2m+3d-3\alpha} > 1.$$

Lastly, coming back to inequality (15), relation (25) provides

$$E'(t) \leq \left(c_5\varepsilon_2 + c_6\varepsilon_2^{1-\frac{2(m+d)}{3\delta}} - c_3\right)\int_\Omega |\nabla V^{\frac{m+d}{2}}|^2 d\mathbf{x} + c_7\Big(\int_\Omega V^m d\mathbf{x}\Big)^\alpha + c_8\Big(\int_\Omega V^m d\mathbf{x}\Big)^{\alpha\beta}, \tag{26}$$

with

$$c_5 = \Gamma^{\frac{3\delta}{m+d}+\frac{6\alpha}{2m+3d}}\frac{3\alpha d c_4\sigma}{2m+3d}, \qquad c_6 = \frac{3\delta c_4 \Gamma^{\frac{3\delta}{m+d}}}{2(m+d)},$$

and

$$c_7 = \Gamma^{\frac{3\delta}{m+d}}(1-d)\varepsilon_2\sigma c_4, \quad c_8 = \Gamma^{\frac{3\delta}{m+d}+\frac{6\alpha}{2m+3d}}\frac{2m+3d-3\alpha}{2m+3d}\varepsilon_2 c_4 d\sigma.$$

As detailed in Appendix, if this relation is satisfied

$$c_3 \geq c_5\Big(\frac{c_5}{c_6}\Big)^{\frac{-3\delta}{2(m+d)}}\Big(\frac{3\delta}{(2m+2d-3\delta)}\Big)^{-\frac{3\delta}{2(m+d)}}\frac{2(m+d)}{2(m+d)-3\delta},$$

then there exits at least a value of ε_2 such that $c_5\varepsilon_2 + c_6\varepsilon_2^{1-\frac{2(m+d)}{3\delta}} - c_3 \leq 0$; for such a value of ε_2 inequality (26) is simplified to

$$E'(t) \leq c_7\Big(\int_\Omega V^m d\mathbf{x}\Big)^\alpha + c_8\Big(\int_\Omega V^m d\mathbf{x}\Big)^{\alpha\beta},$$

or, by (6),

$$\frac{dE}{c_7E^\alpha + c_8E^{\alpha\beta}} \leq 1.$$

Upon an integration, we have for $t < t^*$

$$t^* \geq \int_{E(0)}^\infty \frac{d\eta}{c_7\eta^\alpha + c_8\eta^{\alpha\beta}},$$

so that the proof is complete. ■

Remark 2 Regarding the aforementioned theorems, let us clarify that although Theorem 2 directly arises from Theorem 3 setting $k_2 = 0$, the first and principal result has been derived by Shafer in 2008; subsequently, the detailed proof given above, uses also ideas of Schaefer [22].

Let us underline some aspects concerning the solution u of (P_1), or (P_2), in terms of the E-energy defined in (6):

- If $E(t)$ is unbounded at finite time t^*, there exists a time t_1 (that might be also 0) such that $E(t_1) = E(0)$ and $E(t) > E(t_1)$, for $t \in (t_1, t^*)$ (see Fig. 2); subsequently, it is possible to express the estimate of t^* only in terms of the initial energy $E(0)$, as in (7) and (8).
- If $E(t)$ is unbounded at $t = t^*$, then $||u||_{L^\infty(\Omega)}$ is also unbounded at $t = t^*$; in other words, if u blows up in thc E-cncrgy (rclation (6)) it also docs in thc scnsc of relation (5).
- If the E-energy decreases in time or reaches a constant value (see Fig. 3) we say that u is bounded in time in the sense of the E-energy and, for editing reasons which will be clear later, we improperly set "$t^* = \infty$".

Remark 3 Qualitatively, the difference $k_1 u^p - k_2 |\nabla u|^q$ models a sort of competition between a source effect (represented by the power term), that increases the internal energy of the system and therefore accelerates the blow-up time, and a damping effect (represented by the gradient term), that breaks such an energy and,

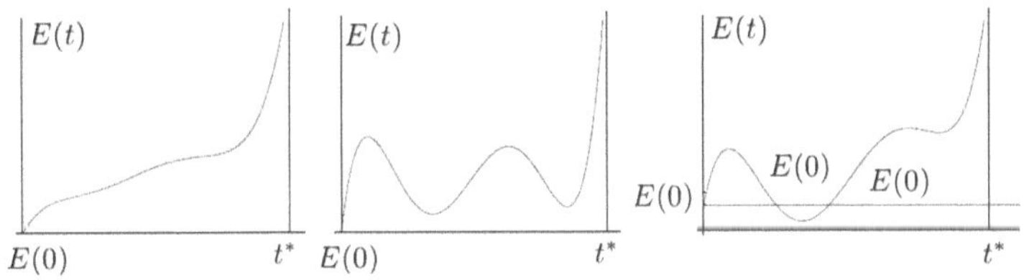

Fig. 2 Possible behaviors of the E-energy in terms of time, once $E(t)$ is supposed to be unbounded at finite time t^*

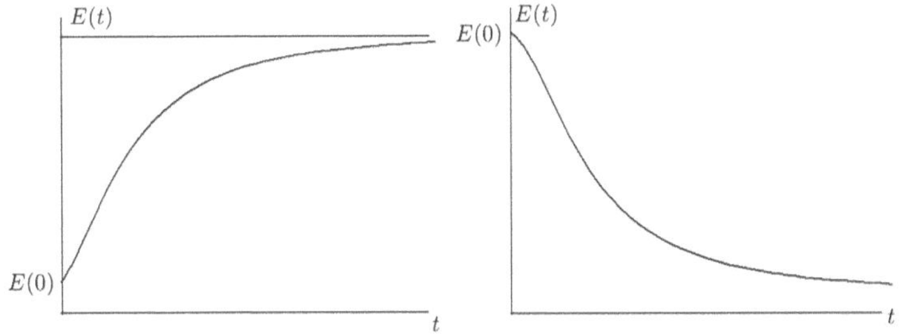

Fig. 3 Possible behaviors of the E-energy in terms of time, once $E(t)$ is supposed to be bounded

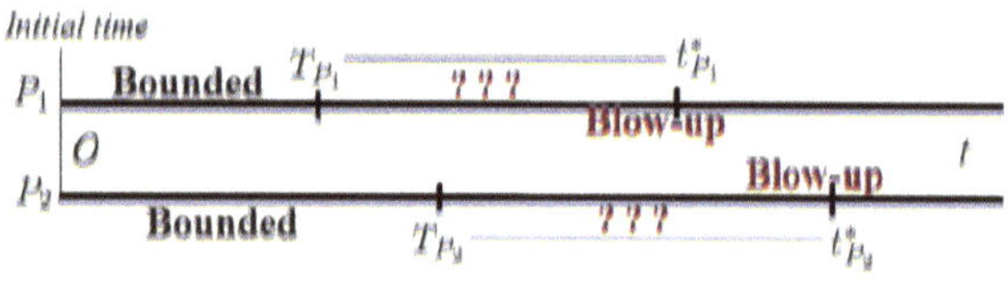

Fig. 4 Analysis of lower bounds and blow-up times of solutions to problems (P_1) and (P_2)

subsequently, contrasts this growth, works against it and slows down the blow-up time.

In terms of the two main problems, the situation discussed in Remark 3 is sketched in Fig. 4: for problem (P_1) (respectively, (P_2)), if the E-energy defined in (6) is finite, the solution u is bounded in such an energy for any t belonging to $[0, T_{P_1})$ (respectively, $[0, T_{P_2})$). In addition, the interval $[0, T_{P_1})$ (respectively, $[0, T_{P_2})$) is not maximal and it is not known *a priori* the length of the interval $(T_{P_1}, t^*_{P_1})$ (respectively, $(T_{P_2}, t^*_{P_2})$). On the contrary, if the E-energy is unbounded at $t^*_{P_1}$ (respectively, $t^*_{P_2}$), it is known that $T_{P_1} < T_{P_2}$ and that $t^*_{P_1} < t^*_{P_2}$.

4 Numerical Resolution Method and Examples

In this section a resolution technique for both problems (P_1) and (P_2), based on a mixed semi-discrete in space and a single-step method in time, is presented. Exactly, in order to derive a general numerical procedure, let us formulate such problems jointly:

$$\begin{cases} u_t = \Delta(u^m) + f_p^q(u, |\nabla u|), & \mathbf{x} \in \Omega, t \in (0, t^*), \\ u = 0, & \mathbf{x} \in \partial\Omega, t \in (0, t^*), \\ u = u_0(\mathbf{x}) \geq 0, & \mathbf{x} \in \Omega, \end{cases} \tag{27}$$

where $f_p^q(u, |\nabla u|)$ will be defined below.

4.1 Finite Element Method: Semi-Discretization in Space

Let Ω be a bounded and regular domain of $\mathbb{R}^n$, with $n \geq 1$. If a mesh of Ω is fixed, and N represents the total number of nodes of Ω, let $\mathcal{U}$ be the numerical approximation of the solution u of (27): therefore,

$$\mathcal{U}(\mathbf{x}, t) = \sum_{i=1}^{N} u^i(t)\varphi^i(\mathbf{x}), \tag{28}$$

where $\varphi^i(\mathbf{x}) \in H_0^1(\Omega)$ is the standard hat basis at the vertex $\mathbf{x}^i$, for $i = 1, \ldots, N$.

Thanks to the divergence theorem, multiplying the differential equation in (27) by a generic test function $\varphi^j(\mathbf{x})$ and taking into consideration the homogeneous boundary condition, for any $j = 1, \ldots, N$, $t \geq 0$, this variational formulation in space is achieved

$$(\mathscr{U}_t, \varphi^j) + (\nabla \mathscr{U}^m, \nabla \varphi^j) = (f_p^q(u, |\nabla u|), \varphi^j); \tag{29}$$

in the equation above, $(\cdot, \cdot)$ stands for the usual L^2 inner product.

In order to compute the evolution in time of the coefficients u^i appearing in (28), let $\Delta t = t_{k+1} - t_k$ be a given time step, with $k = 0, 1, 2, \ldots$ $(t_0 = 0)$, and $\mathscr{U}_k$ the approximation of $\mathscr{U}(\mathbf{x}, t)$ at time t_k. By applying the forward Euler finite difference approximation to system (29), it is seen that

$$(\frac{\mathscr{U}_{k+1} - \mathscr{U}_k}{\Delta t}, \varphi^j) + (\nabla \mathscr{U}_k^m, \nabla \varphi^j) = (f_p^q(\mathscr{U}_k, |\nabla \mathscr{U}_k|), \varphi^j),$$

i.e., taking into account (28),

$$\boldsymbol{M}\frac{\boldsymbol{u}_{k+1} - \boldsymbol{u}_k}{\Delta t} + \boldsymbol{K}\boldsymbol{u}_k^m = \mathscr{F}_p^q(\boldsymbol{u}_k), \tag{30}$$

with

$$\begin{cases} \boldsymbol{M} \in \mathbb{R}^{N \times N} : M_{ij} = \int_\Omega \varphi^i(\mathbf{x})\varphi^j(\mathbf{x})d\mathbf{x}, \\ \boldsymbol{K} \in \mathbb{R}^{N \times N} : K_{ij} = \int_\Omega \nabla\varphi^i(\mathbf{x}) \cdot \nabla\varphi^j(\mathbf{x})d\mathbf{x}, \end{cases}$$

and $\mathscr{F}_p^q(\boldsymbol{u}_k) \in \mathbb{R}^N$ such that

$$\mathscr{F}_p^q(\boldsymbol{u}_k)_j = k_1 \int_\Omega (\sum_{i=1}^N u_k^i \varphi^i(\mathbf{x}))^p \varphi^j(\mathbf{x})d\mathbf{x},$$

if $f_p^q(u, |\nabla u|) = k_1 u^p$ (corresponding to (P_1)) and

$$\mathscr{F}_p^q(\boldsymbol{u}_k)_j = k_1 \int_\Omega (\sum_{i=1}^N u_k^i \varphi^i(\mathbf{x}))^p \varphi^j(\mathbf{x}) - k_2(\sum_{i=1}^N u_k^i |\nabla\varphi^i(\mathbf{x})|)^q \varphi^j(\mathbf{x})d\mathbf{x},$$

if $f_p^q(u, |\nabla u|) = k_1 u^p - k_2 |\nabla u|^q$ (corresponding to (P_2)).

Let us note that we have set $\boldsymbol{u}_k = (u_k^1, \ldots, u_k^N)^T$, where T represents the transposition operator. In these circumstances, u_k^i is the approximation of the solution u of problem (27) at time t_k, for $k = 0, 1, 2, \ldots$, and at space point $\mathbf{x}^i$, for $i = 1, 2, \ldots, N$.

With regards to the estimate of the blow-up time t^*, measured in the sense of Definition 2, the following numerical resolution algorithm is proposed. Let ε_0 be a fixed threshold: once the initial datum $\boldsymbol{u}_0$ and an integration step Δt are given, $\boldsymbol{u}_1$ is computed from (30). Successively, $\boldsymbol{u}_1$ is used to actualize $\boldsymbol{u}_2$, and so on. Moreover, according to (6), we exit the loop when the numerical approximation of the E-energy at step k

$$\varepsilon_k = \int_\Omega \Big(\sum_{i=1}^{N} u_k^i \varphi^i(\mathbf{x}) \Big)^{m(p-1)} d\mathbf{x}, \tag{31}$$

is greater than the initial threshold ε_0 (*Stopping Criterion*); eventually, $t^* \approx k\Delta t$ (see the scheme in Table 1).

Remark 4 It is well known that the forward Euler method is an explicit method which presents only a linear accuracy with respect to the step size, and its local truncation error is $O((\Delta t)^2)$; hence it is a first order method. Moreover, it may manifest undesirable numerical instabilities.

On the other hand, we could obtain stability by means of implicit methods, for instance through the forward Euler method, or others that have higher order (see [9]). Anyway, such approach is more expensive to be implemented since (30) should be replaced by

$$\boldsymbol{M}\frac{\boldsymbol{u}_{k+1} - \boldsymbol{u}_k}{\Delta t} + \boldsymbol{K}\boldsymbol{u}_{k+1}^m = \mathscr{F}_p^q(\boldsymbol{u}_k),$$

and, hence, $\boldsymbol{u}_{k+1}$ should be computed solving an implicit and nonlinear equation.

Nevertheless, since it is very simple to implement and also very intuitive, we prefer to impose specific conditions on the time step size capable to make the

Table 1 Computation of the blow-up time t^*. The necessary input data are the threshold ε_0, the time step Δt and the initial datum $\boldsymbol{u}_0$; successively, it is possible to calculate the sequence $\boldsymbol{u}_k$ and ε_k and, therefore, to calculate t^*

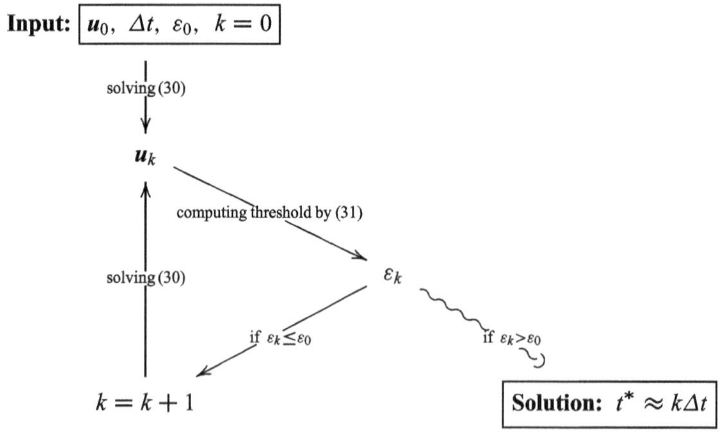

explicit Euler method stable. Specifically, if h represents the diameter of the largest element in the mesh, by choosing $\Delta t/h^2$ small enough the forward method is immune to oscillations and totally suitable to compute the solution as close to the exact one (see again [9] for details). We believe that such approach is totally appropriate to the aims of this research.

4.2 Numerical Simulations in $\mathbb{R}^2$

In the following examples we want to investigate the influence of p, q, m and u_0 on the solution u of (P_1) and (P_2), and specially on the blow-up time t^*. In such sense, in order to carry out an accurate analysis of the efficiency of the presented algorithm and of the obtained results, we have to take in mind Remark 3. In addition, we also rely on relation (2) which suggests that the blow-up time t^* has to increases with m increasing.

4.2.1 Examples

In this section we solve system (27) in order to discuss some aspects corresponding to solutions of (P_1) and (P_2). More exactly, for both problems we focus on

- the analysis of the value of t^* with p, q and m varying;
- the influence of the initial data u_0 on t^*.

Remark 5 Let us point out that these numerical simulations have been obtained by means of the software package FreeFem++ (see [8]). This is a free programming language based in the Finite Element Method, focused on solving partial differential equations. FreeFem++ is implemented in terms of the variational formulations of the corresponding problems, so that it is straightforward to address problems involving PDEs.

The results in Tables 2 and 3 have been computed in the domain $Q = B_1(\mathbf{0}) \times \mathbb{R}_0^+$, being $B_1(\mathbf{0}) = \{\mathbf{x} = (x, y) \in \mathbb{R}^2 : x^2 + y^2 - 1 < 0\}$. Moreover, $\Delta t = 0.001$, $k_1 = k_2 = 1$, $\varepsilon_0 = 10^7$ and $u_0(\mathbf{x}) = \alpha(1 - x^2 - y^2)$, where $\alpha > 0$ was used in order to change the initial E-energy $E_0 = E(u_0)$.

Table 2 emphasizes how the blow-up time t^* of unbounded solutions u to problem (P_1) decreases with p increasing, once m and E_0 are given and also with E_0 increasing, for m and p fixed; the first case can be checked by comparing the first

Table 2 Analysis of problem (P_1)

m	1.2	1.2	1.3	1.3	1.3
p	2.3	2.4	2.4	2.4	2.4
$E(u_0)$	10.99	10.99	10.99	12.57	9.42
$t^*_{P_1}$	0.186	0.108	0.157	0.101	∞

Table 3 Analysis of problem (P_2)

m	1.2	1.2	1.4	1.4
p	2.4	2.4	2.6	2.6
q	1.4	1.3	1.3	1.3
$E(u_0)$	10.99	10.99	10.99	12.57
$t^*_{P_2}$	∞	0.294	0.107	0.062

and second columns and the second by observing the third, fourth and fifth ones. In particular, the last column shows that if E_0 is not big enough, $t^* = \infty$ so that u is globally bounded in E-energy (recall convention). Finally, t^* decreases with the porosity coefficient m (second and third columns).

On the other hand, as explained through the paper, Table 3 also highlights how for m, p and E_0 given, an increasing of q corresponds to a higher dumping effect of the term $|\nabla u|^q$, which results in a decreasing of the blow-up time t^* of unbounded solutions u; it is seen in the first two columns (the data of the first one even return a bounded solution). In addition, if m, p and q are fixed, a higher initial energy E_0 corresponds to solutions whose blow-up time is smaller than solutions associated to smaller E_0 (third and fourth columns).

5 Conclusions

This paper studies the bounded and unbounded (blowing-up) solutions of two nonlinear parabolic problems defined in a bounded and regular domain of $\mathbb{R}^n$, with $n \geq 1$. The equations contain the diffusive term associated to the laplacian of a power of the solution and in a case a reaction term (a power of the solution), which represents a source, and in the other also a power of the gradient of the solution, which models damping effect; moreover Dirichlet boundary conditions are fixed. First we review partial theoretical results concerning existence and boundedness and unboundedness properties of positive solutions to such problems, and then we give lower bound estimates for the blow-up time of the blowing-up solutions in a three-dimensional domain. In addition, we propose and employ a procedure capable to numerically calculate these solutions; this algorithm is achieved by applying a mixed semi-discretization in space and a single-step method in time to both problems. Furthermore, the problems are numerically solved in two-dimensional cases; in particular, the analysis of the results shows that:

- The numerical method is coherent with respect to the expected results since the solutions obey natural laws and expectations.
- The problems are sensitive with respect to small variations of its data, in fact, initial conditions or parameters slightly different each other can return both blowing-up or bounded solutions.

Appendix

For completeness of the reader, we emphasize some details used in the proof of Theorem 3.

Proposition 1 *Let the coefficients c_i $(i = 1, \ldots, 6)$ of Theorem 3 satisfy*

$$c_3 \geq c_5\Big(\frac{c_5}{c_6}\Big)^{\frac{-3\delta}{2(m+d)}}\Big(\frac{3\delta}{(2m+2d-3\delta)}\Big)^{-\frac{3\delta}{2(m+d)}}\frac{2(m+d)}{2(m+d)-3\delta}. \tag{32}$$

Then there exits at least a $\xi \in (0, \infty)$ such that

$$c_5\xi + c_6\xi^{1-\frac{2(m+d)}{3\delta}} - c_3 \leq 0. \tag{33}$$

Proof For any $\xi \in (0, \infty)$, function $\Phi(\xi) := c_5\xi + c_6\xi^{1-\frac{2(m+d)}{3\delta}}$ attains its minimum at the point

$$\xi_m = \Big(\frac{3\delta c_5}{c_6(2m+2d-3\delta)}\Big)^{\frac{-3\delta}{2(m+d)}}.$$

Therefore, since

$$\Phi(\xi_m) = c_5\Big(\frac{c_5}{c_6}\Big)^{\frac{-3\delta}{2(m+d)}}\Big(\frac{3\delta}{(2m+2d-3\delta)}\Big)^{-\frac{3\delta}{2(m+d)}}\frac{2(m+d)}{2(m+d)-3\delta},$$

and (32) holds, relation (33) is proven.

In addiction, let us give this

Remark 6 Relation (32) can be explicitly written as

$$\frac{1}{k_1}\Big(\frac{k_2}{k_1}\Big)^{\frac{\gamma-1}{1-\mu}} \geq \frac{\Gamma^{\frac{3\delta}{m+d}}s^2(m+d)^2}{4m(ms-1)}\Big[\frac{\gamma-\mu}{\gamma-1}\Big(\frac{2\sqrt{\lambda_1}}{ms+q-1}\Big)^{q}\Big]^{-\frac{\gamma-1}{1-\mu}}\Sigma,$$

being

$$\Sigma = \frac{1-\mu}{\gamma-\mu}\Big(\frac{6(m+d)\Gamma^{\frac{6\alpha}{2m+3d}}d\alpha\sigma}{(2m+3d)(2m+2d-3\delta)}\Big)^{1-\frac{3\delta}{2(m+d)}}.$$

Therefore, once m, p, q and Ω are fixed in (P_2), relation (32) is satisfied for k_2 (respectively, k_1) big (respectively, small) enough. Since k_2 is the coefficient associated to $-|\nabla u|^q$, which contrasts the explosion, and k_1 the one associated to u^p, which stimulates it, this effect of coefficients k_1 and k_2 is coherent in terms of estimate (8). In fact, t^* increases when constants c_7 and c_8 decreasing, which in turn decrease with k_2 (respectively k_1) increasing (respectively, decreasing).

Acknowledgements The author is member of the *Gruppo Nazionale per l'Analisi Matematica, la Probabilitá e le loro Applicazioni (GNAMPA)* of the *Istituto Nazionale di Alta Matematica (INdAM)*. The author also gratefully acknowledges Sardinia Regional Government for the financial support (P.O.R. Sardegna, F.S.E. 2007–2013). This work is also supported by the research group *IFQM315-Análisis Teórico y Numérico de Modelos de las Ciencias Experimentales*, of the Department of Mathematics of the University of Cadiz (Spain).

References

1. Acosta, G., Durán, R.G.: An optimal Poincaré inequality in L^1 for convex domains. Proc. Am. Math. Soc. **132**, 195–202 (2004)
2. Aronson, D.G., Crandall, M.G, Peletier, L.A.: Stabilization of solutions of a degenerate diffusion problem. Nonlinear Anal. Theor. **6**, 1001–1022 (1982)
3. Bandle, C., Brunner, H.: Blow-up in diffusion equations: a survey. J. Comput. Appl. Math. **97**, 3–22 (1983)
4. Barenblatt, G.I.: On some unsteady motions of a liquid or a gas in a porous medium. Appl. Math. Mech. **16**(1), 67–78 (1952)
5. Brändle, C., Quirós, F., Rossi, J.D.: Non-simultaneous blow-up for a quasilinear parabolic system with reaction at the boundary. Comm. Pure Appl. Anal. **4**, 523–536 (2005)
6. Farina, M.A., Marras, M., Viglialoro, G.: On explicit lower bounds and blow-up times in a model of chemotaxis. Discret. Contin. Dyn. Syst. Suppl. **2015**, 409–417 (2015)
7. Galaktionov, V.A., Vázquez, J.L.: The problem of blow up in nonlinear parabolic equations. Discret. Contin. Dyn. Syst. **8**, 399–433 (2002)
8. Hecht, F., Pironneau, O., Le Hyaric, A., Ohtsuda, K.: FreeFem++ (Third Edition, Version 3.19). Laboratoire Jacques-Louis Lions, Université Pierre et Marie Curie, Paris. http://www.freefem.org/ff++/
9. Larsson, S., Thomee, V.: Partial Differential Equations with Numerical Methods. Springer, Heidelberg (2003)
10. Leibenzon, L.S.: The motion of a gas in a porous medium. Complete Works, vol. 2, Acad. Sci. URSS, Moscow (1953) (in Russian)
11. Levine, H.A.: Nonexistence of global weak solutions to some properly and improperly posed problems of mathematical physics: the method of unbounded Fourier coefficients. Math. Ann. **329**(2), 205–220 (1975)
12. Marras, M., Vernier-Piro, S., Viglialoro, G.: Estimate from below of blow-up time in a parabolic system with gradient term. Int. J. Pure Appl. Math. **93**(2), 297–306 (2014)
13. Marras, M., Vernier-Piro, S., Viglialoro, G.: Lower bounds for blow-up time in a parabolic problem with a gradient term under various boundary conditions. Kodai Math. J. **3**, 532–543 (2014)
14. Marras, M., Vernier Piro, S., Viglialoro, G.: Lower bounds for blow-up in a parabolic-parabolic Keller-Segel system. Discret. Contin. Dyn. Syst. Suppl. **2015**, 809–916 (2015)
15. Marras, M., Vernier Piro, S., Viglialoro, G.: Blow-up phenomena in chemotaxis systems with a source term. Math. Method Appl. Sci. (2015). doi:http://dx.doi.org/10.1002/mma.3728
16. Muskat, M.: The Flow of Homogeneous Fluids Through Porous Medium. McGraw-Hill, New York (1937)
17. Pattle, R.E.: Diffusion from an instantaneous point source with a concentration-dependent coefficient. Q. J. Mech. Appl. Math. **12**(4), 407–409 (1959)
18. Payne, L.E., Schaefer, P.W.: Lower bound for blow-up time in parabolic problems under Neumann conditions. Appl. Anal. **85**, 1301–1311 (2006)
19. Payne, L.E., Philippin, G.A., Schaefer, P.W.: Blow-up phenomena for some nonlinear parabolic problems. Nonlinear Anal. Theor. **69**(10), 3495–3502 (2008)

20. Quittner, R., Souplet, P.: Superlinear Parabolic Problems. Blow-Up, Global Existence and Steady States. Birkhäuser Advanced Texts. Birkhäuser, Basel (2007)
21. Sacks, P.E.: The initial and boundary value problem for a class of degenerate parabolic equations. Commun. Partial Differ. Equ. **8**(7), 693–733 (1983)
22. Schaefer, P.W.: Lower bounds for blow-up time in some porous medium problems. Proc. Dyn. Syst. Appl. **5**, 442–445 (2008)
23. Souplet, P.: Recent results and open problems on parabolic equations with gradient nonlinearities. Electron. J. Differ. Equ. **2001**, 1–19 (2001)
24. Stinner, Ch., Winkler, M.: Finite time vs. infinite time gradient blow-up in a degenerate diffusion equation. Indiana Univ. Math. J. **57**(5), 2321–2354 (2008)
25. Straughan, B.: Explosive Instabilities in Mechanics. Springer, Berlin (1998)
26. Talenti, G.: Best constant in Sobolev inequality. Ann. Mat. Pura Appl. **110**, 353–372 (1976)
27. Vázquez, J.L.: The Porous Medium Equation: Mathematical Theory. Oxford Mathematical Monographs. Clarendon Press, Oxford (2006)
28. Viglialoro, G.: On the blow-up time of a parabolic system with damping terms. C. R. Acad. Bulg. Sci. **67**(9), 1223–1232 (2014)
29. Viglialoro, G.: Blow-up time of a Keller-Segel-type system with Neumann and Robin boundary conditions. Differ. Integral Equ. **29**(3–4), 359–376 (2016)

Numerical Assessment of the Energy Efficiency of an Open Joint Ventilated Façade for Typical Meteorological Months Data in Southern Spain

Antonio Domínguez-Delgado, Carlos Domínguez-Torres, and José Iñesta-Vaquera

Abstract A numerical evaluation of the energy efficiency of an open joint ventilated façade under climatic conditions operating in Southern Spain is made for typical meteorological month data for each month in the year. Results from CFD computation suggest that the combined effect of the shading of the external wall and the ventilation by the natural convection into the air gap may result in a significative reduction of the heat load during the summer period and a reduction in global energy consumption to get internal comfort in the building when compared with an unventilated one, although the rate of energy savings achieved is relatively sensitive to the combination of environmental conditions. The obtained results seem to indicate that for the whole year, the use of the studied ventilated façade could provide a global energy saving up to 13 % when compared with the use of a standard non-ventilated façade within the orientation and climatic conditions framework considered.

1 Introduction

Over the last years, interest in the development of passive systems for heating and cooling has experienced a remarkable rise because of the need to decrease the energetic costs in the thermal conditioning of buildings.

In hot climates, the main advantage attributed to ventilated façades is the reduction of cooling load for the building climatization. This reduction is achieved by the combination of two factors: ventilation induced by natural convection in the ventilated chamber and protection from solar radiation provided by the external

A. Domínguez-Delgado (✉)
Department of Applied Mathematics 1, University of Sevilla, Avda.Reina Mercedes 2, 41012 Sevilla, Spain
e-mail: domdel@us.es

C. Domínguez-Torres • J. Iñesta-Vaquera
Higher Technical School of Architecture, University of Sevilla, Avda.Reina Mercedes 2, 41012 Sevilla, Spain
e-mail: cardomtor@us.es; joseinesta@hotmail.com

F. Ortegón Gallego et al. (eds.), *Trends in Differential Equations and Applications*, SEMA SIMAI Springer Series 8, DOI 10.1007/978-3-319-32013-7_10

layer of the façade. This way, ventilated façades, if well designed, can significantly reduce the energetic demand for cooling, especially in situations of high solar irradiation.

Previous studies show that the energy efficiency of this type of façade depends strongly on the local weather conditions. Thereby some authors, Ciampi et al. [4] and Patania et al. [13], describe the energetic efficiency decline of these façades when the ambient temperature is high together with a reduction of the benefits of its use in winter due to the penalization that ventilated façades produces in order to take advantage of the solar irradiation. Both situations are found in the climatic context of this study. Hence a careful evaluation of the energy efficiency for this kind of façades when compared with standard non ventilated façades must be carried out in order to determine the suitability of their use in terms of energy efficiency.

In this work we study the thermodynamic behavior of an open joint ventilated façade during actual operating conditions in Southern Spain through the typical meteorological month day data for each month of the year. This way an approximation to the whole year energetic balance is made. This balance is compared to the equivalent one for a standard unventilated façade.

Specifically, we consider the climatic monthly typical values for each month of the year for the city of Seville.

The numerical simulation of the air flow has been performed by using the Navier-Stokes equations for thermodynamic flows and numerical simulations have been carried out with a 2D Finite Element approach by using the *FreeFem++* software [2].

2 The Opaque Open Joint Ventilated Façade

The studied ventilated façade falls into the category of opaque open joint ventilated façades. Basically it has been modeled as a two-dimensional system with a composite inner wall and an outer layer. Thus, between both surfaces an air gap is created.

The inner wall consists of successive layers of gypsum, brick and insulation from inside to outside Fig. 1. The external coating is made of ceramic slabs of dimensions 0.33×0.66 m. Between the slabs there are vertical and horizontal joints of 0.005 m width.

The air gap has a width of 0.04 m and it extends from the floor to the roof of the building, along the entire façade. Communication with the external environment takes place through openings located between the slabs, according to the considered geometry.

Dimensions and physical characteristics of the different layers forming the studied façade are described in Table 1.

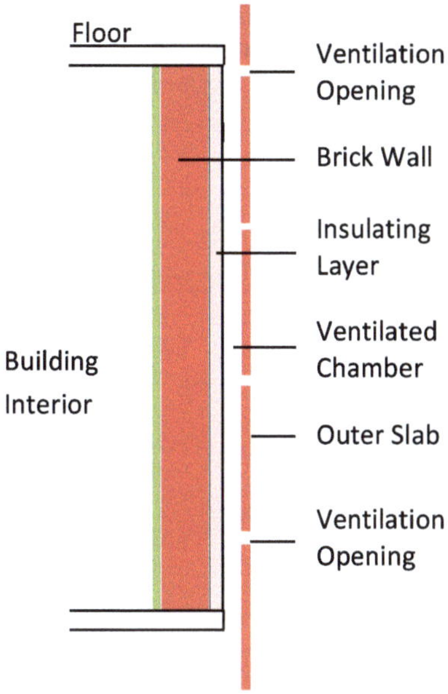

Fig. 1 Schematic section of the open joint ventilated façade

Table 1 Thermophysical characteristics of the ventilated façade

Layer	Description	Thickness	Density	Specific Heat	Conductivity
		(m)	(kg/m^3)	(J/(kgK))	(W/(mK))
1(Ext.)	Ceramic slabs	0.01	2800	1000	3.5
2	Air (ventilation duct)	0.04	1.184	1005	0.0255
3	Insulation	0.03	40	1674	0.029
4	Perforated bricks	0.12	1800	840	0.52
5 (Int.)	Plastering	0.001	1800	831	0.81

3 The Physical Model

In this section it is made a generic description of the physical problem involving the heat transfer and the movement of the air mass in the open joint ventilated façade (OJVF). In the OJFV, the three basic mechanisms of heat transfer are present: radiation, convection and conduction. Specifically, as it is shown in Fig. 2 the heat transfer in the façade is determined by:

- Heat gain on the outer slab due to solar irradiation.
- Heat exchange by radiation between the outer surface and the environment.
- Heat exchange by radiation between the outer surface and the sky.
- Heat exchange by convection between the outer surface and the ambient air.
- Heat transfer by conduction through the outer slab.
- Radiative heat exchange between the surfaces which delimit the ventilated channel.

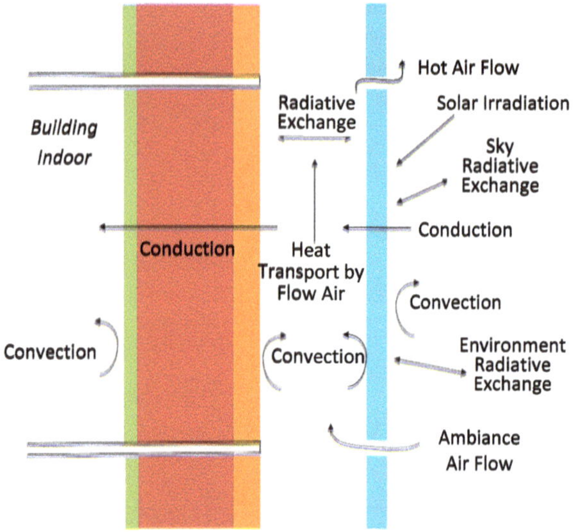

Fig. 2 Heat transfer in the open joint ventilated façade

- Convective heat exchange between the surfaces of the ventilated channel and the air flowing inside it.
- Heat transfer by conduction through the inner wall.
- Heat exchange by convection and radiation between the internal surface of the inner wall and the interior of the building.

The thermal behavior of the ventilated façade can be summarized as follows: at daylight hours the external layer receives direct and diffuse solar radiation plus the radiation of solar origin reflected by the environment, basically by the ground. Part of the solar radiation is absorbed and part is reflected. Moreover, throughout the whole day it takes place a thermal long-wave radiative exchange among the external layer, the ground, the environment and the sky. Simultaneously the external surface of the outer layer exchanges heat by convection with the circulating external air flow whose temperature is determined by the ambient temperature and the heat convective exchange with the surface of the ground in front of the façade. All these contributions result in a heat flow by conduction through the outer wall.

Inside the duct, radiative exchange between the surfaces inside the ventilated channel as well as convective heat transfer between these surfaces an the air flow take place.

The air flow speed through the ventilated chamber is conditioned by the natural convection phenomenon which happens inside the chamber and by the air flow coming in through the openings of the façade, which in turn is influenced by the speed and temperature of exterior air.

Additionally, the velocity of the external air determine the convective heat transfer between the external surface of the outer layer and the ambiance air. This external convective heat transfer is also influenced by the heat transport made by air circulation that is conditioned by the temperatures of the ground surface, the outer layer and the air incoming the domain.

The heat transfer is completed with the conduction through the wall and with the heat transfer that happens on the wall inner surface. This last heat transfer is determined by the convection with the air inside the room and by the thermal radiation with the other walls and objects in the room.

In order to establish the physical model, it is necessary to take into account that air flowing in the interior channel removes or adds heat to the walls of the channel at a rate fundamentally determined by the air flow speed through the chamber and by the difference of temperatures between the channel walls and the air.

Therefore the equations that describe the air flow, the equation for energy transport by the air flow, and the heat transfer equations through the walls, slabs and ground must be solved in every time step. Likewise the radiative exchanges must be computed each time step in order to adequately approximate the heat transfer through the façade.

4 Mathematical Formulation

In this section the governing equations for the air flow and for the thermal conduction through wall, slabs and ground are given.

4.1 Equations that Govern the Fluid

For the air flow we consider the domain Ω which it is described in Sect. 8 and it is showed in Fig. 3. The governing equations for the fluid are the thermodynamic Navier-Stokes equations with a Boussinesq approximation for the buoyancy. These equations can be written as:

- Conservation of momentum:

$$\partial_t \mathbf{U} + \mathbf{U} \cdot \nabla \mathbf{U} - \nabla \cdot (\nu \nabla \mathbf{U}) + \nabla p = \mathbf{b} \qquad \text{in} \quad \Omega \times [0, t_f]. \tag{1}$$

- Continuity:

$$\nabla \cdot \mathbf{U} = 0 \qquad \text{in} \quad \Omega \times [0, t_f]. \tag{2}$$

- Conservation of energy:

$$\partial_t T + \mathbf{U} \cdot \nabla T - \alpha \Delta T = 0 \qquad \text{in} \quad \Omega \times [0, t_f] \tag{3}$$

where the unknowns are $\mathbf{U} = (u, v)$, the velocity for the directions x and y respectively; p is the pressure and T is the temperature of the fluid. ν and α are respectively the cinematic viscosity and the thermal diffusivity of the air. Finally

$$\mathbf{b} = \begin{bmatrix} 0 \\ -g\beta(T - T_a) \end{bmatrix} \tag{4}$$

represents the force of buoyancy due to natural convection, being g the gravitational acceleration and β the coefficient of thermal expansion that can be approximated by $\beta = 1/T_a$ under the hypothesis of ideal gases, where T_a is the ambiance air temperature.

Concerning to the air flow, the condition of non-slip velocity is imposed on all the solid surfaces; in the air inlet to the computational domain the velocity and the temperature of the air is fixed; we take slip condition at the top of the computational domain and finally free outflow. Boundary values for temperature on the building and ground surfaces are given by the energy balance equations described in Sect. 4.3. The initial conditions are fixed from the environmental values of the respective variables.

4.2 Thermal Conduction Through the Wall, Slabs and Ground

Heat conduction through the inner wall is modeled by the equation

$$\frac{\partial T}{\partial t} - \nabla \cdot (\alpha \nabla T) = 0 \tag{5}$$

where the diffusivity coefficient α takes a value corresponding to each material of the various layers of the wall. The same equation is used for thermal conduction through the outer slab where now α is the thermal diffusivity of the slab material.

For the external surfaces of the outer slabs the boundary condition for Eq. (5) is given by the energy balance equation corresponding to each slab external surface as it is explained in Sect. 4.3.

For the slabs and the insulating surface facing the duct, boundary conditions are given by the energy balance equation corresponding to each surface as it is explained in Sect. 4.4.

For the inner layer of the mass wall the boundary condition is imposed with a fixed indoor temperature and a combined convection-radiation heat transfer coefficient of 8 $W/m^2\,K$ taken from several energy building standards.

Thermal conduction through the ground is governed by the same Eq. (5), with a diffusivity coefficient

$$\alpha = 0.5\,10^{-6} m^2/s$$

as it is recommended in [8]. The boundary conditions for the ground are: the monthly average 4 m deep temperature provided by the climatic files from Energy Plus and the energy balance equation at the ground surface as it is explained in Sect. 4.3.

4.3 Energy Balance on the External Surfaces

For the external surfaces of the outer slab and the ground the energy balance is:

$$Q_{c,ext} + Q^{SW} + Q^{LW} - k\frac{\partial T}{\partial n} = 0 \tag{6}$$

where k is the conductivity of the slab or soil and $Q_{c,ext}$ is the convective heat flux between the surface and the air flow. This flux is given by

$$Q_{c,ext} = h_{c,ext}(T_a - T)$$

where T_a is the reference air temperature, T is the surface temperature and $h_{c,ext}$ is the convective heat transfer coefficient described in 4.5. Finally, Q^{SW} and Q^{LW} are respectively the balance of radiative flux of solar origin and the balance of thermal long-wave radiation on the surfaces described in Sects. 4.3.1 and 4.3.2.

4.3.1 Solar Radiative Flux Balance

The radiative flux of solar origin on every exterior surface is given by

$$Q^{SW} = \alpha^s \cdot (I^b + I^d + I^r), \quad (\text{W/m}^2), \tag{7}$$

where α^s is the solar absorptivity of the surface, I^b, I^d are the incident direct and diffuse solar radiation on the surface and I^r is the shortwave radiation of solar origin reflected for the surrounding surfaces and that it is incident on the considered surface.

The calculation of Q^{SW} is explained in Sect. 8.3.1.

4.3.2 Long-Wave Radiative Flux Balance

The long-wave radiative flux balance between the external surfaces, the ambiance and the sky is calculated using the Stefan-Boltzmann's law. For the outer slabs and ground, the long-wave radiation heat flux emitted by every surface is given by

$$E^{LW} = \epsilon\, \sigma\, T^4 \tag{8}$$

where ϵ and T are the emissivity and temperature (given in Kelvin degrees) of the surface and $\sigma = 5.670 \times 10^{-8}$ (W/m^2) is the Stefan-Boltzmann's constant.

In order to evaluate the long-wave radiation exchange with the ambiance, we have considered surrounding objects 50 m away from the ventilated façade with height equal to the façade and a emissivity of 0.85, typical for non-metallic surfaces. For this surrounding we consider a surface temperature equal to the ambiance air temperature.

Furthermore, the relative contribution of the sky downwelling radiation to the long-wave radiative flux on the outer slabs surfaces depends of the fractional part the sky occupies in the field of view of the façade. The estimation of this radiation is described in Sect. 5

Therefore we have four sources of thermal long-wave radiation: the external surfaces of outer slabs, the surface of ground, the surroundings and the sky. Then, the net long wavelength (thermal) radiation flux exchange Q^{LW} in Eq. (6) is the balance for each surface of the emitted long-wave radiation (8) and the absorbed long-wave radiation that hits the considered surface from the other surfaces. The calculation of Q^{LW} is detailed in Sect. 8.3.2.

4.4 Energy Balance on the Duct Surfaces

The energy balance for the surfaces into the duct is:

$$Q_{c,duct} + Q^{LW} - k\frac{\partial T}{\partial n} = 0 \qquad (9)$$

being now k the different materials conductivity of the surfaces facing the duct. $Q_{c,duct}$ is the convective heat transfer of every surface to the air flowing into de duct and Q^{LW} is the long-wave radiative flux balance among the surfaces inside the duct.

Now the surfaces sources of thermal long-wave radiation are the external surface of the insulating layer, the internal faces of the slabs, the floor at the bottom of the duct and the sky at the top. Also the ventilation openings must be taken into account for the global balance of the long-wave radiation into the duct. The radiative exchange Q^{LW} is calculated following the same guidelines developed in Sect. 8.3.2.

To calculate $Q_{c,duct}$ we observe that previous numerical computations [13] show that mean velocity of the flow does not exceed 0.5 m/s into the duct, which is confirmed by our computations. This implies a Reynolds number around 1200 and a laminar behavior of the air flow into the duct. So, the usual Gnielinsky correlation often used to determine the convective heat transfer coefficient in ducts for fully developed turbulent flows is non indicated.

Instead, following Zhai et al. [16] we do a direct calculation of $Q_{c,duct}$. This way we computed the convective heat transfer by

$$Q_{c,duct} = k_{air}(T_{air}(x_8) - T_w)$$

where k_{air} is the air conductivity, T_w is the surface temperature and $T_{air}(x_\delta)$ is the air temperature at a point inside the thermal laminar boundary layer of the flow. The calculation of $Q_{c,duct}$ is explained in Sect. 8.4.

4.5 Convective Heat Transfer Coefficients

Convective heat transfer coefficient $h_{c,ext}$ for external building surfaces is essential in order to calculate heat gains and losses from building façades to the ambiance air. Following the recommendations of Mirsadeghi et al. [12] for low rise buildings, we have considered the value

$$h_{c,ext} = 6.31 V_{loc} + 3.32, \quad (\mathrm{W/m^2\,K})$$

proposed by Liu and Harris [11] for vertical façades. Here V_{loc} is the velocity measured 0.5 m away from the wall surface. For the ground horizontal surface we used the correlations based on Jurges's wind tunnel measurements [9]. The correlation is as follows:

$$h_{e,ext} = 4.1 V_{loc} + 5.8, \quad (\mathrm{W/m^2\,K}).$$

5 Estimation of the Downward Long-Wave Radiation of the Sky

Total sky irradiance onto surfaces on Earth includes the shortwave radiation from the Sun and the thermal long wave radiation from the sky. Solar shortwave radiation takes place only during daylight hours, but thermal downwelling radiation is present throughout the whole day. So, although this radiation is normally named nocturnal radiation, it takes place even during daylight hours. Thereby, building exterior surface temperatures cannot be calculated accurately if the sky long-wave radiation is not considered [10]. In fact, the sky can be used as a heat sink for building radiating surfaces in such a way that if the emitted radiation of a surface exceeds the absorbed radiation, the surface will cool down.

The downward long-wave radiation of the sky, Q_{sky}, is usually approached by using two different concepts: "sky emissivity" or "effective sky temperature".

The effective sky temperature, T_{sky}, is defined as the temperature of the sky when supposing that sky emits long-wave radiation as a blackbody. This way Q_{sky} can be computed as

$$Q_{sky}^{LW} = \sigma T_{sky}^4, \tag{10}$$

where σ is the Stefan-Boltzmann constant.

The other way, sky is assumed to act as a grey body having the temperature equal to the absolute ambiance air temperature T_a and with a global sky emissivity ϵ_{sky}. So, Q_{sky} can be computed as

$$Q_{sky} = \epsilon_{sky}\, \sigma\, T_a^4, \tag{11}$$

Both effective sky temperature and sky emissivity depend on several factors, but the most significant ones are outdoor temperature, relative humidity of the environment and cloud cover.

The effect of cloud cover on downwelling long-wave radiation is complex. Essentially, clouds absorb outgoing IR radiation and emit thermal IR radiation to a temperature higher that emitted by a clear sky. Thus, a cloudy day thermal downwelling sky irradiance can increase over 34 % regarding the sky irradiance of a clear sky.

Walton [15] and Clark et al. [5] estimated that sky emissivity ϵ_{sky} of a cloudy sky can be approximated by

$$\epsilon_{sky} = (.787 + .764 \ln(\frac{T_{dp}}{273}))(1 + \frac{224}{10^4}n - \frac{35}{10^4}n^2 + \frac{28}{10^5}n^3)$$

where T_{dp} is the absolute dewpoint temperature and n is the opaque sky cover in tenths. This model has been used in this work to estimate the downward long-wave radiation of the sky.

6 Studied Façades Features

We compared the energy behavior of the open join ventilated façade (OJVF) with a standard unventilated one (NVF).

The studied OJVF is placed in a building of 5.05 m height with South orientation. The external coating consists of 15 ceramic slabs of dimensions described belove. The Spanish Technical Building Code [6] points out that the ventilated air gap width should be between 30 and 100 mm for ventilated façades. Although some authors [4] point out that the optimum energy efficiency for ventilated façades it is achieved for a camera about 15 cm in width, we stick to the values set by the Technical Code and the studied ventilated façade has a chamber 4 cm in width.

Inside the air gap an aluminium structure support the slabs. This structure is composed of vertical profiles that coincide with the vertical joints blocking their aperture to the external ambiance. So the effective air circulation between external ambiance and the gap is carried out through the horizontal openings.

Dimensions and thermophysical characteristics of the studied façade are listed in Table 1.

For the external surface of the outer layer, an solar radiation absorptivity value equal to 0.3 is considered. The emissivity coefficient of the two surfaces of the external layer has been taken as 0.9 and for the inner wall surface facing the ventilated chamber, the emissivity coefficient has been taken equal to 0.8. The ground in front of the façade is composed of small stones block light colored and an absorptivity value of 0.5 and an emissivity of 0.9 have been considered for it. These values are taken according to the technical specifications for the materials considered [1, 6].

The non ventilated façade has an usual layout consisting of plastering, insulation, perforated bricks with the same characteristics and dimensions than the ventilated façade and a external rough-coated of 15 mm.

7 Climatic Conditions

We have considered a set of environmental conditions that try to reproduce the most yearly relevant features in Southern Spain. These conditions are characterized by high temperatures as well as by relatively high levels of radiation during daytime hours in summer and moderate temperatures in winter. Winds usually are not strong but can range from absolute calm to a relatively moderate wind, which may significantly increase the heat sensation in the hot season. In this work only a wind velocity equal to 1 m/s has been considered for the sake of brevity. Some effects of different wind velocities on the energetic performance of the ventilated façades are showed in [7].

The climatic data used in the computation are been taken from the Energy Plus weather data. For the typical day of every month, hourly values of solar radiation, ambient temperature and downwelling radiation are been used.

8 Numerical Simulation

In this section some aspects related with the numerical simulation are briefly described.

8.1 Computational Domain and Meshing

For the numerical solution of the set of equations describing the thermodynamic air flow, we have started from a two-dimensional computational domain Ω that includes both the ventilated façade and a wide region outside it, in order to adequately simulate air flow in the front and top of the building containing the façade. The incoming region in the front of the studied façade is $10\,H$ wide, and the height of

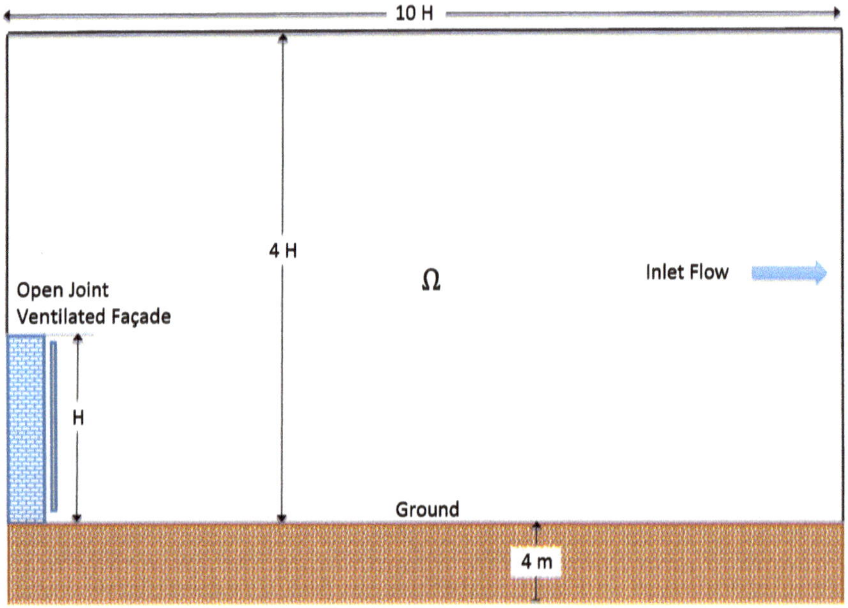

Fig. 3 Non-scaled sketch of the computational domain

the considered computational domain is $4H$, where H is the height of the ventilated façade. A non-scaled sketch of the whole computational domain is showed in Fig. 3.

The domain Ω has been meshed by using triangular elements to perform a discretization of the problem by the Finite Element Method (FEM). Different densities of the grid for Ω have been used since the areas inside the ventilated channel and around the ventilation openings require much more precision than the incoming region in front of the building. The minimum size mesh into the higher precision areas is taken equal to 0.001 m.

The mesh for the wall is conditioned for the thickness of the different layers making up the wall. For the wall the minimum size mesh is taken for the plastering layer where four knots are placed in the x direction. For the outer ceramic slabs we have used ten knots in the x direction, enough to achieve a good precision for the heat conduction equation. The mesh for the ground offers no special difficulty.

8.2 *Numerical Resolution*

The numerical resolution of Eqs. (1)–(3) is made by using mixed a $\mathbb{P}2 - \mathbb{P}1$ Finite Element approximation for the velocity and pressure and a $\mathbb{P}1$ approximation for the temperature.

For time discretization first the partial differential equations are semi-discretized in time. Total derivatives are discretized thanks to the method of characteristics $X^n(x) \approx x - \mathbf{U}^n(x)\,\Delta t$. The nonlinear term is discretized using a semi-implicit formula whereas the other linear terms are discretized implicitly:

$$\frac{(T^{n+1}(x) - T^n \circ X^n(x))}{\Delta t} - \nabla \cdot (\alpha\, \nabla T^{n+1}(x)) = 0 \quad \text{in} \quad \Omega$$

$$\frac{(\mathbf{U}^{n+1}(x) - \mathbf{U}^n \circ X^n(x))}{\Delta t} - \nabla \cdot (\nu\, \nabla \mathbf{U}^{n+1}(x)) + \nabla p^{n+1}(x) = \mathbf{b}^{n+1}(x) \quad \text{in} \quad \Omega$$

$$\nabla \cdot \mathbf{U}^{n+1} = 0 \quad \text{in} \quad \Omega$$

where $\mathbf{b}^{n+1}(x) = \begin{bmatrix} 0 \\ -g\beta(T^{n+1}(x) - T_a) \end{bmatrix}$.

The Eq. (5) for heat transfer through the inner wall, the outer layer and ground have been solved by a $\mathbb{P}1$ Finite Element approximation. The time discretization for these equations are made by using an implicit Euler finite difference scheme:

$$\frac{(T^{n+1}(x) - T^n(x))}{\Delta t} - \nabla \cdot (\alpha\, \nabla T^{n+1}(x)) = 0$$

where now α is the diffusivity of each material.

For the temperature on the solid surfaces, the borders conditions are chosen from the energy balance equations as it is described in Sects. 8.3 and 8.4.

In each time step the radiative balance on the surfaces is calculated and then the balance energy on the solid surfaces is used to get the border conditions for the temperature.

The *FreeFem++* software [2] has been used for the computing implementation of the considered discretizations.

8.3 External Surfaces Energy Balance Calculation

In this section the calculation of solar and long-wave radiative balances on every external surface is described.

8.3.1 Solar Radiative Flux Balance Calculation

Let $i = 1 \ldots N$ be the index of the triangle faces on the exterior surfaces. The balance of the radiative flux of solar origin for every face i is

$$Q_i^{SW} = \alpha_i^s\,(I_i^b + I_i^d + \sum_{j=1}^{N} J_j^{SW} F_{i,j}), \quad (\mathrm{W/m^2}), \tag{12}$$

where I_i^b, I_i^d are the incident direct and diffuse solar radiation on the face i. α_i^s and J_i^{SW} are respectively the solar absorptivity and the solar radiosity of the face i and finally $F_{i,j}$ is the view factor based on i, between faces i and j [3].

The values of the radiosity are calculated by solving the system

$$\mathbf{A}^{\mathbf{SW}}\,\mathbf{J}^{\mathbf{SW}} = \mathbf{E}^{\mathbf{SW}}, \tag{13}$$

with

$$\mathbf{A}^{\mathbf{SW}} = \begin{pmatrix} 1-\rho_1^s F_{1,1} & -\rho_1^s F_{1,2} & \cdots & -\rho_1^s F_{1,N} \\ -\rho_2^s F_{2,1} & 1-\rho_2^s F_{2,2} & \cdots & -\rho_2^s F_{2,N} \\ \cdots & \cdots & \cdots & \cdots \\ -\rho_N^s F_{N,1} & -\rho_N^s F_{N,2} & \ldots & 1-\rho_N^s F_{N,N} \end{pmatrix}$$

$$\mathbf{J}^{\mathbf{SW}} = \begin{pmatrix} J_1^{SW} \\ J_2^{SW} \\ \cdots \\ J_N^{SW} \end{pmatrix}, \quad \mathbf{E}^{\mathbf{SW}} = \begin{pmatrix} \rho_1^s(I_1^b + I_1^d) \\ \rho_2^s(I_2^b + I_2^d) \\ \cdots \\ \rho_n^s(I_N^b + I_N^d) \end{pmatrix}$$

where ρ_i^s is the solar reflectance of face i.

8.3.2 Long-Wave Radiative Flux Balance Calculation

We consider now a total of $N+2$ surfaces. The first N surfaces are the above described faces of the outer slabs and ground. To these faces, we add the $N+1$ face corresponding to the surrounding as described in Sect. 4.3.2 and the $N+2$ corresponding to the sky.

The long-wave radiation heat flux emitted by each face i for $i = 1 \ldots N+2$, is given by

$$E_i^{LW} = \epsilon_i \sigma T_i^4 \tag{14}$$

where ϵ_i and T_i are the emissivity and temperature (given in Kelvin degrees) of the face i.

This way, the balance for long-wave radiation flux on face i, for $i = 1, \ldots, N+2$ is given by

$$Q_i^{LW} = \sum_{j=1}^{N+2} J_j^{LW} F_{i,j} - J_i^{LW}, \quad (\mathrm{W/m^2}), \tag{15}$$

where J_i^{LW} is the long-wave radiosity of the face i and finally $F_{i,j}$ is the view factor.

The values of the long-wave radiosity J_i^{LW} are calculated by solving the system

$$\mathbf{A}^{\mathbf{LW}}\,\mathbf{J}^{\mathbf{LW}} = \mathbf{E}^{\mathbf{LW}}, \tag{16}$$

with

$$\mathbf{A}^{\mathbf{LW}} = \begin{pmatrix} 1-\rho_1 F_{1,1} & -\rho_1 F_{1,2} & \cdots & -\rho_1 F_{1,N+2} \\ -\rho_2 F_{2,1} & 1-\rho_2 F_{2,2} & \cdots & -\rho_2 F_{2,N+2} \\ \vdots & \vdots & \cdots & \vdots \\ -\rho_{N+2} F_{N+2,1} & -\rho_{N+2} F_{N+2,2} & \cdots & 1-\rho_{N+2} F_{N+2,N+2} \end{pmatrix}$$

$$\mathbf{J}^{\mathbf{LW}} = \begin{pmatrix} J_1^{LW} \\ J_2^{LW} \\ \cdots \\ J_{N+2}^{LW} \end{pmatrix}, \quad \mathbf{E}^{\mathbf{LW}} = \begin{pmatrix} E_1^{LW} \\ E_2^{LW} \\ \cdots \\ E_{N+2}^{LW} \end{pmatrix}.$$

Where ρ_i is the long-wave reflectance of the face i and E_i^{LW} is computed by using (14).

For the sky the values of ρ_{N+2}, E_{N+2}^{LW} and T_{N+2} are described in Sect. 5.

8.4 Duct Surfaces Energy Balance Calculation

To implement (9), the radiative exchange Q^{LW} is calculated following the same guidelines showed in Sect. 8.3.2. To calculate the convective heat transfer $Q_{c,duct}$ we use [16]:

$$Q_{c,duct} = k_{air}(T_{air}(x_p) - T_w)$$

where $T_{air}(x_p)$ is the nearest grid knot x_p to the surface.

In order to have a good approximation of the heat transfer between the surfaces of the duct and the air flow, the grid must have some knots inside the thermal laminar boundary layer of the flow. For that, it is enough that the distance dx_p from x_p to the surface verifies

$$dx_p < \delta \tag{17}$$

where δ is the thickness of thermal boundary layer for natural convection [14]. For the geometry considered and the mean velocity found, we used a size mesh of 0.001 m in the x direction in the duct, thereby (17) is verified.

9 Results

The results from the numerical simulations show the expected qualitative behavior for ventilated an unventilated façades. Thus, higher levels of solar radiation and outside temperature produce higher heat flux into the building for both façades but less for ventilated façade than for the unventilated one. Nevertheless, the opposite behavior is observed in the cold season. It is observed that lower levels of solar radiation and outside temperature produce higher heat flux out the building for both façades, but less for the OJVF than for the NVF.

Other significant facts that numerical simulations highlight is the key role that the reflected solar radiation from the ground and the downwelling radiation from the sky play in the energetic behavior of the ventilated façade.

Computations show that for every month the higher slabs reach lower temperatures than slabs located at the bottom. We have considered two floors in the building to estimate the influence of this fact in the heat transfer to both floors. In the following figures we can observe the differences between the heat flux into each floor. In Figs. 4, 5, 6, 7 and 8, the hourly flux into the two floors of the building are showed for typical months of Winter, Spring, Summer and Autumn.

In these figures it can be observed the most important factor in the heat transfer through the façade is the ambient temperature combined with the radiative influence.

Another important fact observed is the time lag between the maximum of the heat flux into the building and the maximum of the ambient temperature and solar irradiation. This lag can be exploited to achieve indoor comfort by using some passive techniques.

In Fig. 9 the monthly flux into the building is showed. For all months in the cold season the heat transfer outward through the OPVJ is lower than through the NVF.

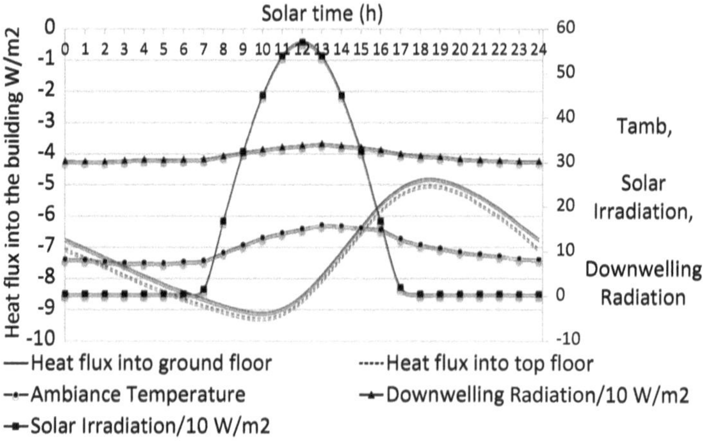

Fig. 4 Heat flux through the ventilated façade in January

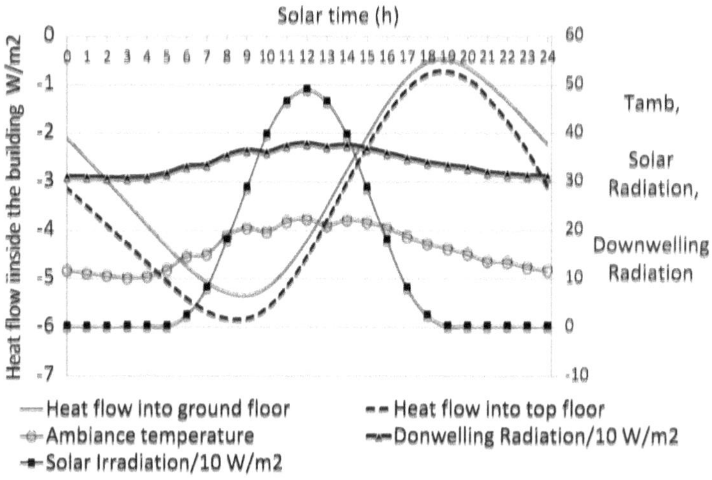

Fig. 5 Heat flux through the ventilated façade in April

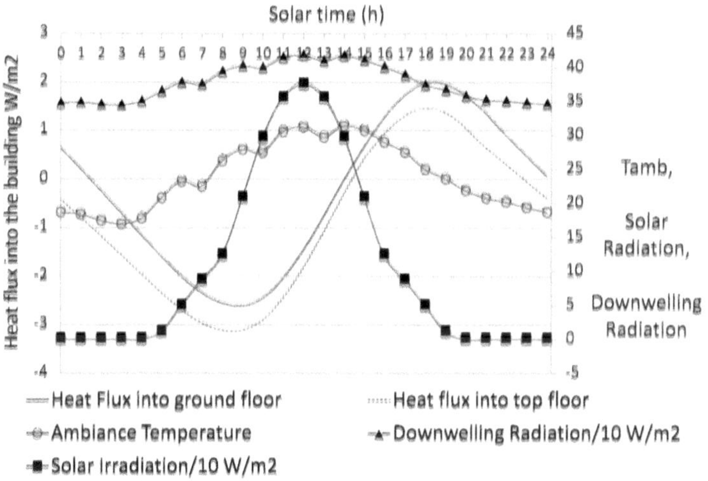

Fig. 6 Heat flux through the ventilated façade in June

Also it is shown in this figure that heat gain for all months in the hot season is lower for the OJVF than for the NVF.

Finally, Fig. 10 displays the yearly global energy behavior for both façades. In the heating season the OJVF reduces the heat loss by approximately 5 % when compared to the NVF. In the cooling season the reduction of heat gain is about 34 %. This implies a yearly saving reduction around 13 % for the OJVF when compared to the standard NVF.

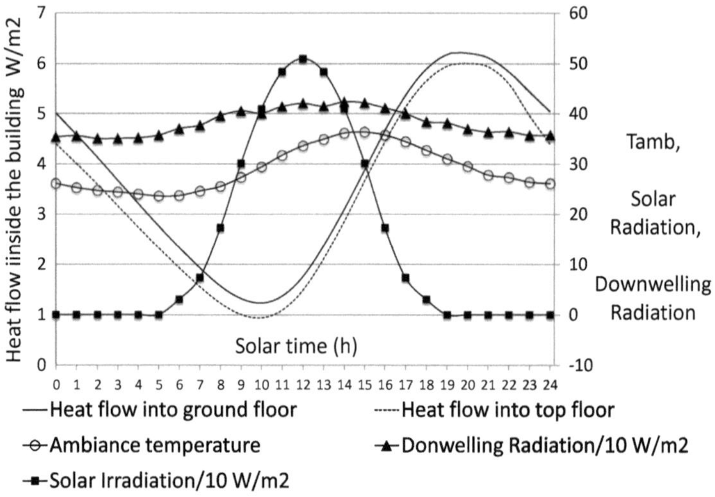

Fig. 7 Heat flux through the ventilated façade in August

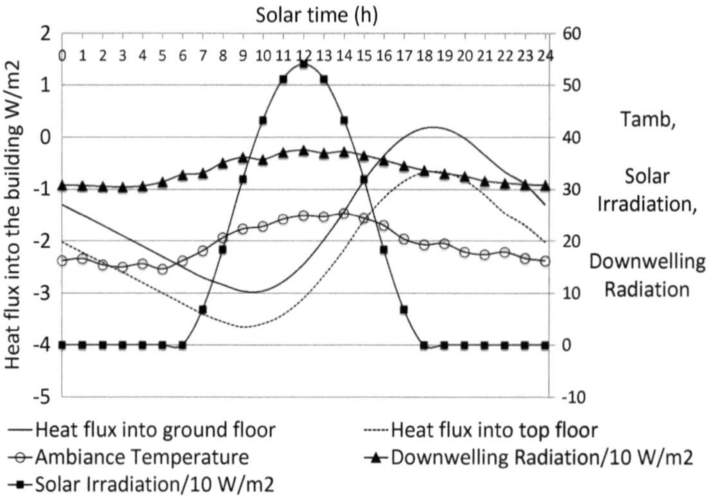

Fig. 8 Heat flux through the ventilated façade in October

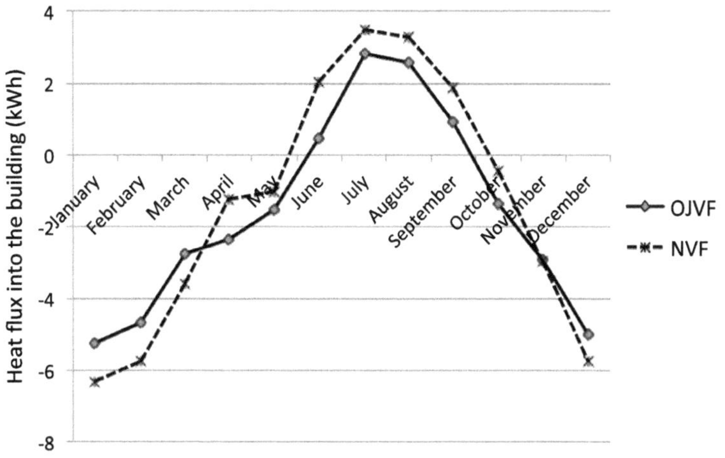

Fig. 9 Monthly heat transfer into the building (kWh) for OJVF and NVF

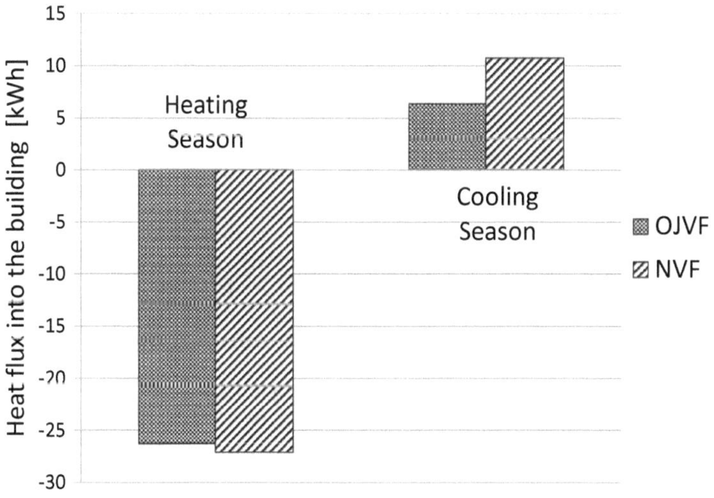

Fig. 10 Whole year energy balance for OJVF and NVF in cooling and heating season

10 Conclusions

A numerical code for simulating the energetic performance of a open joint ventilated façade has been developed. The code has been used to draw conclusions about the thermodynamic behavior of a specific ventilated façade, comparing its efficiency with another non-ventilated façade for typical meteorological month data for each month in the year.

The results seem to indicate that the studied ventilated façade has a better behavior in terms of energetic efficiency compared to non-ventilated façade. The considered ventilated façade could provide energy savings rate of around 13 % for the whole year. In summer this saving rate could reach 34 %.

We conclude that under climatic conditions in southern Spain, the use of the studied ventilated façade could allow a major reduction in the heat load of the building in relation to a non-ventilated façade with the same construction features and could provide a significant reduction in the yearly energetic consumption.

In future research it would be interesting to do a sensitivity analysis and devote a further study to the parameters and closure terms that add an important degree of uncertainty to the analyzed problem.

Acknowledgements The research of Carlos Domínguez Torres was supported by Ministerio de Educación, Ciencia y Deporte, under a Grant of Collaboration with the Departament of Applied Mathematics I of the University of Sevilla, Spain.

References

1. Ashrae Handbook: Heating, Ventilating and Air-Conditioning Applications Ed. Amer Society of Heating (1999)
2. Auliac, S., Le Hyaric, A., Morice, J., Hecht, F., Ohtsuka, K., Pironneau, O.: FreeFem++. Third Edition. Version 3.31–2 (2014). http://www.freefem.org/ff++/ftp/freefem++doc.pdf
3. Bejan, A.: Heat Transfer. Wiley, New York (1992)
4. Ciampi, M., Leccese, F., Tuoni,G.: Ventilated façades energy performance in summer cooling of buildings. Sol. Energy **75**, 491–502 (2003)
5. Clark, G., Allen, C.: The estimation of atmospheric radiation for clear and cloudy skies. In: Proc. of 2nd National Passive Solar Conference (AS/ISES), pp. 675–678 (1978)
6. Código Técnico de la Edificación, Rd 314/2006 de 17 de Marzo. BOE núm. 74, Spain (2006)
7. Domínguez Delgado, A., Durand Neyra, P., Domínguez Torres, C.A.: Estudio del enfriamiento pasivo por fachadas ventiladas en el sur de España. Actas del I Congreso Internacional de Construcción Sostenible y Soluciones Ecoeficientes, Sevilla (2013)
8. EnergyPlus Engineering Reference. U.S. Departament of Energy (2014)
9. Jurges, W.: Heat transfer at a plane wall. Gesund. Ing. **19** (1924)
10. Kehrer, M., Schmidt, T.: Radiation effect on exterior surfaces design. In: Proc. of 8th Nordic Symposium on Building Physics, Denmark (2008)
11. Liu, Y., Harris, D.J.: Full-scale measurements of convective coefficient on external surface of a low-rise building in sheltered conditions. Build. Environ. **42**, 2718–2736 (2007)
12. Mirsadeghi, M., Costola, D., Blocken, B., Hensen, J.L.M.: Review of external convective heat transfer coefficient models in building energy simulation programs: implementation and uncertainty Appl. Therm. Eng. **56**, 134–151 (2013)
13. Patania, F., Gagliano, A., Nocera, F., Ferlito, A., Galesi, A.: Thermofluid-dynamic analysis of ventilated façades. Energy Build. **42**, 1148–1155 (2010)
14. Schlichting, H., Gersten, K.: Boundary-Layer Theory. Springer, Berlin (2000)
15. Walton, G.N.: Thermal Analysis Research Program Reference Manual. NBSSIR 83-2655. National Bureau of Standards, p. 21 (1983)
16. Zhai, Z., Chen, Q.: Numerical determination and treatment of convective heat transfer coefficient in the coupled building energy and CFD simulation. Build. Environ. **36**, 1000–1009 (2004)

Planning Ecotourism Routes in Nature Parks

Eva Barrena, Gilbert Laporte, Francisco A. Ortega, and Miguel A. Pozo

Abstract The main objective of the Nature Parks is to preserve the diversity and integrity of biotic communities for present and future use. Additionally, the Nature Parks can contribute to the invigoration of the sustainable development and culture heritage of its neighboring regions, as well as to the strengthening of the environmental education for visitors by means of direct experiences. From this double point of view, ecotourism is gaining acceptance as a tool for sustainable development since the income of visitors to protected areas can contribute significantly to support the economy of these areas and of the rural communities. This article proposes different methodologies for determining efficient routes of ecotourism where the main objective is the maximization of the cultural transmission experienced by the visitors along the path traveled. The models are formulated by using integer linear programming and its potential applicability is illustrated in the context of the Doñana National Park, Spain.

E. Barrena
Department of Statistics and Operations Research, University of Granada, C/ Santander, 1. 52071 Melilla, Spain
e-mail: ebarrena@ugr.es

G. Laporte
Interuniversity Research Center on Network Enterprise, Logistics and Transportation (CIRRELT), HEC Montreal, 3000 Chemin de la Côte-Sainte-Catherine, Montréal, Canada H3T 2A7
e-mail: gilbert.laporte@cirrelt.ca

F.A. Ortega (✉)
Department of Applied Mathematics I, University of Seville, Higher Technical School of Architecture, Avda. Reina Mercedes 2, 41012 Seville, Spain
e-mail: riejos@us.es

M.A. Pozo
Faculty of Mathematics, Department of Statistics and Operational Research, University of Seville, C/ Tarfia s/n, 41012 Seville, Spain
e-mail: miguelpozo@us.es

F. Ortegón Gallego et al. (eds.), *Trends in Differential Equations and Applications*, SEMA SIMAI Springer Series 8, DOI 10.1007/978-3-319-32013-7_11

1 Introduction

In recent years there has been a major growth of tourist itineraries in every area of the planet. Several classical tourism proposals centered on the local visit of a specific destination have evolved towards more dynamic formulas where interactive journeys, that highlight certain remarkable aspects along routes, are suggested to the traveler [7].

In the design and implementation of these routes one usually offers the clients a pathway where a specific category of heritage predominates: cultural, archaeological, historical, artistic or natural. Additionally, in order to differentiate it from other proposals in the same segment, new attractions (like natural monuments, literary heroes, movie studios, architectural paradigms, etc.) are incorporated along the route. Reference [17] lists several thematic routes in Spain, which have been consolidated in the tourism market because of their interest from different perspectives: gourmet experiences, wine, literature, film, history, geology, etc. Additionally, that document details how the introduction of legends, myths and fictional characters are able to improve the marketing of the tourism products.

The recognition of the scenic value of routes emerged in U.S.A. during the 1920s. Following approval of the Act Scenic Byways in the U.S.A. Congress in 1989, forty-six U.S.A. states conducted initiatives of landscape protection which resulted in the creation of the National Scenic Byways Program [18]. Basically, a route must satisfy two requirements to obtain the recognition needed to become a member of the National Scenic Byways Program (NSB): it must contain at least one of six intrinsic qualities (scenic, recreational, natural, cultural, historical or archaeological) and its territorial extension must cover more than one state. The All-American Roads designation is even more elitist, since it is assumed that the route contains attractions of all these qualities with a degree of uniqueness sufficient to constitute by itself a tourist destination [8]. In November 2010, there were 120 members of NSB registered in the U.S.A. and 31 routes cataloged of All-American Roads that satisfied the above requirements.

Outside North America we also find a large number of tourist itineraries, national or international, which base their appeal on the scenic beauty of a route, additionally supplemented with cultural and anthropological aspects (traditional architecture, folklore, marketing of natural products, handicrafts, etc.). Although the list is extensive, let us cite as examples the Camino de Santiago in Europe [20], the Turquoise Trail in New Mexico [25] and the Red Interlagos in Chile [13].

Following this trend, in recent years we have witnessed around the world a progressive increase in the number of visitors to protected areas, which is a good indicator of the interest for such locations as places of leisure. The combination of rural and natural attractions is the basis of a large number of recreational and tourist activities which, if they are planned and rigorously managed, can generate economic, social and cultural benefits for the local population. The nature tourism is a strategy that has emerged in these protected areas and their surroundings, with the dual aim of supporting the conservation of nature and of creating income opportunities for those communities who live in rural areas [32].

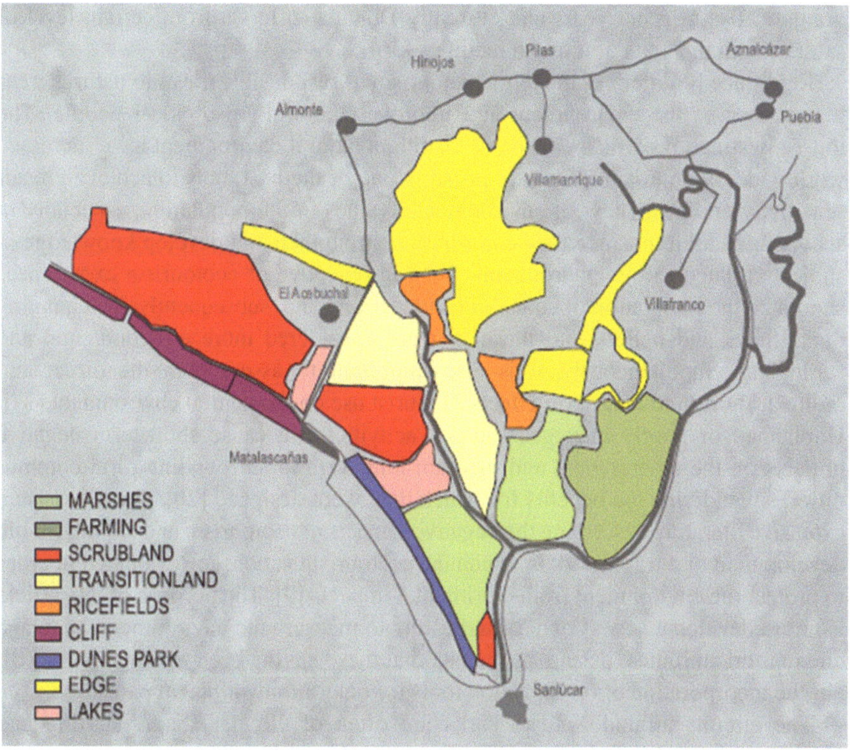

Fig. 1 Doñana region and these different ecosystem types

In the last years we have witnessed in Spain a progressive increase in the number of visitors to protected areas, which is a good indicator of the general interest in natural areas as a place of leisure. At the same time, this trend represents an opportunity for developing an active environmental education along nature trails designed for such a purpose. The Doñana region, the area studied in this research, is located near the mouth of the Guadalquivir River, in the provinces of Huelva, Seville and Cadiz (Fig. 1). It comprises a large area of marsh with lagoons and streams, dunes, pine forests, shrubs and grasses and a rich and varied agriculture.

Since 1978, about 50,000 hectares are protected under the denomination of National Park. Doñana was declared a Biosphere Reserve in 1981 and a World Heritage Site in 1994. Doñana is also part of the list of Wetlands of International Importance (Ramsar Convention), since this area represents a strategic scale in the migratory bird transit between Europe and Africa [1]. Tourism inside and around the Doñana National Park plays an important role in spreading the image of the region internationally and in the economic development of the area. Doñana receives over 400.000 visitors each year. Essentially, the main purpose of these visits is to provide interpretation services in environmental and sustainability education and in other recreational activities managed by the park administration, all by service concessions. The number of visitor centers and information points has increased dramatically in recent years and currently Doñana suffers a from certain level of saturation in this type of tourism facilities [36].

Reference [24] defines the term ecotourism as responsible travel to natural areas that conserves the environment and improves the welfare of local people. The tourist destinations which consist of visiting a natural environment have increased worldwide. Ecotourism routes are possibly one of the best tools to achieve greater sensitivity and awareness of environmental values in the population, especially in school age children, since they can serve to strengthen and develop knowledge in subjects related to the environment. A basic objective of ecotourism experiences should be the facilitation of education and learning and, subsequently, the changing of attitudes and beliefs into those that are considered more environmental and ecological. Simultaneously, it is also important to assess both the direct and indirect, short- and long-term effects of tourist use on the natural environment [27]. Unplanned or poorly managed tourism activities can cause short-term negative impacts on the environment and medium-long term in surrounding rural communities, squandering the benefits for which they were designed [26]. There exists an extensive literature describing the negative impacts of ecotourism and calling for the development of a framework in which the ecotourism actions are evaluated in order to protect the environment from detrimental impacts [9]. The purpose of this article is not to develop a new system of indicators to measure the environmental damage due to tourism routes (despite of the fact that these having been reasonably traced), but the incorporation of optimization tools for determining efficient paths.

The circuits around Natural Parks are often of circular shape, starting and finishing in a center for the environmental interpretation [10, 19]. The length of an itinerary should not exceed a certain threshold so that its realization does not constitute a major effort for people and that the ecosystem does not become affected. The circular routes must flow through different sites since this facilitates the acquisition of new knowledge. The route design should also incorporate a high level of biodiversity. The optimal number of stops for including comments to the visitors is usually close to five, and the duration of each stop should not exceed 10–15 minutes in order to maintain the visitors' attention. Travel time may be increased when people move on foot (hiking in silence) to reach an additional destination that cannot be accessed using a motorized vehicle.

The optimal location of a cycle in a graph is prevalent in areas such as transport and telecommunications and the basis for formulating this problem is commonly inspired in the well-known Travelling Salesman Problem (TSP). Reference [22] classifies the problems of locating cycles in two main categories: problems of Hamiltonian tours, where all the nodes of the graph must necessarily be visited, and non-Hamiltonian problems, where only a subset of nodes must be visited along the cycle.

Arcs Routing Problems (ARP) consist of finding the least-cost route through some edges or arcs of a graph, being subject to certain restrictions [11]. The Orienteering Problem (OP) [35] is the problem of finding a tour maximizing the collected profit and such that the travel cost does not exceed a given value. The Tourist Trip Design Problem (TTPD) can be viewed as a variant of the orienteering problem on a directed graph in which a path must be determined to maximize the utility value derived from the selection of places to be visited without violating budget constraint [33]. This model has already been successfully applied in the field tourism in order to calculate personalised walking routes in historic cities [29, 30, 34] and bicycle itineraries [31].

The Orienteering Tour Problem (OTP) is a specific version where the start and end nodes of the tour are identical [28]. Following this idea of imposing a circular shape in the solutions, the Bus Touring Problem (BTP), introduced by [12] consists of determining the optimal subset of tourist sites to be visited and scenic routes to be traversed between a start point and an end point that coincide, such that the total attractiveness of the tour can be maximized for a given constraints on the total touring time, cost or total distance traveled. The sites and road segments of the geographical region (i.e., vertices and arcs of the graph) are weighted with a non-negative value of attractiveness that denotes the amount of enjoyment derived from visiting a tourist site or traversing a scenic road segment, respectively.

The OTP and TSP models adapted for making decisions based on the selection of heterogeneous arcs, will provide a methodological support for the construction of solutions to the problem of designing ecotourism routes, particularly in environmentally protected areas like National Park of Doñana (Huelva, Spain).

The remainder of this paper is organized as follows. Section 2 describes models that can be applied to determine efficient routes under different objectives and constraints. In Sect. 3 these models are redefined in order to fit the specific context of a Natural Park. Conclusions follow in Sect. 4.

2 Selecting Adequate Models for Our Proposal

As mentioned above, the main objective of this research is to develop a methodology in order to generate optimal routes for visiting natural areas with a high degree of protection, assuming that the routes must be effective from the viewpoint of visibility and for the transmission of information in relation to the existing ecosystems and their diversity. Since the perception of landscape is formed through serial images, the structure of feasible solutions should be treated abstractly as closed paths (cycles) through the set of edges of a graph and, additionally, the optimality condition will be attained by the maximum value of a linear function that accumulates the utility perceived by the observer along the whole route, section by section.

There are not many articles in which the assessment the designed route is carried out on the basis of accumulating the benefit derived of traveling along their arcs, instead of visiting their nodes. In fact, some relevant contributions have already been previously cited:

- References [2–4, 14] propose models that maximize the total utility of the built routes, without violating the vehicle capacity or the time limit.
- Reference [12] determines a transport route attractive in a tourist area.

- Reference [31] provides an heuristic model to optimize the planning of cycling in East Flanders.
- Reference [29] obtains personalised itineraries for each tourist that include a series of activities to carry out sorted in time. A practical application of the developed Tourist Support System is tested in the Autonomous Region of Andalusia.

Assume initially that the territory under analysis is partitioned into polygonal areas in which the prevalence of a particular ecosystem type (monochrome area) has been identified. Subsequently, we can conclude that the underlying space to be investigated is a union of monochromatic polygons (Fig. 2).

If terrain elevation reduces visibility inside a polygon, it could be partitioned into several subpolygons (or triangular cells) in order to guarantee full visibility from the each edge towards the interior of its corresponding cells (Fig. 3). Another reason for splitting the territory into triangular cells is the existence of interesting sites that would deserve to be considered as potential destinations along the itinerary. Such sites must be incorporated as vertices in the final triangulation. Subsequently, we assume that the territory under analysis is already divided into monochromatic triangular areas. Edges that determine the territorial fragmentation can be associated with a single ecosystem or with two ecosystems whenever they are a boundary between two habitats.

Recall that the number of significant features required for a road to be a member of the NSR program is six (specifically: archaeological, cultural, historic, natural, recreational, and scenic qualities). This same level of variety exists in Doñana in terms of territorial biodiversity, because there are six identified habitats: dunes, beach, bank, forest, bushes, and marsh. If the observation made by the visitor of the different biotopes was exclusively concentrated in the nodes of the graph, the design objective of the route could be defined as that of determining a minimum cost tour containing at least one vertex from each cluster (ecosystem). Following this idea, a generalized TSP (GTSP) model turns out to be useful for designing cyclical paths where one instance at least of the existing ecosystems must be collected along the journey.

2.1 *Formulating the GTSP*

The GTSP [21] is a variation of the TSP in which the set of nodes is partitioned into clusters and the objective is to find a minimum cost Hamiltonian cycle passing through at least one node from each cluster. The GTSP and its variants arise in real-life applications such as computer operations, manufacturing logistics, distribution of goods by sea to the potential harbors [23]. An integer linear programming model to formulate the GTSP follows.

Fig. 2 A division of Doñana territory according to the different ecosystem types

Let $G = (V, E)$ be a graph, where $V = \{v_1, v_2, \ldots, v_n\}$ is the set of nodes and $E = \{e = (v_i, v_j) : v_i, v_j \in V, i < j\}$ is the set of feasible edges interconnecting pairs of nodes. The set V is partitioned into m disjoint clusters $V = V_1 \cup V_2 \cup \cdots \cup V_m$. A nonnegative cost $c_e \geq 0$ is associated with each edge $e \in E$ and it is assumed that

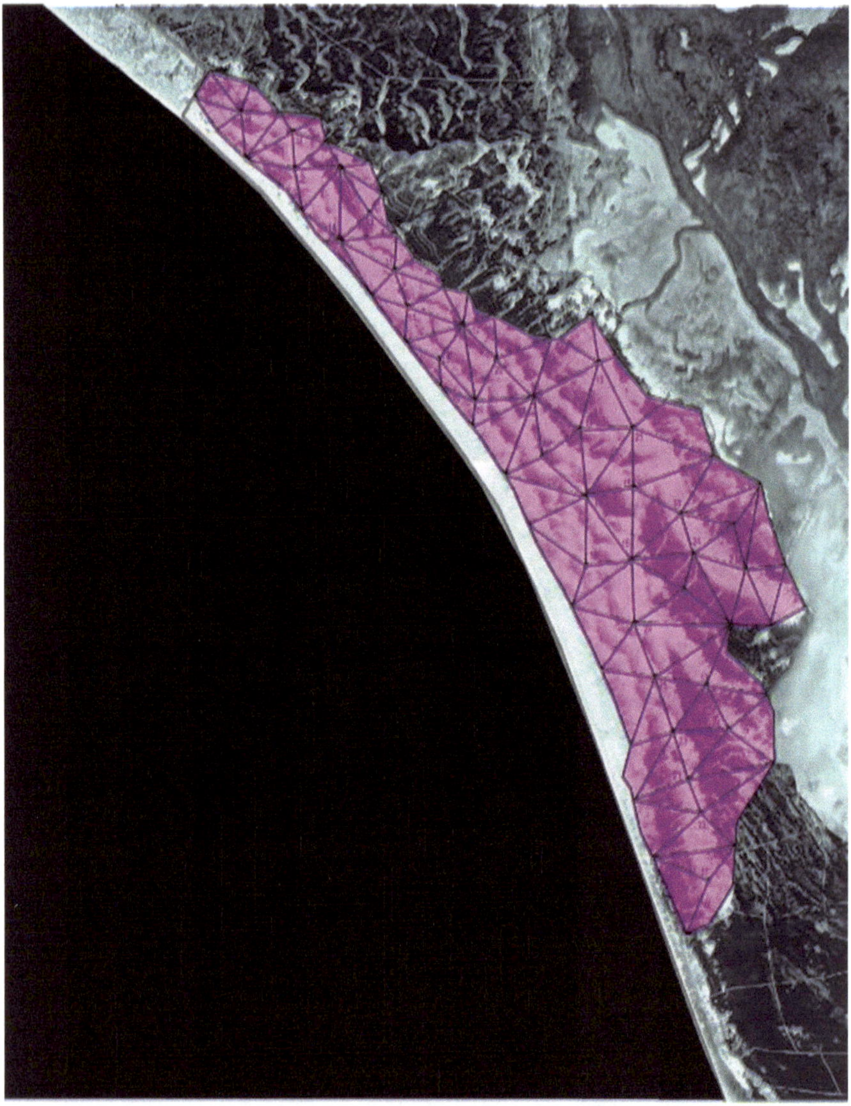

Fig. 3 A triangular fragmentation of territory hosting the dune ecosystem

the triangle inequality is satisfied. We denote by $\delta(i) = \{(v_i, v_j) : v_j \in V,\ i < j\}$ the set of edges incident with node v_i. Two families of variables are used in the model:

x_e: a binary variable equal to 1 if and only if edge e is in the solution path. We denote by $x(\delta(i))$ the sum of values x_e for the edges $e \in \delta(i)$.
y_i: a binary variable equal to 1 if and only if node v_i is in the solution.

The problem is then

$$\text{minimize} \sum_{e \in E} c_e x_e \tag{1}$$

subject to

$$x(\delta(i)) = 2\, y_i \quad (v_i \in V) \tag{2}$$

$$x(\delta(S)) \geq 2(y_i + y_j - 1) \\ (S \subset V,\ 2 \leq |\, S \,| \leq n-2,\ v_i \in S,\ v_j \in V \setminus S) \tag{3}$$

$$\sum_{v_i \in V_h} y_i \geq 1 \quad (h = 1, \ldots, m) \tag{4}$$

$$x_e \in \{0, 1\} \quad (e \in E) \tag{5}$$

$$y_i \in \{0, 1\} \quad (v_i \in V). \tag{6}$$

Objective (1) minimizes the cost of the cycle. Constraints (2) establish the requirement that each node in the solution has two incident edges, and no incident edge otherwise. Inequalities (3) are connectivity constraints which specify that the solution is connected. Inequalities (4) force every vertex subset V_h to be visited at least once. Constraints (5) and (6) impose conditions on the variables.

The cycle in Fig. 4 is of minimum cost and contains at least one vertex from each cluster (different nodes have been assigned to the same cluster if they are instances that correspond to the same ecosystem).

If the utility perceived by the user is not due to the visit to a point but to the travel along a path, the optimization model is clearly different because the objective must now be defined to maximize the profit captured by the traveler along the cycle, which must necessarily be restricted in terms of total length (cost, travel time, time of presence in the park).

2.2 Formulating the GOTP

Assume that the set of edges E can be partitioned in m different clusters: $E = E_1 \cup E_2 \cup \cdots \cup E_m$. The condition of forcing the cycle to contain at least one edge belonging to each of the groups in question can be achieved by incorporating the following restrictions:

$$\textstyle\sum_{e \in E_h} x_e \geq 1 \ (h = 1, \ldots, m). \tag{7}$$

This ensures that the model generates solutions with group diversity. Consistent with the nomenclature coined in the literature, we will refer to this model as the

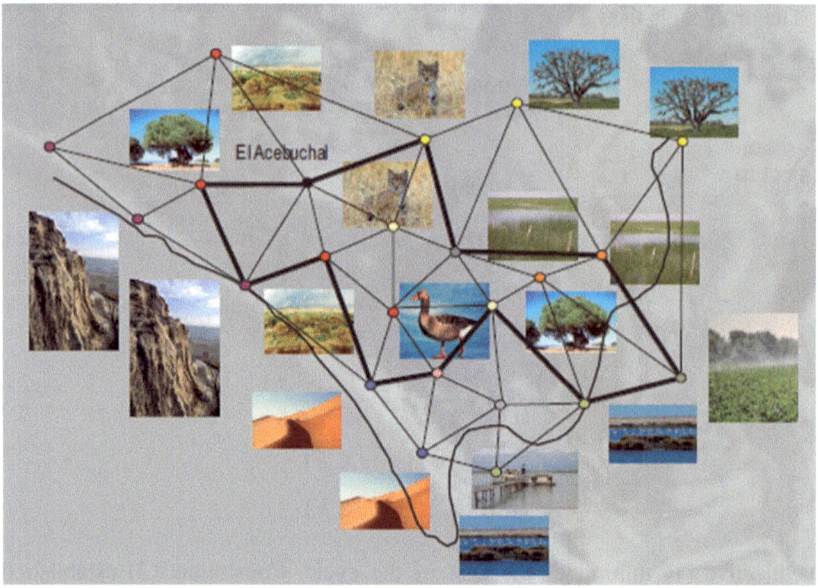

Fig. 4 A GTSP-solution applied to Doñana Natural Park

Generalized Orienteering Tour Problem, introduced by [5]. Nonnegative parameters $p_{ij} \geq 0$ represent the profit obtained by the traveler when traversing each edge. In assessing these parameters, features of visibility and the intensity of its attractions will be taken into account. The problem is then

$$\text{maximize} \sum_{(i,j)\in E} p_{ij}x_{ij} \tag{8}$$

subject to

$$\sum_{(i,j)\in E} c_{ij}x_{ij} \leq c_{\max} \tag{9}$$

$$x(\delta(i)) = 2\,y_i \quad (v_i \in V) \tag{10}$$

$$x(\delta(S)) \geq 2(y_i + y_j - 1); \quad (S \subset V,\ 2 \leq\mid S\mid\leq n-2,\ v_i \in S,\ v_j \in V \setminus S) \tag{11}$$

$$\sum_{(i,j)\in E_h} x_{ij} \geq 1 \quad (h = 1, \ldots, m) \tag{12}$$

$$x_{ij} \in \{0, 1\} \quad ((i,j) \in E) \tag{13}$$

$$y_i \in \{0, 1\} \quad (v_i \in V). \tag{14}$$

The objective (8) maximizes the utility of the design cycle. Constraint (9) means that the solution cost cannot exceed a maximum value. Constraints (10) impose the requirement that each node has two incident edges if it belongs to the solution, and has no incident edge otherwise. Inequalities (12) force every edge cluster to be visited at least once.

3 Reformulating the Problem

Note that, as was previously pointed out, profit (special interest for the traveler) could be located both along arcs as at vertices. Therefore, a new reformulation of the context is required.

Let $G = (\{O\} \cup \{D\} \cup V, A)$ be a graph, where $V = \{v_1, v_2, \ldots, v_n\}$ is the set of nodes, O and D are special nodes called respectively origin and destination that represent starting and ending points in the tour. Moreover, $A = \{a_{ij} = (v_i, v_j) : v_i \in \{O\} \cup V,\ v_j \in \{D\} \cup V,\ v_i \neq v_j\}$ is the set of feasible directed arcs interconnecting pairs of vertices. Assume set A to be partitioned into $m + 1$ disjoint clusters, $A = A_0 \cup A_1 \cup \cdots \cup A_m$, where A_0 is the set of arcs without ecological interest for tourism and A_k is the set of arcs where ecosystem i can be observed ($k = 1, \ldots, m$).

Each arc a_{ij} can be of exactly one of six exclusive types:

- Type 0: $a_{ij} \in A_0$; i.e., this arc has no tourist interest along the path (including the final node v_j).
- Type 1: $a_{ij} \in A_0$, but there exists a point of tourist interest at node v_j that does not deserve a visit.
- Type 2: $a_{ij} \in A_0$, but there exists some tourist interest at node v_j that deserves a visit.
- Type 3: $a_{ij} \in A_0$, but there exists a facility for relaxing travelers at node v_j.
- Type 4: $a_{ij} \in A_k$; this arc has ecological interest since the itinerary runs ecosystem k. Node v_j does not deserve a visit.
- Type 5: $a_{ij} \in A_k$; this arc has ecological interest since the itinerary runs ecosystem k. Node v_j deserves a visit.

Each arc a_{ij} has also the following quantitative parameters:

- c_{ij}: is the time spent in traversing arc $a_{ij} \in A$.
- c_j: is the time spent in visiting node j.
- p_{ij}: is the profit perceived by travelers when crossing arc $a_{ij} \in A$ ($p_{ij} \in [0, 1]$).
- p_j: is the profit acquired by travelers when visiting node j ($p_j \in [0, 1]$).

Eight different options for each arc can be found according to whether the last three parameters c_j, p_{ij} and p_j are zero or positive.

The idea of using graph transformations to solve our routing problem is inspired by [6]. The first step of this transformation consists of generating virtual arcs $a_{ij}^{(1)}, a_{ij}^{(2)}, \ldots, a_{ij}^{(8)}$, in accordance with the characteristics of the segment and the final vertex of real arc $a_{ij} \in A$. For these new virtual arcs, it is necessary to redefine values for the parameters of cost and profit in accordance with the type of arc under consideration:

- Type 0: $c_{ij}^0 = c_{ij}; p_{ij}^0 = 0$.
- Type 1: $c_{ij}^1 = c_{ij}; p_{ij}^1 = p_j$.
- Type 2: $c_{ij}^2 = c_{ij} + t_j; p_{ij}^2 = p_j$.
- Type 3: $c_{ij}^3 = c_{ij} + t_r; p_{ij}^3 = 0$.
- Type 4: $c_{ij}^4 = c_{ij}; p_{ij}^4 = p_{ij}$.
- Type 5: $c_{ij}^5 = c_{ij} + t_j; p_{ij}^5 = p_{ij} + p_j$.

Here, t_r is the time spent during relaxing at node v_j and t_j is the time required for visiting node j.

In this way, the original Generalized Orienteering Tour Problem (GOTP) on G can be transformed into an equivalent directed Generalized Travelling Salesman Problem (GTSP) on a new graph H of virtual arcs.

Since any instance of node routing problem can be efficiently solved (for instance, as is shown in [16] where a branch-and-price process is used), reference [15] propose carrying out a transformation of Arc Routing Problem instances into Node Routing models by using a compact method which guarantees that the number of nodes in the final graph is equal to the number of demanded edges in the arc routing graph, plus one (the depot). This promising methodology helped obtain solutions for some large size instances of Arc Routing Problems.

4 Conclusions

Different models were proposed in this paper for determining efficient routes of ecotourism, where the main objective is the maximization of the cultural transmission experienced by the visitors along the followed itinerary. These models make use of integer linear programming and their potential was illustrated in the context of the Doñana National Park (Spain). Finally, a graph transformation was described in order to provide additional perspectives for solving the resulting instance of Arc Routing Problem which is inherent to this setting.

Acknowledgements This research work was partially supported by the Spanish Ministerio de Ciencia e Innovación under grant MTM2010-19576-C02-01 and MTM2012-37048, by the Excellence program of the Andalusian Government, under grant P09-TEP-5022 and FQM-5849, by the FEDER funds of the European Union, and by the Natural Sciences and Engineering Research Council of Canada (NSERC) under grant 2015-06189.

References

1. Amezaga, J.M., Santamaria, L., Green, A.J.: Biotic wetland connectivity-supporting a new approach for wetland policy. Acta Oecol. **23**, 213–222 (2002)
2. Archetti, C., Feillet, D., Hertz, A., Speranza M.G.: The undirected capacitated arc routing problem with profit. Comput. Oper. Res. **37**, 1860–1869 (2006)
3. Araoz, J., Fernández, E., Zoltan, C.: Privatized rural postman problems. Comput. Oper. Res. **33**, 886–896 (2006)
4. Araoz, J., Fernández, E., Meza, O.: Solving the prize-collecting rural postman problem. Eur. J. Oper. Res. **196**, 3432–3449 (2009)
5. Barrena, E., Ortega, F.A., Pozo, M.A., Ternero, I.: Assessing optimal routes in the natural park of Doñana, Spain. EURO 2012-25th European Conference on Operational Research (2012). Vilnius, Lithuania
6. Blais, M., Laporte, G.: Exact solution of the generalized routing problem through graph transformations. J. Oper. Res. Soc. **54**, 906–910 (2003)
7. Briedenhann, J., Wikens, E.: Tourism routes as a tool for the economic development of rural areas. Vibrant hope or impossible dream. Tour. Manag. **57**, 1–9 (2003)
8. Brown, G.: A method for assessing highway qualities to integrate values in highway planning. J. Transp. Geogr. **11**, 271–283 (2003)
9. Buckley, R.: Environmental Impact of Ecotourism. CAB International, Oxfordshire (2008)
10. Connell, J., Page, S.J.: Exploring the spatial patterns of car-based tourist travel in Loch Lomond and Trossachs National Park, Scotland. Tour. Manag. **29**, 561–580 (2008)
11. Corberán, A., Laporte, G. (eds.) Arc Routing: Problems, Methods and Applications. MOS-SIAM Series on Optimization. Society for Industrial and Applied Mathematics, Philadelphia, USA (2014)
12. Deitch, R., Ladany, S.: The one-period bus touring problem: solved by an effective heuristic for the orienteering tour problem and improvement algorithm. Eur. J. Oper. Res. **127**, 69–77 (2000)
13. Farrell, B., Runyan, D.: Ecology and tourism. Ann. Tour. Res. **18**, 26–40 (1991)
14. Feillet, D., Dejax, P., Gendreau, M.: The profitable arc tour problem: solution with a branch-and-price algorithm. Transp. Sci. **39**, 539–552 (2005)
15. Foulds, L., Longo, H., Martins, J.: A compact transformation of arc routing problems into node routing problems. Ann. Oper. Res. **226**, 177–200 (2015)
16. Fukasawa, R., Longo, H., Lysgaard, J., Reis, M., Uchoa, E., Werneck, R.F.: Robust branch-and-cut-and-price for the capacitated vehicle routing problem. Math. Program. **106(3)**, 491–511 (2006)
17. Hernández-Ramírez, J.: Los caminos del patrimonio. Rutas turisticas e itinerarios culturales. PASOS. Revista de Turismo y Patrimonio Cultural **9(2)**, 225–236 (2011)
18. Kelley, W.J.: National scenic byways. diversity contributes to success. Transp. Res. Rec. J. Transp. Res. Board **1880**, 174–180 (2004)
19. Kinglake National Park Master Plan (2011). Document available at http://parkweb.vic.gov.au
20. Kunaeva, M.: Sustainable tourism management along the Camino de Santiago pilgrimage routes. Master's Thesis Degree Programme in Tourism (2012). HAAGA-HELIA
21. Laporte, G., Nobert, Y.: Generalized traveling salesman problem through n sets of nodes: An integer programming approach. Infor. **21**, 61–75 (1983)
22. Laporte, G., Rodriguez-Martín, I.: Locating a cycle in a transportation or a telecommunications. Networks **50**, 92–108 (2007)
23. Laporte, G., Asef-Vaziri A., Sriskandarajah, C.: Some applications of the generalized traveling salesman problem. J. Oper. Res. Soc. **47**, 1461–1467 (1996)
24. Lindberg, K., Hawkins, D.E.: Ecotourism: A Guide for Planners and Managers. Ecotourism Society, vol. 1. North Bennington, Vermont, USA (1993)
25. Maruyama, N.U., Yen, T.-H., Stronza, A.: Perception of authenticity of tourist art among native american artists in Santa Fe, New Mexico. Int. J. Tour. Res. **10**, 453–466 (2008)

26. Mathieson, A., Wall, G.: Tourism: Economic, Physical and Social Impacts. Longman Group, Essex, England (1982)
27. Orams, M.B.: Towards a more desirable form of ecotourism. Tour. Manag. **16**, 3–8 (1995)
28. Ramesh, R., Yong-Seok, Y., Karwan, M.H.: An optimal algorithm for the orienteering problem. ORSA J. Comput. **4**(2), 155–165 (1992)
29. Rodriguez, B., Molina, J., Perez, F., Caballero, R.: Interactive design of personalised tourism routes. Tour. Manag. **33**, 926–940 (2012)
30. Souffriau, W., Vansteenwegen, P., Vertommen, J., Vanden Berghe, G., Van Oudheusden, D.: A personalized tourist trip design alogrithm for mobile tourisit guides. Appl. Artif. Intell. **22**, 964–985 (2008)
31. Souffriau, W., Vansteenwegen, P., Vanden Berghe, G., Van Oudheusden, D.: The planning of cycle trips in the province of East Flanders. Omega **39**, 209–213 (2011)
32. Tsaur, S.-H., Lin, Y.-C., Lin, J.-H.: Evaluating ecotourism sustainability from the integrated perspective of resource, community and tourism. Tour. Manag. **27**, 640–653 (2006)
33. Tsiligirides, T.: Heuristic methods applied to orienteering. J. Oper. Res. Soc. **35**(9), 797–809 (1984)
34. Vansteenwegen, P., Van Oudheusden, D.: The mobile tourist guide: an OR opportunity. OR Insight **20**, 21–27 (2007)
35. Vansteenwegen, P., Souffriau, W., Van Oudheusden, D.: The orienteering problem: a survey. Eur. J. Oper. Res. **209**(1), 1–10 (2011)
36. Voth, A.: Overcoming National Park Conflicts by Regional Development: Experiences from the Doñana Area in Southern Spain. Exploring the Nature of Management. Proceedings of the Third International Conference on Monitoring and Management of Visitor Flows in Recreational and Protected Areas (2006), pp. 155–160. University of Applied Sciences Rapperswil, Switzerland

Isometries of the Hamming Space and Equivalence Relations of Linear Codes Over a Finite Field

M. Isabel García-Planas and M. Dolors Magret

Abstract Detection and error capabilities are preserved when applying to a linear code an isomorphism which preserves Hamming distance. We study here two such isomorphisms: permutation isometries and monomial isometries.

1 Introduction

Most of the important codes are special types of the so-called linear codes. There are simple encoding and decoding procedures for them. In linear network coding theory, it is usual to consider sets of subspaces of a given linear space over a finite field, in general, and sets of subspaces of a given dimension, in particular.

Since vector space endomorphisms, not even isomorphisms, do not preserve Hamming distance, which is an essential property of each code, we will restrict to consider those isomorphisms which preserve Hamming distance. They map codes onto codes with the same detection and correcting capabilities. These isomorphisms are usually referred to as isometries of the Hamming space. Two examples of such isomorphisms are those given by permutation and monomial matrices, and will be called *permutation isomorphisms* and *monomial isomorphisms*, respectively.

Given a code, we can consider the set of all codes which can be obtained from this one applying different isometries, the isometry class of a code. It makes sense to find invariants describing each isometry class in order to compare codes among each other. Isometry classes can be seen as equivalence classes under suitable equivalence relations (permutation equivalence and monomial equivalence) and as orbits under suitable group actions. This fact allows us to compute the number of non-isometric linear codes of a given dimension. The code equivalence problem has been studied by different authors though it can be a hard problem, specially for *qgeq* 5, (see [3, 4]).

M.I. García-Planas • M.D. Magret (✉)
Universitat Politècnica de Catalunya, Barcelona, Spain
e-mail: maria.isabel.garcia@upc.edu; m.dolors.magret@upc.edu

F. Ortegón Gallego et al. (eds.), *Trends in Differential Equations and Applications*, SEMA SIMAI Springer Series 8, DOI 10.1007/978-3-319-32013-7_12

The structure of the paper is as follows.

Section 2 contains some generalities about linear codes. In Sects. 3 and 4, isometry classes of codes of a given dimension under permutation and monomial equivalence, respectively, are studied. Section 5 is devoted to the computation of the number of permutation and monomial non-isometric classes.

Throughout the paper, we will denote by $\mathbb{F}_p$ the finite field of p elements, p a prime number, $p \neq 2$.

2 Preliminaries

In this section we will recall the basic definitions of linear codes which will be used in the following sections.

A linear (n,k)-code C over $\mathbb{F}_p$ is a k-dimensional vector subspace of $\mathbb{F}_p^n$. Its elements are called codewords. A (n,k)-code has p^k codewords.

Let us denote by $V(k,n,p)$ the set of all k-dimensional subspaces in $\mathbb{F}_p^n$. It is known that their number is equal to:

$$|V(k,n,p)| = \binom{n}{k}_p = \frac{(p^n-1)(p^n-p)\dots(p^n-p^{k-1})}{(p^k-1)(p^k-p)\dots(p^k-p^{k-1})}.$$

Example 1 There are 13 2-dimensional vector subspaces in $\mathbb{F}_3^3$ (therefore 13 $(3,2)$-linear codes over $\mathbb{F}_3^3$):

$$\begin{array}{l}C_1 = [e_1,e_2],\ C_4 = [e_1,e_2+e_3],\ C_7 = [e_1,e_2+2e_3],\\ C_2 = [e_1,e_3],\ C_5 = [e_2,e_1+e_3],\ C_8 = [e_2,e_1+2e_3],\\ C_3 = [e_2,e_3],\ C_6 = [e_3,e_1+e_2],\ C_9 = [e_3,e_1+2e_2],\end{array}$$

$$\begin{array}{l}C_{10} = [e_1+e_2,e_1+e_3],\ C_{11} = [e_1+e_2,e_2+e_3],\\ C_{12} = [e_1+e_3,e_2+e_3],\ C_{13} = [e_1+2e_2,e_2+2e_3].\end{array}$$

The Hamming distance between two codewords $x = (x_1,\dots,x_n)$, $y = (y_1,\dots,y_n)$ is defined by:

$$d(x,y) = |\{i \in \{1,\dots,n\} \mid x_i \neq y_i\}|$$

and the Hamming weight of a codeword $x = (x_1,\dots,x_n)$ is defined by:

$$w(x) = |\{i \in \{1,\dots,n\} \mid x_i \neq 0\}|.$$

The distance or minimum distance of a code C is the minimum number of positions in which any two distinct codewords differ. It is denoted by $d(C)$. The weight of code C is defined as the smallest of the weights of non-zero codewords of C and is denoted by $w(C)$.

The metric space $\mathbb{F}_p^n$ with the Hamming distance d is called Hamming space and denoted by $H(n,p)$.

3 Permutation Isometry Classes

Let us denote by $\mathscr{P}_n(\mathbb{F}_p)$ the group of all permutation matrices. We will denote by $P(i_1,\ldots,i_n)$ the permutation matrix associated to the permutation $i_1\ldots i_n$; that is to say, the permutation matrix in which the non-zero components are in columns $i_1,\ldots,i_n$.

Let us consider, for all $P \in \mathscr{P}_n(\mathbb{F}_p)$, the mapping:

$$\begin{aligned} f_P : V(n,k,p) &\longrightarrow V(n,k,p) \\ C &\longrightarrow CP^t. \end{aligned}$$

This map is an isometry. We will refer to it as the *permutation isometry* associated to the permutation matrix P.

Definition 1 Two linear (n,k)-codes C and C' are called *permutation isometric* if there exists a permutation isometry f_P such that $C' = f_P(C)$, for some permutation matrix $P \in \mathscr{P}_n(\mathbb{F}_p)$.

That is to say, if for all codeword $w' \in C'$ there exists a codeword $w \in C$ such that $w' = wP^t$ (and conversely).

Remark 1 This is an equivalence relation. Given a (n,k)-code C, its equivalence class is $\bar{C}^{\mathscr{P}} = \{f_P(C) \mid P \in \mathscr{P}_n(\mathbb{F}_p)\} = \{CP^t \mid P \in \mathscr{P}_n(\mathbb{F}_p)\}$.

Example 2 Let us consider $p = 3$ and the $(2,3)$-code C_7 (notations as in Example 1) which consists of codewords:

$$\{000, 100, 200, 012, 021, 112, 221, 121, 212\}.$$

Straightforward calculations show that

$$\begin{array}{lll} C_7P(1,2,3) = C_7 & C_7P(2,1,3) = C_8 & C_7P(3,2,1) = C_9 \\ C_7P(1,3,2) = C_7 & C_7P(2,3,1) = C_8 & C_7P(3,1,2) = C_9. \end{array}$$

The permutation class of code C_7 is: $\bar{C}^{\mathscr{P}} = \{C_7, C_8, C_9\}$.

Fixed points or invariant subspaces under all isomorphisms f_P coincide with those codes whose equivalence class consists of only one element.

Example 3 In the same conditions as in Examples 1 and 2 above ($p = 3$), straightforward calculations show that the only code which is invariant under f_P, for all $P \in \mathscr{P}_3(\mathbb{F}_3)$, is C_{13}. More concretely, all 13 vector subspaces are invariant under f_{I_3}, but, in the case of the other permutation matrices, the vector invariant subspaces are those listed below.

$$\begin{aligned}
&P(2,1,3) : C_1, C_6, C_9, C_{12}, C_{13}\\
&P(3,2,1) : C_2, C_5, C_8, C_{11}, C_{13}\\
&P(1,3,2) : C_3, C_4, C_7, C_{10}, C_{13}\\
&P(2,3,1) : C_{13}\\
&P(3,1,2) : C_{13}.
\end{aligned}$$

Therefore the only equivalence class with an only element is: $\bar{C}_{13}^{\mathscr{P}} = \{C_{13}\}$.

Equivalence permutation classes of linear codes coincide with the orbits under the group action

$$\begin{aligned}
\alpha : \mathscr{P}_n(\mathbb{F}_p) \times V(k,n,p) &\longrightarrow V(k,n,p)\\
(P, C) &\longrightarrow CP^t.
\end{aligned}$$

That is to say, denoting the orbit of a k-dimensional vector subspace C by this action by $\mathscr{O}_\alpha(C)$, we have that $\mathscr{O}_\alpha(C) = \overline{C}^{\mathscr{P}}$.

4 Monomial Isometry Classes

As anticipated in the Introduction, isomorphisms given by monomial matrices are also isometries. We recall first some well-known properties of monomial matrices.

Definition 2 A *monomial matrix* of order n is a regular $n \times n$-matrix which has in each row and in each column exactly one non-zero component.

Monomial matrices form a group. The product of monomial matrices is again a monomial matrix. The inverse of a monomial matrix is again a monomial matrix.

Unlike permutation matrices, monomial matrices are not necessarily orthogonal.

The following property of monomial matrices is well-known and will be useful for our purposes.

Lemma 1 *Every monomial matrix is a product of a diagonal matrix with a permutation matrix.*

In general, we will make use of the following notation. Any monomial matrix will be written as:

$$M(a_1, \ldots, a_n; i_1, \ldots, i_n) = \text{diag}\,(a_1, \ldots, a_n)P(i_1, \ldots, i_n).$$

Example 4

$$M(a_1, a_2, a_3; 3, 1, 2) = \begin{pmatrix} 0 & 0 & a_1 \\ a_2 & 0 & 0 \\ 0 & a_3 & 0 \end{pmatrix} = \begin{pmatrix} a_1 & 0 & 0 \\ 0 & a_2 & 0 \\ 0 & 0 & a_3 \end{pmatrix} \begin{pmatrix} 0 & 0 & 1 \\ 1 & 0 & 0 \\ 0 & 1 & 0 \end{pmatrix}$$

$$= \text{diag}\,(a_1, a_2, a_3)P(3, 1, 2).$$

Let us denote by $\mathscr{M}_n(\mathbb{F}_p)$ the group of all monomial matrices. Let us consider, for all $M \in \mathscr{M}_n(\mathbb{F}_p)$ the mapping:

$$\begin{aligned} g_M : V(k, n, p) &\longrightarrow V(k, n, p) \\ C &\longrightarrow CM^t. \end{aligned}$$

This map is an isometry.

Definition 3 Two linear (n, k)-codes C and C' are called *monomially isometric* if there exists an isometry g_M such that $g_M(C) = C'$, for some monomial matrix $M \in \mathscr{M}_n(\mathbb{F}_p)$.

That is to say, if for all codeword $w' \in C'$ there exists a codeword $w \in C$ such that $w' = wM^t$ (and conversely).

Remark 2 This is an equivalence relation. Given a linear (n, k)-code C its equivalence class is: $\overline{C}^{\mathscr{M}} = \{g_M(C) \mid M \in \mathscr{M}_n(\mathbb{F}_p)\} = \{CM^t \mid M \in \mathscr{M}_n(\mathbb{F}_p)\}$.

Obviously, $\overline{C}^{\mathscr{P}} \subseteq \overline{C}^{\mathscr{M}}$.

Example 5 Let us consider, as in Example 2, and with the same notations as in Example 1, the $(3, 2)$-code:

$$C_7 = \{000, 100, 200, 012, 021, 112, 221, 121, 212\} = [e_1, e_2 + 2e_3].$$

Then:

$$\begin{array}{ll}
C_7M(a_1,a_2,a_3;1,2,3) = C_7 & \text{if } a_2 = a_3 \\
C_7M(a_1,a_2,a_3;1,2,3) = C_4 = [e_1,e_2+e_3] & \text{if } a_2 \neq a_3 \\
C_7M(a_1,a_2,a_3;1,3,2) = C_7 & \text{if } a_2 = a_3 \\
C_7M(a_1,a_2,a_3;1,3,2) = C_4 & \text{if } a_2 \neq a_3 \\
C_7M(a_1,a_2,a_3;2,1,3) = C8 & \text{if } a_1 = a_3 \\
C_7M(a_1,a_2,a_3;2,1,3) = C_5 = [e_2,e_1+e_3] & \text{if } a_1 \neq a_3 \\
C_7M(a_1,a_2,a_3;3,1,2) = C_8 & \text{if } a_1 = a_3 \\
C_7M(a_1,a_2,a_3;3,1,2) = C_5 & \text{if } a_1 \neq a_3 \\
C_7M(a_1,a_2,a_3;2,3,1) = C_9 & \text{if } a_1 = a_2 \\
C_7M(a_1,a_2,a_3;2,3,1) = C_6 = [e_3,e_1+e_2] & \text{if } a_1 \neq a_2 \\
C_7M(a_1,a_2,a_3;3,2,1) = C_9 & \text{if } a_1 = a_2 \\
C_7M(a_1,a_2,a_3;3,2,1) = C_6 & \text{if } a_1 \neq a_2
\end{array}$$

and then $\overline{C_7}^{\mathcal{M}} = \{C_4, C_5, C_6, C_7, C_8, C_9\}$.

Fixed points or invariant subspaces under isomorphisms g_M, for all $M \in \mathcal{M}_n(\mathbb{F}_p)$, correspond to those codes having monomial equivalence class consisting only of one element.

Example 6 Straightforward calculations lead to the following list of the 2-dimensional codes in $V(2,3,3)$ which are invariant for the different monomial isometries:

$$\begin{array}{lll}
M(a_1,a_2,a_3;1,2,3): & C_1,C_2,C_3,C_4,C_5,C_6,C_7, & \\
 & C_8,C_9,C_{10},C_{11},C_{12},C_{13} & \text{if } a_1 = a_2 = a_3 \\
 & C_1,C_2,C_3,C_6,C_9 & \text{if } a_1 = a_2 \neq a_3 \\
 & C_1,C_2,C_3,C_5,C_8 & \text{if } a_1 = a_3 \neq a_2 \\
 & C_1,C_2,C_3,C_4,C_7 & \text{if } a_2 = a_3 \neq a_1 \\
M(a_1,a_2,a_3;2,1,3): & C_1,C_6,C_9,C_{12},C_{13} & \text{if } a_1 = a_2 = a_3 \\
 & C_1,C_6,C_9,C_{10},C_{11} & \text{if } a_1 = a_2 = 2a_3 \\
 & C_1 & \text{if } a_1 \neq a_2 \\
M(a_1,a_2,a_3;3,2,1): & C_2,C_5,C_8,C_{11},C_{13} & \text{if } a_1 = a_2 = a_3 \\
 & C_2,C_5,C_8,C_{10},C_{12} & \text{if } a_1 = a_3 = 2a_2 \\
 & C_2 & \text{if } a_1 \neq a_3 \\
M(a_1,a_2,a_3;1,3,2): & C_3,C_4,C_7,C_{10},C_{13} & \text{if } a_1 = a_2 = a_3 \\
 & C_3,C_4,C_7,C_{11},C_{12} & \text{if } a_2 = a_3 = 2a_1 \\
 & C_3 & \text{if } a_2 \neq a_3
\end{array}$$

$$\begin{aligned} M(a_1,a_2,a_3;2,3,1): &\ C_{10} \ \text{if } a_1 = a_3 = 2a_2 \\ &\ C_{11} \ \text{if } a_1 = a_2 = 2a_3 \\ &\ C_{12} \ \text{if } a_2 = a_3 = 2a_1 \\ &\ C_{13} \ \text{if } a_2 = a_3 = a_1 \\ M(a_1,a_2,a_3;3,1,2): &\ C_{10} \ \text{if } a_1 = a_2 = 2a_3 \\ &\ C_{11} \ \text{if } 2a_1 = a_2 = a_3 \\ &\ C_{12} \ \text{if } a_1 = a_3 = 2a_2 \\ &\ C_{13} \ \text{if } a_1 = a_2 = a_3. \end{aligned}$$

According to the list above, we conclude that there are no codes with only one element in its monomial equivalence class because there are no invariant subspaces for all monomial matrices.

Monomial isometry classes coincide with the orbits with respect to the group action of the group of monomial matrices $\mathscr{M}_n(\mathbb{F}_p)$ on the set of vector subspaces of a given dimension.

$$\begin{aligned} \beta : \mathscr{M}_n(\mathbb{F}_p) \times V(k,n,p) &\longrightarrow V(k,n,p) \\ (M,C) &\longrightarrow CM^t. \end{aligned}$$

The orbit of a k-dimensional vector subspace C under this action is: $\mathscr{O}_\beta(C) = \overline{C}^{\mathscr{M}}$.

5 Number of Isometry Classes

The main tool to compute the number of isometry classes is Burnside's Lemma, which can be applied in our case because the equivalence relations considered (permutation and monomial isometry equivalences) are such that isometry equivalence classes coincide with orbits under suitable group actions, as seen in previous Sections.

Let us denote by $\mathscr{S}_n$ the symmetric group on n symbols. Recall that two permutations $\pi_1, \pi_2 \in \mathscr{S}_n$ are conjugate if there exists $\sigma \in \mathscr{S}_n$ such that $\pi_2 = \sigma\pi_1\sigma^{-1}$.

The cycle type of a cycle is the data of how many cycles of each length are present in the cycle decomposition of the cycle into disjoint cycles. If the cycle is a product of m_1 k_1-cycles, m_2 k_2-cycles, ..., m_r k_r-cycles ($0 \leq m_1 \leq m_2 \leq \cdots \leq m_r$), then we will write that its cycle type is $m_1 + m_2 + \cdots + m_r$. With the notations above,

$$\sum_{j=1}^{n} jm_j = n.$$

The following result is well known.

Theorem 1 *Let $\pi_1, \pi_2 \in \mathcal{S}_n$ be two permutations in the symmetric group $\mathcal{S}_n$. Then π_1 and π_2 are conjugate if and only if they have the same cycle type.*

We list below all the permutations of the symmetric groups of n elements for $n = 2$, $n = 3$ and $n = 4$, expressing the decomposition of the cycle in disjoint cycles and the cycle type. For $n > 4$, analogous tables can be constructed.

For $n = 2$ and $n = 3$:

Permutation	Disjoint cycles	Cycle type
1 2	(1)(2)	$1+1$
2 1	(1,2)	2

Permutation	Disjoint cycles	Cycle type
1 2 3	(1)(2)(3)	$1+1+1$
2 1 3	(1,2)(3)	$1+2$
3 2 1	(1,3)(2)	$1+2$
1 3 2	(1)(2,3)	$1+2$
2 3 1	(1,2,3)	3
3 1 2	(1,3,2)	3

For $n = 4$:

Permutation	Disjoint cycles	Cycle type
1 2 3 4	(1)(2)(3)(4)	$1+1+1+1$
2 1 3 4	(1,2)(3)(4)	$1+1+2$
3 2 1 4	(1,3)(2)(4)	$1+1+2$
4 2 3 1	(1,4)(2)(3)	$1+1+2$
1 3 2 4	(2,3)(1)(4)	$1+1+2$
1 4 3 2	(3,4)(1)(3)	$1+1+2$
1 4 3 2	(3,4)(1)(3)	$1+1+2$
3 1 2 4	(1,3,2)(4)	$1+3$
2 3 1 4	(2,3,1)(4)	$1+3$
4 1 3 2	(4,1,2)(3)	$1+3$
2 4 3 1	(2,4,1)(3)	$1+3$
4 2 1 3	(3,4,1)(2)	$1+3$
3 2 4 1	(3,4,1)(2)	$1+3$
1 4 2 3	(4,2,3)(1)	$1+3$
1 3 4 2	(3,4,2)(1)	$1+3$
2 1 4 3	(3,4)(1,2)	$2+2$
3 4 1 2	(2,4)(1,3)	$2+2$
4 3 2 1	(2,3)(1,4)	$2+2$
4 1 2 3	(1,4)(2,3)	4
2 3 4 1	(2,3,4,1)	4
2 4 1 3	(2,4,1,3)	4
3 1 4 2	(3,1,4,2)	4
3 4 2 1	(3,4,2,1)	4
4 3 1 2	(4,3,1,2)	4

Before applying it, we recall the statement of Burnside's Lemma.

Lemma 2 (Burnside's Lemma) *Let $\mathcal{G}$ be a finite group which acts on a set X. For each g in $\mathcal{G}$ we denote by $\sigma(g)$ the set of elements in X that are fixed by g (or g-invariant). Then the number of orbits is*

$$|X/\mathcal{G}| = \frac{1}{|\mathcal{G}|}\sum\nolimits_{g\in\mathcal{G}}\sigma(g).$$

Burnside's Lemma, applied to our particular set-up, yields the following statement.

Theorem 2

(a) The number of permutation isometry classes of codes of a given dimension k is equal to:

$$\frac{1}{n!}\sum\nolimits_{P\in\mathcal{P}_n(\mathbb{F}_p)}\sigma(P)$$

where $\sigma(P)$ is the number of k-dimensional vector subspaces which are invariant under the permutation isomorphism f_P of $V(n,k,p)$.

(b) The number of monomial isometry classes is equal to:

$$\frac{1}{n!}\sum\nolimits_{M\in\mathcal{M}_n(\mathbb{F}_p)}\sigma(M)$$

where $\sigma(M)$ is the number of k-dimensional vector subspaces which are invariant under the monomial isomorphism g_M of $V(n,k,p)$.

Example 7 We consider the set of all 2-dimensional vector subspaces of $\mathbb{F}_3^3$ (see Example 1). In Example 3 we have seen which codes are invariant under the permutation isometries of $V(2,3,3)$. All 13 subspaces are f_P-invariant for $P = I_3$. In the cases where the other permutation matrices are considered, the number of invariant subspaces under f_P is: 5 when $P = P(2,1,3)$, $P = P(3,2,1)$ or $P = P(1,3,2)$ and 1 when $P = P(2,3,1)$ or $P = P(3,1,2)$.

Then the number of permutation equivalence classes is:

$$\frac{1}{3!}(13+5+5+5+1+1) = 5.$$

In the case of monomial isometry equivalence the number of monomial isometry classes, using Example 6, are:

$$\frac{1}{3!}(2\cdot 13+6\cdot 5+3(4\cdot 5+4\cdot 1)+2(8\cdot 1)) = 24.$$

It may be tedious to do this computation in all cases. However, we can restrict to consider the isomorphisms associated to one permutation matrix or monomial matrix in each conjugacy class.

Lemma 3

(a) Let $P(i_1,\dots,i_n)$, $P(j_1,\dots,j_n)$ be two permutation matrices. Assume that $i_1\dots i_n$, $j_1\dots j_n$ have the same cycle type. Then:

$$\sigma(P(i_1,\dots,i_n)) = \sigma(P(j_1,\dots,j_n)).$$

(b) Let $M(a_1,\dots,a_n;i_1,\dots,i_n)$, $M(\alpha_1,\dots,\alpha_n;j_1,\dots,j_n)$ be two monomial matrices, conjugated in $\mathcal{M}_n(\mathbb{F}_p)$. Then:

$$\sigma(M(a_1,\dots,a_n;i_1,\dots,i_n)) = \sigma(M(\alpha_1,\dots,\alpha_n;j_1,\dots,j_n)).$$

We can now state the main result.

Theorem 3

(a) The number of permutation classes is equal to:

$$\frac{1}{n!}\sum\nolimits_{P\in\mathcal{P}_n(\mathbb{F}_p)}|V(k,n,p)^P| = \frac{1}{n!}\sum\nolimits_{\overline{P}^{\mathcal{P}}} s(P)\sigma(P)$$

where $s(P)$ is the number of permutation matrices having the same cycle type.

(b) The number of monomially equivalence classes is equal to:

$$\frac{1}{n!}\sum\nolimits_{M\in\mathcal{M}_n(\mathbb{F}_p)}|V(k,n,p)^M| = \frac{1}{(p-1)^n}\frac{1}{n!}\sum\nolimits_{\overline{M}^{\mathcal{M}}} s(M)\sigma(M)$$

where $s(P)$ is the number of elements in each conjugate class.

Example 8 In Example 7 above, the only computations which had to be done are the following ones.

Cycle type	Permutation matrix	Number of f_P-invariant subspaces
$1+1+1$	I_3	13
$1+2$	$P(2,1,3)$	5
3	$P(2,3,1)$	1

since $\sigma(P(3,2,1)) = \sigma(P(1,3,2)) = \sigma P(2,1,3))$ and $\sigma(P(3,1,2)) = \sigma(P(2,3,1))$. This suffices to compute the number of permutation isomorphism classes:

$$\frac{1}{3!}(13+3\cdot 5+2\cdot 1) = 5.$$

An analogous simplification can be done in the case of monomial isomorphism classes, (see [1]).

To compute the number of vector subspaces of a given dimension which are invariant under permutation and monomial isomorphisms ($\sigma(P)$ and $\sigma(M)$, for all permutation matrices P and monomial matrices M), it is useful to know the decomposition of a vector space into primary components and that of each primary component as a direct sum of cyclic subspaces. We briefly recall this decomposition.

Let f be a linear operator on $\mathbb{F}_p^n$, with associated matrix A in a given basis of $\mathbb{F}_p^n$. We will write $Q_A(t) = \det(A - tI_n)$ the characteristic polynomial of f and denote by $M_A(t)$ the minimal annihilating polynomial of f (the monic polynomial of least degree which annihilates all vectors in $\mathbb{F}_p^n$). Note that they do not depend on the choice of the basis of $\mathbb{F}_p^n$.

Consider the decomposition of the minimal annihilating polynomial of f into irreducible factors:

$$M_f(t) = M_1(t)^{\mu_1} \cdot \dots \cdot M_s(t)^{\mu_s}$$

where $\mu_1, \dots, \mu_s \geq 0$.

The vector subspaces $V_i = \{x \in V \mid M_i(f)^{\mu_i}(x) = 0\}$, $1 \leq i \leq s$, are f-invariant and $V = V_1 \oplus \dots \oplus V_s$ (primary decomposition of V). Moreover, each primary subspace is a direct sum of cyclic subspaces:

$$V_i =< v_i^1 > \oplus \dots \oplus < v_i^{m_i} >, 1 \leq i \leq s$$

where $< v >= [v, f(v), \dots, f^{d-1}(v)]$ being d the least degree of f such that $\dim[v, f(v), \dots, f^{d-1}(v)] = \dim[v, f(v), \dots, f^{d-1}(v), f^d(v)]$. Then we can write

$$V =< v_1^1 > \oplus \dots \oplus < v_1^{m_1} > \oplus \dots \oplus < v_s^1 > \oplus \dots \oplus < v_s^{m_s} > .$$

This is known as the decomposition into cyclic subspaces.

Note that the decomposition of each primary subspace V_i, $1 \leq i \leq s$, as a direct sum of cyclic subspaces is not unique. Nevertheless, two such decompositions have the same sequence of numbers which are the different dimensions of the cyclic subspaces and which will be referred to as *cyclic-primary numbers*.

Example 9 Let us consider f the linear endomorphism of $\mathbb{F}_5^3$ having as associated matrix, in the natural basis of $\mathbb{F}_5^3$, the permutation matrix $P = P(1, 3, 2)$. The characteristic and minimal polynomials are: $Q_P(t) = (t + 4)^2(t + 1)$, $M_P(t) = (t + 4)(t + 1)$.

Primary decomposition: $\mathbb{F}_3^3 = E_1 \oplus E_2$, with $E_1 = \ker(f + 4I_3) = [e_1, e_2 + e_3]$, $E_2 = \ker(f + I_3) = [e_2 + 4e_3]$.

Decomposition in cyclic subspaces: $\mathbb{F}_5^3 = (E_1^1 \oplus E_1^2) \oplus E_2$, with $E_1^1 =< e_1 >$, $E_1^2 =< e_2 + e_3 >$, $E_2 =< e_2 + 4e_3 >$.

Note that these decompositions depend on the finite field.

Example 10 Let us consider f the linear endomorphism of $\mathbb{F}_p^3$ having as associated matrix, in the natural basis of $\mathbb{F}_p^3$ the permutation matrix $P = P(2, 3, 1)$.

The characteristic polynomial is: $Q_P(t) = (t-1)(t^2+t+1)$.

If $p = 5$, t^2+t+1 is irreducible over $\mathbb{F}_5[t]$ and $M_P(t) = (t+1)(t^2+t+1)$.

The primary decomposition is: $\mathbb{F}_5^3 = E_1 \oplus E_2$, with $E_1 = \ker(f + 4I_3) = [e_1+e_2+e_3]$, $E_2 = \ker(f^2+f+I_3) = [e_2+4e_1, e_1+4e_3]$.

The decomposition in cyclic subspaces is:

$$\mathbb{F}_5^3 =< e_1+e_2+e_3 > \oplus < e_1+4e_3 > .$$

But if $p = 7$, $t^2+t+1 = (t+3)(t+5)$ and the minimal annihilating polynomial is: $M_P(t) = (t+3)(t+5)(t+6)$.

The primary decomposition is: $\mathbb{F}_7^3 = E_1 \oplus E_2 \oplus E_3$, with $E_1 = \ker(f+3I_3) = [e_1+4e_2+2e_3]$, $E_2 = \ker(f+5I_3) = [e_1+2e_2+4e_3]$, $E_3 = \ker(f+6I_3) = [e_1+e_2+e_3]$.

The decomposition in cyclic subspaces is:

$$\mathbb{F}_7^3 =< e_1+2e_2+4e_3 > \oplus < e_1+4e_2+2e_3 > \oplus < e_1+e_2+e_3 > .$$

The starting point to obtain this decomposition is obtaining the minimal annihilating polynomial of the endomorphism. In Appendix this polynomial is obtained in the case where the endomorphism is a permutation isomorphism or a monomial isomorphism.

Appendix: Minimal Annihilating Polynomial of Permutation and Monomial Isometries

The minimal annihilating polynomial of a permutation isometry can be determined by the decomposition of the permutation in disjoint cycles. More concretely, if P_1, P_2 are two permutation matrices associated to two permutations σ_2, σ_2 with the same cycle type (conjugate in the symmetric group), the minimal annihilating polynomials $M_{P_1}(t)$ and $M_{P_2}(t)$ coincide.

Proposition 1 *Let P be a permutation matrix associated to a permutation with cycle type $1+\overset{m_1}{\smile\ldots}+1+2+\overset{m_2}{\smile\ldots}+2+3+\overset{m_3}{\smile\ldots}+3+\cdots+r+\overset{m_r}{\smile\ldots}+r$. Then*

$$M_P(t) = MCM\{(t-1)^{n_1}, (t^2-1)^{n_2}, \ldots, (t^3-1)^{n_3}, \ldots, (t^r-1)^{n_r}\}$$

where $n_i = 0$ if $m_i = 0$ and $n_i = 1$ if $m_i > 0$, $1 \le i \le r$ (see [2]).

Example 11 Let us consider $P = P(2,3,4,1,6,5)$. Then $234165 = (2,3,4,1)(6,5)$ has cycle type $2+4$ and

$$M_P(t) = MCM\{(t^2-1)^6,(t^4-1)\} = t^2-1.$$

For any monomial matrix, $M = M(a_1,\ldots,a_n;i_1\ldots i_n)$, the characteristic polynomial of the isomorphism g_M can be obtained from the coefficients $a_1,\ldots,a_n$ and the cycle type of the permutation $i_1\ldots i_n$.

Lemma 4

(a) The characteristic polynomial of M is a product of factors, each of them corresponding to one of the disjoint cycles in the decomposition of $i_1\ldots i_k$.
(b) Given a matrix $M - M(a_1,\ldots,a_k;j_1,\ldots,j_k)$ with $j_1,\ldots,j_k = (j_1,\ldots,j_k)$ a cycle of length k, the characteristic polynomial of M is $t^k - a_1\cdots a_k$.

Example 12

$$M = M(1,1,2,2,2,2,2;2,3,1,5,6,7,4) = \begin{pmatrix} 0&1&0&0&0&0&0\\ 0&0&1&0&0&0&0\\ 2&0&0&0&0&0&0\\ 0&0&0&0&2&0&0\\ 0&0&0&0&0&2&0\\ 0&0&0&0&0&0&2\\ 0&0&0&2&0&0&0 \end{pmatrix}$$

$2\ 3\ 1\ 5\ 6\ 7\ 4 = (2,3,1)(5,6,7,4)$ and then $Q_M(t) = (t^3-2)(t^4-16)$.

In $\mathbb{F}_5[t]$: $Q_M(t) - (t^3-2)(t^4-1)$.
In $\mathbb{F}_7[t]$: $Q_M(t) = (t^3-2)(t^4-2)$.
In $\mathbb{F}_{11}[t]$: $Q_M(t) = (t^3-2)(t^4-5)$.

Let us denote by $G_k(t)$ the GCD of all factors in the characteristic polynomial of $M = M(a_1,\ldots,a_n;i_1,\ldots,i_n)$ of degree k corresponding to cycles of length k, $1\le k\le n$.

Proposition 2 *The minimal annihilating polynomial of monomial matrix $M = M(a_1,\ldots,a_n;i_1,\ldots,i_n)$ is:*

$$P_M(t) = LCM(G_1(t),\ldots,G_k(t)).$$

Example 13 Let us consider

$$M = M(2,3,1,1,3,4,1,1,4;2,3,1,5,4,7,6,8,9)$$
$$= \begin{pmatrix} 0&2&0&0&0&0&0&0&0\\ 0&0&3&0&0&0&0&0&0\\ 1&0&0&0&0&0&0&0&0\\ 0&0&0&0&1&0&0&0&0\\ 0&0&0&3&0&0&0&0&0\\ 0&0&0&0&0&0&1&0&0\\ 0&0&0&0&0&1&0&0&0\\ 0&0&0&0&0&0&0&1&0\\ 0&0&0&0&0&0&0&0&4 \end{pmatrix} \in M_9(\mathbb{F}_7)$$

$2\ 3\ 1\ 5\ 4\ 7\ 6\ 8\ 9 = (2,3,1)(5,4)(7,6)(8)(9)$.
$Q_M(t) = (t^3-6)(t^2-3)(t^2-1)(t-1)(t-4)$.
$G_1(t) = (t-1)(t-4)$, $G_2(t) = (t^2-3)(t^2-1)$, $G_3(t) = t^3-6$.
$P_M(t) = (t-1)(t-3)(t-4)(t-5)(t-6)(t^2-3)$.

An example of invariant subspaces are those spanned by eigenvectors. We can determine the set of eigenvectors of a permutation isomorphism from the decomposition of the permutation associated to it, into disjoint cycles and that of a monomial isomorphism from the decomposition into disjoint cycles and the coefficients in the matrix. The proofs are based on straightforward computations.

Proposition 3 *Let P be a permutation matrix associated to a permutation which is a disjoint product of cycles, and $\lambda \in \mathbb{F}_p$ an eigenvalue of P, λ an mth-root of unity, the vector $(\lambda^{j_k},\dots,\lambda^{j_1})$, where $j_1,\dots,j_k$ is the result of re-ordering the indices of the cycle $(i_1,\dots,i_k)$ in such a way that $j_1 \leq \cdots \leq j_k$. Then $(\lambda_{j_k},\dots,\lambda_{j_1})$ is the eigenvector associated to the eigenvalue λ.*

Example 14 We will consider $p = 5$.

1. Let us consider the 2-cycle $(2,1)$ and the 2×2-matrix associated to it. Then the eigenvector for the eigenvalue $\lambda \in \mathbb{F}_5$, $\lambda^2 = 1$, is $(\lambda,1)$. Since the roots of $\lambda^2 = 1$ are 1 and 4, there are two linearly independent eigenvectors: $(1,1)$ and $(4,1)$.
2. If the permutation 3×3-matrix is associated to the 2-cycle $(2,3,1)$, the eigenvector corresponding to the eigenvalue $\lambda \in \mathbb{F}_5$, $\lambda^3 = 1$, is $(\lambda,\lambda^2,1)$. The equation $\lambda^3 = 1$ has only one root, 1, and therefore there is an unique eigenvector is: $(1,1,1)$.

   If we consider the 3×3-permutation matrix is associated to the 2-cycle $(3,2,1)$, there is also an unique eigenvector: $(1,1,1)$.
3. Let us consider now the case of 4×4-permutation matrices associated to 4-cycles. Let λ be a 4*th*-root of unity (there are four 4*th*-roots of unity: $\lambda_1 = 1$, $\lambda_2 = 2$,$\lambda_3 = 3$ and $\lambda_4 = 4$).

   (i) If the 4-cycle is $(1,2,3,4)$, the eigenvector corresponding to the eigenvalue $\lambda \in \mathbb{F}_5$, is $(\lambda,\lambda^2,\lambda^3,1)$. That is to say, there are four linearly independent eigenvectors:

$$(1, 1, 1, 1), (2, 4, 3, 1), (3, 4, 2, 1), (4, 1, 4, 1).$$

(ii) If the 4-cycle is $(1, 3, 4, 2)$, the eigenvector corresponding to the eigenvalue $\lambda \in \mathbb{F}_5$, $\lambda^4 = 1$, is $(\lambda, \lambda^3, \lambda^2, 1)$. That is to say, there are four linearly independent eigenvectors:

$$(1, 1, 1, 1), (2, 3, 4, 1), (3, 2, 4, 1), (4, 4, 1, 1).$$

(iii) If the 4-cycle is $(1, 2, 3, 4)$, the eigenvector corresponding to the eigenvalue $\lambda \in \mathbb{F}_5$, $\lambda^4 = 1$, is $(\lambda, 1, \lambda^3, \lambda^2)$. That is to say, there are four linearly independent eigenvectors:

$$(1, 1, 1, 1), (2, 1, 3, 4), (3, 1, 2, 4), (4, 1, 4, 1).$$

(iv) If the 4-cycle is $(1, 3, 2, 4)$, the eigenvector corresponding to the eigenvalue $\lambda \in \mathbb{F}_5$, $\lambda^4 = 1$, is $(\lambda, \lambda^2, 1, \lambda^3$. That is to say, there are four linearly independent eigenvectors:

$$(1, 1, 1, 1), (2, 4, 1, 3), (3, 4, 1, 2), (4, 1, 1, 4).$$

(v) If the 4-cycle is $(1, 4, 2, 3)$, the eigenvector corresponding to the eigenvalue $\lambda \in \mathbb{F}_5$, $\lambda^4 = 1$, is $(\lambda, 1, \lambda^2, \lambda^3)$. That is to say, there are four linearly independent eigenvectors:

$$(1, 1, 1, 1), (2, 1, 4, 3), (3, 1, 4, 2), (4, 1, 1, 4).$$

(vi) If the 4-cycle is $(1, 2, 4, 3)$, the eigenvector corresponding to the eigenvalue $\lambda \in \mathbb{F}_5$, $\lambda^4 = 1$, is $(\lambda, \lambda^3, 1, \lambda^2)$. That is to say, there are four linearly independent eigenvectors:

$$(1, 1, 1, 1), (2, 3, 1, 4), (3, 2, 1, 4), (4, 4, 1, 1).$$

4. Let us consider the permutation matrix associated to a cycle of type $2+2+4+8$. Then the minimal annihilating polynomial is: $(t^4 + 1)(t^2 + 1)(t + 1)(t - 1)$. The eigenvalues in $\mathbb{F}_5$ are: $\lambda_1 = 1$, $\lambda_2 = 4$, $\lambda_3 = 2$ and $\lambda_4 = 3$, being the algebraic multiplicities 4,4,1,1, respectively.

Let us assume, for example, that the 2-cycles are: $(9, 16)$ and $(13, 15)$, the 4-cycle is $(1, 3, 5, 14)$ and the 8-cycle is $(2, 4, 6, 7, 8, 10, 11, 12)$. Then the following linearly independent eigenvectors are obtained.

$$\begin{array}{ll}
\lambda_1 = 1 & (0,0,0,0,0,0,0,0,0,0,0,0,1,0,1,0) \\
 & (0,0,0,0,0,0,0,0,1,0,0,0,0,0,0,1) \\
 & (1,0,1,0,1,0,0,0,0,0,0,0,0,1,0,0) \\
 & (0,1,0,1,0,1,1,1,0,1,1,1,0,0,0,0) \\
\lambda_2 = 4 & (0,0,0,0,0,0,0,0,0,0,0,0,4,0,1,0) \\
 & (0,0,0,0,0,0,0,0,4,0,0,0,0,0,0,1) \\
 & (4,0,1,0,4,0,0,0,0,0,0,0,0,1,0,0) \\
 & (0,4,0,1,0,4,1,4,0,1,4,1,0,0,0,0) \\
\lambda_3 = 2 & (2,0,4,0,3,0,0,0,0,0,0,0,0,1,0,0) \\
\lambda_4 = 3 & (3,0,4,0,2,0,0,0,0,0,0,0,0,1,0,0)
\end{array}$$

Let us consider now a monomial matrix Let $M = M(a_1, \dots, a_n; i_1, \dots, i_n)$.

Proposition 4 *The eigenvalues of M are the roots of the polynomials* $t^k - a_{j_1} \cdots a_{j_k}$ *for each cycle* $j_1 \dots j_k$ *of length k in the decomposition of the permutation* $i_1 \dots i_n$ *into disjoint cycles, being* $a_{j_1}, \dots, a_{j_k}$ *the coefficients of M in columns* $j_1, \dots, j_k$.

Example 15 Let us consider

$$M = M(1, 1, 2, 2, 2, 2, 2; 2, 3, 1, 5, 6, 7, 4) = \begin{pmatrix} 0&1&0&0&0&0&0 \\ 0&0&1&0&0&0&0 \\ 2&0&0&0&0&0&0 \\ 0&0&0&0&2&0&0 \\ 0&0&0&0&0&2&0 \\ 0&0&0&0&0&0&2 \\ 0&0&0&2&0&0&0 \end{pmatrix}$$

We have: $2\ 3\ 1\ 5\ 6\ 7\ 4 = (2, 3, 1)(5, 6, 7, 4)$ and therefore $Q_M(t) = (t^3 - 2)(t^4 - 16)$.

In $\mathbb{F}_5[t]$: $Q_M(t) = (t^3 - 2)(t^4 - 1)$ and the eigenvalues of M are: $3(2), 1, 2, 4$.
In $\mathbb{F}_7[t]$: $Q_M(t) = (t^3 - 2)(t^4 - 2)$ and the eigenvalues of M are: $2, 5$.
In $\mathbb{F}_{11}[t]$: $Q_M(t) = (t^3 - 2)(t^4 - 5)$ and the eigenvalues of M are: $7, 2, 9$.

Assume that the permutation $i_1, \dots, i_n$ splits into m_1 cycles of length k_1,..., m_l cycles of length k_l. For any irreducible cycle (j) of length 1, e_j is an eigenvector. We can generalize this as follows.

Proposition 5 *Let* $a_{j_1} \dots a_{j_k}$ *be a cycle of length* $k \geq 2$ *in the decomposition of the characteristic polynomial of M into irreducible factors and* $t^k - a_{j_1} \dots a_{j_k}$ *the corresponding factor in* $Q_M(t)$. *For each root* λ *(in the case where there exists any)* *of this polynomial we obtain an eigenvector:*

$$(\lambda^{k-1}, a_{j_2}a_{j_3}\dots a_{j_k}, \lambda a_{j_3}\dots a_{j_k}, \dots \lambda^{k-3}a_{j_k-1}a_{j_k}, \lambda^{k-2}a_{j_k})$$

Example 16

$$M = M(a_1, a_2, a_3, a_4, a_5, a_6, a_7, a_8, a_9; 2, 3, 4, 1, 6, 7, 8, 9, 5) = \begin{pmatrix} 0 & a_1 & 0 & 0 & 0 & 0 & 0 & 0 & 0 \\ 0 & 0 & a_2 & 0 & 0 & 0 & 0 & 0 & 0 \\ 0 & 0 & 0 & a_3 & 0 & 0 & 0 & 0 & 0 \\ a_4 & 0 & 0 & 0 & 0 & 0 & 0 & 0 & 0 \\ 0 & 0 & 0 & 0 & 0 & a_5 & 0 & 0 & 0 \\ 0 & 0 & 0 & 0 & 0 & 0 & a_6 & 0 & 0 \\ 0 & 0 & 0 & 0 & 0 & 0 & 0 & a_7 & 0 \\ 0 & 0 & 0 & 0 & 0 & 0 & 0 & 0 & a_8 \\ 0 & 0 & 0 & 0 & a_9 & 0 & 0 & 0 & 0 \end{pmatrix}$$

the eigenvectors are:

- $(\lambda^3, a_2a_3a_4, \lambda a_3a_4, \lambda^2 a_4, 0, 0, 0, 0, 0)$.
- $(\mu^4, a_6a_7a_8a_9, \mu a_7a_8a_9, \mu^2 a_8a_9, \mu^3 a_9)$.

For each root λ of $t^4 - a_1a_2a_3a_4$ and μ of $t^5 - a_5a_6a_7a_8a_9$ in $\mathbb{F}_p$.

References

1. Fripertinger, H.: Enumeration of the semilinear isometry classes of linear codes. Bayrether Mathematische Schriften **74**, 100–122 (2005)
2. García-Planas, M.I., Magret, M.D.: Eigenvalues and eigenvectors of permutation matrices. Adv. Pure Math. **5**, 390–394 (2015)
3. Sendrier, N., Simos, D.E.: How easy is code equivalence over $\mathbb{F}_q$? WCC 2013-International Workshop on Coding and Cryptography, Bergen (2013)
4. Sendrier, N., Simos, D.E.: The Hardness of Code Equivalence over $\mathbb{F}_q$ and Its Application to Code-Based Cryptography. Post-Quantum Cryptography, pp. 203–216 (2013)

Advances in the Study of Singular Semilinear Elliptic Problems

Daniela Giachetti, Pedro J. Martínez-Aparicio, and François Murat

Abstract In this paper we deal with some results concerning semilinear elliptic singular problems with Dirichlet boundary conditions. The problem becomes singular where the solution u vanishes. The model of this kind of problems is

$$\begin{cases} u \geq 0 & \text{in } \Omega, \\ -div\, A(x)Du = F(x,u) & \text{in } \Omega, \\ u = 0 & \text{on } \partial\Omega, \end{cases}$$

where Ω is a bounded open set of $\mathbb{R}^N$, $N \geq 1$, A is a coercive matrix with coefficients in $L^\infty(\Omega)$ and $F : (x,s) \in \Omega \times [0,+\infty[\to F(x,s) \in [0,+\infty]$ is a Carathéodory function which is singular at $s = 0$.

Our aim is to study the meaning of the assumptions made on the singular function $F(x,s)$ in the papers [Giachetti et al., J. Math. Pures Appl. (2016, in press); Giachetti et al., Definition, existence, stability and uniqueness of the solution to a semilinear elliptic problem with a strong singularity at $u = 0$ (Preprint, 2016); Giachetti et al., Homogenization of a Dirichlet semilinear elliptic problem with a strong singularity at $u = 0$ in a domain with many small holes (Preprint, 2016)], to extend some uniqueness results of the solution given in the same papers, and to prove the L^∞-regularity of the solutions under some regularity assumption on the data.

D. Giachetti
Facoltà di Ingegneria Civile e Industriale, Dipartimento di Scienze di Base e Applicate per l'Ingegneria, Sapienza Università di Roma, Via Scarpa 16, 00161 Roma, Italy
e-mail: daniela.giachetti@sbai.uniroma1.it

P.J. Martínez-Aparicio (✉)
Departamento de Matemática Aplicada y Estadística, Universidad Politécnica de Cartagena, Paseo Alfonso XIII 52, 30202 Cartagena, Murcia, Spain
e-mail: pedroj.martinez@upct.es

F. Murat
Laboratoire Jacques-Louis Lions, Université Pierre et Marie Curie et CNRS, Boîte Courrier 187, 75252 Paris Cedex 05, France
e-mail: murat@ann.jussieu.fr

F. Ortegón Gallego et al. (eds.), *Trends in Differential Equations and Applications*, SEMA SIMAI Springer Series 8, DOI 10.1007/978-3-319-32013-7_13

1 Introduction

In the papers [3, 4] we study the problem of finding a function u which satisfies, in a convenient sense, the following semilinear singular (in the u variable) problem

$$\begin{cases} u \geq 0 & \text{in } \Omega, \\ -div\, A(x)Du = F(x,u) & \text{in } \Omega, \\ u = 0 & \text{on } \partial\Omega. \end{cases} \tag{1}$$

More precisely in [3] we deal with the case of *mild singularities* (see assumption (4) below). This means that the singular term $F(x,s)$ grows at most like $\dfrac{1}{s^\gamma}$ with $0 < \gamma \leq 1$ near $s = 0$.

In [3] Ω is an open bounded set of $\mathbb{R}^N$, $N \geq 1$ (no regularity is assumed on the boundary $\partial\Omega$ of Ω), the matrix A satisfies

$$\begin{cases} A(x) \in L^\infty(\Omega)^{N\times N}, \\ \exists\alpha > 0,\ A(x) \geq \alpha I \quad \text{a.e. } x \in \Omega, \end{cases} \tag{2}$$

the function F satisfies

$$\begin{cases} F : (x,s) \in \Omega \times [0,+\infty[\to F(x,s) \in [0,+\infty] \text{ is a Carathéodory function,} \\ \text{i.e. } F \text{ satisfies} \\ i) \text{ for a.e. } x \in \Omega,\ s \in [0,+\infty[\to F(x,s) \in [0,+\infty] \text{ is continuous,} \\ ii)\ \forall s \in [0,+\infty[,\ x \in \Omega \to F(x,s) \in [0,+\infty] \text{ is measurable,} \end{cases} \tag{3}$$

and

$$\begin{cases} \exists\gamma, \exists h \text{ with} \\ i)\ 0 < \gamma \leq 1, \\ ii)\ h(x) \geq 0 \text{ a.e. } x \in \Omega,\ h \in L^r(\Omega), \\ \text{with } r = \frac{2N}{N+2} \text{ if } N \geq 3,\ r > 1 \text{ if } N = 2,\ r = 1 \text{ if } N = 1, \\ \text{such that} \\ iii)\ 0 \leq F(x,s) \leq h(x)\left(\dfrac{1}{s^\gamma} + 1\right) \text{ a.e. } x \in \Omega, \forall s > 0. \end{cases} \tag{4}$$

In [3] existence of at least one nonnegative solution in the sense of Definition 1 (given in Sect. 2 below) is proved; moreover uniqueness is also proved if $F(x,s)$ is nonincreasing or more generally "*almost nonincreasing*" in the s variable in the sense that

$$\begin{cases} \text{there exists } \lambda,\ 0 \le \lambda < \lambda_1, \text{ such that} \\ F(x,s) - \lambda s \le F(x,t) - \lambda t \text{ a.e. } x \in \Omega,\ \forall s, \forall t,\ 0 \le t \le s, \end{cases} \tag{5}$$

where λ_1 is the first eigenvalue of the operator $-div\, {}^{s}\!A(x)D$ in $H_0^1(\Omega)$ and ${}^{s}\!A(x) = (A(x) + {}^{t}\!A(x))/2$ is the symmetrized part of the matrix $A(x)$.

Finally, the homogenization of these equations posed in a sequence of domains Ω^ε obtained by removing many small holes from a fixed domain Ω is also considered in [3].

In [4, 5] we study the case of *strong singularities*, which means that hypothesis (4) is replaced by

$$\begin{cases} i)\ \exists h, h(x) \ge 0 \text{ a.e. } x \in \Omega, h \in L^r(\Omega), \\ \text{with } r = \frac{2N}{N+2} \text{ if } N \ge 3,\ r > 1 \text{ if } N = 2,\ r = 1 \text{ if } N = 1, \\ ii)\ \exists \Gamma : s \in [0,+\infty] \to \Gamma(s) \in [0,+\infty[,\ \Gamma \in C^1([0,+\infty[), \\ \text{with } \Gamma(0) = 0,\ \Gamma'(s) > 0\ \forall s > 0, \\ iii)\ 0 \le F(x,s) \le \dfrac{h(x)}{\Gamma(s)} \text{ a.e. } x \in \Omega, \forall s > 0. \end{cases} \tag{6}$$

Remark 1 Note that condition (6) includes condition (4) if $\Gamma(s) = s^\gamma/(s^\gamma + 1)$ with $0 < \gamma \le 1$.

Note also that in (6) the growth of $F(x,s)$ at the singularity $s = 0$ is more general because it includes powerlike growth conditions of the type (4 *iii*) for any $\gamma > 0$ and also more general growth conditions like in the following example

$$F(x,s) = \frac{h(x)}{\exp(-\frac{1}{s})}\left(2 + \sin\left(\frac{1}{s}\right)\right) \text{ a.e. } x \in \Omega,\ \forall s > 0. \tag{7}$$

Note finally that (6 *ii*) implies that the function Γ is increasing and satisfies $\Gamma(s) > 0$ for every $s > 0$, as well as $\Gamma \in \mathrm{Lip}_{\mathrm{loc}}([0,+\infty[)$ with

$$0 < \inf_{a \le t \le b} \Gamma'(t) \le \sup_{a \le t \le b} \Gamma'(t) < +\infty\ \forall\, a, b,\ 0 < a \le b < +\infty.$$

□

An existence result of at least one nonnegative solution in the sense of Definition 4 (given in Sect. 2 below) is proved in [4]. Uniqueness is also proved in [4] if $F(x,s)$ is nonincreasing in the s variable.

In [5] we consider the homogenization in perforated domains for problems with strong singularities.

The works [3, 4] were inspired by the paper [2] of L. Boccardo and L. Orsina where they prove existence and regularity as well as non existence results.

L. Boccardo and J. Casado-Díaz also proved in [1] an uniqueness result of the solutions obtained by approximation and studied the stability of the solution with respect to the G-convergence for a sequence of matrices $A^\varepsilon(x)$ which are equicoercive and equibounded.

In the first part of the present paper we make some remarks about the growth condition (6). More precisely we prove that (6) is equivalent to a family of bounds from above for the function $F(x,s)$ on the sets $\{s \geq k\}$, $k > 0$. In addition we point out by an example that a function F which satisfies the growth condition from above (6) does not in general satisfy a similar growth condition from below.

In the second part of the present paper we extend the uniqueness results given in [3, 4]. In the case of a mild singularity [assumption (4)] we improve condition (5) by allowing $\lambda = \lambda_1$, where λ_1 is the first eigenvalue of the operator $-div\,{}^sA(x)D$ in $H_0^1(\Omega)$, and we prove the uniqueness of the solution under the further assumption that the function $F(x,s)$ does not coincide wit $\lambda_1 + c(x)$ for any function $c(x)$ in any set of the type $\{x \in \Omega,\ s_-(x) \leq s \leq s_+(x)\}$. In the case of a strong singularity of the type $\Gamma(s) = s^\gamma/(s^\gamma + 1)$ with $\gamma > 1$ [assumption (6)] we weaken the condition on the monotonicity of $F(x,s)$ given in [4], requiring that $F(x,s)$ is only "*γ-almost nonincreasing*" in the s variable in the sense that

$$\begin{cases} \text{there exists } \lambda,\ 0 \leq \lambda < 4\dfrac{\gamma+3}{(\gamma+4)^2}\lambda_1, \text{ such that} \\ F(x,s) - \lambda s \leq F(x,t) - \lambda t \ \text{ a.e. } x \in \Omega,\ \ \forall s, \forall t,\ 0 \leq t \leq s. \end{cases} \tag{8}$$

Under this condition we prove that the problem can not have two different solutions u_1 and u_2 satisfying $u_1 - u_2 \in L^\infty(\Omega)$.

In the third part of the present paper we prove, assuming that the function h in (6) belongs to $L^t(\Omega)$, $t > \frac{N}{2}$ if $N \geq 2$, $t = 1$ if $N = 1$, that any solution u to problem (1) (not necessarily obtained by approximation) in the sense of Definition 4 below belongs to $L^\infty(\Omega)$. Note that this L^∞-regularity result holds true under the general growth condition (6) which allows one to consider functions F like (7).

The plan of the paper is the following. In Sect. 2 we recall the definitions of the solution and the existence results given in [3, 4]. In Sect. 3 we study some features of the assumptions (4) and (6). Sect. 4 is devoted to extend the uniqueness results given in [3, 4]. In Sect. 5 we give an L^∞-regularity result for the solutions under a stronger assumption on $F(x,s)$ in the x variable.

Notation

We denote as usual by $\mathscr{D}(\Omega)$ the space of the functions $C^\infty(\Omega)$ whose support is compact and included on Ω, and by $\mathscr{D}'(\Omega)$ the space of distributions on Ω.

For every $s \in \mathbb{R}$ and every $k > 0$ we define as usual

$$s^+ = \max\{s, 0\},\ \ s^- = \max\{0, -s\},$$

$$T_k(s) = \max\{-k, \min\{s,k\}\},\quad G_k(s) = s - T_k(s).$$

2 Definitions of the Solution and Existence Results

In this section we recall the definitions of the solutions to problem (1) that we used in the papers [3] (mild singularity), [4, 5] (strong singularity) and we recall the statements of the existence results in both cases. In order to introduce the notion of solution in the case of a strong singularity, we also need to recall the definition of the space $\mathcal{V}(\Omega)$ of test functions and a formal duality (see (14) below).

2.1 *Mild Singularities*

In the paper [3], in the case of mild singularities, we used the following definition of a solution to problem (1).

Definition 1 ([3]) Assume that the matrix A and the function F satisfy (2), (3) and (4). We say that u is a solution to problem (1) if u satisfies

$$u \in H_0^1(\Omega), \tag{9}$$

$$u \geq 0 \quad \text{a.e. in } \Omega, \tag{10}$$

and

$$\begin{cases} \forall \varphi \in H_0^1(\Omega),\ \varphi \geq 0, \text{ one has} \\ \int_\Omega F(x,u)\varphi < +\infty, \\ \int_\Omega A(x)DuD\varphi = \int_\Omega F(x,u)\varphi. \end{cases} \tag{11}$$

□

The existence result that we proved in [3] is the following.

Theorem 1 ([3]) *Assume that the matrix A and the function F satisfy* (2), (3) *and* (4). *Then there exists at least one solution u to problem* (1) *in the sense of Definition 1.*

2.2 *Strong Singularities*

In order to introduce the notion of solution to problem (1) that we use in [4], in the case of strong singularities, we first define the following space $\mathcal{V}(\Omega)$ of test functions and a new notation (see (14) below).

Definition 2 ([4]) We define the space $\mathscr{V}(\Omega)$ as the space of the functions v which satisfy

$$v \in H_0^1(\Omega) \cap L^\infty(\Omega), \tag{12}$$

$$\begin{cases} \exists I \text{ finite}, \ \exists \hat{\varphi}_i, \exists \hat{g}_i, i \in I, \exists \hat{f}, \text{ with} \\ \hat{\varphi}_i \in H_0^1(\Omega) \cap L^\infty(\Omega), \hat{g}_i \in (L^2(\Omega))^N, \hat{f} \in L^1(\Omega), \\ \text{such that} \\ -div\, {}^t\!A(x)Dv = \sum_{i\in I} \hat{\varphi}_i(-div\, \hat{g}_i) + \hat{f} \text{ in } \mathscr{D}'(\Omega). \end{cases} \tag{13}$$

□

In the definition of $\mathscr{V}(\Omega)$ we use the notation $\hat{\varphi}_i$, $\hat{g}_i$, and $\hat{f}$ to help the reader to identify the functions which enter in the definition of the functions of $\mathscr{V}(\Omega)$.

Note that $\mathscr{V}(\Omega)$ is a vector space.

Definition 3 ([4]) When $v \in \mathscr{V}(\Omega)$ with

$$-div\, {}^t\!A(x)Dv = \sum_{i\in I} \hat{\varphi}_i(-div\, \hat{g}_i) + \hat{f} \quad \text{in } \mathscr{D}'(\Omega),$$

where I, $\hat{\varphi}_i$, $\hat{g}_i$ and $\hat{f}$ are as in (13), and when z satisfies

$$z \in H^1_{loc}(\Omega) \cap L^\infty(\Omega) \text{ with } \varphi z \in H_0^1(\Omega), \ \forall \varphi \in H_0^1(\Omega) \cap L^\infty(\Omega),$$

we will use the following notation

$$\langle\langle -div\, {}^t\!A(x)Dv, z\rangle\rangle_\Omega = \sum_{i\in I} \int_\Omega \hat{g}_i D(\hat{\varphi}_i z) + \int_\Omega \hat{f} z. \tag{14}$$

□

We now give the definition of a solution to problem (1) that we used in [4].

Definition 4 ([4]) Assume that the matrix A and the function F satisfy (2), (3) and (6). We say that u is a solution to problem (1) if u satisfies

$$\begin{cases} i)\ u \in L^2(\Omega) \cap H^1_{\mathrm{loc}}(\Omega), \\ ii)\ u(x) \geq 0 \ \text{ a.e. } x \in \Omega, \\ iii)\ G_k(u) \in H_0^1(\Omega) \quad \forall k > 0, \\ iv)\ \varphi DT_k(u) \in (L^2(\Omega))^N \ \forall k > 0, \ \ \forall \varphi \in H_0^1(\Omega) \cap L^\infty(\Omega), \end{cases} \tag{15}$$

$$\begin{cases} \forall v \in \mathscr{V}(\Omega),\ v \geq 0, \\ \text{with } -div\,{}^t\!A(x)Dv = \sum_{i\in I} \hat{\varphi}_i(-div\,\hat{g}_i) + \hat{f} \text{ in } \mathscr{D}'(\Omega), \\ \text{where } \hat{\varphi}_i \in H_0^1(\Omega) \cap L^\infty(\Omega), \hat{g}_i \in (L^2(\Omega))^N, \hat{f} \in L^1(\Omega), \\ \text{one has} \\ i)\ \int_\Omega F(x,u)v < +\infty, \\ ii)\ \int_\Omega {}^t\!A(x)DvDG_k(u) + \sum_{i\in I}\int_\Omega \hat{g}_i D(\hat{\varphi}_i T_k(u)) + \int_\Omega \hat{f}T_k(u) = \\ = \langle -div\,{}^t\!A(x)Dv, G_k(u)\rangle_{H^{-1}(\Omega),H_0^1(\Omega)} + \langle\langle -div\,{}^t\!A(x)Dv, T_k(u)\rangle\rangle_\Omega = \\ = \int_\Omega F(x,u)v\ \ \forall k > 0. \end{cases} \tag{16}$$

□

This is a definition of a solution by transposition in the spirit of those introduced by J.-L. Lions and E. Magenes and by G. Stampacchia.

Remark 2 We prove in [4] that every solution to problem (1) in the sense of Definition 4 satisfies

$$\beta(T_k(u)) \in H_0^1(\Omega)\ \forall k > 0,$$

and that the following estimate holds true

$$\alpha\|D\beta(T_k(u))\|^2_{(L^2(\Omega))^N} < \|h\|_{L^1(\Omega)}\ \forall k > 0,$$

where the function $\beta : s \in [0,+\infty[\to \beta(s) \in [0,+\infty[$ is defined by

$$\beta(s) = \int_0^s \sqrt{\Gamma'(\tau)}d\tau.$$

It is easy to show that in the case where $\Gamma(s) = s^\gamma/(s^\gamma + 1)$ with $\gamma \geq 1$ one can actually prove [with the techniques of [4] but using now, in the formal computation, the test function $(T_k(u))^\gamma$ in place of $\Gamma(T_k(u))$] that

$$(T_k(u))^{\frac{\gamma+1}{2}} \in H_0^1(\Omega), \tag{17}$$

which is a slightly different result. The estimate in this case is the following

$$\alpha\frac{4\gamma}{(\gamma+1)^2}\|DT_k(u)^{\frac{\gamma+1}{2}})\|^2_{(L^2(\Omega))^N} \leq \|h\|_{L^1(\Omega)}(1 + k^\gamma)\ \forall k > 0.$$

□

In this case of strong singularities we proved in [4] the following existence result.

Theorem 2 ([4]) *Assume that the matrix A and the function F satisfy* (2), (3) *and* (6). *Then there exists at least one solution u to problem* (1) *in the sense of Definition 4.*

3 Remarks on the Growth Assumptions of the Singular Term

In this section we discuss some issues concerning assumption (6).

In the following Proposition 1 we prove that condition (6) [and therefore (4))] can be rewritten (in an equivalent way) through a family of bounds from above for the function F on the sets $\{s \geq k\}$, $k > 0$.

Proposition 1 *Assumption* (6) *is equivalent to the following one*

$$\begin{cases} \forall k > 0, \ \exists h_k \in L^r(\Omega), \\ \text{with } r = \frac{2N}{N+2} \text{ if } N \geq 3, \ r > 1 \text{ if } N = 2, \ r = 1 \text{ if } N = 1, \\ \text{such that } 0 \leq F(x,s) \leq h_k(x) \text{ a.e. } x \in \Omega, \forall s \geq k. \end{cases} \tag{18}$$

Proof It is obvious that (6) implies (18) since we can take $h_k(x) = \dfrac{h(x)}{\Gamma(k)}$ for $s \geq k$.

Now we prove that (18) implies (6).

Choosing $k = \frac{1}{n}$ we deduce from (18) that

$$\begin{cases} \forall n \in \mathbb{N}, \ n \geq 1, \ \exists \overline{h}_n \in L^r(\Omega), \\ \text{such that } 0 \leq F(x,s) \leq \overline{h}_n(x) \text{ a.e. } x \in \Omega, \forall s \geq \frac{1}{n}. \end{cases} \tag{19}$$

We can always assume that, for $n \geq 1$, one has $\overline{h}_{n+1}(x) \geq \overline{h}_n(x)$ almost everywhere in Ω, and we define c_n by

$$c_n = \frac{1}{\|\overline{h}_n\|_{L^r(\Omega)}} \frac{1}{2^n} \quad \forall n \geq 1.$$

Therefore we have $c_{n+1} \leq \frac{1}{2} c_n$ and

$$c_n \to 0 \text{ as } n \to +\infty. \tag{20}$$

Now we define $\overline{h} = \displaystyle\sum_2^\infty c_n \overline{h}_n$; then $\overline{h} \in L^r(\Omega)$ since $c_n \|\overline{h}_n\|_{L^r(\Omega)} = \dfrac{1}{2^n}$ and since the series $\displaystyle\sum_2^\infty \frac{1}{2^n}$ is convergent.

Let us define a function g by setting $g\left(\frac{1}{n}\right) = c_{n+1}$ for $n \geq 1$, by defining g as the linear interpolation between c_{n+2} and c_{n+1} for $\frac{1}{n+1} \leq s \leq \frac{1}{n}$, and by setting $g(0) = 0$. This function g is increasing, piecewise affine and continuous (see (20)) on $[0, 1]$.

Using inequality (19) for $s \geq \frac{1}{n+1}$ and the fact that g is increasing, we have for $\frac{1}{n+1} \leq s \leq \frac{1}{n}$ and $n \geq 1$

$$0 \leq F(x,s) \leq \overline{h}_{n+1}(x) = c_{n+1}\overline{h}_{n+1}(x)\frac{1}{c_{n+1}} \leq \overline{h}(x)\frac{1}{c_{n+1}} = \overline{h}(x)\frac{1}{g(\frac{1}{n})} \leq \overline{h}(x)\frac{1}{g(s)}.$$

We have proved that (6 *iii*) is satisfied for $0 \leq s \leq 1$ with the functions $h = \overline{h}$ and $\Gamma = g$, where $g \in C^0([0,1])$, g piecewise affine, $g(0) = 0$ and $g(s) > 0$ and $g'(s) > 0$ for $s > 0$.

For what concerns $s \geq 1$, we take an increasing and concave function $g \in C^1([1,+\infty]) \cap \mathrm{Lip}([1,+\infty])$ such that $g(1) \leq g(s) \leq 2g(1)$ for $s \geq 1$. Using (18) for $s \geq 1$ and the latest inequality we have for $s \geq 1$

$$F(x,s) \leq h_1(x) = \frac{g(1)h_1(x)}{g(1)} \leq \frac{2g(1)h_1(x)}{g(s)}.$$

Setting $h(x) = \sup\left\{\overline{h}(x), 2g(1)h_1(x)\right\}$, we have proved that (6 *iii*) is satisfied for $s \geq 0$ with the functions h and g, where $h \in L^r(\Omega)$, and where g satisfies (6 *ii*) but does not belong neither to $C^1([0,+\infty])$ nor to $\mathrm{Lip}([0,+\infty])$.

It is easy to replace the function g by $\overline{g}$ defined by

$$\overline{g}(s) = \int_0^s \inf\{1, g'(t)\}dt;$$

the function $\overline{g}$ is increasing, piecewise affine and continuous on $[0,1]$, C^1 on $[1,+\infty]$ and satisfies

$$\begin{cases} \overline{g}(0) = 0, \ \overline{g}(s) > 0 \ \forall s > 0, \ \overline{g} \in \mathrm{Lip}([0,+\infty]), \\ 0 < \inf_{a \leq t \leq b} \overline{g}'(t) \leq \sup_{0 \leq t \leq +\infty} \overline{g}'(t) < +\infty \ \forall\, a, b, \ 0 < a \leq b < +\infty, \end{cases} \tag{21}$$

as well as $\overline{g} \leq g$; a more technical process allows one to build a function Γ with $\Gamma \leq \overline{g}$ which still satisfies (21) and belongs to $C^1([0,+\infty])$.

We have (almost completely) proved that (18) implies (6). □

In the next Remark we point out that condition (6 *iii*) prescribes only a growth from above on the function F but no growth from below.

Remark 3 In this remark we give an example of function F satisfying condition (6 *iii*) with $\Gamma(s) = s^2$ and $h = 1$, i.e. such that

$$0 \le F(x,s) \le \frac{1}{s^2} \quad \forall s > 0$$

which does not satisfy

$$F(x,s) \ge \frac{C}{s^2} \text{ for any } C > 0. \tag{22}$$

Let $0 < \theta < 1$ and define

$$s_n = \theta^{(2^n - 1)} \text{ and } t_n = \theta^{(2^{(n+1)}-2)},\ \forall n \ge 0.$$

Then one has

$$\begin{cases} 0 < s_n < 1,\ s_{n+1} = \theta t_n,\ t_n = s_n^2,\ \forall n \ge 0, \\ 0 < s_{n+1} < t_n < s_n < 1,\ \forall n < 0, \\ s_n \to 0 \text{ as } n \to +\infty. \end{cases}$$

We now define a function F such that

$$\begin{cases} F(s) \text{ is continuous for every } s > 0, \\ F(s) = 1 \text{ for every } s \ge 1, \\ F(s) = \dfrac{1}{(s_n)^2} \text{ for every } s \text{ such that } t_n \le s \le s_n,\ \forall n \ge 0. \end{cases}$$

Then

$$F(s_n) = \frac{1}{(s_n)^2},\ F(t_n) = \frac{1}{t_n},\ \forall n \ge 0.$$

It remains to define F in the intervals $s_{n+1} \le s \le t_n$. Since $\dfrac{1}{s} < \dfrac{1}{(s)^2}$ for $s < 1$, we can choose the function F such that

$$\frac{1}{s} \le F(s) \le \frac{1}{(s)^2} \text{ for every } s \text{ such that } s_{n+1} \le s \le t_n,\ \forall n \ge 0.$$

Then the function F satisfies

$$\frac{1}{s} \le F(s) \le \frac{1}{(s)^2} \text{ for every } s,\ s \le 1.$$

But $F(t_n) = \dfrac{1}{t_n}$ for every $n \ge 0$, and t_n tends to 0 as n tends to $+\infty$, which proves that there is no $C > 0$ such that (22) holds true. □

4 New Results About the Uniqueness of the Solution

In this section we prove some further results about the uniqueness of solution which complete the results proved in [3, 4]. For the convenience of the reader we include here the results of [3, 4].

We denote by λ_1 and ϕ_1 the first eigenvalue and the first eigenfunction of the operator $-div\, {}^sA(x)D$ in $H_0^1(\Omega)$, where ${}^sA(x) = (A(x) + {}^tA(x))/2$ is the symmetrized part of the matrix $A(x)$, namely

$$\begin{cases} \phi_1 \in H_0^1(\Omega),\ \phi_1 \geq 0,\ \displaystyle\int_\Omega |\phi_1|^2 = 1, \\ -div\, {}^sA(x)D\phi_1 = \lambda_1 \phi_1 \text{ in } \mathscr{D}'(\Omega). \end{cases} \tag{23}$$

4.1 Uniqueness in the Case of a Mild Singularity

We first recall the uniqueness result that we proved in [3] in the case of a mild singularity.

Theorem 3 ([3]) *Assume that the matrix A and the function F satisfy* (2), (3) *and* (4). *Assume moreover that the function $F(x,s)$ is "almost nonincreasing" in s, i.e. that*

$$\begin{cases} \textit{there exists } \lambda,\ 0 \leq \lambda < \lambda_1 \textit{ such that} \\ F(x,s) - \lambda s \leq F(x,t) - \lambda t \text{ a.e. } x \in \Omega,\ \forall s, \forall t,\ 0 \leq t \leq s. \end{cases} \tag{24}$$

Then the solution to problem (1) *in the sense of Definition 1 is unique.*

Remark 4 Note that (24) holds with $\lambda = 0$ when $F(x,s)$ is assumed to be nonincreasing in the s variable.

Note also that if in place of (24) one assumes that the function

$$s \in [0, +\infty] \to F(x,s) - \lambda_1 s \text{ is nonincreasing,} \tag{25}$$

uniqueness of the solution to problem (1) in the sense of Definition 1 in general does not hold true.

Indeed, consider the case where the matrix A satisfies (2) and is symmetric and where the function F is defined by

$$F(x,s) = \lambda_1 T_k(s) \quad \forall s \geq 0, \tag{26}$$

where T_k is the truncation at height $k > 0$, for some k fixed, and where λ_1 and ϕ_1 are defined by (23).

The function F defined by (26) satisfies assumptions (3), (4) and (25).

Recall that ϕ_1, the unique solution to (23), belongs to $L^\infty(\Omega)$. Then for every t with $0 \le t \le k/\|\phi_1\|_{L^\infty(\Omega)}$, the function

$$u = t\phi_1$$

is a solution to (1) in the classical sense, and therefore in the sense of Definition 1.

This proves that uniqueness does not hold if assumption (24) is replaced by the weaker assumption (25). □

This counterexample, even if naive, indicates the mechanism of the possible non uniqueness of the solution when hypothesis (24) is replaced by hypothesis (25): indeed, if there are two different solutions to (1), then one has for almost every $x \in \Omega$

$$F(x,s) = \lambda_1 s + c(x) \text{ for } s_-(x) \le s \le s_+(x),$$

as stated in the next Theorem. In other terms, uniqueness holds true for the solution to (1) in the sense of Definition 1 when $F(x,s)$ satisfies (25) and does not coincide with $\lambda_1 s + c(x)$ for any $c(x)$ in any set of the type $\{x \in \Omega,\ s_-(x) \le s \le s_+(x)\}$.

Theorem 4 *Assume that the matrix A and the function F satisfy* (2), (3) *and* (4). *Assume moreover that F satisfies* (25).

If there exist two different solutions $\hat{u}$ *and* u *to* (1) *in the sense of Definition 1, the function F satisfies*

$$F(x,s) = \lambda_1 s + c(x) \ \ a.e.\ x \in \Omega,\ \ \forall s,\ s_-(x) \le s \le s_+(x), \tag{27}$$

where $c(x) \in L^1_{loc}(\Omega)$ *and where*

$$s_+(x) - s_-(x) = t\phi_1(x) \ \ a.e.\ x \in \Omega, \tag{28}$$

for some $t > 0$ *which does not depend on x (recall that* $\phi_1(x) > 0$ *for almost every* $x \in \Omega$*).*

Proof Assume indeed that $\hat{u}$ and u are two solutions to (1) in the sense of Definition 1 such that $\hat{u} \ne u$. Using $(\hat{u}-u)^+$ and $(\hat{u}-u)^-$ as test functions in the equations satisfied by u and $\hat{u}$ and subtracting, one obtains

$$\int_\Omega A(x)D(\hat{u}-u)D(\hat{u}-u) = \int_\Omega (F(x,\hat{u}) - F(x,u))(\hat{u}-u),$$

or equivalently

$$\begin{cases} \displaystyle\int_\Omega A(x)D(\hat{u}-u)D(\hat{u}-u) - \lambda_1 \int_\Omega |(\hat{u}-u)|^2 = \\ \displaystyle= \int_\Omega ((F(x,\hat{u}) - \lambda_1\hat{u}) - (F(x,u) - \lambda_1 u))(\hat{u}-u). \end{cases} \tag{29}$$

Assumption (25) then implies that the right-hand side of (29) is nonpositive, which in turn implies, by the uniqueness of the first eigenfunction ϕ_1 up to a multiplicative constant, that

$$\hat{u} - u = t\phi_1 \text{ for some } t \in \mathbb{R}. \tag{30}$$

Exchanging if necessary $\hat{u}$ and u one can assume $t > 0$. Taking Ψ^+ and Ψ^- as test functions in (11) for every $\Psi \in H_0^1(\Omega)$ proves that any solution u to problem (1) in the sense of Definition 1 satisfies

$$F(x,u) \in L^1_{\text{loc}}(\Omega), \quad -div\, A(x)Du = F(x,u) \text{ in } \mathscr{D}'(\Omega). \tag{31}$$

From (31) applied to $\hat{u}$ and to u, and from (30) one deduces that

$$F(x, u(x) + t\phi_1(x)) = F(x, u(x)) + t\lambda_1\phi_1(x) \text{ a.e. } x \in \Omega,$$

for the parameter t defined above. Since the function $F(x,s) - \lambda_1 s$ is assumed to be nondecreasing in s and since $\phi_1(x) > 0$ for almost every $x \in \Omega$, this implies that

$$F(x, u(x) + r\phi_1(x)) = F(x, u(x)) + r\lambda_1\phi_1(x) \text{ a.e. } x \in \Omega \ \forall r,\ 0 \le r \le t.$$

This proves (27) and (28) with

$$c(x) = F(x, u(x)),\ s_-(x) = u(x),\ s_+(x) = u(x) + t\phi_1(x);$$

note that $c(x) \in L^1_{\text{loc}}(\Omega)$ in view of (31). □

Remark 5 The result of Theorem 4 also holds true in the case where the function $F(x,s)$ is non singular at $s = 0$, is not assumed to be nonnegative and satisfies

$$|F(x,s)| \le h(x) \text{ a.e. } x \in \Omega,\ \ \forall s \in \mathbb{R},$$

where $h(x)$ satisfies (4 *ii*). As far as we know, the result is new also in this case. □

4.2 Uniqueness in the Case of a Strong Singularity

In [4] we proved the following uniqueness result.

Theorem 5 ([4]) *Assume that the matrix A and the function F satisfy* (2), (3) *and* (6). *Assume moreover that the function* $F(x,s)$ *is nonincreasing with respect to s, i.e. that*

$$F(x,s) \le F(x,t) \text{ a.e. } x \in \Omega,\ \forall s, \forall t,\ 0 \le t \le s. \tag{32}$$

Then the solution to problem (1) *in the sense of Definition 4 is unique.*

Remark 6 If we compare the two uniqueness results Theorems 3 and 5, we note that in Theorem 3, in the case of a mild singularity, we assumed that $F(x,s)$ is "almost nonincreasing" in the s variable [see (24)] while in Theorem 5, in the case of a strong singularity, we assumed the stronger condition that $F(x,s)$ is nonincreasing in the s variable [see (32)]. □

Now, we prove a Comparison Principle in the case of a strong singularity under the weaker assumption that $F(x,s)$ is "γ-almost nonincreasing" in s. Unfortunately we are not able to prove a Comparison Principle which is completely general, since we have to assume that $(u_1 - u_2)^+ \in L^\infty(\Omega)$ where u_1 and u_2 are the solutions we want to compare. Moreover our proof deals only with the case where in (6 *iii*) the function Γ is given by $\Gamma(s) = s^\gamma/(s^\gamma + 1)$ with $\gamma > 1$.

Proposition 2 (Comparison Principle) *Assume that the matrix A satisfies* (2). *Assume that the two functions $F_1(x,s)$ and $F_2(x,s)$ satisfy* (3) *and* (6) *with $\Gamma(s) = s^\gamma/(s^\gamma+1)$, $\gamma > 1$. Assume moreover that either $F_1(x,s)$ or $F_2(x,s)$ is "γ-almost nonincreasing", i.e. satisfies*

$$\begin{cases} \textit{there exists } \lambda,\ 0 \le \lambda < 4\dfrac{\gamma+3}{(\gamma+4)^2}\lambda_1 \textit{ such that} \\ F(x,s) - \lambda s \le F(x,t) - \lambda t \ \text{ a.e. } x \in \Omega,\ \forall s, \forall t,\ 0 \le t \le s, \end{cases} \tag{33}$$

and that

$$F_1(x,s) \le F_2(x,s) \ \textit{a.e.}\ x \in \Omega, \quad \forall s \ge 0. \tag{34}$$

Let u_1 and u_2 be any solutions in the sense of Definition 4 to problem $(1)_1$ *and* $(1)_2$, *where* $(1)_1$ *and* $(1)_2$ *are* (1) *with $F(x,s)$ replaced respectively by $F_1(x,s)$ and $F_2(x,s)$. Assume also that*

$$(u_1 - u_2)^+ \in L^\infty(\Omega). \tag{35}$$

Then one has

$$u_1(x) \le u_2(x) \ \textit{a.e.}\ x \in \Omega.$$

Remark 7 Given $F_1(x,s)$ and $F_2(x,s)$ which satisfy (6), there is no loss of generality to assume that $F_1(x,s)$ and $F_2(x,s)$ satisfy (6) with the same h and Γ. Indeed if $0 \le F_i(x,s) \le \dfrac{h_i(x)}{\Gamma_i(s)}$, $i = 1, 2$, setting $h(x) = \sup\{h_1(x), h_2(x)\}$ and $\Gamma(s) = \inf\{\Gamma_1(s), \Gamma_2(s)\}$, one has $\dfrac{h_i(x)}{\Gamma_i(s)} \le \dfrac{h(x)}{\Gamma(s)}$. □

Proof In this proof we set

$$\varphi = ((u_1 - u_2)^+)^m, \quad \text{with} \quad m \geq 1 + \frac{\gamma + 1}{2}.$$

First Step In this first step we will prove that

$$\varphi^2 = ((u_1 - u_2)^+)^{2m} \in \mathscr{V}(\Omega) \text{ for } m \geq 1 + \frac{\gamma + 1}{2}. \tag{36}$$

Since u_1, u_2 belong to $H^1_{\text{loc}}(\Omega)$ and $m > 1$, using (35), we have

$$((u_1 - u_2)^+)^m \in H^1_{\text{loc}}(\Omega) \cap L^\infty(\Omega), \tag{37}$$

with

$$D(((u_1 - u_2)^+)^m) = m((u_1 - u_2)^+)^{m-1} D(u_1 - u_2) \text{ in } \mathscr{D}'(\Omega).$$

Choosing $k \geq \|(u_1 - u_2)^+\|_{L^\infty(\Omega)}$, we have $(u_1 - u_2)^+ = T_k(u_1 - u_2)^+$ and therefore, since $u_2 \geq 0$,

$$\begin{cases} \left|D(((u_1 - u_2)^+)^m)\right| - m(T_k(u_1 - u_2)^+)^{m-1}|D(u_1 - u_2)| \leq \\ \leq m(T_k(u_1))^{m-1}(|Du_1| + |Du_2|). \end{cases}$$

Since $m - 1 \geq \frac{\gamma+1}{2}$ and since $(T_k(u_1))^{\frac{\gamma+1}{2}} \in H^1_0(\Omega) \cap L^\infty(\Omega)$ for any $k > 0$ [see (17)], we have $(T_k(u_1))^{m-1} \in H^1_0(\Omega) \cap L^\infty(\Omega)$. Therefore for $i = 1, 2$, and for every $j > 0$, since

$$\begin{cases} 0 \leq (T_k(u_1))^{m-1}|Du_i| = (T_k(u_1))^{m-1}|DT_j(u_i) + DG_j(u_i)| \leq \\ \leq (T_k(u_1))^{m-1}|DT_j(u_j)| + (T_k(u_1))^{m-1}|DG_j(u_i)|, \end{cases}$$

we have, by (15 *iv*) and (15 *iii*), $(T_k(u_1))^{m-1}|Du_i| \in L^2(\Omega)$, $i = 1, 2$. Since $0 \leq ((u_1 - u_2)^+)^m \leq (T_k(u_1))^m \in H^1_0(\Omega) \cap L^\infty(\Omega)$ it follows that

$$\varphi = ((u_1 - u_2)^+)^m \in H^1_0(\Omega) \cap L^\infty(\Omega) \text{ for } m \geq 1 + \frac{\gamma + 1}{2}.$$

Finally since for $\phi \in H^1_0(\Omega) \cap L^\infty(\Omega)$ one has

$$-\, div\, {}^t\!A(x)D\phi^2 = -2\, div(\phi\, {}^t\!AD\phi) = -2\, {}^t\!AD\phi D\phi + 2\phi(-div\, {}^t\!AD\phi), \tag{38}$$

it is not difficult to see that $\phi^2 \in \mathscr{V}(\Omega)$ when $\phi \in H^1_0(\Omega) \cap L^\infty(\Omega)$.

This completes the proof of (36).

Second Step Since $\varphi^2 = ((u_1 - u_2)^+)^{2m} \geq 0$, we can take φ^2 as test function in (16 *ii*) because of (36). We obtain

$$\begin{cases} \langle -div\, {}^tA(x)D\varphi^2, G_k(u_i)\rangle_{H^{-1}(\Omega),H_0^1(\Omega)} + \langle\langle -div\, {}^tA(x)D\varphi^2, T_k(u_i)\rangle\rangle_\Omega = \\ = \int_\Omega F_i(x,u_i)\varphi^2, \ \ i = 1,2. \end{cases}$$

In view of (14) and of (38) with $\phi = \varphi$, this is nothing but

$$\begin{cases} 2\int_\Omega \varphi\, {}^tAD\varphi DG_k(u_i) - 2\int_\Omega {}^tAD\varphi D\varphi T_k(u_i) + 2\int_\Omega {}^tAD\varphi D(\varphi T_k(u_i)) = \\ = \int_\Omega F_i(x,u_i)\varphi^2, \ \ i = 1,2, \end{cases}$$

which by an easy computation gives

$$2\int_\Omega \varphi\, {}^tAD\varphi Du_i = \int_\Omega F_i(x,u_i)\varphi^2, \ \ i = 1,2.$$

Taking the difference between these two equations it follows that

$$2\int_\Omega \varphi\, {}^tA(x)D\varphi D(u_1 - u_2) = \int_\Omega (F_1(x,u_1) - \int_\Omega F_2(x,u_2))\varphi^2. \tag{39}$$

Since $(u_1 - u_2)^+ \in H^1_{\text{loc}}(\Omega) \cap L^\infty(\Omega)$ by (37), one has in $L^1_{\text{loc}}(\Omega)$

$$\begin{cases} \varphi\, {}^tA(x)D\varphi D(u_1 - u_2) = \\ = m((u_1-u_2)^+)^m((u_1-u_2)^+)^{m-1}\, {}^tA(x)D(u_1-u_2)^+D(u_1-u_2) = \\ = \dfrac{m}{(m+\frac{1}{2})^2}\, {}^sA(x)D\big(((u_1-u_2)^+)^{(m+\frac{1}{2})}\big)\, D\big(((u_1-u_2)^+)^{(m+\frac{1}{2})}\big). \end{cases} \tag{40}$$

Subtracting the term $\lambda \int_\Omega ((u_1 - u_2)^+)^{2m+1}$ in both sides of (39) and using (40), we get

$$\begin{cases} \dfrac{2m}{(m+\frac{1}{2})^2}\int_\Omega {}^sA(x)D\big(((u_1-u_2)^+)^{\frac{2m+1}{2}}\big)\, D\big(((u_1-u_2)^+)^{\frac{2m+1}{2}}\big) + \\ -\lambda \int_\Omega ((u_1-u_2)^+)^{2m+1} \leq \\ \leq \int_\Omega \big(F_1(x,u_1) - F_2(x,u_2) - \lambda(u_1-u_2)\big)((u_1-u_2)^+)^{2m}. \end{cases}$$

By the characterization of the first eigenvalue λ_1 of ${}^s\!A$ and setting

$$G_1(x,s) = F_1(x,s) - \lambda s \text{ and } G_2(x,s) = F_2(x,s) - \lambda s,$$

we have

$$\begin{cases} \left(\dfrac{2m}{(m+\frac{1}{2})^2}\lambda_1 - \lambda\right)\displaystyle\int_\Omega ((u_1-u_2)^+)^{2m+1} \le \\ \le \displaystyle\int_\Omega (G_1(x,u_1) - G_2(x,u_2))((u_1-u_2)^+)^{2m}, \end{cases}$$

in which we choose $m = 1 + \frac{\gamma+1}{2}$. This implies

$$\begin{cases} \left(\dfrac{4(\gamma+3)}{(\gamma+4)^2}\lambda_1 - \lambda\right)\displaystyle\int_\Omega ((u_1-u_2)^+)^{2m+1} \le \\ \le \displaystyle\int_\Omega (G_1(x,u_1) - G_2(x,u_2))((u_1-u_2)^+)^{2m}. \end{cases} \tag{41}$$

Third Step We want to show that

$$(G_1(x,u_1) - G_2(x,u_2))((u_1-u_2)^+)^{2m} \le 0 \text{ a.e. in } x \in \Omega. \tag{42}$$

This will imply, by (41), that $u_1 \le u_2$.

Recall that as a consequence of (38), $\phi^2 \in \mathscr{V}(\Omega)$ when $\phi \in H_0^1(\Omega) \cap L^\infty(\Omega)$, and therefore that $\psi^2 \in \mathscr{V}(\Omega)$ when $\psi \in \mathscr{D}(\Omega)$. In view of (16 *i*), this implies that $F_i(x,u_i)\psi^2$, $i = 1,2$, belongs to $L^1(\Omega)$, and therefore that $F_i(x,u_i)$, $i = 1,2$, is finite almost everywhere. This fact excludes almost everywhere any indeterminacies of the type $0 \times \infty$ and of the type $\infty - \infty$ in the computations below.

Consider first the case where $G_1(x,s)$ is nonincreasing in s.

In view of (6 *iii*) and of (34) one has

$$-\lambda u_2 \le F_1(x,u_2) - \lambda u_2 = G_1(x,u_2) \le G_2(x,u_2),$$

which implies that $G_1(x,u_2)$ is finite almost everywhere. Moreover, using the fact that $G_1(x,s)$ is nonincreasing in s and then (34), one has

$$\begin{cases} (G_1(x,u_1) - G_2(x,u_2))((u_1-u_2)^+)^{2m} \le \\ \le (G_1(x,u_2) - G_2(x,u_2))((u_1-u_2)^+)^{2m} \le 0 \\ \text{on the set } \{x \in \Omega : u_1(x) > u_2(x)\}, \end{cases} \tag{43}$$

while on the other hand one has

$$\begin{cases}(G_1(x,u_1)-G_2(x,u_2))((u_1-u_2)^+)^{2m}=0\\ \text{on the set } \{x\in\Omega : u_1(x)\le u_2(x)\}.\end{cases} \tag{44}$$

Collecting (43) and (44) proves (42) in this first case.

Consider now the case where $G_2(x,s)$ is nonincreasing in s in this first case.

In view of (6 *iii*) and since $G_2(x,s)$ is nonincreasing in s one has

$$-\lambda u_1 \le F_2(x,u_1)-\lambda u_1 = G_2(x,u_1)\le G_2(x,u_2) \text{ on the set } \{x\in\Omega : u_1(x)>u_2(x)\},$$

which implies that $G_2(x,u_1)$ is finite almost everywhere. Moreover, using (34) and then the fact that $G_2(x,s)$ is nondecreasing in s, one has

$$\begin{cases}(G_1(x,u_1)-G_2(x,u_2))((u_1-u_2)^+)^{2m}\le\\ \le (G_2(x,u_1)-G_2(x,u_2))((u_1-u_2)^+)^{2m}\le 0\\ \text{on the set } \{x\in\Omega : u_1(x)>u_2(x)\},\end{cases} \tag{45}$$

while (44) still holds true.

Collecting together (45) and (44) proves (42) in this second case.

The proof of Proposition 2 is complete. □

Now we can state the following uniqueness result.

Theorem 6 *Assume that the matrix A and the function F satisfy* (2), (3) *and* (6) *with* $\Gamma(s)=s^\gamma/(s^\gamma+1)$ *for some* $\gamma>1$. *Assume moreover that the function* $F(x,s)$ *is "*γ*-almost nonincreasing" with respect to s, i.e. satisfies assumption* (33). *Then if* u_1 *and* u_2 *are two solutions to problem* (1) *in the sense of Definition 4 which are such that* $(u_1-u_2)\in L^\infty(\Omega)$, *one has* $u_1=u_2$.

Proof Applying the Comparison Principle to the case where $F_1(x,s)=F_2(x,s)$ with $F(x,s)$ satisfying (2), (3), (6) with $\Gamma(s)=s^\gamma/(s^\gamma+1)$ for some $\gamma>1$ immediately proves the uniqueness Theorem 6.

Remark 8 As in Proposition 2 and Theorem 6, we deal in this Remark with functions $F(x,s)$ which satisfy (6) with $\Gamma(s)=s^\gamma/(s^\gamma+1)$ for some $\gamma>1$.

Note that in this setting, the definition (33) of a function "γ-almost increasing" depends on the value of γ. This is not the case when $0<\gamma\le 1$, since in this case the definition (24) of a function "almost increasing" does not depend on the value of γ. Note also that the limit as $\gamma>1$ tends to 1 of condition (33) is different of (and stronger than) condition (24), since the limit $(4/5)^2$ of the constant $4(\gamma+3)/(\gamma+4)^2$ which appears in (33) is strictly smaller than the constant 1 which appears in (24). □

5 L^∞-Regularity of the Solutions

In this section we prove that any solution u to problem (1) in the sense of Definition 4 belongs to $L^\infty(\Omega)$ (with an a priori estimate in this space) if the function h in (6 *iii*) belongs to $L^t(\Omega)$, $t > \frac{N}{2}$ if $N \geq 2$, $t = 1$ if $N = 1$, and not only to $L^r(\Omega)$.

The L^∞-regularity of the solutions obtained by approximation has been proved in [2]. In [1] (see also [3]) the authors proved the L^∞-regularity for general solutions (not necessarily obtained by approximation) in the case of a mild singularity. Our result below is concerned with any solution in the sense of Definition 4 to problem (1) for any function F satisfying the general growth condition (6), which includes in particular mild singularities but also strong singularities with powerlike growth conditions of the type (4 *iii*) for any $\gamma > 0$ or even like the one of example (7).

Specifically we prove the following regularity result.

Proposition 3 (**$L^\infty(\Omega)$ Regularity**) *Assume that the matrix A and the function F satisfy* (2), (3) *and* (6). *Assume moreover that*

$$h \in L^t(\Omega),\ t > \frac{N}{2} \ \text{if}\ N \geq 2,\ t = 1 \ \text{if}\ N = 1. \tag{46}$$

Then every u solution to problem (1) *in the sense of Definition 4 satisfies*

$$u \in L^\infty(\Omega), \quad \|u\|_{L^\infty(\Omega)} \leq 1 + \frac{1}{\alpha\Gamma(1)} C(|\Omega|, N, t)\|h\|_{L^t(\Omega)}, \tag{47}$$

for a constant $C(|\Omega|, N, t)$ which depends only on $|\Omega|$, N and t and is nondecreasing in $|\Omega|$.

Proof We will first prove that

$$\int_\Omega A(x)DG_k(u)DG_k(u) = \int_\Omega F(x,u)G_k(u) \quad \forall k > 0. \tag{48}$$

Note that we can not use $G_k(u)$ as test function in (16 *ii*) since $G_k(u)$ does not belong to $\mathcal{V}(\Omega)$.

Following the proof of Proposition 5.1 in [4] we define for every k and n with $0 < k < n$ the function $S_{k,n}$ as

$$S_{k,n}(s) = \begin{cases} 0 & \text{if } 0 \leq s \leq k, \\ s - k & \text{if } k \leq s \leq n, \\ n - k & \text{if } n \leq s. \end{cases}$$

We can prove as in the first two steps of the proof of Proposition 5.1 in [4] that $S_{k,n}(u) \in \mathcal{V}(\Omega)$ and that

$$\int_\Omega A(x)DG_k(u)DS_{k,n}(u) = \int_\Omega F(x,u)S_{k,n}(u). \tag{49}$$

We can also prove as in the third step of the proof of Proposition 5.1 in [4] that $S_{k,n}(u)$ is bounded in $H_0^1(\Omega)$ for $k > 0$ fixed independently of $n > k$. This allows us to pass to the limit in n in the left-hand side of (49) up to a subsequence of n. On the other hand we can apply Fatou's Lemma to the right-hand side of (49) getting

$$\int_\Omega F(x,u)G_k(u) < +\infty.$$

Since $0 \le S_{k,n}(s) \le G_k(s)$ for $n > k$, Lebesgue's dominated convergence allows us order to pass to the limit in the right-hand side of (49) as n tends to infinity.

This proves (48).

Using now the coercivity (2) and the growth condition (6 *iii*) in (48) we have

$$\begin{cases} \alpha \displaystyle\int_\Omega |DG_k(u)|^2 \le \int_\Omega h(x)\frac{G_k(u)}{\Gamma(u)} = \\ = \displaystyle\int_\Omega h(x)\frac{G_k(u)}{\Gamma(u)}\chi_{\{u\ge k\}} \le \int_\Omega h(x)\frac{G_k(u)}{\Gamma(1)}\chi_{\{u\ge k\}} \quad \forall k \ge 1. \end{cases}$$

Define for $k \ge 0$

$$\varphi(k) = \text{meas}\{x \in \Omega : u(x) \ge k\}.$$

When $N \ge 3$ (the proof is analogous when $N = 2$ and $N = 1$), using Sobolev's inequality and Hölder's inequality with p defined by

$$\frac{1}{t} + \frac{1}{2^*} + \frac{1}{p} = 1,$$

and processing as in [6] one obtains

$$\varphi(h) \le \frac{1}{(h-k)^{2^*}}\left(\frac{C_S^2}{\alpha\Gamma(1)}\|h\|_{L^t(\Omega)}\right)^{2^*}\varphi(k)^{\frac{2^*}{p}} \quad \forall h,k,\ \ h > k \ge 1,$$

where $2^* = \frac{2N}{N-2}$ and where the Sobolev's constant C_S is defined by

$$\|v\|_{L^{2^*}(\Omega)} \le C_S\|Dv\|_{(L^2(\Omega))^N} \quad \forall v \in H_0^1(\Omega) \text{ when } N \ge 3.$$

Since $\frac{2^*}{p} > 1$, Lemma 4.1 of [6] implies that $\varphi(1+d) = 0$, or in other terms that

$$u(x) \leq 1 + d \quad \text{a.e. } x \in \Omega,$$

where

$$d = \frac{C_S^2}{\alpha\Gamma(1)} \|h\|_{L^t(\Omega)} |\Omega|^{\frac{2^*}{p}-1} 2^{\frac{2^*}{2^*-p}},$$

since $\varphi(1) \leq |\Omega|$. This immediately gives (47).

Remark 9 Observe that when F satisfies the growth condition (4), equality (48) is trivial because $G_k(u) \in H_0^1(\Omega)$ is an admissible test function in (11) since in this case $u \in H_0^1(\Omega)$. □

Acknowledgements The authors would like to thank their own institutions (Dipartimento di Scienze di Base e Applicate per l'Ingegneria della Facoltà di Ingegneria Civile e Industriale di Sapienza Università di Roma, Departamento de Matemática Aplicada y Estadística de la Universidad Politécnica de Cartagena, Laboratoire Jacques-Louis Lions de l'Université Pierre et Marie Curie Paris VI et du CNRS) for providing the support of reciprocal visits which allowed them to perform the present work. The work of Pedro J. Martínez-Aparicio has been partially supported by the grant MTM2015-68210-P of the Spanish Ministerio de Economía y Competitivided (MINECO), the FQM-116 grant of the Junta de Andalucía and the grant Programa de Apoyo a la Investigación de la Fundación Séneca-Agencia de Ciencia y Tecnología de la Región de Murcia 19461/PI/14.

References

1. Boccardo, L., Casado-Díaz, J.: Some properties of solutions of some semilinear elliptic singular problems and applications to the G-convergence. Asymptot. Anal. **86**, 1–15 (2014)
2. Boccardo, L., Orsina, L.: Semilinear elliptic equations with singular nonlinearities. Calc. Var. Partial Differential Equations **37**, 363–380 (2010)
3. Giachetti, D., Martínez-Aparicio, P.J., Murat, F.: A semilinear elliptic equation with a mild singularity at $u = 0$: existence and homogenization. J. Math. Pures Appl. (2016). doi:10.1016/j.matpur.2016.04.007
4. Giachetti, D., Martínez-Aparicio, P.J., Murat, F.: Definition, existence, stability and uniqueness of the solution to a semilinear elliptic problem with a strong singularity at $u = 0$. Preprint (2016)
5. Giachetti, D., Martínez-Aparicio, P.J., Murat, F.: Homogenization of a Dirichlet semilinear elliptic problem with a strong singularity at $u = 0$ in a domain with many small holes. Preprint (2016)
6. Stampacchia, G.: Le problème de Dirichlet pour les équations elliptiques du second ordre à coefficients discontinus. Ann. Inst. Fourier Grenoble **15**, 189–258 (1965)

Weighted Extrapolation Techniques for Finite Difference Methods on Complex Domains with Cartesian Meshes

A. Baeza, P. Mulet, and D. Zorío

Abstract The design of numerical boundary conditions in high order schemes is a challenging problem that has been tackled in different ways depending on the nature of the problem and the scheme used to solve it numerically. In this paper we propose a technique to extrapolate the information from the computational domain to ghost cells for schemes with structured Cartesian Meshes on complex domains. This technique is based on the application of Lagrange interpolation with weighted filters for the detection of discontinuities that permits a data dependent extrapolation, with high order at smooth regions and essentially non oscillatory properties near discontinuities. This paper is a sequel of Baeza et al. (J Sci Comput, 2015), where a boundary extrapolation procedure with Boolean filters was developed. We show that weighted extrapolation can tackle discontinuities more robustly than the procedure introduced in Baeza et al. (J Sci Comput, 2015).

1 Introduction

Hyperbolic conservation laws have become the focus of many research lines in the last decades. Very few analytic solutions are known for these equations, and thus numerical methods to approximate them have been developed and improved along the years.

The main motivation of this work is to extend some of the methods focused on Cartesian meshes, whose use was relegated to rectangular domains and low order boundary conditions, to problems with complex domains using high order boundary conditions with extrapolation techniques at ghost cells capable to cope with discontinuities in weak solutions that may approach the boundary.

Some authors have approached this problem from different perspectives, such as in [9], where a technique based on second order Lagrange interpolation with limiters is developed, [10, 11], with a high order approach but problem dependent and relatively high computational cost. In [1], an extrapolation technique based

A. Baeza • P. Mulet • D. Zorío (✉)
Departament de Matemàtica Aplicada, Universitat de València, Valencia, Spain
e-mail: antonio.baeza@uv.es; mulet@uv.es; david.zorio@uv.es

F. Ortegón Gallego et al. (eds.), *Trends in Differential Equations and Applications*, SEMA SIMAI Springer Series 8, DOI 10.1007/978-3-319-32013-7_14

on Boolean filters for the detection of discontinuities is developed, but with the drawback of having a tuning parameter and lack of robustness for some demanding problems (simulation failure for a certain threshold values range).

Our approach can be understood as an extension of [9] and [1] in the sense that it is based on Lagrange extrapolation with weights akin to the WENO procedure, but without imposing limitations on the order of the method or the number of ghost cells and agnostic about the equation.

The organization of the paper is the following: In Sect. 2 we present the equations and the numerical methods that we consider in this paper. The details of the procedure for meshing complex domains with Cartesian meshes are explained in Sect. 3. In Sect. 4 we expound how we perform extrapolations with the method for the detection of singularities. Some numerical results that are obtained with this methodology are presented in Sect. 5, with some simple tests in 1D to illustrate the correct behavior of the proposed techniques and some more complex ones in 2D. Finally, some conclusions are drawn in Sect. 6.

2 Numerical Schemes

The equations that will be considered throughout this paper are hyperbolic systems of m two-dimensional conservation laws

$$u_t + f(u)_x + g(u)_y = 0, \quad u = u(x, y, t), \tag{1}$$

defined on an open and bounded spatial domain $\Omega \subseteq \mathbb{R}^2$, with Lipschitz boundary $\partial\Omega$ given by a finite union of piece-wise smooth curves, $u : \Omega \times \mathbb{R}^+ \to \mathbb{R}^m$, and fluxes $f, g : \mathbb{R}^m \to \mathbb{R}^m$. These equations are supplemented with an initial condition, $u(x, y, 0) = u_0(x, y)$, $u_0 : \Omega \to \mathbb{R}^m$, and different boundary conditions that may vary depending on the problem.

Although the techniques that will be expounded in this paper are applicable to other numerical schemes, we use here Shu-Osher's finite difference conservative methods [8] with a WENO5 (*Weighted Essentially Non-Oscillatory*) [5] spatial reconstruction, Donat-Marquina's flux-splitting [4] and the RK3-TVD ODE solver [7] in a method of lines fashion that we briefly describe here for the sake of completeness. This combination of techniques was proposed in [6].

3 Meshing Procedure

We define our mesh starting from a reference vertical line, $x = \overline{x}$ and a horizontal one $y = \overline{y}$. Let $h_x > 0$ and $h_y > 0$ be the horizontal and vertical spacings of the mesh, so that the vertical lines in the mesh are determined by: $x = x_r := \overline{x} + rh_x$, $r \in \mathbb{Z}$ and the horizontal ones by $y = y_s := \overline{y} + sh_y$, $s \in \mathbb{Z}$. The cell with center

(x_r, y_s) is defined by:

$$[x_r - \frac{h_x}{2}, x_r + \frac{h_x}{2}] \times [y_s - \frac{h_y}{2}, y_s + \frac{h_y}{2}].$$

The computational domain is then given by

$$\mathscr{D} := \{(x_r, y_s) : \quad (x_r, y_s) \in \Omega, \quad r, s \in \mathbb{Z}\} = (\overline{x} + h_x\mathbb{Z}) \times (\overline{y} + h_y\mathbb{Z}) \cap \Omega.$$

Notice that $\mathscr{D}$ is finite since Ω is bounded.

3.1 Ghost Cells

We recall that WENO schemes of order $2k - 1$ use an stencil (consecutive indexes) of $2k$ points, therefore k additional cells are needed at both sides of each horizontal and vertical mesh line in order to perform a time step. These additional cells are usually named *ghost cells* and, in terms of their centers, are given by:

$$\mathscr{GC} := \mathscr{GC}_x \cup \mathscr{GC}_y,$$

where

$$\mathscr{GC}_x := \{(x_r, y_s) : \quad 0 < d\,(x_r, \; \Pi_x\,(\mathscr{D} \cap (\mathbb{R} \times \{y_s\}))) \leq kh_x, \quad r, s \in \mathbb{Z}\},$$
$$\mathscr{GC}_y := \left\{(x_r, y_s) : \quad 0 < d\left(y_s, \; \Pi_y\,(\mathscr{D} \cap (\{x_r\} \times \mathbb{R}))\right) \leq kh_y, \quad r, s \in \mathbb{Z}\right\},$$

where Π_x and Π_y denote the projections on the respective coordinates and,

$$d(a, B) := \inf\{|b - a| : \; b \in B\},$$

for given $a \in \mathbb{R}$ and $B \subseteq \mathbb{R}$. Notice that $d(a, \emptyset) = +\infty$, since, by convention, $\inf \emptyset = +\infty$.

3.2 Normal Lines

There are many ways in which a numerical boundary extrapolation can be done, but not all of them are suitable for the stability of the method or provide accurate results. The motivation of the choice of the nodal disposition that we will next introduce has been explained in [1].

We focus now on the two-dimensional setting and boundaries with prescribed Dirichlet conditions, e.g., reflective boundary conditions for the Euler equations. In this situation, it seems reasonable that the extrapolation at a certain ghost cell P be based on the prescribed value at the nearest boundary point. It can be proven that a point $N_0 = N(P) \in \partial\Omega$ satisfying

$$\|P - N_0\|_2 = \min\{\|P - B\|_2 : \quad B \in \partial\Omega\}$$

also satisfies that the line determined by P and N_0 is normal to the curve $\partial\Omega$ at N_0, if $\partial\Omega$ is differentiable at N_0. Uniqueness of N_0 holds whenever P is close enough to the boundary, so we will henceforth denote $N(P) = N_0$.

We refer to [1] for further details.

3.2.1 Choice of Nodes on Normal Lines

If we wish to formally preserve a certain precision in the resulting scheme, it is necessary to extrapolate the information from the interior of the domain in an adequate manner. Therefore, if the basic numerical scheme has order r it is reasonable to use extrapolation of this order at least. For the sake of clarity, we will not distinguish between interpolation or extrapolation when these take place at the interior of the domain.

We proceed in a similar fashion as in [9] and [1]. Let $P \in \mathscr{GC}$ and consider the corresponding point in $\partial\Omega$ at minimal distance, $N(P)$.

At first place, one needs to obtain data from the information in $\mathscr{D}$ at a set of points $\mathscr{N}(P) = \{N_1, \ldots, N_R\}$, with $R \geq r$, on the line determined by the points P and $N_0 = N(P)$. By a CFL stability motivation, we will do the selection with a spacing between them of at least the distance between P and $N(P)$. The justification of this fact was done in [1]. We will choose the nodes depending on the slope of the normal line, so that the use of interior information is maximized. See [1] for further details.

We denote by $v = (v_1, v_2)$ the vector determined by P and $N(P)$, and $\mathscr{S}_q = \{N_{q,1}, \ldots, N_{q,R}\}$ the closest set of points to N_q from the computational domain sharing the same coordinate than N_q, whose horizontal or vertical 1D disposition depends on the angle of the normal line as explained in [1].

Figure 1 shows graphical examples of the boundary extrapolation setup for a certain ghost cell P.

Since we are now concerned on performing a weighted extrapolation and the computation of smoothness indicators to build the weights can be very computationally expensive if the data is not equally spaced, as it will generally happen on Dirichlet boundaries, we perform an additional step before extrapolating the value at the ghost cell in order to generate a new stencil such that, together with the boundary node, the global stencil is formed by equally spaced points.

Therefore, if Dirichlet conditions are prescribed, we use the data obtained in N_q, $1 \leq q \leq R$ to perform 1D interpolations at the points P_q, $1 \leq q \leq R-1$, where $P_q = (P_0^x + qh_x, P_0^y + q\frac{v_2}{v_1}h_y)$, $0 \leq q \leq R-1$ if $|v_1| \geq |v_2|$ or $P_q =$

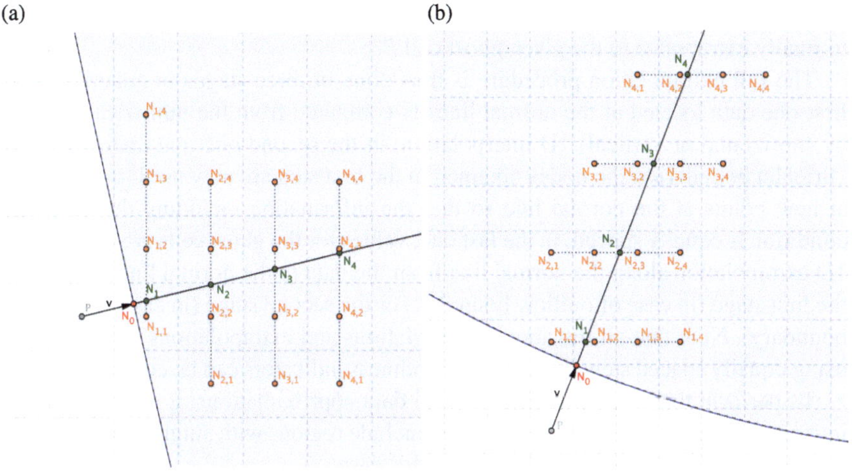

Fig. 1 Examples of choice of stencil for Neumann boundary conditions: (**a**) vertical arrangement of nodes in $\mathscr{S}_q$; (**b**) horizontal arrangement of nodes in $\mathscr{S}_q$

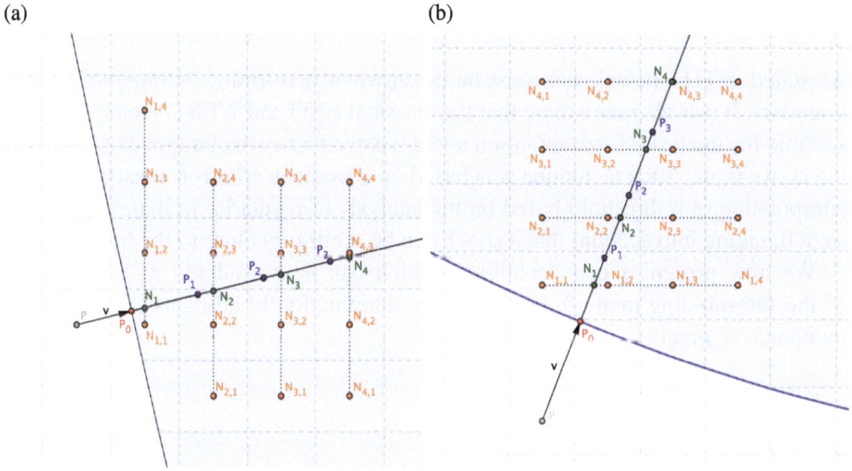

Fig. 2 Examples of choice of stencil for Dirichlet boundary conditions: (**a**) vertical arrangement of nodes in $\mathscr{S}_q$; (**b**) horizontal arrangement of nodes in $\mathscr{S}_q$

$(P_0^x + q\frac{v_1}{v_2}h_x, P_0^y + qh_y)$, $0 \leq q \leq R-1$, otherwise, and use the data from the stencil $\mathscr{S}(P) = \{P_0, P_1, \ldots, P_{R-1}\}$ to extrapolate it at the ghost cell P. In case of outflow conditions, we extrapolate directly the data from the stencil $\mathscr{S}(P) = \{N_1, N_2, \ldots, N_R\}$. See Fig. 2 for graphical examples.

The above procedure for the selection of the interpolation nodes at the normal lines and their corresponding sets $\mathscr{S}_q$ is performed only once at the beginning of the simulation as long as the boundary does not change. With an adequate use of this data structure, one can reconstruct data at order r (in case of smoothness) at the points $N_1, \ldots, N_R$ on the normal line. Once these values are obtained, they are used to finally extrapolate to the given ghost cell P.

The full extrapolation procedure is thus done in three stages in general: in the first one data located at the normal lines is computed from the numerical solution by (horizontal or vertical) 1D interpolation; in the second one, only performed on Dirichlet boundaries, the nodes obtained in the first step are now used to interpolate at new points at the normal line so that the information including the boundary condition is equally spaced; in the last one, values for the ghost cells are obtained by 1D extrapolation along the normal line from the data in the normal line obtained in the first stage (in case of outflow boundary) or the second stage (in case of Dirichlet boundary). Note that all the above interpolations and extrapolations are performed using equally spaced stencils, so that smoothness indicators can be computed easily.

To perform these two one-dimensional data approximations, it should be taken into account that the selected stencils can include regions with singularities. We will see in the next section how to proceed in this case.

4 Extrapolation

As stated in [1], special care must be taken when performing extrapolation at the boundary. It was also seen there that the classical ENO and WENO methods are not suitable for the task of extrapolation and developed a new technique that overcame the above issue. Such technique was based on a Boolean criterion consisting on the computation of a threshold based on the analysis of regularity in the extrapolation stencil, taking into account that a shock can be arbitrarily close to the boundary.

We now present a new technique, which can be considered as an evolution of the thresholding method, based on the computation of adimensional and scale independent weights.

4.1 *Weighted Extrapolation*

We define inductively the following set of indexes:

$$J_0 = \{j_0\}, \text{ and } X_0 = \{x_j\}_{j \in J_0} = \{x_{j_0}\}$$

where

$$j_0 = \underset{j \in J}{\operatorname{argmin}} |x_j - x_*|.$$

That is, X_0 is the one point set including the closest node to the extrapolation point.

Assume we have defined J_{k-1}, then J_k is defined by

$$J_k = J_{k-1} \cup \{j_k\} \text{ and } X_k = \{x_j\}_{j \in J_k}$$

where

$$x_{j_k} = \underset{j \in J \setminus J_{k-1}}{\operatorname{argmin}} |x_j - x_*|.$$

That is, we add the closest node to x_* from the remaining nodes to choose as we increase k. X_k and I_k can be defined for $0 \leq k \leq r$.

By construction, it is clear that such sets can be written as a sequence of nodes with successive indexes, i.e., a stencil:

$$X_k = \{x_{i_k+j}\}_{j=0}^{k}$$

for some $0 \leq i_k \leq r - k$, $0 \leq k \leq r$.

Now, for each k, $0 \leq k \leq r$, we define p_k the interpolating polynomial of degree at most k such that $p_k(x_{i_k+j}) = u_{i_k+j}$, $\forall j$, $0 \leq j \leq k$.

And given $\{\omega_k\}_{k=1}^{r}$ a set of weights, $0 \leq \omega_k \leq 1$, we define the following recurrence:

$$\begin{aligned} u_*^{(0)} &= p_0(x_*) = u_{i_0}, \\ u_*^{(k)} &= (1 - \omega_k) u_*^{(k-1)} + \omega_k p_k(x_*), \quad 1 \leq k \leq r. \end{aligned}$$

We define the final result of the weighted extrapolation as

$$u_* := u_*^{(r)},$$

which will be taken as an approximation for the value $u(x_*)$.

The chosen weights should verify that $\omega_k \approx 0$ if the stencil J_k crosses a discontinuity and $\omega_k \approx 1$ if the data from the stencil is smooth. We will show below a weight construction that verifies that property as well as the capability of preserving the accuracy order of the extrapolation in case of smoothness.

From now on, we will assume that the nodes X are equally spaced and define $h = x_{i+1} - x_i$.

For each $1 \leq k \leq r$, we define a slight modification of the Jiang-Shu smoothness indicator associated to the stencil J_k as the following value:

$$I_k = \frac{1}{r} \sum_{\ell=1}^{k} \int_{x_0}^{x_r} h^{2\ell-1} p_k^{(\ell)}(x)^2 dx + \varepsilon,$$

where $\varepsilon > 0$ is a small positive number (in all our experiments, we take $\varepsilon = 10^{-100}$).

Now, given $1 \leq r_0 \leq E\left(\frac{r}{2}\right)$, where $E(x) = \max \mathbb{Z} \cap (-\infty, x]$, we will seek for a smoothness zone along the stencils of $r_0 + 1$ points as a reference.

This procedure will work if there is only one discontinuity in the stencil, and the restriction $r_0 \leq E\left(\frac{r}{2}\right)$ is set in order to avoid a stencil overlapping, since a discontinuity might eventually be in the overlapping zone and thus none of the stencils would include smooth data.

4.1.1 Original Weights (OW)

We define

$$IS_k = \min_{0 \leq j \leq r-k} \frac{1}{r} \sum_{\ell=1}^{r_0} \int_{x_0}^{x_r} h^{2\ell-1} q_{k,j}^{(\ell)}(x)^2 dx + \varepsilon, \quad 1 \leq k \leq r_0,$$

where $q_{k,j}$ is the polynomial of degree at most k such that $q_{k,j}(x_{j+i}) = u_{j+i}$ for $0 \leq i \leq k$, $0 \leq j \leq r-k$.

Now, the weights are defined as follows

$$\begin{aligned} \omega_k &= 1 - \left(1 - \left(\frac{IS_k}{I_k}\right)^{s_1}\right)^{s_2}, \quad 1 \leq k \leq r_0, \\ \omega_k &= \min\left\{1 - \left(1 - \left(\frac{IS_{r_0}}{I_k}\right)^{s_1}\right)^{s_2}, 1\right\}, \quad r_0 + 1 \leq k \leq r. \end{aligned} \tag{2}$$

The parameter s_1 enforces the convergence to 0 when the stencil is not smooth, while the parameter s_2 enforces the convergence to 1 when it is smooth.

It can be shown that for a smooth stencil, if there exists some $1 \leq k_0 \leq r_0$ such that $|u^{(k_0)}| >> 0$ around the stencil, then

$$\omega_k = 1 - \mathcal{O}(h^{s_2})$$

and if the stencil crosses a discontinuity, then

$$\omega_k = \mathcal{O}(h^{2s_1}).$$

Taking into account these considerations, and assuming the above hypothesis, it can be proven that if $s_2 \geq 1$ then

$$u_* = u(x^*) + \mathcal{O}(h^{r+1}).$$

4.1.2 Unique Weight Extrapolation (UW)

Since we seek for robustness combined with efficiency, the above extrapolation method can be replaced in practice by a simpler one, based in the computation of only one weight, that we will now explain in detail.

Such simplification is performed in the following sense: Instead of gradually increasing the degree of the interpolating polynomials, we will just average the constant extrapolation ($k = 0$) and maximum degree extrapolation ($k = r$), that is, we will consider

$$u_* = (1-\omega)p_0(x_*) + \omega p_r(x_*) = (1-\omega)u_{i_0} + \omega p_r(x_*),$$

where

$$\omega = \min\left\{1 - \left(1 - \left(\frac{IS_{r_0}}{I_r}\right)^{s_1}\right)^{s_2}, 1\right\}.$$

In order to lower even more the computational cost and ensure that $0 < \omega \leq 1$ without having to bound it artifically by 1 when $r_0 > 1$, we can replace the definition of I_r, which is a smoothness indicator of the whole $r+1$ points stencil, by the average of all smoothness indicators of the substencils of $r_0 + 1$ points, i.e.:

$$I_r^* := \frac{1}{r - r_0 + 1} \sum_{j=0}^{r-r_0} I_{r_0,j},$$

where

$$I_{r_0,j} = \frac{1}{r_0} \sum_{\ell=1}^{r_0} \int_{x_j}^{x_{r_0+j}} h^{2\ell-1} q_{r_0,j}^{(\ell)}(x)^2 dx + \varepsilon. \tag{3}$$

Then one can define

$$\omega = 1 - \left(1 - \left(\frac{IS_{r_0}}{I_r^*}\right)^{s_1}\right)^{s_2},$$

which in this case it clearly verifies $0 \leq \omega \leq 1$.

After a similar analysis as performed for OW, under the hypothesis $\exists k_0 \in \mathbb{N}$, $1 \leq k_0 \leq r_0$ such that $|u^{(k_0)}| >> 0$ around the stencil, then

$$u_* = u(x^*) + \mathcal{O}(h^{r'+1}),$$

where $r' = s_2(r_0 - k_0 + 1)$.

We will use in our experiments the above extrapolation technique taking $R = 10$ (a set of 10 points will be used to look for smooth zones), $r = 4$ (stencils of 5 points for extrapolation), $r_0 = s_1 = 2$, $s_2 = 4$ in order to achieve fifth order convergence provided that the two first derivatives do not annihilate simultaneously.

Unlike OW, this extrapolation method does not provide the optimal order in presence of a discontinuity, since the convergence order decays to 1 regardless of the position of the discontinuity. However, this is not a major issue and in practice the results are good.

In this sense, we have to take into account as well that when a discontinuity passes through the boundary, there will be a moment when only one cell will be available, leading into an essentially constant extrapolation, regardless of the method used, which yields again a first order accurate approximation.

5 Numerical Experiments

5.1 *One-Dimensional Experiments*

In this section we present some one-dimensional numerical experiments where both the accuracy of the extrapolation method for smooth solutions and its behavior in presence of discontinuities will be tested and analyzed for the UW extrapolation method.

This approach will illustrate that the accuracy order will still be the expected one in the smooth case and that the extrapolation method shows good performance in the non-smooth case.

5.1.1 Linear Advection

The problem statement for this test is the same as in [10] and [1]. We consider the linear advection equation

$$u_t + u_x = 0, \quad \Omega := (-1, 1),$$

with initial condition given by $u(x, 0) = 0.25 + 0.5 \sin(\pi x)$ and boundary condition $u(-1, t) = 0.25 - 0.5 \sin(\pi(1+t))$, $t \geq 0$. We apply a numerical outflow condition at $x = 1$, where Dirichlet boundary conditions cannot be imposed due to the direction of propagation of the information.

It is immediately checked that the unique (smooth) solution to this problem is

$$u(x, t) = 0.25 + 0.5 \sin(\pi(x - t)).$$

In order to numerically test the order of accuracy we perform tests at resolutions given by $n = 20 \cdot 2^j$ points, $j = 1, \ldots, 5$. The cell centers are $x_j := -1 + (j + \frac{1}{2})h_x$, with $h_x := \frac{2}{n}$. We recall that the set of all cell centers which are interior to Ω is

$$\mathscr{D} := \left\{x_j : j \in \{0, \ldots, n-1\}\right\}.$$

Since we use WENO5 reconstruction, we require 3 extra cells at each side of the boundary, where extrapolation from the interior will take place.

- $x = -1$: x_j, $-3 \leq j \leq -1$.
- $x = 1$: x_j, $n \leq j \leq n+2$.

Given that the ODE solver is third order accurate, in order to attain fifth order accuracy in the overall scheme, we need to select a time step given by $\Delta t = \left(\frac{2}{n}\right)^{\frac{5}{3}}$, with corresponding Courant numbers $\Delta t / h_x = (2/n)^{2/3} \leq 1/20^{2/3}$.

Since the left boundary conditions are time dependent, we also have to take into account that a specific approximation is needed in each of the 3 stages in each RK3-TVD time step. In general, if the inflow condition is given by some function $g(t)$ which is at least twice continuously differentiable, we have to use the following values at the boundary to preserve third order accuracy [3]:

- First stage: $g(t_k)$.
- Second stage: $g(t_k) + \Delta t g'(t_k)$.
- Third stage: $g(t_k) + \frac{1}{2}\Delta t g'(t_k) + \frac{1}{4}\Delta t^2 g''(t_k)$.

Taking into account all the previous considerations, we execute the simulation until $t = 1$ for all the previously specified resolutions and we study the errors in the 1 and ∞ norms, together with the order deduced from them. We consider different modalities of boundary extrapolation:

- By thresholding, taking $\delta = \delta' = 0.99$ (Table 1).
- Weighted, using the unique weight modality (Table 2).

From the Tables 1 and 2, it can be appreciated that the behaviour of the weighted extrapolation method is better than the thresholding technique for lower resolutions, while it is essentially as good as the thresholding method for higher resolutions.

Table 1 Extrapolation by thresholding, $\delta = 0.99$

n	Error $\|\cdot\|_1$	Order $\|\cdot\|_1$	Error $\|\cdot\|_\infty$	Order $\|\cdot\|_\infty$
40	5.45E−5	–	3.81E−4	–
80	3.06E−6	4.15	3.65E−5	3.38
160	1.34E−8	7.83	2.10E−7	7.44
320	2.64E−10	5.67	6.95E−10	8.93
640	8.26E−12	5.00	2.13E−11	5.03

Table 2 Weighted extrapolation

n	Error $\|\cdot\|_1$	Order $\|\cdot\|_1$	Error $\|\cdot\|_\infty$	Order $\|\cdot\|_\infty$
40	9.99E−6	–	2.44E−5	–
80	2.79E−7	5.16	7.35E−7	5.05
160	8.51E−9	5.03	2.31E−8	4.99
320	2.65E−10	5.00	6.95E−10	5.06
640	8.26E−12	5.00	2.13E−11	5.03

5.1.2 Linear Advection, Discontinuous Solution

We illustrate with this experiment the behavior of the schemes when discontinuities are present and the entailed improvement with respect to using Lagrange extrapolation with no filters. We consider the same meshing and data as in Sect. 5.1.1 for the previous problem, but now the boundary condition is:

$$u(-1,t) = g(t) = \begin{cases} 0.25 \text{ if } t \le 1 \\ -1 \text{ if } t > 1 \end{cases}$$

With this definition, the unique (weak) solution to this problem has a moving discontinuity and is given by:

$$u(x,t) = \begin{cases} -1 & \text{if} \quad x < t-2 \\ 0.25 & \text{if } t-2 \le x \le t-1 \\ 0.25 + 0.5\sin(\pi(x-t)) & \text{if} \quad x \ge t-1 \end{cases}$$

In Fig. 3 we check the graphical results that correspond to the simulation until $t = 1.5$, first using Lagrange extrapolation with no filters, afterwards with a filter with $\delta = 0.75$ and finally weighted extrapolation. As it can be seen in Fig. 3, Lagrange extrapolation without filters leads to spurious oscillations around the left side of the discontinuity, while thresholding and weighted extrapolation remove them.

5.2 *Two-Dimensional Experiments*

The equations that will be considered in this section are the two-dimensional Euler equations for inviscid gas dynamics

$$u_t + f(u)_x + g(u)_y = 0, \quad u = u(x,y,t),$$

$$u = \begin{bmatrix} \rho \\ \rho v^x \\ \rho v^y \\ E \end{bmatrix}, \quad f(u) = \begin{bmatrix} \rho v^x \\ p + \rho (v^x)^2 \\ \rho v^x v^y \\ v^x(E+p) \end{bmatrix}, \quad g(u) = \begin{bmatrix} \rho v^y \\ \rho v^x v^y \\ p + \rho (v^y)^2 \\ v^y(E+p) \end{bmatrix}. \tag{4}$$

In these equations, ρ is the density, (v^x, v^y) is the velocity and E is the specific energy of the system. The variable p stands for the pressure and is given by the equation of state:

$$p = (\gamma - 1)\left(E - \frac{1}{2}\rho((v^x)^2 + (v^y)^2)\right),$$

where γ is the adiabatic constant, that will be taken as 1.4 in all the experiments.

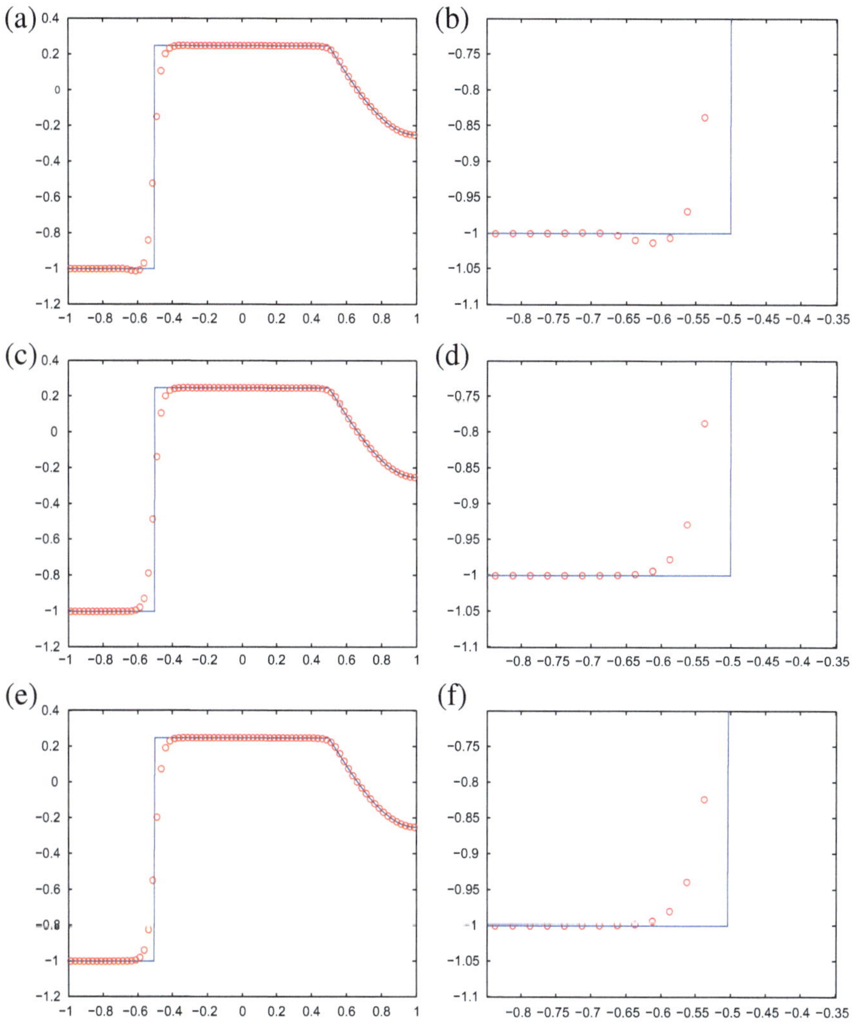

Fig. 3 Comparison of different extrapolations for the linear advection test with discontinuous solution. (**a**) Lagrange extrapolation; (**b**) Lagrange extrapolation (zoom); (**c**) extrapolation with thresholds; (**d**) extrapolation with thresholds (zoom); (**e**) weighted extrapolation; (**f**) weighted extrapolation (zoom)

5.2.1 Double Mach Reflection

This experiment uses the Euler equations to model a vertical right-going Mach 10 shock colliding with an equilateral triangle. By symmetry, this is equivalent to a collision with a ramp with a slope of 30 degrees with respect to the horizontal line, which is how we will model the simulation to halve the computational cost.

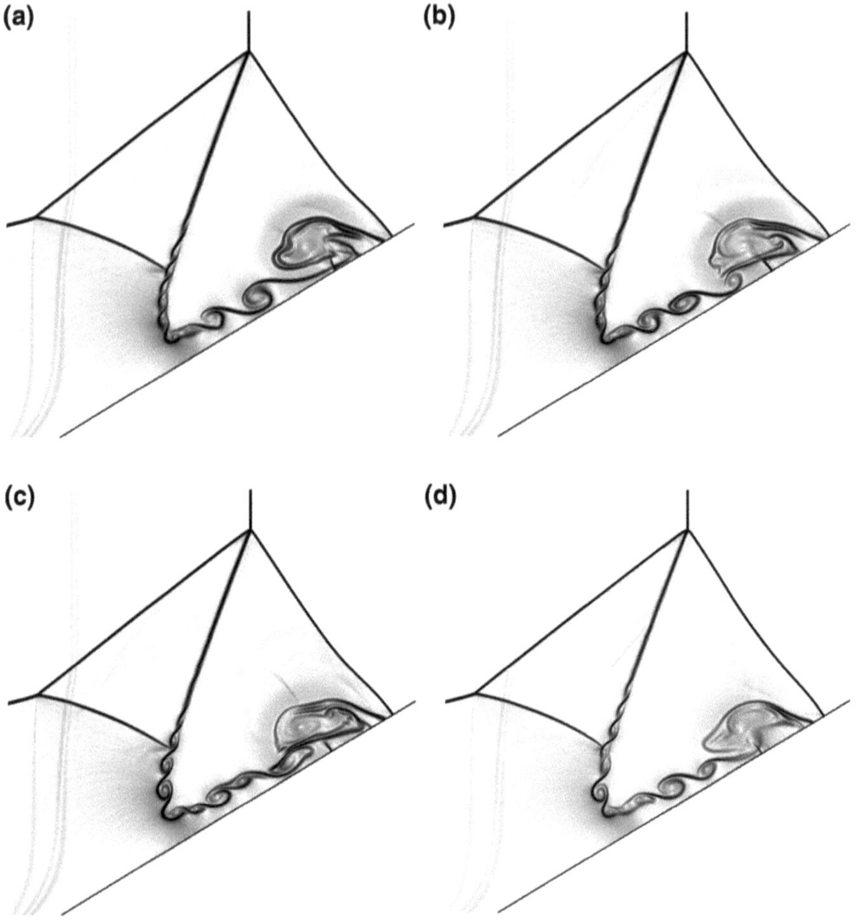

Fig. 4 Schlieren plots of the density field at $t = 0.2$. (**a**) Density field: thresholding, $\delta = 0.9$; (**b**) density field: thresholding, $\delta = 0.5$; (**c**) density field: thresholding, $\delta = 0.35$; (**d**) density field: weighted extrapolation

The initial conditions are the following:

$$u = (\rho, v^x, v^y, E) = (8.0, 8.25, 0, 563.5) \text{ if } x \leq \hat{x}$$
$$u = (\rho, v^x, v^y, E) = (1.4, 0, 0, 2.5) \qquad \text{if } x > \hat{x}$$

where $\hat{x}$ is the point where the ramp starts.

We perform the simulation until $t = 0.2$. The experiment consists in different simulations with different threshold values and the weighted extrapolation. In Fig. 4 we present a Schlieren plot of the result for the density ρ at a resolution of $h_x = h_y = \frac{\sqrt{3}}{2}\frac{1}{640}$, both for thresholding and weighted extrapolation.

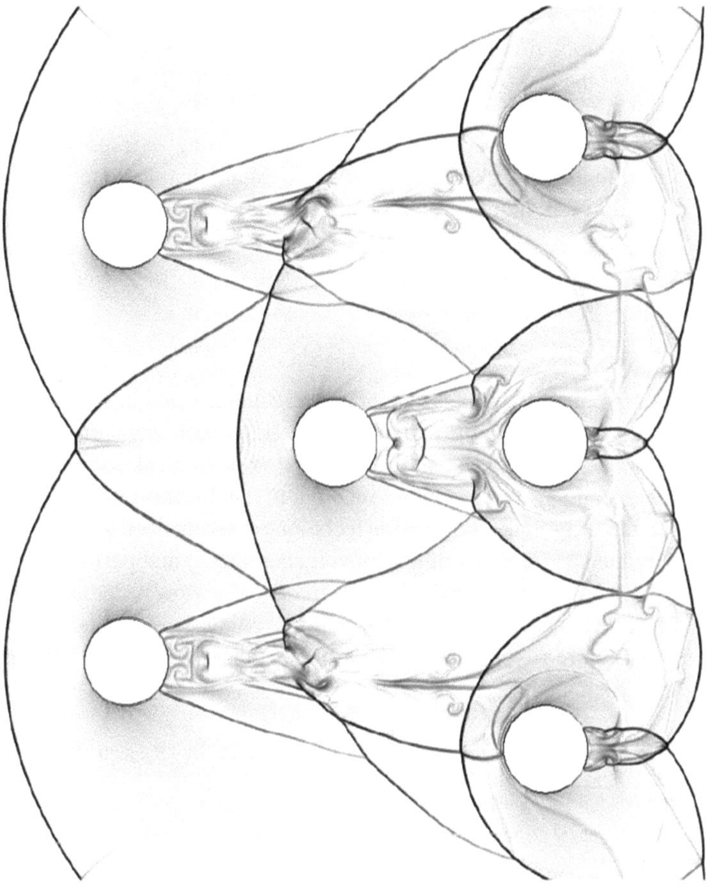

Fig. 5 Density field: $t = 0.5$

The results show that the weighted extrapolation proposed in this paper produces a result with a similar or even better quality than the ones obtained by the thresholding technique without having to adjust additional tuning parameters. Moreover, for simulations attempted to be performed with threshold values lower than 0.2, the numerical simulation failed due to the existence of unphysical quantities for the pressure, produced by oscillations near a shock, which illustrates that the weighted extrapolation procedure is more robust than the thresholding technique.

5.2.2 Interaction of a Shock with Multiple Circular Obstacles

We repeat the previous experiment by adding multiple circles in the domain as shown in Fig. 5. This test can also be found in [2]. In this case, we run the simulation until $t = 0.5$ and a mesh size of $h_x = h_y = \frac{1}{512}$ on the whole domain, using the weighted extrapolation technique. As in the previous experiment, we present a Schlieren plot for the last time step in Fig. 5. These results are again consistent with those obtained in [2].

It must be remarked that in this case, thresholding extrapolation fails for threshold lower than about 0.99, which illustrates again that weighted extrapolation is more robust for more demanding problems like this one.

6 Conclusions

In this paper we have compared a new weighted extrapolation for boundary conditions with a thresholding technique. We have seen both theoretically and through numerical experiments that weighted extrapolation entails an improvement that overcomes some of the drawbacks inherent to the thresholding method.

Moreover, the weighted extrapolation technique does not need a tuning parameter (except the exponents for the weights convergence speed) and permits a successful detection for discontinuities, in which case it reduces to a low order method in order to avoid the appearance of spurious oscillations. Numerical results have reported as well that it is more robust than thresholding extrapolation in some complex and demanding problems.

Acknowledgements This research was partially supported by Spanish MINECO grants MTM2011-22741 and MTM2014-54388.

References

1. Baeza, A., Mulet, P., Zorío, D.: High order boundary extrapolation technique for finite difference methods on complex domains with Cartesian meshes. J. Sci. Comput. **66**, 761–791 (2016)
2. Boiron, O., Chiavassa, G., Donat, R.: A high-resolution penalization method for large Mach number flows in the presence of obstacles. Comput. Fluids **38**, 703–714 (2009). doi:10.1016/j.compfluid.2008.07.003
3. Carpenter, M., Gottlieb, D., Abarbanel, S., Don, W.S.: The theoretical accuracy of Runge-Kutta time discretizations for the initial boundary value problem: a study of the boundary error. SIAM J. Sci. Comput. **16**, 1241–1252 (1995)
4. Donat, R., Marquina, A.: Capturing shock reflections: an improved flux formula. J. Comput. Phys. **125**, 42–58 (1996)
5. Jiang, G.S., Shu, C.W.: Efficient implementation of Weighted ENO schemes. J. Comput. Phys. **126**, 202–228 (1996)
6. Marquina, A., Mulet, P.: A flux-split algorithm applied to conservative models for multicomponent compressible flows. J. Comput. Phys. **185**, 120–138 (2003)
7. Shu, C.W., Osher, S.: Efficient implementation of essentially non-oscillatory shock-capturing schemes. J. Comput. Phys. **77**, 439–471 (1988)
8. Shu, C.W., Osher, S.: Efficient implementation of essentially non-oscillatory shock-capturing schemes, II. J. Comput. Phys. **83**(1), 32–78 (1989)

9. Sjogreen, B., Petersson, N.: A Cartesian embedded boundary method for hyperbolic conservation laws. Commun. Comput. Phys. **2**, 1199–1219 (2007)
10. Tan, S., Shu, C.W.: Inverse Lax-Wendroff procedure for numerical boundary conditions of conservation laws. J. Comput. Phys. **229**, 8144–8166 (2010)
11. Tan, S., Wang, C., Shu, C.W., Ning, J.: Efficient implementation of high order inverse Lax-Wendroff boundary treatment for conservation laws. J. Comput. Phys. **231**(6), 2510–2527 (2012)

High Order Nyström Methods for Transmission Problems for Helmholtz Equation

Víctor Domínguez and Catalin Turc

Abstract We present super-algebraic compatible Nyström discretizations for the four Helmholtz boundary operators of Calderón's calculus on smooth closed curves in 2D. These discretizations are based on appropriate splitting of the kernels combined with very accurate product-quadrature rules for the different singularities that such kernels present. A Fourier based analysis shows that the four discrete operators converge to the continuous ones in appropriate Sobolev norms. This proves that Nyström discretizations of many popular integral equation formulations for Helmholtz equations are stable and convergent. The convergence is actually super-algebraic for smooth solutions.

1 Introduction

The design of robust discretizations of the boundary integral equations in 2D has been an active research topic in the last decades. The analysis of Galerkin discretizations of boundary integral equations is by now well understood in the case of smooth boundaries and boundary data. Indeed, their stability can be established based on the coercivity of the principal parts of the boundary integral operators featured in the integral formulations (a first result along these lines can be traced back to [12]), and compact perturbation analysis arguments. On the other hand, although Nyström/collocation methods are simpler to implement, their analysis is somewhat more complicated. Given that for 2D problems boundary integral operators can be thought of as periodic pseudo-differential operators, the analysis of discretization schemes for boundary integral equations relies on Fourier analysis. Galerkin as well as Nyström/collocation methods for periodic integral equations have been fully analyzed for many periodic integral equations and these techniques

V. Domínguez (✉) • C. Turc
Departamento de Ingeniería Matemática e Informática, Universidad Pública de Navarra, Avda Tarazona s/n, 31500 Tudela, Spain

Department of Mathematical Sciences and Center for Applied Mathematics, New Jersey Institute of Technology, University Heights, 323 Dr. M.L. King Jr. Blvd, Newark, NJ 07102, USA
e-mail: catalin.c.turc@njit.edu; victor.dominguez@unavarra.es

F. Ortegón Gallego et al. (eds.), *Trends in Differential Equations and Applications*, SEMA SIMAI Springer Series 8, DOI 10.1007/978-3-319-32013-7_15

have been also used to derive new methods as qualocation schemes, cf. [13] and references therein.

Boundary integral formulations of Helmholtz equations in a certain domain rely on single and double layer acoustic potentials and their Dirichlet and Neumann traces on the boundary of that domain. These traces lead to the natural definition of four boundary integral operators which are referred to as the Helmholtz boundary integral operators of Calderon's calculus. In this paper we focus on Nyström methods based on suitable quadrature rules for the discretization of the four Helmholtz boundary integral operators that feature in Calderon's calculus. These provide a means of defining fully discrete versions of these operators which can be used easily to discretize complicate formulations involving rather complex compositions of different boundary operators. Moreover, these discretizations can be easily used in conjunction with iterative solvers based on Krylov subspace methods.

The aim of this paper is not to propose new discretizations of the Helmholtz boundary integral operators. Actually, most of those considered here can be found and have been thoroughly analyzed in the literature, mostly by Kress (cf [5, 6] and references therein). Our objective is therefore different: we want to propose *compatible* discretizations of the four Helmholtz boundary integral operators that lead to super-algebraic schemes for most of the boundary integral formulations of the Helmholtz equation in 2D.

Helmholtz transmission problems for smooth interfaces provide a sufficiently complex environment for testing our discretizations as they feature all of the four Helmholtz boundary integral operators in Calderon's calculus. Discretizations of integral formulations of other types of boundary conditions can be readily produced and analyzed with the methods we present in this paper.

Some of the formulations considered in this paper are direct, i.e., the unknowns are physical quantities of the problem (typically the trace and the normal derivative of the solution), others are indirect. Some of the indirect formulations considered in this text could be more economical from a computational point of view. Besides, some more sophisticated integral formulations lead to matrices with clustered eigenvalues, which usually ensures a faster convergence of Krylov methods such as GMRES. Demanding better spectral properties requires working with more complex formulations whose discretization could seem challenging at first sight. We will show that the discrete boundary layer operators can be used as black boxes in such a way that the discretization of any integral formulation, however complicated, is in fact straightforward. Moreover, for smooth data, we prove that the numerical solutions converge super-algebraically, that is, faster than any negative power of N, the number of degrees of freedom.

The paper is structured as follows: in Sect. 2 we discuss briefly the Helmholtz transmission problem and introduce the boundary layer potentials and operators for the Helmholtz equation. In Sect. 3 we reformulate these mappings as integral operators acting on spaces of $2\pi-$periodic functions via a parameterization of the interface. We present also their numerical discretizations and analyze their convergence. We conclude by showing in Sect. 4 how these compatible discretizations can be applied to solve numerically several boundary integral formulations of the original Helmholtz transmission problem. Well-posedness and convergence estimates are derived for the integral equations considered in this paper. Some numerical experiments are presented in the final Sect. 6.

2 Helmholtz Transmission Problems and Boundary Integral Operators

We start introducing the domain of the transmission problem (see Fig. 1). Let D^- be a compact domain with smooth boundary Γ which for simplicity we will assume to be simply connected. Denote also $D^+ := \mathbb{R}^2 \setminus \overline{D^-}$. We will write γ for the trace operator and ∂_n for the unit normal derivative on Γ pointing toward D^+. Given two wavenumbers k_+, k_- that are complex numbers with non-negative imaginary part, we consider the following Helmholtz transmission problem:

$$\begin{aligned} \Delta u^+ + k_+^2 u^+ &= 0 \qquad \text{in } D^+ \\ \Delta u^- + k_-^2 u^- &= 0 \qquad \text{in } D^- \\ \gamma u^+ - \gamma u^- &= -u_{\text{inc}}, \quad \partial_n u^+ - \nu \partial_n u^- = -\partial_n u_{\text{inc}} \\ \partial_r u^+ - \mathrm{i} k_+ u^+ &= o(|\mathbf{r}|^{-1/2}). \end{aligned} \tag{1}$$

Here ∂_r is the partial derivative on the radial direction and u_{inc} is an incident wave that is a solution of the Helmholtz problem for k_+ on a neighborhood of $\overline{D^-}$. We assume that the transmission problem above together with its adjoint, that is the transmission problem defined by taking $k_\pm$ in $D^\mp$, are uniquely solvable. For instance, if $k_\pm$ are real and $\nu > 0$ these hypotheses are known to be satisfied. We refer to [4] for more comprehensive sets of values of $k_\pm$ and ν fulfilling these hypotheses.

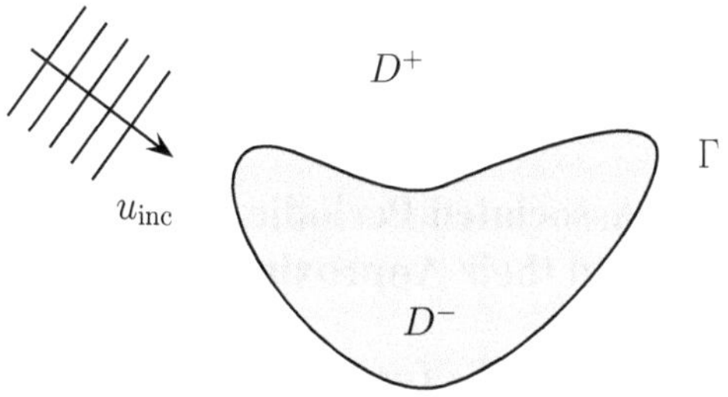

Fig. 1 Sketch of the domain of the transmission problem

Let

$$\Phi_k(\mathbf{x}) := \frac{\mathrm{i}}{4} H_0^{(1)}(k|\mathbf{x}|)$$

($H_0^{(1)}$ is the Hankel function of first kind and order 0) be the outgoing fundamental solution of the Helmholtz equation in $\mathbb{R}^2$. The single and double layer operators are defined as follows

$$SL_k\varphi := \int_\Gamma \Phi_k(\cdot - \mathbf{y})\varphi(\mathbf{y})\mathrm{d}\sigma(\mathbf{y}), \qquad DL_k g := \int_\Gamma \frac{\partial \Phi_k(\cdot - \mathbf{y})}{\partial \mathbf{n}(\mathbf{y})} g(\mathbf{y})\mathrm{d}\sigma(\mathbf{y}). \tag{2}$$

We stress that for any density, the layer operators define solutions of the Helmholtz equation in $\mathbb{R}^2 \setminus \Gamma$ which satisfy, in addition, the radiation condition at infinity (last condition in (1)). Moreover, the third Green formula states

$$u^{\pm} = \mp SL_{k_\pm} \partial_n u_\pm \pm DL_{k_\pm} \gamma u_\pm. \tag{3}$$

Let us denote by $\gamma^\pm$, $\partial_n^\pm$ the trace and respectively the normal derivative taken from $D^\pm$. We have then the jump properties

$$\begin{aligned} \gamma^\pm SL_k &= V_k & \partial_n^\pm SL_k &= \mp\tfrac{1}{2}I + K_k^\top \\ \gamma^\pm DL_k &= \pm\tfrac{1}{2}I + K_k^\top & \partial_n^\pm DL_k &= H_k \end{aligned} \tag{4}$$

where I denotes the identity, V_k is the single layer operator, K_k and $K_k^\top$ are the double layer and adjoint double layer operator, and H_k is the hypersingular operator.

We can now proceed as follows: (a) we can use (4) and the transmission conditions stated in (1) to compute the Cauchy data of the solution $(u^\pm, \partial_n^\pm u)$ and reconstruct these functions using (3); (b) we can try to write the $u^\pm$ in terms of some unknown densities associated with the potentials (2) and solve for these densities via equations obtained from (4). Approach (a) leads to the so-called direct methods whereas schemes obtained from (b) are known as indirect methods.

3 Associated Periodic Integral Operators and their Approximation

3.1 *Periodic Integral Operators*

Let us consider a smooth regular 2π-periodic parameterization of the curve Γ given by $\mathbf{x} : \mathbb{R} \to \Gamma$. First we set the transmission data

$$h(s) := -(\gamma_\Gamma u_{\rm inc} \circ \mathbf{x})(s), \quad \eta(s) := -(\partial_n u_{\rm inc} \circ \mathbf{x})(s)\, |\mathbf{x}'(s)|. \tag{5}$$

We follow the same rule to reformulate layer potentials and boundary integral operators as 2π–periodic integral operators in the following sense: for SL_k and the associated boundary integral operators V_k and $K_k^\top$, the norm of the parameterization $|\mathbf{x}'(t)|$ (t is the integration variable) is incorporated in the density function φ in (2), whereas for DL_k, and the corresponding boundary integral operators K_k and H_k this term is incorporated in the kernels of these operators. In addition, the operators $K_k^\top$ and H_k are multiplied by $|\mathbf{x}(s)|$, where s will be used henceforth as the variable corresponding to the target point in all of the integral operators considered in this text. With these conventions, we write the single, double and adjoint double layer operator as follows

$$[\mathrm{V}_k\varphi](s) = \int_0^{2\pi} A(s,t)\log\sin^2\tfrac{s-t}{2}\varphi(t)\mathrm{d}t + \int_0^{2\pi} B(s,t)\varphi(t)\mathrm{d}t \tag{6}$$

$$[\mathrm{K}_k g](s) = \int_0^{2\pi} C(s,t)\sin^2\tfrac{s-t}{2}\log\sin^2\tfrac{s-t}{2}g(t)\mathrm{d}t + \int_0^{2\pi} D(s,t)g(t)\mathrm{d}t \tag{7}$$

$$[\mathrm{K}_k^\top\varphi](s) = \int_0^{2\pi} C(t,s)\sin^2\tfrac{t-s}{2}\log\sin^2\tfrac{t-s}{2}\varphi(t)\mathrm{d}t + \int_0^{2\pi} D(t,s)\varphi(t)\mathrm{d}t \tag{8}$$

with

$$A(s,t) = -\frac{1}{4\pi}J_0(k|\mathbf{x}(s)-\mathbf{x}(t)|)$$

$$C(s,t) = -\frac{k(\mathbf{x}(s)-\mathbf{x}(t))\cdot(x_2'(t),-x_1'(t))}{|\mathbf{x}(s)-\mathbf{x}(t)|^2}\,\frac{J_1(k|\mathbf{x}(s)-\mathbf{x}(t)|)}{|\mathbf{x}(s)-\mathbf{x}(t)|}\,\frac{|\mathbf{x}(s)-\mathbf{x}(t)|^2}{\sin^2\frac{s-t}{2}}$$

$$B(s,t) = \frac{\mathrm{i}}{4}H_0^1(k|\mathbf{x}(s)-\mathbf{x}(t)|) - A(s,t)\log\sin^2\tfrac{s-t}{2}$$

$$D(s,t) = \frac{\mathrm{i}k}{4}H_1^1(k|\mathbf{x}(s)-\mathbf{x}(t)|)\frac{(\mathbf{x}(s)-\mathbf{x}(t))\cdot(x_2'(t),-x_1'(t))}{|\mathbf{x}(s)-\mathbf{x}(t)|}$$
$$-C(s,t)\sin^2\tfrac{s-t}{2}\log\sin^2\tfrac{s-t}{2}.$$

(Observe that K_k and $\mathrm{K}_k^\top$ are transpose to each other). Very well known properties of the Bessel functions imply that the functions A, B, C, D are smooth functions if so is the map $\mathbf{x}$, as we have already assumed above.

Regarding the parameterized version of the hypersingular operator, the integration-by-parts like formula due to Maue [10] (see also [11]) allows to write H_k as the integro-differential operator

$$[\mathrm{H}_k g]\,(s) = [\mathrm{D}\mathrm{V}_k\mathrm{D}\,g]\,(s) - \mathrm{i}k^2\left[\mathrm{V}_k((\mathbf{x}'(s)\cdot\mathbf{x}'(\cdot))g\right](s). \tag{9}$$

Here, $\mathrm{D}\varphi := \varphi'$ is simply the differentiation operator.

3.2 Nyström Discretization

The structure of the kernels introduced in the previous section leads to tackle, apart from the derivative operator, the evaluation of integrals as

$$[\Psi\varphi](s) := \int_0^{2\pi} \psi(s-t)a(s,t)\varphi(t)\,\mathrm{d}t, \tag{10}$$

where a, ψ are $2\pi-$periodic, with a being smooth and ψ, in principle, singular at 0. The operators defined in Eq. (10) are $2\pi-$periodic pseudo-differential operators (cf. [13, Chap. 7]).

3.2.1 Trigonometric Interpolation

Let us denote

$$\mathbb{T}_N := \mathrm{span}\,\langle e_n \,:\, -N < n \leq N\rangle, \quad \text{with} \quad e_n(t) := \exp(\mathrm{i}nt), \quad (n \in \mathbb{Z})$$

the space of trigonometric polynomials of degree N. On $\mathbb{T}_N$ we consider the trigonometric interpolation problem on the uniform grid $\{j\pi/N\}$:

$$\mathbb{T}_N \ni P_N g \quad \text{s.t.} \quad (P_N g)(\tfrac{j\pi}{N}) = g(\tfrac{j\pi}{N}),\ j = 0,\ldots,2N-1.$$

The solution of the interpolating problem is given by

$$\sum_{n=-N+1}^{N}\left[\frac{1}{2N}\sum_{m=-N+1}^{N} g(\tfrac{j\pi}{N})e_n(-\tfrac{\mathrm{i}m\pi}{N})\right]e_n(\mathrm{i}nt) \tag{11}$$

which can be computed in $\mathcal{O}(n\log n)$ operations using FFT.

3.2.2 Discrete Operators

We now introduce

$$[\Psi_N\varphi](s) := \int_0^{2\pi} \psi(s-t)P_N\left[a(s,\cdot)\varphi\right](t)\,\mathrm{d}t \approx [\Psi\varphi](s) \tag{12}$$

as discrete approximations of (10). Clearly $\Psi_N\varphi$ depends only on the pointwise values of the density at the grid points, which justifies the use of the term "discrete" when referring to these operators.

Obviously, we are just working with a product-integration rule and the applicability of such procedure relies on being able to compute

$$\hat{\psi}(n) := \frac{1}{2\pi}\int_0^{2\pi} \psi(t)e_{-n}(t)\,\mathrm{d}t, \quad n \in \mathbb{Z}$$

i.e., the Fourier coefficients of the weight function ψ. Fortunately, for the weight functions featured above, these Fourier coefficients can be computed explicitly. Indeed, for $\psi_1 := \log\sin^2\frac{t}{2}$ we have

$$\begin{aligned}\hat{\psi}_1(n) &= \tfrac{1}{2\pi}\int_0^{2\pi} \log(\sin^2\tfrac{t}{2})\, e_{-n}(t)\mathrm{d}t = \tfrac{1}{2\pi}\int_0^{2\pi} \log(\sin^2\tfrac{t}{2})\, \cos(nt)\mathrm{d}t \\ &= \begin{cases} -2\log 4, & n = 0, \\ -2|n|^{-1}, & \text{otherwise,}\end{cases}\end{aligned}$$

whereas for $\psi_2 := \sin^2\frac{t}{2}\log\sin^2\frac{t}{2}$ straightforward calculations yield

$$\begin{aligned}\hat{\psi}_2(n) &= \tfrac{1}{2\pi}\int_0^{2\pi} \sin^2\tfrac{t}{2}\log(\sin^2\tfrac{t}{2})\, e_{-n}(t)\mathrm{d}t \\ &= \tfrac{1}{8\pi}\int_0^{2\pi} \log(\sin^2\tfrac{t}{2})\ (2\cos(nt) - \cos(n-1)t - \cos(n+1)t)\,\mathrm{d}t \\ &= \begin{cases} \frac{1}{2}, & n = 0, \\ -\frac{3}{8}, & |n| = 1, \\ \frac{1}{4}\Big[\frac{1}{|n+1|} + \frac{1}{|n-1|} - \frac{2}{|n|}\Big], & \text{otherwise.}\end{cases}\end{aligned}$$

We stress that the calculation in the case of the weight ψ_1 can be traced back to [8, 9] (see also [6]). For the remaining case, $\psi_0 \equiv 1$, the same approach gives us (see (11))

$$\begin{aligned}\int_0^{2\pi} (P_N g)(t)\,\mathrm{d}t &= \sum_{n=-N+1}^{N}\left[\frac{1}{2N}\sum_{m=-N+1}^{N} g(\tfrac{j\pi}{N})\exp(-\tfrac{imn\pi}{N})\right]\int_0^{2\pi}\exp(int)\,\mathrm{d}t \\ &= \frac{\pi}{N}\sum_{j=0}^{2N-1} g(\tfrac{j\pi}{N}),\end{aligned}$$

i.e., the trapezoidal rule. Therefore, for $\psi \equiv 1$, we simply have

$$[\Psi_N\varphi](s) = \frac{\pi}{N}\sum_{j=0}^{2N-1} a\big(s,\tfrac{j\pi}{N}\big)\varphi\big(\tfrac{j\pi}{N}\big).$$

3.2.3 Discrete Helmholtz Boundary Integral Operators

For the single layer operator we work with two types of discretizations. The first one, proposed originally by Kress (cf. [6] and references therein) is simply

$$[\mathrm{V}_{k,N}\varphi](s) := \int_0^{2\pi} \psi_1(s-t)\,[P_N A(s,\cdot)\varphi]\,(t)\,\mathrm{d}t + \int_0^{2\pi} [P_N B(s,\cdot)\varphi](t)\mathrm{d}t. \qquad (13)$$

One can use the same approach for the double layer operator and obtain

$$[\mathrm{K}_{k,N}\varphi](s) := \int_0^{2\pi} \psi_1(s-t)[P_N C(s,\cdot)\sin^2\tfrac{s-\cdot}{2}\,\varphi](t)\mathrm{d}t + \int_0^{2\pi} [P_N D(s,\cdot)\varphi]\,(t)\mathrm{d}t. \qquad (14)$$

The operator $\mathrm{K}^\top_{k,N}$ can be defined accordingly.

Alternatively, we can proceed in a different way and define the *more accurate* approximation

$$[\widetilde{\mathrm{K}}_{k,N}g](s) := \int_0^{2\pi} \psi_2(s-t)P_N[C(s,\cdot)g](t)\,\mathrm{d}t + \int_0^{2\pi} [P_N D(s,\cdot)g](t)\mathrm{d}t. \qquad (15)$$

The operator $\widetilde{\mathrm{K}}^\top_{k,N}$ can be obviously defined in the same manner.

We can actually use the same approach for the single layer operator V_k. Indeed, let us write first

$$A(s,t) = -\frac{1}{4\pi} + \frac{1 - J_0(k|\mathbf{x}(s)-\mathbf{x}(t)|)}{4\pi\sin^2\frac{s-t}{2}}\sin^2\tfrac{s-t}{2} =: -\frac{1}{4\pi} + \widetilde{A}(s,t)\sin^2\tfrac{s-t}{2}. \qquad (16)$$

We point out that function $A(s,t)$ is smooth with

$$\widetilde{A}(s,s) \equiv \frac{k^2}{4\pi}|\mathbf{x}'(s)|.$$

Hence, using the Bessel operator defined as

$$[\Lambda\varphi]\,(s) := -\frac{1}{4\pi}\int_0^{2\pi} \log\sin^2\tfrac{s-t}{2}\,\varphi(t)\,\mathrm{d}t,$$

we have derived the following alternative expression for the single layer operator

$$\mathrm{V}_k\varphi = \Lambda\varphi + \int_0^{2\pi} \widetilde{A}(\,\cdot\,,t)\psi_2(\,\cdot\, - t)\varphi(t)\,\mathrm{d}t + \int_0^{2\pi} B(\,\cdot\,,t)\varphi(t)\,\mathrm{d}t =: \Lambda\varphi + \mathrm{R}_k\varphi,$$

which can be exploited to lead to the following approximation

$$\begin{aligned}\big[\widetilde{\mathrm{V}}_{k,N}\varphi\big](s) &:= \big[\Lambda\varphi\big](s) + \int_0^{2\pi} \psi_2(s-t)\, P_N[\widetilde{A}(s,\cdot)\varphi](t)\,\mathrm{d}t + \int_0^{2\pi} P_N[\widetilde{B}(s,\cdot)\varphi](t)\,\mathrm{d}t \\ &=: \big[\Lambda\varphi\big](s) + \big[\widetilde{\mathrm{R}}_{k,N}\varphi\big](s). \end{aligned} \tag{17}$$

Obviously, $\widetilde{\mathrm{V}}_{k,N}$ can be applied, in principle, only to trigonometric polynomials, since otherwise the first term gives rise to an infinite series. As we will see later, this is not a severe constraint for the numerical approximations we propose.

Finally, applying integration by parts and making use of the same quadrature rules, we have

$$\mathrm{H}_k = \mathrm{D}\Lambda\mathrm{D} + \mathrm{T}_k \tag{18}$$

with

$$\big[\mathrm{T}_k\varphi\big](s) = \int_0^{2\pi} E(s,t)\log\sin^2\tfrac{s-t}{2}\varphi(t)\,\mathrm{d}t + \int_0^{2\pi} F(s,t)\varphi(t)\,\mathrm{d}t$$

where

$$\begin{aligned} E(s,t) &:= -\partial_s\partial_t\widetilde{A}(s,t)\sin^2\tfrac{s-t}{2} + \tfrac{1}{2}\big(\partial_s\widetilde{A}(s,t) - \partial_t\widetilde{A}(s,t)\big)\sin(s-t) \\ &\quad + \tfrac{1}{2}\widetilde{A}(s,t)\cos(s-t) - \mathrm{i}k^2\big(\mathbf{x}'(s)\cdot\mathbf{x}'(t)\big)A(s,t) \\ F(s,t) &:= -\partial_s\partial_t B(s,t) + \tfrac{1}{2}\big(\partial_s\widetilde{A}(s,t) - \partial_t\widetilde{A}(s,t)\big)\sin(s-t) \\ &\quad + \widetilde{A}(s,t)(\tfrac{1}{2} + \cos(s-t)) - \mathrm{i}k^2(\mathbf{x}'(s)\cdot\mathbf{x}'(t))B(s,t). \end{aligned}$$

Then, following the same convention, we can define

$$\mathrm{H}_{k,N}\varphi := \mathrm{D}\Lambda\mathrm{D}\varphi + \mathrm{T}_{k,N}\varphi \tag{19}$$

with

$$\big[\mathrm{T}_{k,N}\varphi\big](s) = \int_0^{2\pi} \psi_1(s-t)\,[P_N E(s,\cdot)\varphi]\,(t)\mathrm{d}t + \int_0^{2\pi} [P_N F(s,\cdot)\varphi](t)\mathrm{d}t$$

Again $\mathrm{H}_{k,N}$ is not a full discrete operators, but when applied to trigonometric polynomials it can be computed exactly which turns out to be enough for our purposes.

3.3 Convergence Analysis

We develop our analysis in periodic Sobolev norms. For any $p \in \mathbb{R}$ we first define the Sobolev norm

$$\|\varphi\|_p^2 := |\hat{\varphi}(0)|^2 + \sum_{n\neq 0} |n|^{2p} |\hat{\varphi}(n)|^2.$$

The periodic Sobolev spaces of order p, denoted in what follows by H^p, can be defined, for instance, as the completion of trigonometric polynomials in this norm.

We are ready to state the main theorem. The proof follows from application of similar ideas to those introduced in [6, Chaps. 12 and 13] (see also [1]). Let us point out that henceforth, for given $A : X \to Y$, we denote by $\|A\|_{X\to Y}$ its operator norm.

Theorem 1 *Let $p > 1/2$ and $q \geq -1$ with $p + q > 1/2$. Then, if $\mathrm{A} \in \{\mathrm{K}_k, \mathrm{K}_k^\top, \mathrm{V}_k, \mathrm{H}_k\}$ and A_N is the corresponding approximation, i.e., $\mathrm{A}_N \in \{\mathrm{K}_{k,N}, \mathrm{K}_{k,N}^\top, \mathrm{V}_{k,N}, \mathrm{H}_{k,N}\}$,*

$$\|\mathrm{A} - \mathrm{A}_N\|_{H^{p+q}\to H^p} \leq C_{p,q} N^{-q-\min\{p,1\}}. \tag{20}$$

On the other hand, for $\mathrm{A} \in \{\mathrm{K}_k, \mathrm{K}_k^\top, \mathrm{V}_k\}$ and $\widetilde{\mathrm{A}}_N \in \{\widetilde{\mathrm{K}}_{k,N}, \widetilde{\mathrm{K}}_{k,N}^\top, \widetilde{\mathrm{V}}_{k,N}\}$ the corresponding discretization, we have for $q \geq -3$ with $p + q > 1/2$.

$$\|\mathrm{A} - \widetilde{\mathrm{A}}_N\|_{H^{p+q}\to H^p} \leq C_{p,q} N^{-q-\min\{p,3\}}. \tag{21}$$

Proof For any function ψ we denote the convolution operator in the usual manner:

$$[\psi * \varphi](s) := \int_0^{2\pi} \psi(s-t)\varphi(t)\, \mathrm{d}t = \sum_{n=-\infty}^{\infty} \hat{\psi}(n)\hat{\varphi}(n)e_n(s).$$

Then it is straightforward to check that for φ smooth enough, see (10),

$$[\Psi\varphi](s) = \int_0^{2\pi} a(s,t)\psi(s-t)\varphi(t)\, \mathrm{d}t = \sum_{n=-\infty}^{\infty} a_n(s)[\psi * (e_n\varphi)](s)$$

where $e_n(s) = \exp(\mathrm{i}ns)$ and

$$a_n(s) := \frac{1}{2\pi}\int_0^{2\pi} a(s,t)e_{-n}(t)\mathrm{d}t$$

is the nth Fourier coefficient of $a(s,\cdot)$. Since function a is assumed to be smooth, then for any P it holds that

$$\sup_{n\in\mathbb{N}} (1+|n|)^P \|a_n\|_{L^\infty(0,2\pi)} < C_P.$$

Let us restrict ourselves to the cases $\psi = \psi_m$, for $m = 0, 1, 2$ (see the beginning of Sect. 3.2.2). Denote then by Ψ_m the corresponding operator and by $\Psi_{m,N}$, its numerical approximation cf. (12). Clearly, the proof of this Theorem can be reduced to studying

$$(\Psi_m - \Psi_{m,N})\varphi = \sum_{n=-\infty}^{\infty} a_n \psi_m * (e_n\varphi - P_N(e_n\varphi)).$$

We will make use of the following results:

(a) For $m = 1, 2$ it holds

$$\|\psi_m * \varphi\|_p \le C_q \|\varphi\|_{p-q}, \quad q \le 2m - 1$$

whereas for $m = 0$

$$\|\psi_0 * \varphi\|_p = 2\pi|\hat{\varphi}(0)| \le 2\pi\|\varphi\|_{p+q}, \quad \forall q \in \mathbb{R}.$$

Indeed, for $m = 1, 2$

$$|\hat{\psi}_m(n)| \le C_m(1 + |n|)^{1-2m}$$

with C_m independent of m which implies

$$\begin{aligned}\|\psi_m * \varphi\|_p^2 &= |\hat{\psi}_m(0)\hat{\varphi}(0)|^2 + \sum_{n=-\infty}^{\infty} |n|^{2p}|\hat{\psi}(n)\hat{\varphi}(n)|^2 \\ &\le C_m^2 \left[|\hat{\psi}_m(0)\hat{\varphi}(0)|^2 + \sum_{n=-\infty}^{\infty} |n|^{2(p+1-2m)}|\hat{\varphi}(n)|^2 \right] = C_m^2 \|\varphi\|_{p-2m+1}^2.\end{aligned}$$

(b) The convergence estimate for the trigonometric interpolant [13, Theorem 8.2.1]

$$\|P_N\varphi - \varphi\|_p \le C_{p,q} N^{-q} \|\varphi\|_{p+q}, \quad \forall p, q \ge 0, \quad p + q > 1/2. \tag{22}$$

(c) The fact that H^p for $p > 1/2$ is an algebra, cf [13, Lemma 5.13.1] and therefore

$$\|a\varphi\|_p \le C_p \|a\|_p \|\varphi\|_p.$$

(d) The obvious bound $\|e_n\|_p \le \max\{1, |n|^p\}$.

We are ready to analyze the approximation error of the discrete operators. First, for $m = 0$, that is, for integral operators with smooth kernel, we have

$$\begin{aligned}
\|(\Psi_0 - \Psi_{0,N})\varphi\|_p &\le C \sum_{n=-\infty}^{\infty} \|a_n\|_p \|\psi_0 * (e_n\varphi - P_N(e_n\varphi))\|_p \\
&= 2\pi C_p \sum_{n=-\infty}^{\infty} \|a_n\|_p \|e_n\varphi - P_N(e_n\varphi)\|_0 \\
&\le C_{p,q} N^{-p-q} \sum_{n=-\infty}^{\infty} \|a_n\|_p \|e_n\|_{p+q} \|\varphi\|_{p+q} \\
&\le C_{p,q} N^{-p-q} \Big[\sum_{n=-\infty}^{\infty} \|a_n\|_p (1+|n|)^{p+q} \Big] \|\varphi\|_{p+q} \\
&\le C'_{p,q} N^{-p-q} \|\varphi\|_{p+q}
\end{aligned}$$

for all $p + q > 1/2$.

Let us examine the case $m = 2$. If $p \ge 3$, we can proceed similarly to conclude

$$\begin{aligned}
\|(\Psi_2 - \Psi_{2,N})\varphi\|_p &\le \sum_{n=-\infty}^{\infty} \|a_n\|_p \|e_n\varphi - P_N(e_n\varphi)\|_{p-3} \\
&\le C_{p,q} N^{-q-3} \sum_{n=-\infty}^{\infty} \|a_n\|_p \|e_n\|_{p+q} \|\varphi\|_{p+q} \le C_{p,q} N^{-q-3} \|\varphi\|_{p+q},
\end{aligned}$$

provided that $p + q > 1/2$ and $q \ge -3$. If $p \in [0, 3]$, we can only get convergence estimates for the interpolator in H^0 (we can not expect faster convergence in weaker norms). Therefore we have instead

$$\|(\Psi_2 - \Psi_{2,N})\varphi\|_p \le \sum_{n=-\infty}^{\infty} \|a_n\|_p \|e_n\varphi - P_N(e_n\varphi)\|_0 \le C_{p,q} N^{-p-q} \|\varphi\|_{p+q}.$$

Collecting these bounds, the result for $m = 3$ follows readily.

Case $m = 1$ is left as exercise for the reader. □

We recall the functional properties of the boundary operators in the Sobolev setting. Define

$$\mathscr{D}_k := \begin{bmatrix} -\mathrm{K}_k & \mathrm{V}_k \\ -\mathrm{H}_k & \mathrm{K}_k^\top \end{bmatrix}.$$

Then, $\mathscr{D}_k : H^{p+1} \times H^p \to H^{p+1} \times H^p$ is continuous for any $p \in \mathbb{R}$. Actually it holds

$$\mathrm{K}_k, \mathrm{K}_k^\top, \mathrm{R}_k : H^p \to H^{p+3}. \tag{23}$$

This extra regularizing property has been repeatedly used in the design and analysis of boundary integral methods for Helmholtz equation.

If we define

$$\mathscr{D}_{k,N} := \begin{bmatrix} -\mathrm{K}_{k,N} & \mathrm{V}_{k,N} \\ -\mathrm{H}_{k,N} & \mathrm{K}_{k,N}^\top \end{bmatrix}, \quad \widetilde{\mathscr{D}}_{k,N} := \begin{bmatrix} -\widetilde{\mathrm{K}}_{k,N} & \widetilde{\mathrm{V}}_{k,N} \\ -\mathrm{H}_{k,N} & \widetilde{\mathrm{K}}_{k,N}^\top \end{bmatrix},$$

the following result can be easily derived from Theorem 1.

Proposition 1 *For any* $p > 1/2$,

$$\mathscr{D}_{k,N}, \widetilde{\mathscr{D}}_{k,N} : H^{p+1} \times H^p \to H^{p+1} \times H^p \tag{24}$$

are uniformly continuous. Moreover, if $p > 1/2$ *and* $q \geq -1$ *with* $p + q > 1/2$,

$$\|\mathscr{D}_{k,N} - \mathscr{D}_k\|_{H^{p+q}\times H^{p+q}\to H^p\times H^p} + \|\widetilde{\mathscr{D}}_{k,N} - \mathscr{D}_k\|_{H^{p+q}\times H^{p+q}\to H^p\times H^p} \leq CN^{-q-\min\{1,p\}}, \tag{25}$$

and, for $q \geq -2$, $p > 1/2$ *and* $p + q > 1/2$,

$$\|\widetilde{\mathscr{D}}_{k,N} - \mathscr{D}_k\|_{H^{p+q+1}\times H^{p+q}\to H^{p+1}\times H^p} \leq CN^{-q-\min\{2,p\}}. \tag{26}$$

Proof Define

$$\mathscr{E}_k := \begin{bmatrix} -\mathrm{K}_k & \mathrm{V}_k \\ -\mathrm{T}_k & \mathrm{K}_k^\top \end{bmatrix}, \quad \mathscr{E}_{k,N} := \begin{bmatrix} -\mathrm{K}_{k,N} & \mathrm{V}_{k,N} \\ -\mathrm{T}_{k,N} & \mathrm{K}_{k,N}^\top \end{bmatrix}. \tag{27}$$

Then

$$\mathscr{D}_{k,N} - \mathscr{D}_k = \mathscr{E}_{k,N} - \mathscr{E}_k. \tag{28}$$

Equation (20) in Theorem 1 proves (25) since

$$\|\mathrm{K}_{k,N} - \mathrm{K}_k\|_{H^{p+q}\to\times H^p} \leq CN^{-q-\min\{1,p\}} \tag{29a}$$

$$\|\mathrm{K}_{k,N}^\top - \mathrm{K}_k^\top\|_{H^{p+q}\to H^p} \leq CN^{-q-\min\{1,p\}} \tag{29b}$$

$$\|\mathrm{V}_{k,N} - \mathrm{V}_k\|_{H^{p+q}\to H^p} \leq CN^{-q-1-\min\{1,p\}} \tag{29c}$$

$$\|\mathrm{T}_{k,N} - \mathrm{T}_k\|_{H^{p+q}\to H^p} \leq CN^{-q-\min\{1,p\}} \tag{29d}$$

which hold for $p+q > 1/2$ and $q \geq -1$. Moreover, from the mapping properties of the continuous operators, these estimates with $q = 0$ imply the first result for $\mathscr{D}_{k,N}$.

For the second estimate, we start now from

$$\widetilde{\mathscr{D}}_{k,N} - \mathscr{D}_k = \widetilde{\mathscr{F}}_{k,N} - \mathscr{F}_k \tag{30}$$

where

$$\mathscr{F}_k := \begin{bmatrix} -\mathrm{K}_k & \mathrm{R}_k \\ -\mathrm{T}_k & \mathrm{K}_k^\top \end{bmatrix}, \quad \widetilde{\mathscr{F}}_{k,N} := \begin{bmatrix} -\widetilde{\mathrm{K}}_{k,N} & \widetilde{\mathrm{R}}_{k,N} \\ -\mathrm{T}_{k,N} & \widetilde{\mathrm{K}}_{k,N}^\top \end{bmatrix}, \tag{31}$$

for which we have the error convergence estimates

$$\|\widetilde{\mathrm{K}}_{k,N} - \mathrm{K}_k\|_{H^{p'+q'} \to H^{p'}} \le C N^{-q' - \min\{3,p'\}}, \quad q' \ge -3 \tag{32a}$$

$$\|\widetilde{\mathrm{K}}_{k,N}^\top - \mathrm{K}_k^\top\|_{H^{p'+q'} \to H^{p'}} \le C N^{-q' - \min\{3,p'\}}, \quad q' \ge -3 \tag{32b}$$

$$\|\widetilde{\mathrm{R}}_{k,N} - \mathrm{R}_k\|_{H^{p'+q'} \to H^{p'}} \le C N^{-q' - \min\{3,p'\}}, \quad q' \ge -3 \tag{32c}$$

$$\|\mathrm{T}_{k,N} - \mathrm{T}_k\|_{H^{p'+q'} \times H^{p'}} \le C N^{-q' - \min\{1,p'\}}, \quad q' \ge -1. \tag{32d}$$

(With the restriction $p' + q' > 1/2$ in all these cases). Choosing $q' = q$ and $p' = p$ in all the estimates in (32) we get (25) which, in particular, implies (24) as a simple consequence. To prove (26), we take $(p', q') = (p+1, q)$ in (32a), $(p', q') = (p, q)$ in (32b), $(p', q') = (p+1, q-1)$ in (32c) and $(p', q') = (p, q+1)$ in (32d).

In short, we have shown in this section two different types of discrete versions of the Helmholtz boundary layer operators. The first type of discretization is simpler and works well for equations stated in $H^p \times H^p$ such as the equations of the second kind where the hypersingular operator is not the leading term, either because it does not appear or because the strong singular part is canceled out. The second type of discretization involving the operators $\widetilde{\mathscr{D}}_{k,N}$ turns out to be more appropriate for formulations in the natural space $H^{p+1} \times H^p$ or for complex formulations where the operators are more involved and/or the operator H_k plays a dominant role. Actually, we could keep $\mathrm{K}_{k,N}$ and $\mathrm{K}_{k,N}^\top$ in $\widetilde{\mathscr{D}}_{k,N}$ and the desired convergence property, namely $\|\widetilde{\mathscr{D}}_{k,N} - \mathscr{D}_k\|_{H^p \times H^{p+1} \to H^p \times H^{p+1}} \to 0$ for any $p > 1/2$, still holds. We have prefered, however, to collect in $\widetilde{\mathscr{D}}_{k,N}$ the more accurate discretization. We will consider several examples of these cases in next section.

4 Boundary Integral Equations for Transmission Problems and their Nyström Discretizations

We consider numerical approximations of several well-posed formulations of the transmission problem (1) presented in Sect. 2. Equipped with the discrete operators introduced and analyzed in the previous section, the stability and convergence of the resulting schemes can be now easily proven.

For the sake of a simpler notation, we will denote in this section only by $\mathrm{V}_\pm$, $\mathrm{H}_\pm$, etc the corresponding layer operators for $k_\pm$. Their discrete versions will be denoted, as before, by simply adding the subscript N.

First we consider the Kress-Roach formulation cf [7]. Defining

$$\mathscr{L}_1\begin{bmatrix} a\\ \varphi\end{bmatrix} := \left(\tfrac{1+\nu}{2}\mathscr{I} + \begin{bmatrix} \nu\mathrm{K}_- - \mathrm{K}_+ & \mathrm{V}_+ - \mathrm{V}_-\\ \nu(\mathrm{H}_- - \mathrm{H}_+) & \nu\mathrm{K}_+^\top - \mathrm{K}_-^\top\end{bmatrix}\right)\begin{bmatrix} a\\ \varphi\end{bmatrix},$$

where $\mathscr{I}$ is the identity operator matrix, this formulation amounts to solving the system of boundary equations

$$\mathscr{L}_1\begin{bmatrix} a\\ \varphi\end{bmatrix} = \begin{bmatrix} f\\ \lambda\end{bmatrix}. \tag{33}$$

It is well known that if $(f,\lambda) = (h, \nu\,\eta)$ cf. (5), then the unique solution is $a = u_\mathrm{t}\circ\mathbf{x}$, $\varphi = |\mathbf{x}'|(\partial_n u_\mathrm{t})\circ\mathbf{x}$ where u_t is exterior part of the total wave: $u_\mathrm{t} = u_+ + u_\mathrm{inc}$. Clearly, once this equation is solved, taking into account the transmission conditions (1), we can evaluate $u^\pm$ by means of (2).

The discrete versions of the operators $\mathscr{L}_1$ are given by

$$\begin{aligned}\mathscr{L}_{1,N} &:= \tfrac{1+\nu}{2}\mathscr{I} + \mathscr{P}_N\begin{bmatrix} \nu\mathrm{K}_{-,N} - \mathrm{K}_{+,N} & \mathrm{V}_{+,N} - \mathrm{V}_{-,N}\\ \nu(\mathrm{H}_{-,N} - \mathrm{H}_{+,N}) & \nu\mathrm{K}_{+,N}^\top - \mathrm{K}_{-,N}^\top\end{bmatrix}\\ &= \tfrac{1+\nu}{2}\mathscr{I} + \mathscr{P}_N\begin{bmatrix} \nu\mathrm{K}_{-,N} - \mathrm{K}_{+,N} & \mathrm{V}_{+,N} - \mathrm{V}_{-,N}\\ \nu(\mathrm{T}_{-,N} - \mathrm{T}_{+,N}) & \nu\mathrm{K}_{+,N}^\top - \mathrm{K}_{-,N}^\top\end{bmatrix}\end{aligned}$$

(recall (18)–(19)) where

$$\mathscr{P}_N = \begin{bmatrix} P_N & \\ & P_N\end{bmatrix}.$$

Thus, the discrete problem is given by

$$\mathscr{L}_{1,N}\begin{bmatrix} a_N\\ \varphi_N\end{bmatrix} = \begin{bmatrix} P_N f\\ P_N\lambda\end{bmatrix}. \tag{34}$$

Observe that the last equation implies that $(a_N, \varphi_N) \in \mathbb{T}_N \times \mathbb{T}_N$ which allows us to reformulate the method as a true Nyström scheme, where the unknowns are the pointwise values of the densities at the grid points $\{\frac{j\pi}{N}\}$.

We will consider next the Costabel-Stephan formulation [2]: Let

$$\mathscr{L}_2 := \begin{bmatrix} -(\mathrm{K}_- + \mathrm{K}_+) & \nu^{-1}\mathrm{V}_+ + \mathrm{V}_- \\ -(\mathrm{H}_- + \nu\mathrm{H}_+) & \mathrm{K}_+^\top + \mathrm{K}_-^\top \end{bmatrix}$$
$$= (1+\nu^{-1})\begin{bmatrix} & \Lambda \\ -\nu \mathrm{D}\Lambda\mathrm{D} & \end{bmatrix} + \begin{bmatrix} -\mathrm{K}_- - \mathrm{K}_+ & \nu^{-1}\widetilde{\mathrm{R}}_+ + \widetilde{\mathrm{R}}_- \\ -(\mathrm{T}_- + \nu\mathrm{T}_+) & \mathrm{K}_+^\top + \mathrm{K}_-^\top \end{bmatrix}$$

and the associated system of integral equations

$$\mathscr{L}_2\begin{bmatrix} a \\ \varphi \end{bmatrix} = \begin{bmatrix} f \\ \lambda \end{bmatrix}. \tag{35}$$

In this case, if we take $(f,\lambda) = (h,\eta)$, then $(a,\varphi) = (u_t \circ \mathbf{x}, |\mathbf{x}'|(\partial_n u_t)\circ \mathbf{x})$ is again the exact solution.

Letting

$$\widetilde{\mathscr{L}}_{2,N} := (1+\nu^{-1})\begin{bmatrix} & \Lambda \\ -\nu \mathrm{D}\Lambda\mathrm{D} & \end{bmatrix} + \mathscr{P}_N\begin{bmatrix} -\widetilde{\mathrm{K}}_{-,N} - \widetilde{\mathrm{K}}_{+,N} & \nu^{-1}\widetilde{\mathrm{R}}_{+,N} + \widetilde{\mathrm{R}}_{-,N} \\ -(\mathrm{T}_{-,N} + \nu\mathrm{T}_{+,N}) & \widetilde{\mathrm{K}}_{+,N}^\top + \widetilde{\mathrm{K}}_{-,N}^\top \end{bmatrix},$$

the method we propose for solving (35) can be written in operational form as follows

$$\widetilde{\mathscr{L}}_{2,N}\begin{bmatrix} a_N \\ \varphi_N \end{bmatrix} = \begin{bmatrix} P_N f \\ P_N \lambda \end{bmatrix}. \tag{36}$$

As before, $(a_N, \varphi_N) \in \mathbb{T}_N \times \mathbb{T}_N$ for any pair (f,λ) on the right hand side. (This can be easily seen by noticing that the leading part in $\widetilde{\mathscr{L}}_{2,N}$ is diagonal in the complex exponential bases).

The so-called regularized combined field integral equation, proposed in [3] will be also analyzed here. Let

$$\mathscr{L}_3 = \frac{1}{\nu+1}\mathscr{L}_1 + \frac{2}{\nu+1}\begin{bmatrix} & \mathrm{V}_\kappa \\ -\nu\mathrm{H}_\kappa & \end{bmatrix}\mathscr{L}_2 = \begin{bmatrix} \frac{1}{2}I + \mathrm{K}_- & -\nu^{-1}\mathrm{V}_- \\ \nu\mathrm{H}_- & \frac{1}{2}I - \mathrm{K}_-^\top \end{bmatrix} + \mathscr{R}_\kappa\mathscr{L}_2$$

with

$$\mathscr{R}_\kappa := \frac{1}{\nu+1}\begin{bmatrix} I & 2\mathrm{V}_\kappa \\ -2\nu\mathrm{H}_\kappa & \nu I \end{bmatrix}.$$

The boundary integral equation is then given by

$$\mathscr{L}_3\begin{bmatrix} a \\ \varphi \end{bmatrix} = \mathscr{R}_\kappa\begin{bmatrix} f \\ \lambda \end{bmatrix}. \tag{37}$$

It can be shown (see [3]) that this system of integral equations admits a unique solution provided that κ is chosen to be a complex number with positive imaginary part. Moreover, this parameter can be adjusted to make eigenvalues cluster around 1. Besides, by construction if we plug (h, η) in the right hand side, the unique solution is $(u_t \circ \mathbf{x}, |\mathbf{x}'|(\partial_n u_t) \circ \mathbf{x})$. In other words, this is a new direct method where $\mathscr{R}_k$ works as some sort of preconditioner for $\mathscr{L}_2$.

The discretization of the regularized equations is done as follows. First, we set

$$\mathscr{R}_{\kappa,N} := \frac{1}{\nu+1}\begin{bmatrix} I & 2\Lambda + P_N\widetilde{\mathrm{R}}_{\kappa,N} \\ -2\nu\mathrm{D}\Lambda\mathrm{D} - 2\nu P_N\mathrm{T}_{\kappa,N} & \nu I \end{bmatrix}$$

and next we define

$$\begin{aligned}\widetilde{\mathscr{L}}_{3,N} &:= \begin{bmatrix} \frac{1}{2}I + P_N\widetilde{\mathrm{K}}_{-,N} & -\nu^{-1}\Lambda - \nu^{-1}P_N\widetilde{\mathrm{R}}_{-,N} \\ \nu\mathrm{D}\Lambda\mathrm{D} + \nu P_N\mathrm{T}_{-,N} & \frac{1}{2}I - P_N\widetilde{\mathrm{K}}^{\top}_{-,N} \end{bmatrix} + \mathscr{R}_{\kappa,N}\widetilde{\mathscr{L}}_{2,N} \\ &= \begin{bmatrix} \frac{1}{2}I & -\nu^{-1}\Lambda \\ \nu\mathrm{D}\Lambda\mathrm{D} & \frac{1}{2}I \end{bmatrix} + \mathscr{P}_N\begin{bmatrix} \widetilde{\mathrm{K}}_{-,N} & \nu^{-1}\widetilde{\mathrm{R}}_{-,N} \\ \nu\mathrm{T}_{-,N} & -\widetilde{\mathrm{K}}^{\top}_{-,N} \end{bmatrix} + \mathscr{R}_{\kappa,N}\widetilde{\mathscr{L}}_{2,N}.\end{aligned}$$

(Observe that the first matrix operator maps $\mathbb{T}_N \times \mathbb{T}_N$ into itself.) The numerical algorithm, in operator form, is given by

$$\widetilde{\mathscr{L}}_{3,N}\begin{bmatrix} a_N \\ \varphi_N \end{bmatrix} = \mathscr{R}_{\kappa,N}\begin{bmatrix} P_N f \\ P_N\lambda \end{bmatrix}. \tag{38}$$

Observe again that the right-hand-sides are trigonometric polynomials, and thus so are the solutions of these discrete problems.

We also investigate an integral formulation based on an indirect method. That is, unlike the formulations considered so far, the unknown is not immediately related to traces on the boundary of the solution of the transmission problem. This integral formulation, has an interesting feature: the solution of the transmission Helmholtz problem can be reconstructed from knowledge of one boundary density only. In other words, this integral equation needs half as many unknowns as the other integral formulations considered in this paper thus far. Let us describe this equation, which was first introduced in [4]. We seek a function μ so that

$$u^- = -2[\mathrm{SL}_-\mu], \quad u^+ = \nu\mathrm{SL}_+(I + 2\mathrm{K}^{\top}_-)\mu - 2\mathrm{DL}_+\mathrm{V}_-\mu$$

($\mathrm{SL}_\pm$ and $\mathrm{DL}_\pm$ are the corresponding parameterized layer potentials). The density μ can be computed by solving the boundary integral *equation*

$$L_4\,\mu := -\frac{\nu+1}{2}\mu + \mathbf{K}\mu - \mathrm{i}\rho\mathbf{V}\mu = f, \tag{39}$$

where $f = \lambda - \mathrm{i}\rho g$. Here ρ is a coupling parameter which must be real and different from zero to ensure the well-posedness of the equation. In the definition of the operator L_4 we used the operators

$$\mathbf{K} := -\mathrm{K}_-^\top(\nu I - 2\mathrm{K}_-^\top) - \nu \mathrm{K}_+^\top(I + 2\mathrm{K}_-^\top) + 2(\mathrm{H}_+ - \mathrm{H}_-)\mathrm{V}_-$$

and

$$\mathbf{V} := -\nu \mathrm{V}_+(I + 2\mathrm{K}_-^\top) - (I - 2\mathrm{K}_+)\mathrm{V}_-.$$

The discretizations of these operators are given by

$$\begin{aligned}\mathbf{K}_N &:= -\mathrm{K}_{-,N}^\top(\nu I - 2\mathrm{K}_{-,N}^\top) - \nu \mathrm{K}_{+,N}^\top(I + 2\mathrm{K}_{-,N}^\top) + 2(\mathrm{T}_{+,N} - \mathrm{T}_{-,N})\mathrm{V}_{-,N}]\\ \mathbf{V}_N &:= -\nu \mathrm{V}_{+,N}(I + 2\mathrm{K}_{-,N}^\top) - (I - 2\mathrm{K}_{+,N})\mathrm{V}_{-,N}.\end{aligned}$$

Thus, we define

$$L_{4,N} := -\tfrac{\nu+1}{2} I + P_N \mathbf{K}_N - \mathrm{i}\rho P_N \mathbf{V}_N,$$

and the discretization of the equation $L_4\mu = f$ is given by

$$L_{4,N}\mu_N = P_N f. \tag{40}$$

Again, $\mu_N \in \mathbb{T}_N$ regardless of the right hand side f.

Theorem 2 *The mappings* (33), (35), (37) *and* (39)

$$\begin{aligned}\mathscr{L}_1 &: H^p \times H^p \to H^p \times H^p, \quad j = 1, 3\\ \mathscr{L}_j &: H^{p+1} \times H^p \to H^{p+1} \times H^p, \quad j = 1, 2, 3\\ L_4 &: H^p \to H^p\end{aligned} \tag{41}$$

are continuous and invertible for all $p \in \mathbb{R}$.

Moreover, for $p > 1/2$,

$$\|\mathscr{L}_1 - \mathscr{L}_{1,N}\|_{H^p\times H^p \to H^p\times H^p} \le C_p N^{-\min\{p,1\}}, \tag{42a}$$

$$\|\mathscr{L}_2 - \widetilde{\mathscr{L}}_{2,N}\|_{H^{p+1}\times H^p \to H^p\times H^{p+1}} \le C_p N^{-\min\{p,2\}}, \tag{42b}$$

$$\|\mathscr{L}_3 - \widetilde{\mathscr{L}}_{3,N}\|_{H^{p+1}\times H^p \to H^p\times H^{p+1}} \le C_p N^{-\min\{p,1\}}, \tag{42c}$$

$$\|\mathscr{L}_3 - \widetilde{\mathscr{L}}_{3,N}\|_{H^{p+1}\times H^{p+1} \to H^{p+1}\times H^{p+1}} \le C_p N^{-\min\{p,1\}}, \tag{42d}$$

$$\|L_4 - L_{4,N}\|_{H^p \to H^p} \le C_p N^{-\min\{p,1\}}. \tag{42e}$$

Furthermore, we have the following convergence results: For all $p > 1/2$ *and* $q \geq 0$, *if* (a^1, φ^1) *denotes the exact solution for* (33) *and* (a_N^1, φ_1^N) *is the corresponding numerical solution of* (34), *it holds*

$$\|a - a_N^1\|_p + \|\varphi - \varphi_N^1\|_p \leq CN^{-q}\left[\|a\|_{p+q} + \|\varphi\|_{p+q}\right]. \tag{43a}$$

Let for $j = 2, 3$ $(\tilde{a}_N^j, \tilde{\varphi}_N^j)$ *the continuous solution of* (35) *and* (37) *and* $(\tilde{a}_N^j, \tilde{\varphi}_N^j)$ *the discrete solution of* (36) *and* (38). *Then we have*

$$\|a - \tilde{a}_N^2\|_{p+1} + \|\varphi - \tilde{\varphi}_N^2\|_p \leq CN^{-q}\left[\|a\|_{p+q+1} + \|\varphi\|_{p+q}\right]. \tag{44a}$$

$$\|a - \tilde{a}_N^3\|_{p+1} + \|\varphi - \tilde{\varphi}_N^3\|_p \leq CN^{-q}\left[\|f\|_{p+q+1} + \|\lambda\|_{p+q}\right]. \tag{44b}$$

$$\|a - \tilde{a}_N^3\|_{p+1} + \|\varphi - \ddot{\varphi}_N^3\|_{p+1} \leq CN^{-q}\left[\|f\|_{p+q+1} + \|\lambda\|_{p+q+1}\right]. \tag{44c}$$

Finally if μ *is the solution of* L_4 *and* μ_N *that given by the numerical scheme* (40),

$$\|\mu - \mu_N\|_p \leq CN^{-q}\|\mu\|_{p+q}, \quad p > 1/2,\ q \geq 0.$$

In the estimates above, $C > 0$ *is independent of* a, φ, f, λ *or* μ, *and* N.

Proof The functional properties stated in (41) are well known and can be easily derived from the functional properties of the operators involved (see Proposition 1).

The proofs for all the convergence estimates share the same ideas. Thus, for the sake of brevity we restrict ourselves to consider a few representative cases to illustrate the kind of techniques used here.

Proof of (42a) **and** (43a) Denote as in (27)

$$\mathscr{E}_\pm := \begin{bmatrix} -\mathrm{K}_\pm & \mathrm{V}_\pm \\ -\mathrm{T}_\pm & \mathrm{K}_\pm^\top \end{bmatrix}.$$

Notice that $\mathscr{E}_\pm : H^p \times H^p \to H^{p+1} \times H^{p+1}$ and therefore, from from (22),

$$\|(\mathscr{P}_N - \mathscr{I})\mathscr{E}_\pm\|_{H^{p+q}\times H^{p+q}\to H^p\times H^p} \leq CN^{-q-1} \tag{45}$$

for any $p \geq 0$, $q \geq -1$ with $p + q > 1/2$. Setting accordingly

$$\mathscr{E}_{\pm,N} := \begin{bmatrix} -\mathrm{K}_{\pm,N} & \mathrm{V}_{\pm,N} \\ -\mathrm{T}_{\pm,N} & \mathrm{K}_{\pm,N} \end{bmatrix}$$

we notice that cf (25) (see also (28))

$$\|\mathscr{E}_k - \mathscr{E}_{k,N}\|_{H^{p+q}\times H^{p+q}\to H^p\times H^p} \leq CN^{-q-\min\{1,p\}}. \tag{46}$$

On the other hand,

$$\mathscr{L}_1 = \tfrac{1+\nu}{2}\mathscr{I} + \begin{bmatrix}1 & \\ & \nu\end{bmatrix}\mathscr{E}_+ - \mathscr{E}_-\begin{bmatrix}\nu & \\ & 1\end{bmatrix},$$

$$\mathscr{L}_{1,N} = \tfrac{1+\nu}{2}\mathscr{I} - \mathscr{P}_N\begin{bmatrix}1 & \\ & \nu\end{bmatrix}\mathscr{E}_{+,N} + \mathscr{P}_N\mathscr{E}_{-,N}\begin{bmatrix}\nu & \\ & 1\end{bmatrix}.$$

Therefore, (45) and (46) yield

$$\begin{aligned}
&\|\mathscr{L}_1 - \mathscr{L}_{1,N}\|_{H^{p+q}\times H^{p+q}\to H^p\times H^p}\\
&\qquad \le \max\{\nu,1\}\Big[\|(\mathscr{P}_N - \mathscr{I})\mathscr{E}_\pm\|_{H^{p+q}\times H^{p+q}\to H^p\times H^p}\\
&\qquad\quad + \|\mathscr{P}_N\|_{H^p\times H^p\to H^p\times H^p}\big[\|\mathscr{E}_\pm - \mathscr{E}_{\pm,N}\|_{H^{p+q}\times H^{p+q}\to H^p\times H^p}\big]\\
&\qquad \le CN^{-q-\min\{1,p\}}.
\end{aligned} \tag{47}$$

In particular, setting $q = 0$ implies (42a). The error estimate for the numerical method is obtained using standard techniques:

$$\begin{aligned}
\left\|\begin{bmatrix}a - a_N^1\\ \varphi - \varphi_N^1\end{bmatrix}\right\|_p &\le C\left\|\mathscr{L}_{1,N}\begin{bmatrix}a - a_N^1\\ \varphi - \varphi_N^1\end{bmatrix}\right\|_p\\
&\le \left\|(\mathscr{L}_{1,N} - \mathscr{L}_1)\begin{bmatrix}a\\ \varphi\end{bmatrix}\right\|_p + \left\|\mathscr{L}_1\begin{bmatrix}a\\ \varphi\end{bmatrix} - \mathscr{L}_{1,N}\begin{bmatrix}a_N\\ \varphi_N\end{bmatrix}\right\|_p\\
&\le \left\|(\mathscr{L}_{1,N} - \mathscr{L}_1)\begin{bmatrix}a\\ \varphi\end{bmatrix}\right\|_p + \left\|(\mathscr{I} - \mathscr{P}_N)\mathscr{L}_1\begin{bmatrix}a\\ \varphi\end{bmatrix}\right\|_p\\
&\le CN^{-q}\big(\|a\|_{p+q} + \|\varphi\|_{p+q}\big).
\end{aligned}$$

Proof of (42b) **and** (44a) For $\mathscr{L}_2$, we proceed in the same fashion with

$$\mathscr{F}_\pm := \begin{bmatrix}-\mathrm{K}_\pm & \mathrm{R}_\pm\\ -\mathrm{T}_\pm & \mathrm{K}_\pm^\top\end{bmatrix},\quad \widetilde{\mathscr{F}}_{\pm,N} := \begin{bmatrix}-\widetilde{\mathrm{K}}_{\pm,N} & \widetilde{\mathrm{R}}_{\pm,N}\\ -\mathrm{T}_{\pm,N} & \widetilde{\mathrm{K}}_{\pm,N}^\top\end{bmatrix}.$$

which allows us to write

$$\mathscr{L}_2 = (1+\nu^{-1})\begin{bmatrix} & \Lambda\\ -\nu\,\mathrm{D}\Lambda\mathrm{D} & \end{bmatrix} + \begin{bmatrix}\nu^{-1/2} & \\ & \nu^{1/2}\end{bmatrix}\mathscr{F}_+\begin{bmatrix}\nu^{1/2} & \\ & \nu^{-1/2}\end{bmatrix} + \mathscr{F}_-,$$

$$\widetilde{\mathscr{L}}_{2,N} = (1+\nu^{-1})\begin{bmatrix} & \Lambda\\ -\nu\,\mathrm{D}\Lambda\mathrm{D} & \end{bmatrix} + \begin{bmatrix}\nu^{-1/2} & \\ & \nu^{1/2}\end{bmatrix}\mathscr{F}_{+,N}\begin{bmatrix}\nu^{1/2} & \\ & \nu^{-1/2}\end{bmatrix} + \mathscr{F}_{-,N}.$$

Since $\mathscr{F}_\pm : H^{p+1} \times H^p \to H^{p+3} \times H^{p+2}$ holds as well, estimate (22) yields

$$\|(\mathscr{P}_N - \mathscr{I})\mathscr{F}_\pm\|_{H^{p+q+1}\times H^{p+q}\to H^{p+1}\times H^p} \leq CN^{-q-2}$$

for any $p \geq 0$ and $q \geq -1$. On the other hand, from (26) (see also (30)),

$$\|\mathscr{F}_k - \widetilde{\mathscr{F}}_{k,N}\|_{H^{p+q+1}\times H^{p+q}\to H^{p+1}\times H^p} \leq CN^{-q-\min\{2,p\}}$$

for any $p > 1/2$ and $q \geq -2$ with $p + q > 1/2$.

Thus

$$\begin{aligned}
&\|\mathscr{L}_2 - \widetilde{\mathscr{L}}_{2,N}\|_{H^{p+q+1}\times H^{p+q}\to H^{p+1}\times H^p} \\
&\qquad \leq \max\{\nu, 1\}\Big[\|(\mathscr{P}_N - \mathscr{I})\mathscr{F}_\pm\|_{H^{p+q+1}\times H^{p+q}\to H^{p+1}\times H^p} \\
&\qquad\quad +\|\mathscr{P}_N\|_{H^{p+1}\times H^p\to H^{p+1}\times H^p}\big[\|\mathscr{F}_\pm - \widetilde{\mathscr{F}}_{\pm,N}\|_{H^{p+q+1}\times H^{p+q}\to H^{p+1}\times H^p}\big]\\
&\qquad \leq CN^{-q-\min\{2,p\}}
\end{aligned} \tag{48}$$

which, with $q = 0$, implies in particular (42b). Estimate (44a) is proved from (48) as in (43a).

Proof of (42d) **and** (44c) Notice first that $\mathscr{F}_\pm : H^p \times H^p \to H^{p+3} \times H^{p+1}$ is continuous and

$$\|\mathscr{F}_\pm - \widetilde{\mathscr{F}}_{\pm,N}\|_{H^{p+q}\times H^{p+q}\to H^{p+2}\times H^p} \leq CN^{-q-\min\{p,1\}}, \quad p, p+q > 1/2,\ q \geq -1$$

which can be deduced from Eqs. (32a) and (32c) with $p' = p + 2$ and $q' = q - 2$ and from Eqs. (32b) and (32d) with $p' = p$ and $q' = q$. Thus, similar arguments as those used above for $\mathscr{L}_2$ can be applied to show a different estimate:

$$\begin{aligned}
&\|\mathscr{L}_2 - \widetilde{\mathscr{L}}_{2,N}\|_{H^{p+q+1}\times H^{p+q+1}\to H^{p+2}\times H^p} \\
&\qquad \leq C\Big[\|(\mathscr{P}_N - \mathscr{I})\mathscr{F}_\pm\|_{H^{p+q+1}\times H^{p+q+1}\to H^{p+2}\times H^p} \\
&\qquad\quad +\|\mathscr{P}_N\|_{H^{p+2}\times H^p\to H^{p+2}\times H^p}\big[\|\mathscr{F}_\pm - \widetilde{\mathscr{F}}_{\pm,N}\|_{H^{p+q}\times H^{p+q}\to H^{p+2}\times H^p}\big]\Big] \\
&\qquad \leq C'N^{-q-2}\|\mathscr{F}_\pm\|_{H^{p+q+1}\times H^{p+q+1}\to H^{p+q+4}\times H^{p+q+2}} + C'N^{-q-\min\{p,1\}} \\
&\qquad \leq C''N^{-q-\min\{p,1\}}
\end{aligned} \tag{49}$$

which holds for $p > 1/2, q \geq -1$ and $p + q > 1/2$.

We are now ready to start analyzing the more complex formulation of this paper, namely $\mathscr{L}_3$ and the corresponding discretization given by $\widetilde{\mathscr{L}}_{3,N}$. Clearly,

$$\begin{aligned}\mathscr{L}_3 - \widetilde{\mathscr{L}}_{3,N} &= \frac{1}{\nu+1}(\mathscr{L}_1 - \widetilde{\mathscr{L}}_{1,N}) + \frac{2}{\nu+1}\begin{bmatrix} & \mathrm{R}_\kappa - \widetilde{\mathrm{R}}_{\kappa,N} \\ -\nu(\mathrm{T}_\kappa - \mathrm{T}_{\kappa,N}) & \end{bmatrix}\mathscr{L}_2 \\ &\quad + \frac{2}{\nu+1}\begin{bmatrix} & \mathrm{V}_{\kappa,N} \\ -\nu \mathrm{H}_{\kappa,N} & \end{bmatrix}(\mathscr{L}_2 - \widetilde{\mathscr{L}}_{2,N}) \\ &=: \mathscr{T}_1 + \mathscr{T}_2 + \mathscr{T}_3. \end{aligned} \tag{50}$$

First term with $\widetilde{\mathscr{L}}_{1,N}$ defined as $\mathscr{L}_{1,N}$ with $\widetilde{\mathrm{V}}_{\pm,N}, \widetilde{\mathrm{K}}_{\pm,N}$ and $\widetilde{\mathrm{K}}^\top_{\pm,N}$ instead, can be analyzed as in (47) to get

$$\|\mathscr{T}_1\|_{H^{p+q+1}\times H^{p+q+1}\to H^{p+1}\times H^{p+1}} \le CN^{-1}. \tag{51}$$

For the second term we emphasize that

$$\begin{aligned}&\|\mathscr{T}_2\|_{H^{p+q+1}\times H^{p+q+1}\to H^{p+1}\times H^{p+1}} \\ &\qquad \le CN^{-q-\min\{p,1\}}\|\mathscr{L}_2\|_{H^{p+q+1}\times H^{p+q+1}\to H^{p+q+1}\times H^{p+q}}. \end{aligned} \tag{52}$$

(We have applied (32c) with $p' = p+1$ and $q' = q-1$ and (32d) with $p' = p+1$ and $q' = q$ and the mapping properties of $\mathscr{L}_2$).

Regarding the third term, using (49) we get

$$\begin{aligned}\|\mathscr{T}_3\|_{H^{p+q+1}\times H^{p+q+1}\to H^{p+1}\times H^{p+1}} &\le C\|\mathscr{L}_2 - \widetilde{\mathscr{L}}_{2,N}\|_{H^{p+q+1}\times H^{p+q+1}\to H^{p+2}\times H^{p}} \\ &\le CN^{-q-\min\{p,1\}}. \end{aligned} \tag{53}$$

Gathering (51)–(53) in (50) we obtain

$$\|\mathscr{L}_3 - \widetilde{\mathscr{L}}_{3,N}\|_{H^{p+q+1}\times H^{p+q+1}\to H^{p+1}\times H^{p+1}} \le CN^{-q-\min\{p,1\}} \tag{54}$$

which implies (42d) by taking $q = 0$.

To prove (44c), we can easily see that, as in (48), we simply have to bound

$$\left\|(\mathscr{L}_3 - \widetilde{\mathscr{L}}_{3,N})\begin{bmatrix} a \\ \varphi \end{bmatrix}\right\|_{p+1}, \quad \left\|\left(\mathscr{P}_N\mathscr{R}_{\kappa,N} - \mathscr{R}_\kappa\right)\begin{bmatrix} f \\ \lambda \end{bmatrix}\right\|_{p+1}.$$

The first term has been already studied in (54). Regarding the second term, we have

$$\begin{aligned}
&\left\| \left(\mathscr{P}_N \mathscr{R}_{\kappa,N} - \mathscr{R}_\kappa\right) \begin{bmatrix} f \\ \lambda \end{bmatrix} \right\|_{p+1} \\
&\quad \leq C\big[\|\mathscr{R}_{\kappa,N} - \mathscr{R}_\kappa\|_{H^{p+q+1}\times H^{p+q+1}\to H^{p+1}\times H^{p+1}} \\
&\qquad + \|(\mathscr{P}_N - \mathscr{I})\mathscr{R}_\kappa\|_{H^{p+q+1}\times H^{p+q+1}\to H^{p+1}\times H^{p+1}}\big]\big[\|f\|_{p+q+1} + \|\lambda\|_{p+q+1}\big] \\
&\quad \leq CN^{-q}\big[\|f\|_{p+q+1} + \|\lambda\|_{p+q+1}\big].
\end{aligned}$$

Notice that, unlike (43a), $\|f\|_{p+q+1}$, $\|\lambda\|_{p+q+1}$ cannot be bounded in terms of $\|a\|_{p+q+1}$ and $\|b\|_{p+q+1}$ because we cannot guarantee that $\mathscr{R}_\kappa$ is invertible. However, it follows that

$$\begin{aligned}
&\|a\|_{p+q+1} + \|b\|_{p+q+1} \\
&\qquad \leq \|\mathscr{L}_3^{-1}\mathscr{R}_\kappa\|_{H^{p+q+1}\times H^{p+q+1}\to H^{p+q+1}\times H^{p+q+1}}\big[\|f\|_{p+q+1} + \lambda\|_{p+q+1}\big]
\end{aligned}$$

which allows us to write the convergence in terms of the regularity of the right-hand-side instead.

The main point of this theorem is that convergence in higher Sobolev space norms of the Helmholtz boundary operators allows to prove easily the stability and convergence of the Nyström discretizations. The higher order discretizations $\{\widetilde{\mathrm{V}}_{k,N}, \widetilde{\mathrm{R}}_{K,N}, \widetilde{\mathrm{R}}_{K,N}^{\top}\}$ guarantee convergence of Nyström discretizations for rather complex formulations whereas the simpler, but less accurate discretizations of second kind integral formulations such as those based on the operators $\mathscr{L}_1$ still converge. The analysis based on the results of Theorem 1, whose details are a bit more subtle, allows us to employ optimal discretizations and norms in which the stability and convergence results hold. Observe that on account of Sobolev embedding theorems, all of the convergence results established above imply convergence in the L^∞ norm.

5 Numerical Experiments

For brevity, we only present numerical results for the Costabel-Stephan formulation $\mathscr{L}_2$. We refer the reader to [1, 3] for extensive numerical results for the other formulations.

The domains we have considered are the geometries depicted in Fig. 2. We have taken $k_+ = 8, k_- = 32$ in the Helmholtz transmission problems (1), with $\nu = 1$ in the transmission conditions across the interface. We have applied the numerical schemes $\widetilde{\mathscr{L}}_{2,N}$ and $\mathscr{L}_{2,N}$. The latter scheme is that defined using $\mathrm{V}_{\pm,N}$, the less accurate approximation for $\mathrm{V}_\pm$. We point out that only the first discretization has been analyzed in this paper.

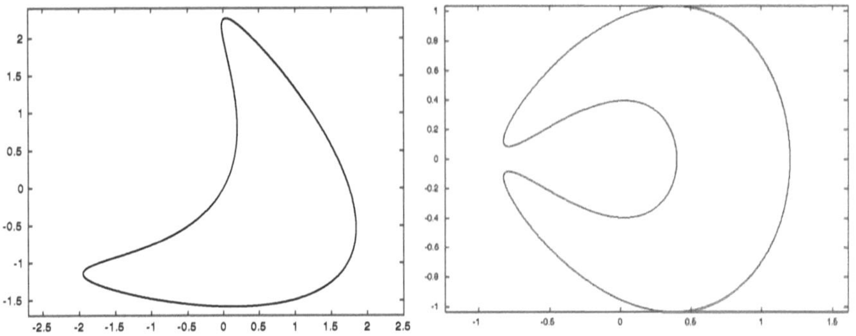

Fig. 2 Kite and cavity geometries considered in the numerical experiments

Table 1 L^∞ error estimate in the far field for the discretizations $\mathscr{L}_{2,N}$ and $\widetilde{\mathscr{L}}_{2,N}$ for the Helmholtz transmission problem in the kite (left) and cavity (right) domains

	Kite		Cavity	
N	$\mathscr{L}_{2,N}$	$\widetilde{\mathscr{L}}_{2,N}$	$\mathscr{L}_{2,N}$	$\widetilde{\mathscr{L}}_{2,N}$
96	3.2E−02	9.1E−03	7.1E−02	1.7E−02
128	4.1E−04	2.5E−05	8.3E−04	6.3E−05
160	5.9E−11	5.8E−12	2.0E−10	3.3E−11

The L^∞ error estimate in the far field for the numerical solutions is shown in Table 1. The *exact* solution has been computed using $\mathscr{L}_{1,N}$ for sufficiently large N, which, in turns, provides an indirect demonstration of the performance of this discretization too.

Both methods converge super-algebraically to the exact solution, although $\widetilde{\mathscr{L}}_{2,N}$ performs better with even a slightly faster convergence. Convergence, and specially stability of $\mathscr{L}_{2,N}$ remains as an open problem and certainly will deserve more research in the future.

Acknowledgements Catalin Turc gratefully acknowledge support from NSF through contract DMS-1312169. Víctor Domínguez is partially supported by Ministerio de Economía y Competitividad, through the grant MTM2014-52859.

This research was partially supported by Spanish MINECO grants MTM2011-22741 and MTM2014-54388.

References

1. Boubendir, Y., Domínguez, V., Turc, C.: High-order Nyström discretizations for the solution of integral equation formulations of two-dimensional Helmholtz transmission problems. IMA J. Numer. Anal. **36**(1), 463–492 (2016)
2. Costabel, M., Stephan, E.: A direct boundary integral equation method for transmission problems. J. Math. Anal. Appl. **106**(2), 367–413 (1985)

3. Domínguez, V., Lyon, M., Turc, C.: High-order Nyström discretizations for the solution of integral equation formulations of two-dimensional Helmholtz transmission on interfaces with corners. Submitted. Preprint available in arXiv:1509.04415 (2015)
4. Kleinman, R.E., Martin, P.A.: On single integral equations for the transmission problem of acoustics. SIAM J. Appl. Math. **48**(2), 307–325 (1988)
5. Kress, R.: On the numerical solution of a hypersingular integral equation in scattering theory. J. Comput. Appl. Math. **61**(3), 345–360 (1995)
6. Kress, R.: Linear Integral Equations, 3rd edn. Springer, New York (2014)
7. Kress, R., Roach, G.F.: Transmission problems for the Helmholtz equation. J. Math. Phys. **19**(6), 1433–1437 (1978)
8. Kussmaul, R.: Ein numerisches Verfahren zur Losung des Neumannschen Aussenraumproblems fur die Helmholtzsche Schwingungsgleichung. Computing **4**, 246–273 (1969)
9. Martensen, E.: Uber eine Methode zum raumlichen Neumannschen Problem mit einer ANwendung fur torusartige Berandungen. Acta Math. **109**, 75–135 (1963)
10. Maue, A.W.: Zur Formulierung eines allgemeinen Beugungsproblems durch eine Integralgleichung. Z. Phys. **126**, 601–618 (1949)
11. Nédélec, J.-C.: Integral equations with nonintegrable kernels. Integr. Equ. Oper. Theory **5**(4), 562–572 (1982)
12. Nédélec, J.-C., Planchard, J.: Une méthode variationnelle d'éléments finis pour la résolution numérique d'un problème extérieur dans $\mathbb{R}^3$. Rev. Française Automat. Informat. Recherche Opérationnelle Sér. Rouge **7**, 105–129 (1973)
13. Saranen, J., Vainikko, G.: Periodic integral and pseudo-differential equations with numerical approximation. Springer, Berlin (2002)

Algebraic Inverse Integrating Factors for a Class of Generalized Nilpotent Systems

Antonio Algaba, Natalia Fuentes, Cristóbal García, and Manuel Reyes

Abstract Usually, the study of differential systems with linear part null is done using quasi-homogeneous expansions of vector fields. Here, we use this technique for analyzing the existence of an inverse integrating factor for generalized nilpotent systems, in general non-integrable, whose lowest-degree quasi-homogeneous term is the Hamiltonian system $y^2\partial_x + x^3\partial y$.

1 Introduction

We consider an autonomous system

$$\dot{\mathbf{x}} = \mathbf{F}(\mathbf{x}) = (P(\mathbf{x}), Q(\mathbf{x}))^T, \ \mathbf{x} \in \mathbb{C}^2, \tag{1}$$

where $\mathbf{F}$ is a formal planar vector field defined in a neighborhood of the origin $U \subset \mathbb{C}^2$ having a singular point at the origin, i.e., $\mathbf{F}(\mathbf{0}) = \mathbf{0}$ and $P, Q \in \mathbb{C}[[x, y]]$ (algebra of the power series in x and y with coefficient in $\mathbb{C}$).

A non-null $\mathscr{C}^1$ class function V is an inverse integrating factor of system (1) (or also of $\mathbf{F}$) on U if it satisfies the linear partial differential equation $L_{\mathbf{F}}V = \text{div}(\mathbf{F})V$, being $L_{\mathbf{F}}V := P\partial V/\partial x + Q\partial V/\partial y$, the Lie derivative of V respect to $\mathbf{F}$, and $\text{div}(\mathbf{F}) := \partial P/\partial x + \partial Q/\partial y$, the divergence of $\mathbf{F}$. This name for V comes from the fact that $1/V$ defines on $U \setminus \{V = 0\}$ an integrating factor of system (1) (which allows to compute a first integral of the system on $U \setminus \{V = 0\}$). So, if system (1) possesses an inverse integrating factor V then it is integrable on $U \setminus \{V = 0\}$. For more details about the relation between the integrability and the inverse integrating factor see [5, 9, 10, 19, 21].

The expressions of V are often simpler than the expressions of the first integrals, see [7, 8]. The domain of definition and the regularity of V usually are larger than the domain and the regularity of the first integral, see [9, 13, 22, 23].

A. Algaba • N. Fuentes • C. García • M. Reyes (✉)
Department of Mathematics, Huelva University, Calle Dr. Cantero Cuadrado, 6, 21071, Huelva, Spain
e-mail: colume@uhu.es

F. Ortegón Gallego et al. (eds.), *Trends in Differential Equations and Applications*, SEMA SIMAI Springer Series 8, DOI 10.1007/978-3-319-32013-7_16

This concept also plays an important role in the study of the existence of limit cycles of a vector field, because the zero-set $\{V = 0\}$, formed by orbits of the system (1), contains the limit cycles of the system (1) which are in U, whenever they exist, see [12, 15, 17]. The zero-set $\{V = 0\}$ also contains the homoclinic and heteroclinic connections between hyperbolic saddle equilibria, see [14]. Moreover, the cyclicity of a limit cycle is related to the vanishing order of V, see [16].

The existence of inverse integrating factors in a neighborhood of a singularity has been studied in some particular cases. We remark, among others, the papers of Enciso and Peralta-Salas [12], Chavarriga et al. [8], Christopher et al. [11] and Algaba et al. [5].

We are concerned to determine what degenerate systems have an algebraic inverse integrating factor over $\mathbb{C}((x, y))$ (which will be named AIIF) where $\mathbb{C}((x, y))$ denotes the quotient field of the algebra of the power series $\mathbb{C}[[x, y]]$. In this sense, the only results we know are Walcher [24] where is claimed its existence for non-degenerate cusp nilpotent singularity, and Algaba et al. [4] where it is characterized all nilpotent systems having an algebraic inverse integrating factor.

Here, we deal with systems whose lowest degree quasi-homogeneous term is $(y^2, x^3)^T$. These systems are a class of generalized nilpotent systems. The integrability problem of these class has been studied by Giné [18]. He proves that the formal first integrals, if any, are of the form $y^k + F(x, y)$ where F starts with terms of order higher than k. Moreover, if the system has a local analytic first integral then it has also a local analytic first integral of the form $y^k + F(x, y)$. This shape of the first integral allows giving necessary conditions of analytic and formal integrability for several families of polynomial systems.

This paper is organized as follows: the following section is devoted to providing an expansion in quasi-homogeneous terms of an orbitally equivalent normal form of systems whose lowest-degree term is a Hamiltonian vector field, Theorem 1.

Our results are presented in Sect. 3. We obtain a suitable normal form to study the existence of an algebraic inverse integrating factor of perturbations of the quasi-homogeneous Hamiltonian system $y^2\partial_x + x^3\partial_y$, see Theorem 2. Theorem 3 gives a normal form of such systems having an algebraic inverse integrating factor. Moreover, we give the shape of them, Theorem 4 states that the existence of a formal inverse integrating factor is equivalent to the analytic integrability for these systems.

Finally, in Sect. 4 we compute the systems with an algebraic inverse integrating factor for a family of planar systems (Theorem 5) and we solve the integrability problem of the family (Theorem 6).

2 Quasi-Homogeneous Normal Forms

For more details on the concepts and definitions we give in this section, see [1].

Given $\mathbf{t} = (t_1, t_2)$ non-null with t_1 and t_2 non-negative integer numbers without common factors, we will denote by $\mathcal{P}_k^{\mathbf{t}}$ to the vector space of quasi-homogeneous polynomials of type $\mathbf{t}$ and degree k, i.e.

$$\mathcal{P}_k^{\mathbf{t}} = \{f \in \mathbb{C}[x,y] : f(\varepsilon^{t_1}x, \varepsilon^{t_2}y) = \varepsilon^k f(x,y)\},$$

and by

$$\mathcal{Q}_k^{\mathbf{t}} = \{\mathbf{F} = (P,Q)^T : P \in \mathcal{P}_{k+t_1}^{\mathbf{t}},\ Q \in \mathcal{P}_{k+t_2}^{\mathbf{t}}\}$$

to the vector space of the quasi-homogeneous polynomial vector fields of type $\mathbf{t}$ and degree k. Any vector field can be expanded into quasi-homogeneous terms of type $\mathbf{t}$ of successive degrees. Thus, the vector field $\mathbf{F}$ can be written in the form

$$\mathbf{F} = \mathbf{F}_r + \mathbf{F}_{r+1} + \cdots,$$

for some $r \in \mathbb{Z}$, where $\mathbf{F}_j = (P_{j+t_1}, Q_{j+t_2})^T \in \mathcal{Q}_j^{\mathbf{t}}$ and $\mathbf{F}_r \not\equiv \mathbf{0}$. Such expansion will be expressed as $\mathbf{F} = \mathbf{F}_r +$ q-h.h.o.t., where "q-h.h.o.t." means "quasi-homogeneous higher order terms."

If we select the type $\mathbf{t} = (1,1)$, we are using in fact the Taylor expansion, but in general, each term in the above expansion involves monomials with different degrees.

The key in the problem of obtaining a normal form of the system (1) is to analyze the effect of a near-identity transformation $\mathbf{x} = \mathbf{y} + \mathbf{P}_k(\mathbf{y})$ and a reparameterization of the time by $\frac{dt}{dT} = 1 + \tau_k(\mathbf{x})$, where $\mathbf{P}_k \in \mathcal{Q}_k^{\mathbf{t}}$ and $\tau_k \in \mathcal{P}_k^{\mathbf{t}}$, with $k \geq 1$.

The quasi-homogeneous terms of the transformed system $\dot{\mathbf{y}} = \mathbf{G}(\mathbf{y})$ agree with the original ones up to degree $r + k - 1$ and for the degree $r + k$ it has

$$\begin{aligned}\mathbf{G}_{r+k} &= \mathbf{F}_{r+k} - (D\mathbf{P}_k\mathbf{F}_r - D\mathbf{F}_r\mathbf{P}_k) + \tau_k\mathbf{F}_r = \mathbf{F}_{r+k} - [\mathbf{P}_k, \mathbf{F}_r] + \tau_k\mathbf{F}_r \\ &= \mathbf{F}_{r+k} - \mathcal{L}_{r+k}(\mathbf{P}_k, \tau_k)\end{aligned}$$

where we have introduced the homological operator under orbital equivalence:

$$\begin{aligned}\mathcal{L}_{r+k} : \mathcal{Q}_k^{\mathbf{t}} \times \mathcal{P}_k^{\mathbf{t}} &\longrightarrow \mathcal{Q}_{r+k}^{\mathbf{t}} \\ (\mathbf{P}_k, \tau_k) &\rightarrow \mathcal{L}_{r+k}(\mathbf{P}_k, \tau_k) = [\mathbf{P}_k, \mathbf{F}_r] - \tau_k\,\mathbf{F}_r.\end{aligned} \tag{2}$$

Following the ideas of the conventional normal form theory, it is enough to choose $(\mathbf{P}_k, \tau_k) \in \mathcal{Q}_k^{\mathbf{t}} \times \mathcal{P}_k^{\mathbf{t}}$ adequately in order to simplify the $(r+k)$-degree quasi-homogeneous term in system (1), by annihilating the part belonging to the range of the linear operator $\mathcal{L}_{r+k}$. In other words, we can achieve that $\mathbf{F}_{r+k}$ belongs to a complementary subspace to the range of $\mathcal{L}_{r+k}$. When this has been done, we say that the corresponding term has been reduced to normal form under orbital equivalence. So, by means of a sequence of time-reparameterizations and near identity transformations (by performing the procedure for $k = 1$, then for $k = 2$ and so on) system (1) can be formally reduced to normal form under orbital equivalence, i.e. the system can be transformed into

$$\dot{\mathbf{y}} = \mathbf{G}(\mathbf{y}) = \mathbf{G}_r(\mathbf{y}) + \mathbf{G}_{r+1}(\mathbf{y}) + \cdots, \tag{3}$$

with $\mathbf{G}_r \not\equiv \mathbf{0}$ and $\mathbf{G}_{r+k} \in \mathrm{Cor}(\mathscr{L}_{r+k}) \subseteq \mathscr{Q}^{\mathbf{t}}_{r+k}$ where $\mathrm{Cor}(\mathscr{L}_{r+k})$ is any complementary subspace to the range of the homological operator $\mathscr{L}_{r+k}$. We note that such space is not unique, in general.

Given $h \in \mathcal{P}^{\mathbf{t}}_{r+|\mathbf{t}|}$, we define the linear operator

$$\begin{aligned} \ell_j : \mathcal{P}^t_{j-r} &\longrightarrow \mathcal{P}^{\mathbf{t}}_j \\ \mu_{j-r} &\longrightarrow \ell_j(\mu_{j-r}) := \frac{\partial h}{\partial x}\frac{\partial \mu_{j-r}}{\partial y} - \frac{\partial h}{\partial y}\frac{\partial \mu_{j-r}}{\partial x}, \end{aligned} \tag{4}$$

(Poisson bracket of h and μ_{j-r}) and denote by $\mathrm{Cor}(\ell_j)$ a complementary subspace to the range of the linear operator ℓ_j (co-range of the operator ℓ_j).

We recall the conservative-dissipative splitting of a quasi-homogeneous vector field. Given a fixed type $\mathbf{t} = (t_1, t_2)$, for each $\mathbf{F}_k \in \mathscr{Q}^{\mathbf{t}}_k$, there exist unique polynomials $\mu_k \in \mathcal{P}^{\mathbf{t}}_k$ and $h_{k+|\mathbf{t}|} \in \mathcal{P}^{\mathbf{t}}_{k+|\mathbf{t}|}$ such that

$$\mathbf{F}_k = \mathbf{X}_{h_{k+|\mathbf{t}|}} + \mu_k \mathbf{D}_0, \tag{5}$$

where $h_{k+|\mathbf{t}|} = \frac{1}{k+|\mathbf{t}|}(\mathbf{D}_0 \wedge \mathbf{F}_k)$ and $\mu_k = \frac{1}{k+|\mathbf{t}|}\mathrm{div}(\mathbf{F}_k)$. ($\mathbf{X}_h$ denotes the Hamiltonian vector field whose Hamiltonian function is h; that is, $\mathbf{X}_h := (-\partial h/\partial y, \partial h/\partial x)^T$). Its proof can be found in [4].

Fixed $h \in \mathcal{P}^{\mathbf{t}}_{r+|\mathbf{t}|}$, we consider the systems of the form

$$\dot{\mathbf{x}} = \mathbf{X}_h + \text{q-h.h.o.t.}, \tag{6}$$

i.e. a class of systems which can be considered as perturbations of a Hamiltonian system $\mathbf{X}_h \in \mathscr{Q}^{\mathbf{t}}_r$, whose Hamiltonian function h is a quasi-homogeneous function.

In what follows, we will denote $\mathbf{D}_0 := (t_1 x, t_2 y)^T \in \mathscr{Q}^{\mathbf{t}}_0$.

Algaba et al. [3, 6] proved the following properties of the operators $\mathscr{L}_{r+k}$ and ℓ_k.

Proposition 1 ([3, 6]) *Consider system (6). For every non-negative integer k, it verifies:*

1. $\mathscr{L}_{r+k}(\mathscr{Q}^{\mathbf{t}}_k \times Cor(\ell_k)) = \mathscr{L}_{r+k}(\mathscr{Q}^{\mathbf{t}}_k \times \mathcal{P}^{\mathbf{t}}_k)$.
2. $Cor(\mathscr{L}_{r+k}) = \mathbf{X}_{S_{r+k+|\mathbf{t}|}} \oplus Cor(\ell_{r+k})\mathbf{D}_0$.
   being $S_{r+k+|\mathbf{t}|}$ *a subspace verifying* $Cor(\ell_{r+k+|\mathbf{t}|}) = S_{r+k+|\mathbf{t}|} \oplus hCor(\ell_k)$.
3. *If h has only simple factors on* $\mathbb{C}[x, y]$, *then* $Cor(\ell_{r+k+|\mathbf{t}|}) = hCor(\ell_k)$, *for all* $k > r$ *with* $\mathcal{P}^{\mathbf{t}}_{k-r} \neq \{0\}$.

Notice that we can to obtain $\mathrm{Cor}(\ell_{r+k+|\mathbf{t}|})$ from the co-range of the scalar linear operator ℓ_k.

Moreover, Algaba et al. [3] give the following property of the sets $\mathcal{P}_k^{\mathbf{t}}$:

Lemma 1 *Fixed* $\mathbf{t} = (t_1, t_2)$, *it has that:*

1. $\mathcal{P}_k^{\mathbf{t}} = \{0\}$, *if* $k \notin \mathscr{I}^{\mathbf{t}}$.
2. *If* $k > t_1 t_2 - |\mathbf{t}|$, *then* $k \in \mathscr{I}^{\mathbf{t}}$, *i.e.* $\mathcal{P}_k^{\mathbf{t}}$ *is a non-trivial space*

being $\mathscr{I}^{\mathbf{t}} = \{k = k_1 t_1 + k_2 t_2 + k_3 t_1 t_2 \in \mathbb{N} : k_1, k_2, k_3 \in \mathbb{N}, k_1 < t_2, k_2 < t_1\}$.

We define the following subsets of $\mathbb{N}_0$:

$$\mathcal{J}_1 = \{j,\ j \leq r\},$$

$$\mathcal{J}_2 = \{j,\ j \geq r+1 \text{ such that } \mathcal{P}_{j-r}^{\mathbf{t}} = \{0\}\}, \tag{7}$$

$$\mathcal{J} = \{j \in \mathcal{J}_1 \cup \mathcal{J}_2,\ S_{r+|\mathbf{t}|+j} \neq \{0\}\}. \tag{8}$$

From Lemma 1, there exists $m_0 := \max\{\mathbb{N}_0 \setminus \mathscr{I}^{\mathbf{t}}\}$. Thus, $j \notin \mathcal{J}_2$, for all $j \geq n_0 := 1 + r + m_0$.
These properties allow to give an expression of $\mathrm{Cor}(\mathscr{L}_{r+k})$, or equivalently, to obtain an orbital equivalent normal form up any order.

Theorem 1 *Consider system (6) with h having only simple factors on* $\mathbb{C}[x,y]$. *An orbital equivalent normal form becomes*

$$\dot{\mathbf{x}} = \mathbf{X}_{h+g} + \mu \mathbf{D}_0, \tag{9}$$

being $g = \sum_{j \in \mathcal{J}} g_{r+|\mathbf{t}|+j}$ *with* $g_{r+|\mathbf{t}|+j} \in S_{r+|\mathbf{t}|+j}$ *and* $\mu = \sum_{j>r} \mu_j$ *with* $\mu_j \in Cor(\ell_j)$ *and* $\mu_{j+r+|\mathbf{t}|} = h\mu_j$ *for all* $j \notin \mathcal{J}_1 \cup \mathcal{J}_2$ *(i.e.* $j \geq n_0$*).*

Consequently, it is enough the computation of the co-ranges of ℓ_j from $r+1$ to $n_0 + r + |\mathbf{t}| - 1$ to provide a normal form.

3 Main Results

We consider the degenerate systems of the form

$$(\dot{x}, \dot{y})^T = (y^2 + \sum_{j \geq 3} P_j(x,y), \sum_{j \geq 3} Q_j(x,y))^T, \tag{10}$$

with P_j and Q_j homogeneous polynomials of degree j and $Q_3(1,0) \neq 0$ (without loss of generality, we can assume $Q_3(1,0) = 1$).

The first quasi-homogeneous polynomial of type (3, 4), according to the degree, are:

$$\begin{array}{lll}\mathcal{P}_3^{(3,4)} = \{x\}, & \mathcal{P}_4^{(3,4)} = \{y\}, & \mathcal{P}_6^{(3,4)} = \{x^2\},\\ \mathcal{P}_7^{(3,4)} = \{xy\}, & \mathcal{P}_8^{(3,4)} = \{y^2\}, & \mathcal{P}_9^{(3,4)} = \{x^3\},\\ \mathcal{P}_{10}^{(3,4)} = \{x^2y\}, & \mathcal{P}_{11}^{(3,4)} = \{xy^2\}, & \mathcal{P}_{12}^{(3,4)} = \{y^3,\ x^4\}.\end{array}$$

We write $P_j(x,y) = \sum_{j=m+n} a_{mn}x^my^n$ and $Q_j(x,y) = \sum_{j=m+n} b_{mn}x^my^n$. The quasi-homogeneous expansion with respect to $\mathbf{t} = (3,4)$ of system (10) is

$$\begin{pmatrix}\dot{x}\\ \dot{y}\end{pmatrix} = \overbrace{\begin{pmatrix}y^2\\ x^3\end{pmatrix}}^{\mathbf{F}_5} + \overbrace{\begin{pmatrix}a_{30}x^3\\ b_{21}x^2y\end{pmatrix}}^{\mathbf{F}_6} + \overbrace{\begin{pmatrix}a_{21}x^2y\\ b_{12}xy^2\end{pmatrix}}^{\mathbf{F}_7} + \overbrace{\begin{pmatrix}a_{12}xy^2\\ b_{03}y^3 + b_{40}x^4\end{pmatrix}}^{\mathbf{F}_8} + \text{q-h.h.o.t.}, \tag{11}$$

i.e., system (6) for $r = 5$, $h = x^4/4 - y^3/3$ (h has only simple factors on $\mathbb{C}[x,y]$).

Thus, $\mathbb{N}_0 \setminus \mathscr{I}^{(3,4)} = \{1,2,5\}$, $m_0 = 5$ and $n_0 = 11$. Also, $\mathfrak{J}_1 = \{1,2,3,4,5\}$, $\mathfrak{J}_2 = \{6,7,10\}$.

Table 1 shows the range and co-range of ℓ_j for $6 \le j \le 22$.

It is straightforward to check that $\mathfrak{J}$ is an empty set. Therefore, g is identically null. So, from Theorem 1, we give the following result:

Theorem 2 *A normal form orbitally equivalent of system (10) is*

$$\begin{aligned}(\dot{x},\dot{y})^T = (y^2,x^3)^T + \sum_{j\ge 0}\big(&\alpha_{12j+6}x^2h^j + \alpha_{12j+7}xyh^j + \alpha_{12j+10}x^2yh^j\\ &+ \alpha_{12j+12}x^2h^{j+1} + \alpha_{12j+15}xh^{j+1} + \alpha_{12j+16}xh^{j+1}\big)\,\mathbf{D}_0,\end{aligned} \tag{12}$$

Table 1 Range and co-range of operator ℓ_j for system (11)

Range(ℓ_6)=span$\{0\}$,	Cor(ℓ_6)=span$\{x^2\}$
Range(ℓ_7)=span$\{0\}$,	Cor(ℓ_7)=span$\{xy\}$
Range(ℓ_8)=span$\{y^2\}$,	Cor(ℓ_8)=span$\{0\}$
Range(ℓ_9)=span$\{x^3\}$,	Cor(ℓ_9)=span$\{0\}$
Range(ℓ_{10})=span$\{0\}$,	Cor(ℓ_{10})=span$\{x^2y\}$
Range(ℓ_{11})=span$\{xy^2\}$,	Cor(ℓ_{11})=span$\{0\}$
Range(ℓ_{12})=span$\{7x^4-12h\}$,	Cor(ℓ_{12})=span$\{h\}$
Range(ℓ_{13})=span$\{x^3y\}$,	Cor(ℓ_{13})=$\{0\}$
Range(ℓ_{14})=span$\{x^2y^2\}$,	Cor(ℓ_{14})=$\{0\}$
Range(ℓ_{15})=span$\{x^3-6xh\}$,	Cor(ℓ_{15})=span$\{xh\}$
Range(ℓ_{16})=span$\{11x^4y-12yh\}$,	Cor(ℓ_{16})=span$\{yh\}$
Range(ℓ_{17})=span$\{x^3y^2\}$,	Cor(ℓ_{17})=$\{0\}$
Range(ℓ_{18})=span$\{13x^6-36x^2h\}$,	Cor(ℓ_{18})=span$\{x^2h\}$
Range(ℓ_{19})=span$\{7x^5-12xyh\}$,	Cor(ℓ_{19})=span$\{xyh\}$
Range(ℓ_{22})=span$\{17x^6y-9x^2yh\}$,	Cor(ℓ_{22})=span$\{x^2yh\}$

with $\mathbf{D}_0 = (3x, 4y)^T$. *The coefficients* α_j *are named coefficients of order* j *of system (12).*

We state the well-known relationship between inverse integrating factors of formally orbital equivalent vector fields.

Proposition 2 *Let* Φ *be a diffeomorphism and* η *a function on* $U \subset \mathbf{R}^2$ *such that* $detD\Phi$ *has no zero on* U *and* $\eta(\mathbf{0}) \neq 0$. *If* $V(\mathbf{x}) \in \mathbb{C}[[x, y]]$ *is an inverse integrating factor of the system* $\dot{\mathbf{x}} = \mathbf{F}(\mathbf{x})$, *then* $\eta(\mathbf{y})(det(D\Phi(\mathbf{y}))^{-1}V(\Phi(\mathbf{y}))$ *is an inverse integrating factor of* $\dot{\mathbf{y}} = \Phi_*(\eta\mathbf{F})(\mathbf{y}) := D\Phi(\mathbf{y})^{-1}\eta(\mathbf{y})\mathbf{F}(\Phi(\mathbf{y}))$.

The following results have been established in Algaba et al. [4]:

Proposition 3 *The functions* $f(h)$ *being* f *a scalar non-constant function of class* $\mathscr{C}^1$ *are first integrals and inverse integrating factors of the Hamiltonian system* $\dot{\mathbf{x}} = \mathbf{X}_h$ *with* $h \in \mathcal{P}^{\mathbf{t}}_{r+|\mathbf{t}|}$.

Proposition 4 *Consider system* $\dot{\mathbf{x}} = \mathbf{X}_h + \mu\mathbf{D}_0$ *with* $h \in \mathcal{P}^{\mathbf{t}}_{r+|\mathbf{t}|}$ *having only simple factors in its factorization on* $\mathbb{C}[x, y]$ *and* $\mu = \sum_{j>r}\mu_j$, $\mu_j \in Cor(\ell_j)$ *and we denote by* $N = min\{j, \ \mu_{r+j} \not\equiv 0\}$. *If* V *is an algebraic inverse integrating factor, then* $V = f(h)^{1+N/(r+|\mathbf{t}|)}$, *being* f *a scalar formal function with* $f(0) = 0$ *and* $f'(0) = 1$, *is the unique algebraic inverse integrating factor, up to a multiplicative constant.*

Proposition 5 *Consider system* $\dot{\mathbf{x}} = \mathbf{X}_h + (\lambda g(h) + \nu)\mathbf{D}_0$ *with* $h \in \mathcal{P}^{\mathbf{t}}_{r+|\mathbf{t}|}$ *having only simple factors in its factorization on* $\mathbb{C}[x, y]$, $\lambda \in Cor(\ell_{r+N}) \setminus \{0\}$, g *a scalar function,* $g(0) = 1$, *and* $\nu = \sum_{j>N}\nu_{r+j}$, $\nu_j \in Cor(\ell_j)$, $\nu_{r+N+l(r+|\mathbf{t}|)} \equiv 0$ *for all non-negative integer* l, *then, under these conditions, the system possesses an algebraic inverse integrating factor if and only if* $\nu \equiv 0$.

Proposition 6 *Consider system* $\dot{\mathbf{x}} = \mathbf{X}_h + \lambda g(h)\mathbf{D}_0$ *where* $h \in \mathcal{P}_{r+|\mathbf{t}|}$, $\lambda \in \mathcal{P}_{r+N}$ *and* g *a* $\mathscr{C}^1$ *class function with* $g(0) = 1$. *Then, the function* $h^{1+N/(r+|\mathbf{t}|)}g(h)$ *is the unique inverse integrating factor of the system, up to a multiplicative constant.*

Next result is the main result of the paper, which characterizes the systems (10) having an algebraic inverse integrating factor.

Theorem 3 *System (10) possesses an algebraic inverse integrating factor if and only if it is orbitally equivalent to one, and only one, of the following systems:*

1.

$$(\dot{x}, \dot{y})^T = (y^2, x^3)^T. \tag{13}$$

In this case, the functions $f(h + \cdots)$ *being* f *a scalar non-constant function of class* $\mathscr{C}^1$ *are inverse integrating factors of system (10).*

2.

$$(\dot{x}, \dot{y})^T = (y^2, x^3)^T + \alpha_{12L+6}x^2h^Lg(h)\mathbf{D}_0. \tag{14}$$

Moreover, the algebraic inverse integrating factor of system (10) is given by $(h+\cdots)^{L+1+1/12}g(h+\cdots)$.

3.
$$(\dot{x},\dot{y})^T = (y^2,x^3)^T + \alpha_{12L+7}xyh^L g(h)\mathbf{D}_0. \tag{15}$$

Moreover, the algebraic inverse integrating factor of system (10) is given by $(h+\cdots)^{L+1+2/12}g(h+\cdots)$.

4.
$$(\dot{x},\dot{y})^T = (y^2,x^3)^T + \alpha_{12L+10}x^2yh^L g(h)\mathbf{D}_0. \tag{16}$$

Moreover, the algebraic inverse integrating factor of system (10) is given by $(h+\cdots)^{L+1+5/12}g(h+\cdots)$.

5.
$$(\dot{x},\dot{y})^T = (y^2,x^3)^T + \alpha_{12L+12}h^{L+1} g(h)\mathbf{D}_0. \tag{17}$$

Moreover, the algebraic inverse integrating factor of system (10) is given by $(h+\cdots)^{L+1+7/12}g(h+\cdots)$.

6.
$$(\dot{x},\dot{y})^T = (y^2,x^3)^T + \alpha_{12L+15}xh^{L+1} g(h)\mathbf{D}_0. \tag{18}$$

Moreover, the algebraic inverse integrating factor of system (10) is given by $(h+\cdots)^{L+1+10/12}g(h+\cdots)$.

7.
$$(\dot{x},\dot{y})^T = (y^2,x^3)^T + \alpha_{12L+16}yh^{L+1} g(h)\mathbf{D}_0 \tag{19}$$

Moreover, the algebraic inverse integrating factor of system (10) is given by $(h+\cdots)^{L+1+11/12}g(h+\cdots)$

with $\mathbf{D}_0=(3x,4y)^T$, $L\geq 0$, *g a scalar function with* $g(0)=1$ *and* α_j *non-null.*

Proof of Theorem 3 We perform the transformation which brings the system (10) into the system (12).

If $\alpha_j = 0$ for all j, system (12) is system (13). From Propositions 2 and 3 it has that the system (10) admits an algebraic inverse integrating factor and they are of the form $f(h+\cdots)$.

Otherwise, we assume that there exists some $\alpha_j \neq 0$.

Let $j_0 = \min\{j,\ \alpha_j \neq 0\}$ be. Suppose, for instance, that $j_0 = 12L+6$ for a certain $L>0$. So, system (12) is of the form $\dot{\mathbf{x}} = \mathbf{X}_h + (\lambda g(h)+\nu)\mathbf{D}_0$, with $\lambda = \alpha_{12L+6}x^2h^L$, $N = 12L+1$, $g(h) = 1+\alpha_{12(L+1)+6}/\alpha_{12L+6}h+\cdots$ and $\nu = \sum_{j>12L+6}\nu_j$ with $\nu_j \in \mathrm{Cor}(\ell_j)$, $\nu_{12j+6} \equiv 0$ for all non-negative integer j. From Proposition 5, system (12) possesses an algebraic inverse integrating factor if and only if $\nu \equiv 0$, i.e. system (12) agrees with system (14). From Proposition 6, the function $V = h^{L+1+1/12}g(h)$ is an algebraic inverse integrating factor and from Proposition 4, it is unique up to a multiplicative constant. Last, undoing the change and by using Proposition 2, it has that system (10) has an algebraic inverse integrating factor of the form $V(h + \cdots)$.

For the cases $j_0 \neq 12L + 6$ it has the remaining systems (15)–(19) since are systems considered in Proposition 5 being λ the polynomial xyh^L, x^2yh^L, h^{L+1}, x^2h^L, xh^{L+1} or yh^{L+1}. So, the proof is completed. □

Next result solves the formal integrability problem for system (10). In the analytical case, according to the result of Mattei and Moussu [20], once the existence of a formal first integral has been established, we can ensure that there is also an analytical first integral.

Theorem 4 *System (10) is integrable (it has a first integral) if and only if it admits a formal inverse integrating factor (which can be zero at origin).*

Proof of Theorem 4 Necessary condition. Assume that system (10) is an integrable system. From Algaba et al. [3], the system is orbitally equivalent to a Hamiltonian system. So, by Theorem 2, system (10) is orbitally equivalent to system (12) with $\alpha_j = 0$, for all j. From Theorem 3, it follows that the functions $f(h + \cdots)$ being f a scalar non-constant function of class $\mathscr{C}^1$ are inverse integrating factors of system (10). Moreover, f may or may not to be zero at origin.

Sufficient condition. Assume that system (10) has an inverse integrating factor. From Theorem 3, system (10) possesses a formal inverse integrating factor if it is orbitally equivalent to the Hamiltonian system (13). So, system (10) is an integrable system. □

Note that the integrable systems (10) are orbital-reversible since are orbitally equivalent to $y^2\partial_x + x^3\partial_y$, which is invariant to $(x, y, t) \to (-x, y, -t)$. On the contrary, the non-integrability does not have any relation with the reversibility. For example, by applying Theorem 3, we check that:

1. System $\dot{x} = y^2 + 3x^2(ay + b(y^3/3 + x^4/4))$, $\dot{y} = x^3 + 4xy(ay + b(y^3/3 + x^4/4))$ is reversible to the involution $(-x, y)$ and does not have an inverse integrating factor.
2. System $\dot{x} = y^2 + 3x^3$, $\dot{y} = x^3 + 4x^2y$ has the inverse integrating factor $(y^3/3 + x^4/4)^{13/12}$ and it is not orbital reversible.

4 Application

We consider the family of generalized nilpotent systems given by

$$(\dot{x}, \dot{y})^T = (y^2 + a_{30}x^3 + a_{21}x^2y, x^3 + b_{21}x^2y + b_{12}xy^2)^T \tag{20}$$

Next result provides systems (20) that have an algebraic inverse integrating factor.

Theorem 5 *System (20) possesses an algebraic inverse integrating factor if and only if it satisfies one and only one of the following series of conditions:*

(i) $3b_{21} - 4a_{30} = a_{21} = b_{12} = 0$, $b_{21} + 3a_{30} \neq 0$.
(ii) $a_{30} = b_{21} = 3b_{12} - 4a_{21} = 0$, $b_{12} + a_{21} \neq 0$.
(iii) $a_{30} = b_{21} = 4b_{12} - 3a_{21} = 0$, $b_{12} + a_{21} \neq 0$.
(iv) $b_{21} + 3a_{30} = b_{12} + a_{21} = 0$.

Proof of Theorem 5 First, we re-write the parameters of the following form:

$$a_{30} = 3d_{20} - c_{31},\ b_{21} = 4d_{20} + 3c_{31},\ a_{21} = 3d_{11} - 2c_{22},\ b_{12} = 4d_{11} + 2c_{22}.$$

System (20) becomes

$$\begin{pmatrix} \dot{x} \\ \dot{y} \end{pmatrix} = \overbrace{\begin{pmatrix} y^2 \\ x^3 \end{pmatrix}}^{\mathbf{F}_5} + \overbrace{X_{c_{31}x^2y} + d_{20}x^2 \begin{pmatrix} 3x \\ 4y \end{pmatrix}}^{\mathbf{F}_6} + \overbrace{X_{c_{22}x^2y^2} + d_{11}xy \begin{pmatrix} 3x \\ 4y \end{pmatrix}}^{\mathbf{F}_7}. \quad (21)$$

The structure of the proof consists into computing successively the coefficients of the dissipative part of the normal form (12) given by Theorem 2.

The effective computation of the normal form in this example is performed following a procedure based on Lie transforms, because then we can exploit the strengths of the computer algebra systems. The details of the procedure can be found in [2], and the main tool is the Lie product of vector fields. Using *Maple* in the computations, we have obtained the expressions for the coefficients of the normal form (12).

The coefficient of order 6 of system (12) is $\alpha_6 = d_{20}$.

We distinguish the following cases:

- **Case** $d_{20} \neq 0$. By means of the change $x = \frac{u}{d_{20}^3}$, $y = \frac{v}{d_{20}^4}$ and rescaling the time by $dt = d_{20}^5 dT$, the system is transformed into

$$\begin{pmatrix} u' \\ v' \end{pmatrix} = \begin{pmatrix} v^2 \\ u^3 \end{pmatrix} + X_{\tilde{c}_{31}u^2v} + u^2 \begin{pmatrix} 3u \\ 4v \end{pmatrix} + X_{\tilde{c}_{22}u^2v^2} + \tilde{d}_{11}uv \begin{pmatrix} 3u \\ 4v \end{pmatrix}. \quad (22)$$

  where $' = \frac{d}{dT}$, $\tilde{c}_{31} = \frac{c_{31}}{d_{20}}$, $\tilde{d}_{11} = \frac{d_{11}}{d_{20}}$, $\tilde{c}_{22} = \frac{c_{22}}{d_{20}}$.

  From Theorem 3, system (22) admits an algebraic inverse integrating factor if and only if its normal form is given by system (14), i.e. the coefficients α_j, with $j \neq 12k + 6$, are zero, for all k.

  The coefficient of order 7 of the normal form for system (22) is $\alpha_7 = 7\tilde{d}_{11} - 13\tilde{c}_{31}$. If $\alpha_7 = 0$ ($\tilde{d}_{11} = \frac{13}{7}\tilde{c}_{31}$), the following coefficients are, up a positive multiplicative constant:

$$
\begin{aligned}
\alpha_{10} &= -\left[\tilde{c}_{22}^2 - (3\tilde{c}_{31}^2 + \tfrac{11}{7}\tilde{c}_{31} - 2)\tilde{c}_{22} + \tfrac{1}{588}\tilde{c}_{31}(1029\tilde{c}_{31}^3 + 896\tilde{c}_{31}^2 - 372\tilde{c}_{31} - 336)\right],\\
\alpha_{12} &= -\left[\tilde{c}_{22}^3 + (\tfrac{15}{2}\tilde{c}_{31}^2 - \tfrac{549}{70}\tilde{c}_{31} - \tfrac{41}{425})\tilde{c}_{22}^2\right.\\
&\quad -(\tfrac{45}{4}\tilde{c}_{31}^4 - \tfrac{1023}{70}\tilde{c}_{31}^3 - \tfrac{45314}{20825}\tilde{c}_{31}^2 + \tfrac{7547}{5950}\tilde{c}_{31} + \tfrac{6}{17})\tilde{c}_{22}\\
&\quad + \tfrac{1}{1166200}\tilde{c}_{31}(5685225\tilde{c}_{31}^5 - 7796880\tilde{c}_{31}^4 - 3296356\tilde{c}_{31}^3\\
&\quad \left. +1077816\tilde{c}_{31}^2 + 637616\tilde{c}_{31} + 117600)\right],\\
\alpha_{15} &= (\tfrac{4368}{55}\tilde{c}_{31} - \tfrac{49296}{935})\tilde{c}_{22}^4 + (-\tfrac{5824}{11}\tilde{c}_{31}^3 + \tfrac{746616}{935}\tilde{c}_{31}^2 + \tfrac{16743064}{32725}\tilde{c}_{31} + \tfrac{40608672}{3108875})\tilde{c}_{22}^3\\
&\quad +(\tfrac{61152}{55}\tilde{c}_{31}^5 - \tfrac{412308}{187}\tilde{c}_{31}^4 - \tfrac{13748072}{6545}\tilde{c}_{31}^3 + \tfrac{1723218744}{21762125}\tilde{c}_{31}^2 + \tfrac{37628458608}{21762125}\tilde{c}_{31}\\
&\quad +\tfrac{3456128}{621775})\tilde{c}_{22}^2 + (-\tfrac{52416}{55}\tilde{c}_{31}^7 + \tfrac{2039856}{935}\tilde{c}_{31}^6 + \tfrac{87060662}{32725}\tilde{c}_{31}^5 - \tfrac{1374070672}{21762125}\tilde{c}_{31}^4\\
&\quad -\tfrac{77026810328}{21762125}\tilde{c}_{31}^3 - \tfrac{56817681856}{152334875}\tilde{c}_{31}^2 + \tfrac{1343749888}{4352425}\tilde{c}_{31} + \tfrac{1477632}{17765})\tilde{c}_{22}\\
&\quad +\tfrac{1456}{5}\tilde{c}_{31}^9 - \tfrac{680706}{935}\tilde{c}_{31}^8 - \tfrac{35263124}{32725}\tilde{c}_{31}^7 - \tfrac{36588708}{1978375}\tilde{c}_{31}^6 + \tfrac{278054454704}{152334875}\tilde{c}_{31}^5\\
&\quad +\tfrac{561705613456}{1066344125}\tilde{c}_{31}^4 - \tfrac{23540044736}{96940375}\tilde{c}_{31}^3 - \tfrac{3968393728}{30466975}\tilde{c}_{31}^2 - \tfrac{2955264}{124355}\tilde{c}_{31}.
\end{aligned}
$$

It is easy to check that the three coefficients α_{10}, α_{12} and α_{15} are zero simultaneously if and only if $\tilde{c}_{22} = \tilde{c}_{31} = 0$. So, system (20) becomes

$$(\dot{x}, \dot{y})^T = (y^2 + 3d_{20}x^3,\ x^3 + 4d_{20}x^2y)^T,$$

i.e. case (i). This system possesses the algebraic inverse integrating factor $(4y^3 - 3x^4)^{13/12}$.

- **Case** $d_{20} = 0$. In this case, $\alpha_7 = d_{11}$.

We first assume that $d_{11} \neq 0$. Therefore, the first coefficient of the normal form different from zero is of order 7. From Theorem 3, system (22) admits an algebraic inverse integrating factor if and only if its normal form is given by system (15), i.e. the coefficients α_j, with $j \neq 12k + 7$, are zero, for all k.

The coefficient α_{10} of system (12) is, up multiplicative constant,

$$\alpha_{10} = d_{11}c_{31}(35c_{31}^2 - 30c_{22} + 3d_{11}).$$

We analyze the following possibilities:

(a) $c_{31} = 0$. The coefficient α_{12} is 0 and

$$\alpha_{15} = d_{11}c_{22}(d_{11} + 2c_{22})\left[18d_{11}^2 - 5c_{22}d_{11} - 10c_{22}^2\right].$$

We again distinguish three cases:

(a.1) $c_{22}=0$**.** In this case, system (20) becomes

$$(\dot{x},\dot{y})^T=(y^2+3d_{11}x^2y,\ x^3+4d_{11}xy^2)^T,$$

i.e. case (ii). This system possesses the algebraic inverse integrating factor $(4y^3-3x^4)^{7/6}$.

(a.2) $c_{22}=-\frac{1}{2}d_{11}\neq 0$**.** In this case, the system is

$$(\dot{x},\dot{y})^T=(y^2+4d_{11}x^2y,\ x^3+3d_{11}xy^2)^T,$$

i.e. case (iii).

Such system has the inverse integrating factor $\left(-4y^3+3x^4-6d_{11}x^2y^2+3d_{11}^2y^4\right)^{7/6}$.

(a.3) $18d_{11}^2-5c_{22}d_{11}-10c_{22}^2=0$**.** The following coefficient $\alpha_j,\ j\neq 12k+7$, non-null is, up positive multiplicative constant,

$$\begin{aligned}\alpha_{27}&=-c_{22}d_{11}(d_{11}+2c_{22})q(d_{11},c_{22})\\ q(d_{11},c_{22})&=101640000c_{22}^8+203280000c_{22}^7d_{11}-1447525800c_{22}^6d_{11}^2\\ &\quad-2349158700c_{22}^5d_{11}^3+2319593710c_{22}^4d_{11}^4+3313232335c_{22}^3d_{11}^5\\ &\quad-6331529024c_{22}^2d_{11}^6-3604918332c_{22}d_{11}^7+10328947200d_{11}^8.\end{aligned}$$

Imposing the condition **(a.3)**, it has that $\alpha_{27}\neq 0$. Thus, the system does not have an algebraic inverse integrating factor.

(b) $c_{22}=\frac{1}{10}d_{11}+\frac{7}{5}c_{31}^2,\ c_{31}\neq 0$**.** In this case,

$$\alpha_{12}=c_{31}^3d_{11}(21d_{11}+17c_{31}^2).$$

Taking $d_{11}=-\frac{17}{21}c_{31}^2$, we get, up positive multiplicative constant,

$$\alpha_{15}=-c_{31}^{10}\neq 0.$$

Therefore, the system does not have an algebraic inverse integrating factor.

Last, we assume that $d_{11}=0$. In such case, the system is a Hamiltonian system whose Hamiltonian function is $-\frac{1}{3}y^3+\frac{1}{4}x^4+c_{31}x^2y+c_{22}x^2y^2$. Such function is also an algebraic inverse integrating factor of the system. This is case (iv). □

The family (20) is a particular case of the family considered in [18, Theorem 7] where the integrability of the system is studied. Giné [18] gives sufficient conditions of integrability. Concretely, it proves that if $b_{21}=a_{30}=a_{21}+b_{12}=0$ then system (20) is integrable.

As a direct consequence of Theorem 4, it has the following result which characterizes the integrability of the systems (20):

Theorem 6 *The integrable systems of the family (20) are*

$$(\dot{x}, \dot{y})^T = (y^2 + a_{30}x^3 + a_{21}x^2y,\ x^3 - 3a_{30}x^2y - a_{21}xy^2)^T,$$

(i.e. $b_{21} + 3a_{30} = a_{21} + b_{12} = 0$*) for any real numbers* a_{30} *and* a_{21}*.*

Acknowledgements This work has been partially supported by *Ministerio de Ciencia y Tecnología, Plan Nacional I+D+I* co-financed with FEDER funds, in the frame of the project MTM2014-56272-C2-02, and by *Consejería de Educación y Ciencia de la Junta de Andalucía* (FQM-276 and P12-FQM-1658).

References

1. Algaba, A., Freire, E., Gamero, E., García, C.: Quasihomogeneous normal forms J. Comput. Appl. Math. **150**, 193–216 (2003)
2. Algaba, A., Freire, E., Gamero, E., García, C.: An algorithm for computing quasi-homogeneous formal normal forms under equivalence. Acta Appl. Math. **80**, 335–359 (2004)
3. Algaba, A., Gamero, E., García, C.: The integrability problem for a class of planar systems. Nonlinearity **22**(2), 95–420 (2009)
4. Algaba, A., García, C., Reyes, M.: Nilpotent systems admitting an algebraic inverse integrating factor over $\mathbb{C}((x, y))$. Qual. Theory Dyn. Syst. **10**(2), 303–316 (2011)
5. Algaba, A., García, C., Reyes, M.: Existence of an inverse integrating factor, center problem and integrability of a class of nilpotent systems. Chaos Solitons Fractals **45**, 869–878 (2012)
6. Algaba, A., Fuentes, N., García, C., Reyes, M.: A class of non-integrable systems admitting an inverse integrating factor. J. Math. Anal. Appl. **420**(2), 1439–1454 (2014)
7. Chavarriga, J.: Integrable systems in the plane with a center type linear part. Appl. Math. Warsaw **22**, 285–309 (1994)
8. Chavarriga, J., Giacomini, H., Giné, J., Llibre, J.: On the integrability of two-dimensional flows. J. Differential Equations **157**, 163–182 (1999)
9. Chavarriga, J., Giacomini, H., Giné, J., Llibre, J.: Darboux integrability and the inverse integrating factor. J. Differential Equations **194**, 116–139 (2003)
10. Christopher, C.J., Llibre, J.: Integrability via invariant algebraic curves for planar polynomial differential systems. Ann. Differential Equations **16**, 5–19 (2000)
11. Christopher, C., Mardesic, P., Rousseau, C.: Normalizable, integrable, and linealizable saddle points for complex quadratic systems in $\mathbb{C}^2$. J. Dyn. Control Syst. **9**, 311–363 (2003)
12. Enciso, A., Peralta-Salas, D.: Existence and vanishing set of inverse integrating factors for analytic vector fields. Bull. London Math. Soc. **41**, 1112–1124 (2009)
13. Ferragut, A., Llibre, J., Mahdi, A.: Polynomial inverse integrating factors for polynomial vector fields. Discrete Contin. Dyn. Syst. **17**, 387–395 (2006)
14. García, I., Grau, M.: A survey on the inverse integrating factor. Qual. Theory Dyn. Syst. **9**(1–2), 115–166 (2010)
15. García, I., Shafer, D.: Integral invariants and limit sets of planar vector fields. J. Differential Equations **217**(2), 363–376 (2005)
16. García, I., Giacomini, H., Grau, M.: The inverse integrating factor and the Poincaré map. Trans. Am. Math. Soc. **362**(7), 3591–3612 (2010)

17. Giacomini, H., Llibre, J., Viano, M.: On the nonexistence, existence and uniqueness of limit cycles. Nonlinearity **9**, 501–516 (1996)
18. Giné, J.: Analytic integrability and characterization of center for generalized nilpotent singular point. Appl. Math. Comput. **148**, 849–868 (2004)
19. Kooij, R.E., Christopher, C.J.: Algebraic invariant curves and the integrability of polynomial systems. Appl. Math. Lett. **6**, 51–53 (1993)
20. Mattei, J.F., Moussu, R.: Holonomie et intégrales premières. Ann. Sci. Ecole Normale Superieure **13**, 469–523 (1980)
21. Mazzi, L., Sabatini, M.: A characterization of centers via first integrals. J. Differential Equations **76**, 222–237 (1988)
22. Prelle, M.J., Singer, M.F.: Elementary first integrals of of differential equations. Trans. Am. Math. Soc. **279**, 215–229 (1983)
23. Singer, M.F.: Liouvillian first integrals of differential equations. Trans. Am. Math. Soc. **333**, 673–688 (1992)
24. Walcher, S.: Local integrating factors. J. Lie Theory **13**, 279–289 (2003)

WENO Schemes for Multi-Dimensional Porous Media Flow Without Capillarity

R. Bürger, F. Guerrero, M.C. Martí, and P. Mulet

Abstract In this work we derive a numerical technique based on finite-difference WENO schemes for the simulation of multi-dimensional multiphase flows in a homogeneous porous medium. The key idea is to define a compatible discretization for the fluxes of the convective term in order to maintain their divergence-free character not only in the continuous setting but also in the discrete setting, ensuring the conservation of the sum of the saturations through time evolution. The one-dimensional numerical technique is derived in detail for the case of neglected capillarity effects. Numerical results obtained with one-dimensional and two-dimensional standard tests of multiphase flow in a homogeneous porous medium are shown.

1 Introduction

Mathematical models for multiphase flow processes in porous media under vertical equilibrium have been used since the end of the nineteenth century in many different physical situations such as oil and gas reservoirs [1, 5], water filtration [13] and enhanced oil recovery [4]. All these applications have arisen in part due to the development of suitable numerical models and efficient computational methods, either using finite differences [2, 16] or finite elements [4].

The physical situation we are interested in study is the following: we consider immiscible and incompressible N-phases fluid, therefore, phase densities and viscosities of each phase are assumed to be constant, flowing through a rigid and

R. Bürger
Departamento de Ingeniería Matemática, CI2MA and Universidad de Concepción, Concepcion, Chile
e-mail: rburger@ing-mat.udec.cl

F. Guerrero • P. Mulet (✉)
Department de Matemàtica Aplicada, Universitat de València, Valencia, Spain
e-mail: Francisco.Guerrero-Cortina@uv.es; mulet@uv.es

M.C. Martí
CI2MA and Universidad de Concepción, Concepcion, Chile
e-mail: mmarti@ci2ma.udec.cl

F. Ortegón Gallego et al. (eds.), *Trends in Differential Equations and Applications*,
SEMA SIMAI Springer Series 8, DOI 10.1007/978-3-319-32013-7_17

homogeneous porous medium, with constant porosity and absolute permeability, with no internal sources or sinks. We consider the phase velocities given by the extension of Darcy's law and we neglect the capillarity effects.

These assumptions lead to a closed system of partial differential equations for the mass conservation that can be written as follows:

$$\frac{\partial \mathbf{u}}{\partial t} + \frac{\partial \mathbf{f}(\mathbf{u})}{\partial x} = \frac{\partial}{\partial x}\left[\mathbf{B}(\mathbf{u})\frac{\partial \mathbf{u}}{\partial x}\right]$$

where $\mathbf{u}$ is the vector of saturations, $\mathbf{f}(\mathbf{u})$ is the fractional vector representing the buoyancy effects, modeling the convective terms, and $\mathbf{B}(\mathbf{u})$ is the capillary diffusion tensor, which models diffusion terms. When capillary effects are neglected or small, discontinuities and/or sharp gradients will develop in the numerical solutions that call for sophisticated techniques to obtain accurate numerical simulation results.

High-Resolution Shock-Capturing (HRSC) schemes have been developed to properly handle discontinuities in numerical simulations while obtaining high order of resolution in smooth regions of the solution. Among the HRSC schemes, the Weighted Essentially Non-oscillatory (WENO) finite-difference schemes have become one of the most popular methods to approximate the solutions of hyperbolic conservation laws. These schemes combine the upwinding framework, in which the direction of propagation of the information on the computational mesh determines the discretization of the equations on that mesh, with the WENO high-order upwind-biased reconstructions, to control the creation and evolution of spurious numerical oscillations.

Donat et al. [8, 9, 11] develop a progressive evolution towards the application of finite-difference WENO schemes to one-dimensional multiphase porous media flow processes. In [8] the one-dimensional vertical equilibrium two-phase model of [6] is considered, assuming vertical equilibrium and using Darcy's law [7], and it is shown that WENO schemes can be successfully used to obtain accurate solutions of the problem when neglecting capillary effects. When capillary effects are considered, the stability restrictions for explicit numerical schemes are very restrictive, so the use of Implicit-Explicit (IMEX) Runge-Kutta schemes is proposed to overcome the stability restrictions on the time step due to the diffusive terms.

In [9, 11] the model is extended to three-phase flow and general N-phase flow, respectively. It is proved that the IMEX-WENO scheme is suitable for solving the physical problem independently of the magnitude of the considered capillary effects.

The model solved in [9, 11] is based on the fact that the assumption of vertical equilibrium allows to explicitly express the derivative of the pressure appearing in the constitutive mathematical model equations, due to Darcy's law, in terms of capillary pressures and other terms depending on the various fluid phases. While this explicit expression of the pressure derivative is readily available in the one-dimensional case, it is not possible to obtain a similar expression for the 2D model, making impossible the extension of Donat et al.'s work to a multi-dimensional framework. Since multiphase flows in porous media are of interest not only in a one-dimensional framework, it is the primary purpose of this contribution to extend the use of finite-difference conservative WENO schemes to a multi-dimensional version of this problem. This avoids using the explicit expression of the derivative of the pressure. To this end, we propose to define a compatible data-dependent discretization for the fluxes, valid for all equations of the system, that allows us to explicitly obtain numerical approximations of the pressure and its derivative.

The paper is organized as follows. In Sect. 2, we derive the model equations for multi-dimensional multiphase flow in porous media and we introduce the idea of our proposed method in a multi-dimensional setting. Section 3 includes the detailed definition of our semi-discrete formulation in the one-dimensional case. Section 4 extends the method to the two-dimensional case. In Sect. 5 we perform some numerical experiments on standard tests on multiphase flow in porous media, both in 1D and 2D, to test the capabilities of the scheme proposed. Finally, in Sect. 6 we collect some conclusions and some proposals for future work.

2 Porous Media Flow

We denote by $u_i(x,t)$, $i = 1, \dots, N$, $x \in \Omega = (0,1)^d$, $t \in \mathbb{R}^+$, the concentration (or saturation) of the ith phase in the pore space, ρ_i its density, assumed constant, and g the gravity acceleration vector, $g = (0, \dots, 0, -9.81)^T \in \mathbb{R}^d$. Then, by Darcy's law, the velocity of the ith phase is given by

$$f_i = \lambda_i(u_i)(\rho_i g - \nabla p_i), \quad f_i \in \mathbb{R}^d, \tag{1}$$

where p_i is the phase pressure and

$$\lambda_i = \frac{k k_i(u_i)}{\phi \mu_i} \geq 0$$

is the (normalized by porosity) relative mobility of the ith phase, assumed to be function of their corresponding phase saturation only, $\lambda_i = \lambda_i(u_i)$, where $k_i(u_i)$ is the relative permeability of phase i, $0 \leq k_i(u_i) \leq 1$, k is the absolute permeability of the porous medium, measuring the ability of the porous material to allow fluids to pass through it, ϕ is the (constant) porosity and μ_i is the viscosity of phase i, assumed constant in our case.

Taking this into account, the continuity equations of all phases can be written as

$$0 = \frac{\partial u_i}{\partial t} + \operatorname{div}\left(\lambda_i(u_i)(\rho_i g - \nabla p_i)\right), \quad i = 1, \dots, N. \tag{2}$$

These equations are supplemented with initial conditions and known normal fluxes at the boundary, $f_i(x,t) \cdot n = q_i(x,t)$, for $x \in \partial\Omega$, with n denoting the unit normal vector to the boundary pointing outwards Ω. In this paper we use $q_i = 0$.

The assumption that the fluid occupies the whole pore space yields that the saturations satisfy $\sum_i u_i = 1$. Therefore, if we sum all the equations in (2) we get:

$$0 = \frac{\partial \sum_i u_i}{\partial t} + \operatorname{div}\left(\sum_{i=1}^{N} \lambda_i(u_i)(\rho_i g - \nabla p_i)\right).$$

If initially the fluid phases saturate the pores, i.e., $\sum_i u_i(x,0) = 1$, for all $x \in \Omega$, then we deduce that this will hold through time evolution, i.e. $\sum_i u_i(x,t) = 1$, for all $x \in \Omega, t \in \mathbb{R}^+$ if and only if

$$0 = \operatorname{div}\left(\sum_{i=1}^{N} \lambda_i(u_i)(\rho_i g - \nabla p_i)\right). \tag{3}$$

The equations in (2) and (3) form a system of $N+1$ equations in the $2N$ unknowns $u_i, p_i, i = 1, \dots, N$, but they can be reduced to N equations in the $2N-1$ unknowns $u_1, \dots, u_{N-1}, p_1, \dots, p_N$ if we take into account that (3) is equivalent to $u_1 + \dots + u_N = 1$, assuming that this holds initially. Therefore, $N-1$ additional equations have to be supplied in order to solve (2) and (3).

For this purpose capillary pressures $\bar{p}_1, \dots, \bar{p}_{N-1}$ are introduced as the pressure differences with respect to a reference *non-wetting* phase, which we take to be the Nth phase:

$$\bar{p}_i = p_N - p_i, \quad i = 1 \dots, N-1.$$

Capillary pressures are specified as functions of the saturations, $\bar{p}_i = \bar{p}_i(u_i)$.

Then, by this assumption, (3) can be written as follows, where the pressure gradients of each fluid phase are expressed in terms of the corresponding capillary pressure:

$$0 = \operatorname{div}\left(\sum_{i=1}^{N} \lambda_i(u_i)(\rho_i g - \nabla p_N + \nabla \bar{p}_i(u_i))\right), \tag{4}$$

$$-\operatorname{div}(\lambda(u)\nabla p_N)) = -\sum_{i=1}^{N} \operatorname{div}(\lambda_i(u_i)\rho_i g) - \sum_{i=1}^{N} \operatorname{div}\left(\lambda_i(u_i)\bar{p}_i'(u_i)\nabla u_i\right),$$

where we define

$$\lambda(u) := \sum_{i=1}^{N} \lambda_i(u_i), \quad u = (u_1, \dots, u_N).$$

With appropriate boundary conditions for p_N, the elliptic equation (4) can be solved for p_N, thus entailing a functional relation $p_N[u]$.

In 1D, (4), under the assumption of zero total boundary flux $\sum_i q_i(t) = 0$, can be explicitly solved as

$$\nabla p_N = \sum_{i=1}^{N} \frac{\lambda_i(u_i)}{\lambda(u)} (\rho_i g + \nabla \bar{p}_i(u_i)).$$

This yields a system of conservation laws for the concentrations u_i, $i = 1, \dots, N$:

$$0 = \frac{\partial u_i}{\partial t} + \operatorname{div}\left(\lambda_i(u_i)\left(\rho_i g - \sum_{j=1}^{N} \frac{\lambda_j(u_j)}{\lambda(u)} (\rho_j g + \nabla \bar{p}_j(u_j)) + \nabla \bar{p}_i(u_i)\right)\right). \tag{5}$$

In [9] it was proved that the diffusive part of (5) is weakly parabolic (i.e., the eigenvalues of the diffusion tensor are non-negative), while the convective part may have non-hyperbolic regions. In that paper, the authors propose the use of a high-order WENO schemes, developed by Liu et al. in [15] and improved by Jiang and Shu in [12], to discretize the convective part and to deal with the steep gradients that may appear during the process. To overcome the severe stability restrictions associated with explicit schemes for parabolic equations, an Implicit-Explicit (IMEX) strategy, where the parabolic terms are handled by an implicit discretization, is proposed and it is shown that it provides highly accurate and efficient numerical solutions.

Unfortunately, the work developed in [8, 9, 11] can not be extended to a multidimensional framework, as Eq. (4) can not be explicitly solved in more than one dimension. In order to extend the use of finite-difference WENO schemes to numerically solve multidimensional multiphase flow problems in porous media, we propose the use of finite-difference WENO schemes to numerically solve Eq. (2), using a compatible discretization for the fluxes f_i in (1) that preserves the divergence-free character of the numerical fluxes, i.e., we require Eq. (3) to be satisfied also in the discrete setting. This property is necessary if one wants to assure that the conservation of the concentration holds during time evolution, i.e., $\sum_i u_i(x,t) = 1,\ \forall t \geq 0$.

For the sake of simplicity, to define the compatible discretization of the fluxes, we will henceforth neglect the effects of capillarity, i.e. we assume zero capillary pressures $\bar{p}_i = 0$, so (4) is written for $p := p_N$ as

$$-\operatorname{div}(\lambda(u)\nabla p)) = -\sum_{i=1}^{N} \operatorname{div}(\lambda_i(u_i)\rho_i g). \tag{6}$$

Specifically, we consider a uniform Cartesian computational mesh on $\Omega = (0,1)^d$, that contains m^d points, and, for $i = 1, \dots, N$, we denote by $v_i(t) \in \mathbb{R}^{m \times \cdots \times m}$

the d-dimensional matrix containing approximations of the sought solution at the mesh points $x_j, \boldsymbol{j} \in \mathbb{N}^d$, $v_{i,j}(t) \approx u_i(x_j, t)$.

Let $D_h[v]$ be a data-dependent discretization of $-$div (like that provided by finite-difference WENO schemes, see the next section for further details), i.e., a linear operator:

$$D_h[v]\colon \left(\mathbb{R}^{m\times\cdots\times m}\right)^d \to \mathbb{R}^{m\times\cdots\times m}.$$

Since the adjoint operator of $-$div is ∇, if we consider $D_h[v] \approx -$div, then the discretization that we propose for $\nabla = (-\mathrm{div})^*$ should be $\nabla_h[v] = D_h[v]^*$,

$$D_h[v]^*\colon \mathbb{R}^{m\times\cdots\times m} \to \left(\mathbb{R}^{m\times\cdots\times m}\right)^d.$$

Since the operator $D_h[v]$ is linear, using the same argument that led us to (6), we obtain a compatible spatial semi-discretization:

$$\begin{aligned} 0 &= v_i'(t) - D_h[v]\tilde{f}_i \\ 0 &= \left(\sum_i v_i(t)\right)' - D_h[v]\sum_i \tilde{f}_i, \end{aligned} \tag{7}$$

$$\tilde{f}_i = \lambda_i(v)(\rho_i g - D_h[v]^* p_h) \approx f_i$$

where $v_i = (v_{i,j}(t)), \boldsymbol{j} \in \mathbb{N}^d$ and p_h is the d-dimensional matrix that approximates the function p on the computational mesh.

If $\sum_i v_{i,j}(0) = 1$, then we deduce that $\sum_i v_{i,j}(t) = 1$, for all $t \in \mathbb{R}^+$ if and only if

$$0 = D_h[v]\sum_i \tilde{f}_i \tag{8}$$

and

$$D_h[v]\Lambda(v)D_h[v]^* p_h = \sum D_h[v]\lambda_i(v)\rho_i g. \tag{9}$$

Here we have used the notation $\Lambda(v)$ for a diagonal matrix such that $(\Lambda(v)q)_j = (\sum_i \lambda_i(v_{i,j}))q_j$, understanding that λ and λ_i act in a pointwise manner on their matrix arguments. Now, (9) is a data-dependent discretization of (6), with a matrix which is symmetric and positive semidefinite, since the elements in the diagonal of $\Lambda(v)$ are non-negative. The goal of this work is to use an ODE solver to solve (7) together with the elliptic pressure-velocity equation (9).

Since there are more details that have to be taken into account, for instance, upwinding and numerical viscosity, we detail in the next section the operator form in a one-dimensional setting.

3 One-Dimensional Porous Media Flows

We consider a uniform grid on [0, 1] defined by the cell centers $x_j = \left(j - \frac{1}{2}\right) h$, $j = 1, \dots, m$, with cell boundaries given by $x_{j+\frac{1}{2}} = x_j + \frac{h}{2}$, where $h = 1/m$ is the uniform grid spacing, so that $0 = x_{\frac{1}{2}}$, $1 = x_{m+\frac{1}{2}}$.

To obtain high-order finite-difference conservative schemes for the approximate solution of (2), we use Shu and Osher's technique [17], for which the spatial derivative in (2) can be exactly obtained by a conservative finite difference formula that involves values of φ_i at the cell boundaries,

$$f_i(u(x))_x = \frac{1}{h}\left(\varphi_i\left(x + \frac{h}{2}\right) - \varphi_i\left(x - \frac{h}{2}\right)\right), \tag{10}$$

where the functions φ_i are implicitly defined as

$$f_i(u(x)) = \frac{1}{h}\int_{x-\frac{h}{2}}^{x+\frac{h}{2}} \varphi_i(\xi)d\xi, \quad i = 1, \dots, N. \tag{11}$$

Then we can discretize the spatial derivative in (2) as:

$$(f_i(u))_x(x_j) = \frac{\hat{f}_i(x_{j+\frac{1}{2}}) - \hat{f}_i(x_{j-\frac{1}{2}})}{h} + \mathscr{O}(h^r) \tag{12}$$

where $\hat{f}_i$ is a highly accurate approximation to φ_i obtained from known grid values of $f_i(u)$ [which are cell averages of φ_i by (11)] on a stencil around $x_{j+\frac{1}{2}}$ such that $\varphi_i\left(x_{j+\frac{1}{2}}\right) = \hat{f}_i\left(x_{j+\frac{1}{2}}\right) + e\left(x_{j+\frac{1}{2}}\right) h^r + \mathscr{O}(h^{r+1})$, for a locally Lipschitz continuous function e.

We will compute the numerical fluxes $\hat{f}_{i,j+\frac{1}{2}} = \hat{f}_i(x_{j+\frac{1}{2}})$, $i = 1, \dots, N$, using a component-wise finite-difference scheme as:

$$\hat{f}_{i,j+\frac{1}{2}} = \mathscr{R}^+(f^+_{i,j-r+1}, \dots, f^+_{i,j+r-1}, x_{j+\frac{1}{2}}) + \mathscr{R}^-(f^-_{i,j-r+2}, \dots, f^-_{i,j+r}; x_{j+\frac{1}{2}}), \tag{13}$$

where the functions $f_i^{\pm}$ define a flux splitting for f_i (necessary for stability purposes), $f^{\pm}_{i,k} = f_i^{\pm}(x_k)$ and $\mathscr{R}^{\pm}$ are upwind-biased $(2r - 1)$-order WENO reconstruction operators, that we next describe.

In a general setting, given some cell averages $\bar{g}_l$ [which are $f^+_{i,l}$ in (13)] of a function g [which corresponds to φ_i in (11)] on a stencil around the point $x_{j+\frac{1}{2}}$, a $(2r-1)$-order WENO reconstruction of $g(x_{j+\frac{1}{2}})$, is determined by a convex combination:

$$q(x_{j+\frac{1}{2}}) = \sum_{k=0}^{r-1} w_{k,j} p^r_{k+j}(x_{j+\frac{1}{2}}),$$

where $p^r_{k+j}(x)$ is the $(r-1)$th degree polynomial reconstruction defined on the stencil $S_k = \{x_{j+k-r+1}, \ldots, x_{j+k}\}$, $k = 0, \ldots, r-1$, i.e.,

$$\int_{x_{l-\frac{1}{2}}}^{x_{l+\frac{1}{2}}} g(x)dx = \bar{g}_l, \quad l = j+k-r+1, \ldots, j+k,$$

satisfying $p^r_{k+j}(x_{j+\frac{1}{2}}) = g(x_{j+\frac{1}{2}}) + \mathcal{O}(h^r)$ and $w_{k,j}$ are weight functions which depend on the smoothness of the function g on the corresponding stencil, so that polynomials corresponding to singularity-crossing stencils should have a negligible contribution to the convex combination.

For instance, for the third-order WENO scheme, named WENO3, that corresponds to $r = 2$, one can write the left-biased WENO reconstruction appearing in (13), as:

$$\begin{aligned}\mathscr{R}^+(f^+_{i,j-1}, f^+_{i,j}, f^+_{i,j+1}, x_{j+\frac{1}{2}}) = q(x_{j+\frac{1}{2}}) = \\ w^+_{0,j}\left(-\frac{1}{2}f^+_{i,j-1} + \frac{3}{2}f^+_{i,j}\right) + w^+_{1,j}\left(\frac{1}{2}f^+_{i,j} + \frac{1}{2}f^+_{i,j+1}\right) = \\ \gamma^+_{-1,j}f^+_{i,j-1} + \gamma^+_{0,j}f^+_{i,j} + \gamma^+_{1,j}f^+_{i,j+1},\end{aligned}$$

where the coefficients $\gamma^+_{l,j}$, $l = -1, 0, 1$ are defined as linear combinations of the corresponding WENO3 weight functions $w^+_{k,j}$, $j = 1, \ldots, m$, $k = 0, 1$ and, for this left-biased reconstruction, $\gamma^+_{2,j} = 0$. The same can be done for the right-biased WENO reconstruction, using the weights $w^-_{k,j}$ to define the coefficients $\gamma^-_{l,j}$, $l = -1, \ldots, 2$, where, in this case, $\gamma^-_{-1,j} = 0$. As it can be readily seen in this example, in general one has

$$\sum_{l=-r+1}^{r} \gamma^{\pm}_{l,j} = \sum_{k=0}^{r-1} w^{\pm}_{k,j} = 1. \tag{14}$$

In order to avoid the phase dependence on the definition of the parameters $\gamma^{\pm}_{l,j}$, we are using weights based on component-wise global smoothness indicators defined by Levy et al. in [14]. These smoothness indicators, proposed to improve the resolution of the scheme near the discontinuities, are defined as an average of the smoothness indicators defined in [12] and are valid for all the components of the system. We will use the values of v instead of the values of $f^{\pm}(v)$ in their computation. We refer the reader to [12, 15] for further details on WENO schemes.

With these considerations, the numerical flux of a component-wise finite-difference $(2r-1)$-WENO scheme is given by

$$\hat{f}_{i,j+\frac{1}{2}} = \sum_{l\in\mathscr{S}} \gamma_{l,j}^{+}\tilde{f}_{i,j+l}^{+} + \gamma_{l,j}^{-}\tilde{f}_{i,j+l}^{-}, \tag{15}$$

with the coefficients $\gamma_{l,j}^{\pm} = \gamma_l^{\pm}(v_{\mathscr{S}+j})$, $\mathscr{S} = \{-r+1,\dots,r\}$, defined as linear combinations of the corresponding $(2r-1)$-WENO weight functions and where $v_{\mathscr{S}+j}$ denotes the $N\times 2r$ matrix whose entries are $(v_{\mathscr{S}+j})_{i,k} = v_{i,j-r+k}$, $i=1,\dots,N$, $k=1,\dots,2r$.

In the definition of the numerical fluxes (15) we employ a Lax-Friedrichs flux splitting:

$$\begin{aligned} \tilde{f}_{i,j}^{\pm} &= \frac{1}{2}\left(f_{i,j}^{\pm} \pm \alpha v_{i,j}\right), \\ f_{i,j}^{\pm} &= \lambda_{i,j}(\rho_i g - \nabla_j^{\pm} p), \\ \nabla_j^{\pm} p = \nabla_h[v]_j^{\pm} p &= (\nabla_h[v]^{\pm} p)_j, \quad \lambda_{i,j} = \lambda_i(v_{i,j}) \end{aligned} \tag{16}$$

(i.e., $\nabla_h[v]_j^{\pm}$ is the jth row of $\nabla_h[v]^{\pm}$) with the parameter α being an upper bound of the eigenvalues of $f'(v)$, and where the linear operators $p \mapsto \nabla_h[v]^{\pm} p$ approximating ∇p remain to be determined.

Using (15) and (16), we can write the numerical flux for each phase as

$$\hat{f}_{i,j+\frac{1}{2}} = \frac{1}{2}\sum_{l=-r+1}^{r}\left(\gamma_{j,l}^{+} f_{i,j+l}^{+} + \gamma_{j,l}^{-} f_{i,j+l}^{-} + \alpha(\gamma_{j,l}^{+} - \gamma_{j,l}^{-})u_{i,j+l}\right). \tag{17}$$

For the computation of some numerical fluxes near the boundary, such as $\hat{f}_{i,\frac{1}{2}}$ or $\hat{f}_{i,m+\frac{1}{2}}$, the knowledge of some values of the fluxes $f_{i,j}$ and saturations $u_{i,j}$ outside the computational domain (for $j \notin \{1,\dots,m\}$) is required. To define these flux values we implement zero-flux boundary conditions $f_i(0) = f_i(1) = 0$ via linear extrapolation, based on [3], as follows:

$$\begin{aligned} f_{i,-j}^{\pm} &= -f_{i,j+1}^{\pm}, \\ f_{i,m+j+1}^{\pm} &= -f_{i,m-j}^{\pm}, \quad j = 0,\dots,r-1. \end{aligned}$$

For the saturations we use outflow boundary conditions:

$$\begin{aligned} v_{i,-j} &= v_{i,j+1}, \\ v_{i,m+j+1} &= v_{i,m-j}, \quad j = 0,\dots,r-1. \end{aligned}$$

With the notation for $m+1$ vectors z_i, $i = 1, \dots, N$

$$z_{i,j} = \hat{f}_{i,j+\frac{1}{2}}, \quad j = 0, \dots, m$$

and (16), (17), the numerical flux can be written in matrix-vector form as follows:

$$\begin{aligned} z_i &= \Gamma^+ E \Lambda_i (\rho_i g e - \nabla_h^+ p) + \Gamma^- E \Lambda_i (\rho_i g e - \nabla_h^- p) + \alpha(\Gamma^+ - \Gamma^-) F v_i \\ &= g(\Gamma^+ + \Gamma^-) E \Lambda_i \rho_i e - (\Gamma^+ E \Lambda_i \nabla_h^+ + \Gamma^- E \Lambda_i \nabla_h^-) p + \alpha(\Gamma^+ - \Gamma^-) F v_i, \end{aligned}$$

where we have introduced $\Gamma^\pm = \Gamma^\pm[v]$ as the $(m+1) \times (m+2r)$ matrices given by

$$\Gamma^\pm_{j+1, j+l+r} = \frac{1}{2} \gamma^\pm_{j,l}, \quad j = 0, \dots, m, \quad l = -r+1, \dots, r. \tag{18}$$

The matrix $\Lambda_i = \Lambda_i[v]$ is the diagonal matrix with $\Lambda_i[v]_{j,j} = \lambda_i(v_{i,j})$ and $e = (1, \dots, 1)^T \in \mathbb{R}^m$.

The $(m+2r) \times m$ matrices E, F have been introduced to implement the boundary conditions for fluxes and saturations respectively in matrix/vector form. These matrices are defined to introduce the values of the boundary fluxes and saturations into the vector of fluxes or saturations, respectively. For example, for WENO3 scheme, that corresponds to $r = 2$, these matrices are

$$E = \left[-e_2, -e_1, e_1, \dots e_m, -e_m, -e_{m-1}\right]^T, \quad F = \left[e_2, e_1, e_1, \dots e_m, e_m, e_{m-1}\right]^T,$$

where e_i is the ith vector in the canonical basis of $\mathbb{R}^m$.

We know that the action of the operator $D_h[v] \approx -\mathrm{div}$ on f_i, using the definition of z_i, is described as follows:

$$(D_h[v] f_i)_j = -\frac{\hat{f}_{i,j+\frac{1}{2}} - \hat{f}_{i,j-\frac{1}{2}}}{h} = \frac{z_{j-1,i} - z_{j,i}}{h}, \quad j = 1, \dots, m. \tag{19}$$

Then, if we introduce $D_m \in \mathbb{R}^{m \times (m+1)}$ as

$$D_m = \frac{1}{h} \begin{bmatrix} -1 & 1 & & & \\ 0 & -1 & 1 & & \\ \cdots & \cdots & \cdots & \cdots & \cdots \\ & & & -1 & 1 \end{bmatrix},$$

and use the definition of the vector z_i, we can rewrite (19) as

$$\begin{aligned} D_h[v] f_i &= -D_m z_i \\ &= -g D_m (\Gamma^+ + \Gamma^-) E \Lambda_i \rho_i e - (-D_m \Gamma^+ E \Lambda_i \nabla_h^+ - D_m \Gamma^- E \Lambda_i \nabla_h^-) p \\ &\quad - \alpha D_m (\Gamma^+ - \Gamma^-) F v_i, \end{aligned}$$

so that the incompressibility equation (8) corresponds to:

$$\begin{aligned}0 =& D_h[v]\sum_i f_i\\ =& -gD_m(\Gamma^+ + \Gamma^-)E\sum_i \Lambda_i\rho_i e - (-D_m\Gamma^+ E\Lambda\nabla_h^+ - D_m\Gamma^- E\Lambda\nabla_h^-)p\\ &- \alpha D_m(\Gamma^+ - \Gamma^-)F\sum_i v_i,\end{aligned}$$

where $\Lambda = \sum_i \Lambda_i$. Taking into account that $\sum_i v_i = e$ and $\Gamma^{\pm}Fe = \frac{1}{2}e$ from (14) and (18), we can write this equation as

$$(-D_m\Gamma^+ E\Lambda\nabla_h^+ - D_m\Gamma^- E\Lambda\nabla_h^-)p = -gD_m(\Gamma^+ + \Gamma^-)E\sum_i \Lambda_i\rho_i e.$$

If we require that the matrix of this system be symmetric and positive semidefinite, then we have to take $\nabla^{\pm} = \nabla^{\pm}[v]$ to be proportional to $(-D_m\Gamma^{\pm}E)^T$. From (18), the right scaling yields:

$$\nabla^{\pm} = 2(-D_m\Gamma^{\pm}E)^T.$$

To summarize, the spatial semidiscretization of the problem can be written as

$$\begin{aligned}v_i' &= b_i[v] - A_i[v]p[v] - \alpha D_m(\Gamma^+ - \Gamma^-)Fv_i,\\ A[v]p[v] &= b[v], \quad A[v] = \sum_i A_i[v], \quad b[v] = \sum_i b_i[v],\end{aligned}$$

where

$$\begin{aligned}A_i[v] &= \nabla^+[v]^T\Lambda_i\nabla^+[v] + \nabla^-[v]^T\Lambda_i\nabla^-[v],\\ b_i[v] &= g\rho_i(\nabla^+[v] + \nabla^-[v])^T\Lambda_i[v]e,\end{aligned}$$

and

$$D_h[v] = (\nabla[v])^T = \left(\frac{\nabla^+[v] + \nabla^-[v]}{2}\right)^T.$$

The spatially-discretized scheme can be solved by an appropriate ODE solver. In this work we use Shu and Osher's TVD Runge-Kutta 3 method proposed in [17], that can be written as:

$$v_i^{(1)} = v_i^n - \Delta t \mathscr{D}(v_i^n),$$
$$v_i^{(2)} = \frac{3}{4}v_i^n + \frac{1}{4}v_i^{(1)} - \frac{1}{4}\Delta t \mathscr{D}\left(v_i^{(1)}\right),$$
$$v_i^{n+1} = \frac{1}{3}v_i^n + \frac{2}{3}v_i^{(2)} - \frac{2}{3}\Delta t \mathscr{D}\left(v_i^{(2)}\right),$$

where $\mathscr{D}(v)_i = b_i[v] - A_i[v]p[v] - \alpha D_m(\Gamma^+ - \Gamma^-)Fv_i$ and $v_i^n \approx v_i(t_n)$, $i = 1,\dots,N$.

4 Two-Dimensional Porous Media Flows

In this section we present the ideas for a two-dimensional extension of the scheme presented in the previous section for 1D. We show the results obtained by the component-wise WENO scheme.

For simplicity, we assume a Cartesian mesh (x_i, y_j) on $\Omega = (0,1)^2$, with $x_i = y_i = (i - \frac{1}{2})h$, $i = 1,\dots,m$, $h = 1/m$. Let us denote $f_i = (f_i^x, f_i^y)$, $g = (0, -9.81) = (g^x, g^y)$:

$$f_i^s = \lambda_i(u_i)\left(\rho_i g^s - \frac{\partial p}{\partial s}\right), \quad s = x, y.$$

Denote by α_x, α_y upper bounds of the eigenvalues of $(f^x)'$ and $(f^y)'$, respectively.

We order the nodes in (Cartesian) row major order, i.e. the node (x_i, y_j) is at position $\overline{(i,j)} := (j-1)m + i$.

From an extended $(m+2r)\times(m+2r)$ matrix v, we define $m(m+1)\times m(m+2r)$ matrices $\Gamma^{x,\pm}$, $\Gamma^{y,\pm}$ whose nonzero entries are given by

$$\Gamma[v]^{x,\pm}_{\overline{(j+1,k)},\overline{(j+l+r,k)}} = \Gamma[v]^{x,\pm}_{j+1+(k-1)(m+1),j+l+r+(k-1)(m+2r)} =$$
$$= \frac{1}{2}\gamma_l^{\pm}(v_{\mathscr{S}+j,k}); \quad j = 0,\dots,m; \quad k = 1,\dots,m;$$
$$\Gamma[v]^{y,\pm}_{\overline{(j,k)},\overline{(j,k+l+1)}} = \Gamma[v]^{y,\pm}_{j+km,j+(k+l+1)m} = \frac{1}{2}\gamma_l^{\pm}(v_{j,\mathscr{S}+k})$$
$$j = 1,\dots,m; \quad k = 0,\dots,m$$

with $l = -r+1,\dots,r$.

The two-dimensional extension of the proposal for the ∇ operators is

$$\nabla^{x,\pm}[v] = 2(-D_m^x \Gamma^{x,\pm}[v]E^x)^T, \quad \nabla^{y,\pm}[v] = 2(-D_m^y \Gamma^{y,\pm}[v]E^y)^T,$$

where

$$D_m^x = I_m \otimes D_m, \quad E^x = I_m \otimes E, \quad D_m^y = D_m \otimes I_m, \quad E^y = E \otimes I_m.$$

Analogously, we define:

$$F^x = I_m \otimes F, \quad F^y = F \otimes I_m.$$

These matrices E^x, E^y, F^x, F^y correspond to the two-dimensional extension of the boundary conditions already shown in Sect. 3 for the one-dimensional case. The boundary conditions for the x-component of the fluxes are $f_i^x(0,y) = f_i^x(1,y) = 0$, then we have:

$$\begin{aligned} f^x_{i;-j,k} &= -f^x_{i;j+1,k}, \\ f^x_{i;m+j+1,k} &= -f^x_{i;m-j,k}, \quad j = 0,\dots,r-1. \end{aligned}$$

The boundary conditions for the y-component of the fluxes are $f_i^y(x,0) = f_i^y(x,1) = 0$. Then we have

$$\begin{aligned} f^y_{i;j,-k} &= -f^y_{i;j,k+1}, \\ f^y_{i;j,m+k+1} &= -f^y_{i;j,m-k}, \quad k = 0,\dots,r-1. \end{aligned}$$

For the saturations we use outflow boundary conditions:

$$\begin{aligned} u_{i;-j,k} &= u_{i;j+1,k}, \\ u_{i;m+j+1,k} &= u_{i;m-j,k}, \quad j = 0,\dots,r-1, \\ u_{i;j,-k} &= u_{i;j,k+1}, \\ u_{i;j,m+k+1} &= u_{i;j,m-k}, \quad k = 0,\dots,r-1. \end{aligned}$$

Therefore, the spatial semi-discretization for $v_{i;\overline{(j,k)}}(t) \approx u_i(x_j, y_k, t)$ is given by

$$\begin{aligned} v_i' &= b_i[v] - A_i[v]p[v] - \alpha_x D_m^x(\Gamma^{x,+} - \Gamma^{x,-})F^x v_i - \alpha_y D_m^y(\Gamma^{y,+}[v] - \Gamma^{y,-}[v])F^y v_i, \\ A[v]p[v] &= b[v], \quad A[v] = \sum_i A_i[v], \quad b[v] = \sum_i b_i[v], \end{aligned} \tag{20}$$

where

$$\begin{aligned} A_i[v] &= \sum_{\text{sign}=\pm} \sum_{\text{var}=x,y} \nabla^{\text{var,sign}}[v]^T \Lambda_i[v] \nabla^{\text{var,sign}}[v], \\ b_i[v] &= \rho_i \sum_{\text{sign}=\pm} \sum_{\text{var}=x,y} g^{\text{var}} \nabla^{\text{var,sign}}[v]^T \Lambda_i[v] e, \end{aligned}$$

with the diagonal $m^2 \times m^2$ matrix $\Lambda_i[v]$ given by

$$\Lambda_i[v]_{\overline{(j,k)},\overline{(j,k)}} = \lambda_i(v_{i;j,k}).$$

5 Numerical Results

5.1 One-Dimensional Experiments

To test the numerical results obtained with the newly developed scheme, we consider the same test problem as in [9, 11], which corresponds to a simulation associated to a problem involving water filtration in a vertical column of porous soil containing oil, gas and water. The initial conditions for the simulation are as follows:

$$u = (u_w, u_g, u_o) = \begin{cases} (1,0,0), & 0 \leq x \leq 0.5; \\ (0.2,0.2,0.6), & 0.5 < x \leq 1.0. \end{cases}$$

The subindices w, g and o refer to water, gas and oil respectively. The viscosities and densities considered along all the numerical tests are $(\mu_w, \mu_g, \mu_o) = (1,1,1)$ and $(\rho_w, \rho_g, \rho_o) = (1, 0.0012, 0.85)$ respectively.

The numerical solutions have been obtained with WENO3 reconstructions and the third-order TVD-Runge-Kutta ODE solver, with a mesh with $m = 512$ nodes. The results obtained by our proposed numerical scheme are compared with those obtained by the scheme developed in [9], in which the pressure gradient is obtained by an analytical expression in terms of the capillary pressures and the relative mobilities of the different fluid phases.

As it can be appreciated in Figs. 1 and 2 the numerical solution obtained when using our proposed scheme, labeled WENO3-RK3, is very similar to the solution obtained with the scheme by Donat et al. [9], labeled DGM 2014, even near the boundaries and sharp profiles.

5.2 Two-Dimensional Experiments

Because of Neumann boundary conditions, the equation $A[v]p[v] = b[v]$ in (20) has many solutions, differing only in the addition of a constant, and therefore they all give the same discretized gradient. One can see that the system resulting of removing one equation and setting to zero the corresponding unknown yields a solution of the original system. We have used the conjugate gradient method with an ILU(0) preconditioner [10] on this reduced system. We are aware that more sophisticated and efficient alternatives (such as those based on multigrid techniques) could be applied, but this is out of the scope of this contribution.

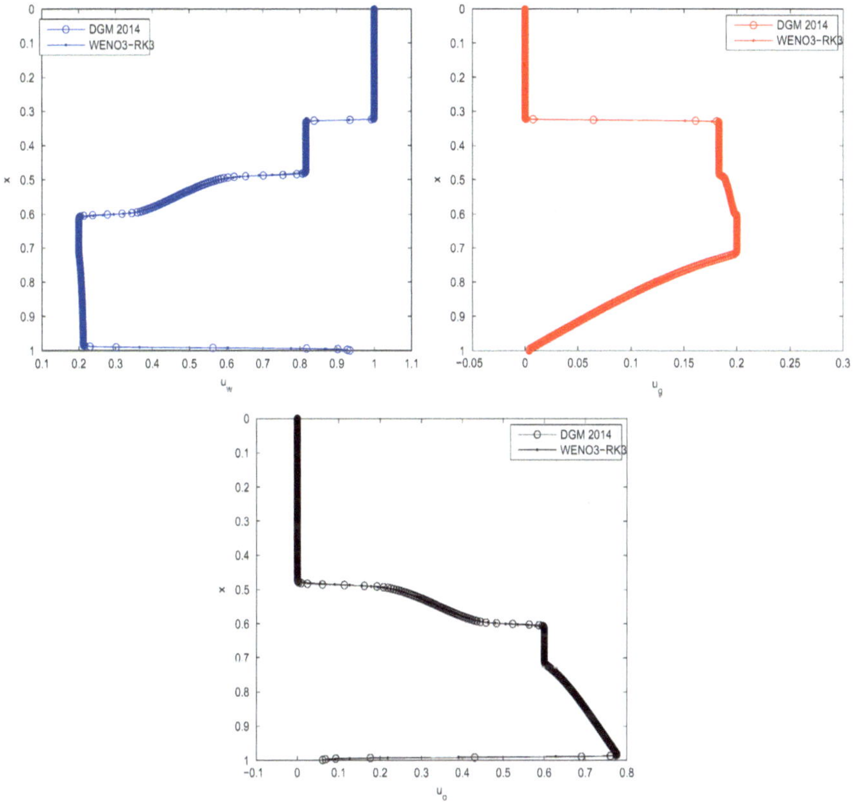

Fig. 1 Water, gas and oil saturation numerical solutions at time $T = 1.0$, $m = 512$, $k/h = 0.5$

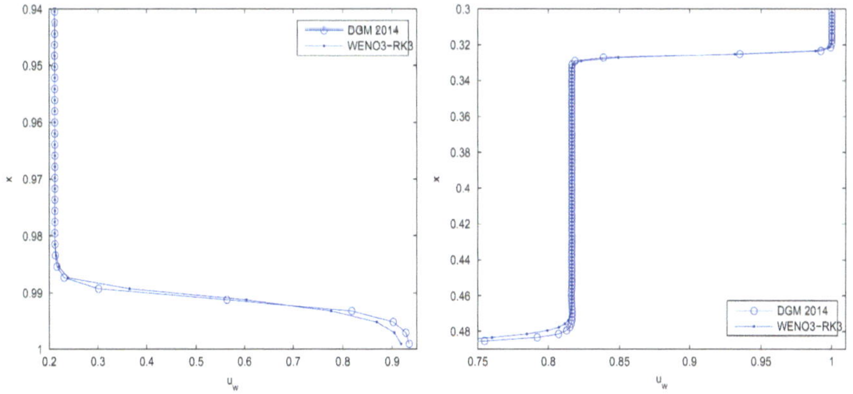

Fig. 2 Enlarged regions of the water saturation numerical solution at time $T = 1.0$, $m = 512$, $k/h = 0.5$

5.2.1 Test Problem 1

To test our two-dimensional WENO scheme we propose the next experiment. It represents the two dimensional version of the test problem shown in [9, 11]. Here $y = 0$ represents the top of the domain and $y = 1$ its bottom.

The initial conditions for water, gas and oil saturations on the domain $\Omega = (0, 1)^2$ are:

$$u = (u_w, u_g, u_o)(x, y) = \begin{cases} (1, 0, 0) & (x, y) \in [0.25, 0.75] \times [0, 0.5] \\ (0.2, 0.2, 0.6) & (x, y) \notin [0.25, 0.75] \times [0, 0.5]. \end{cases}$$

Figure 3 shows the evolution over time given by our two-dimensional WENO scheme. We are using WENO3 with the values of the numerical viscosities $\alpha_x = \alpha_y = 0.8$ empirically determined in order to obtain stable solutions, but not too softened near discontinuities.

The global behavior of the numerical solution seems to be physically correct. Water flows downwards with a widening of the region dominated by water saturation as time increases and an accumulation in the bottom of the domain. We can also observe how the gas is accumulating very fast at the top of the domain as time increases. The behavior of the oil is slightly different: it flows downwards near the top and upwards near the bottom of the domain.

In Table 1 we show the computational times needed to obtain the numerical solution at $T = 1$ for some mesh sizes $m \times m$. A logarithmic least squares adjustment applied to the table yields that the computational time is $\approx 10^{-5.33} m^{3.77}$. Let us analyze this asymptotic computational time: The number of computations in each time step on an $m \times m$ spatial grid is $\mathcal{O}(m^2)+$ the cost of solving the sparse $m^2 \times m^2$ system of linear equations appearing in (20). The computational cost of a direct solver on this system with bandwidth m is $\mathcal{O}(m^3)$ and, on the other side, the computational cost of a multigrid solver could be not lower than $\mathcal{O}(m^2)$. Since there are $\mathcal{O}(m)$ time steps, the computational cost with direct solves would be $\mathcal{O}(m)\big(\mathcal{O}(m^2) + \mathcal{O}(m^3)\big) = \mathcal{O}(m^4)$, whereas for multigrid techniques could be as low as $\mathcal{O}(m)\big(\mathcal{O}(m^2) + \mathcal{O}(m^2)\big) = \mathcal{O}(m^3)$. The exponent 3.77 in the previous adjustment suggests that there is some room for improving the performance of the elliptic solver.

Table 1 Computation times for the WENO3 scheme for test problem 1 at $T = 1.0$

m	CPU time (s)
50	12.9
100	142.5
200	2109.2
400	31, 955.8

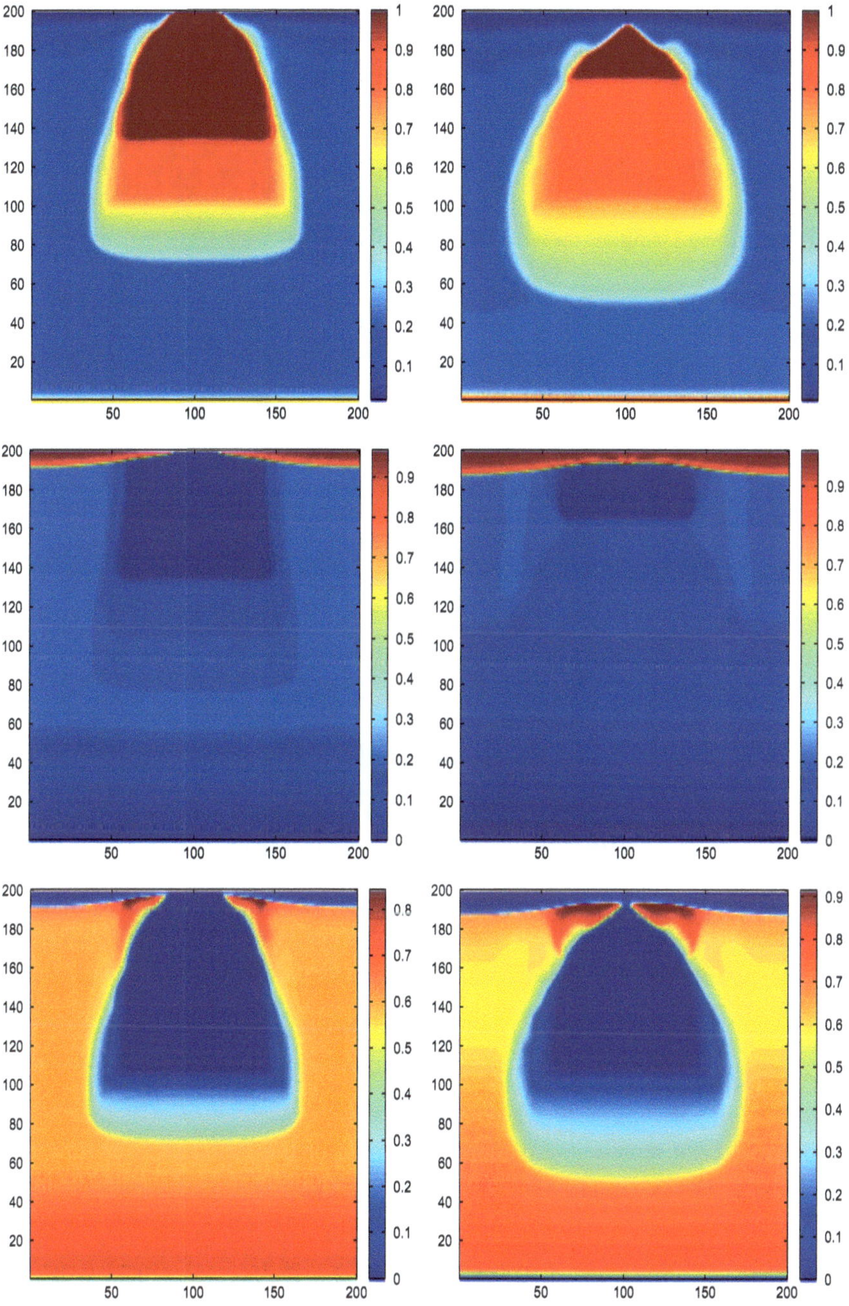

Fig. 3 Water, gas and oil saturations (from *top to bottom*) numerical solution for the test problem 1 at time $T = 1.0$ (*left*) and $T = 2.0$ (*right*), $m = 200$, $k/h = 0.5$

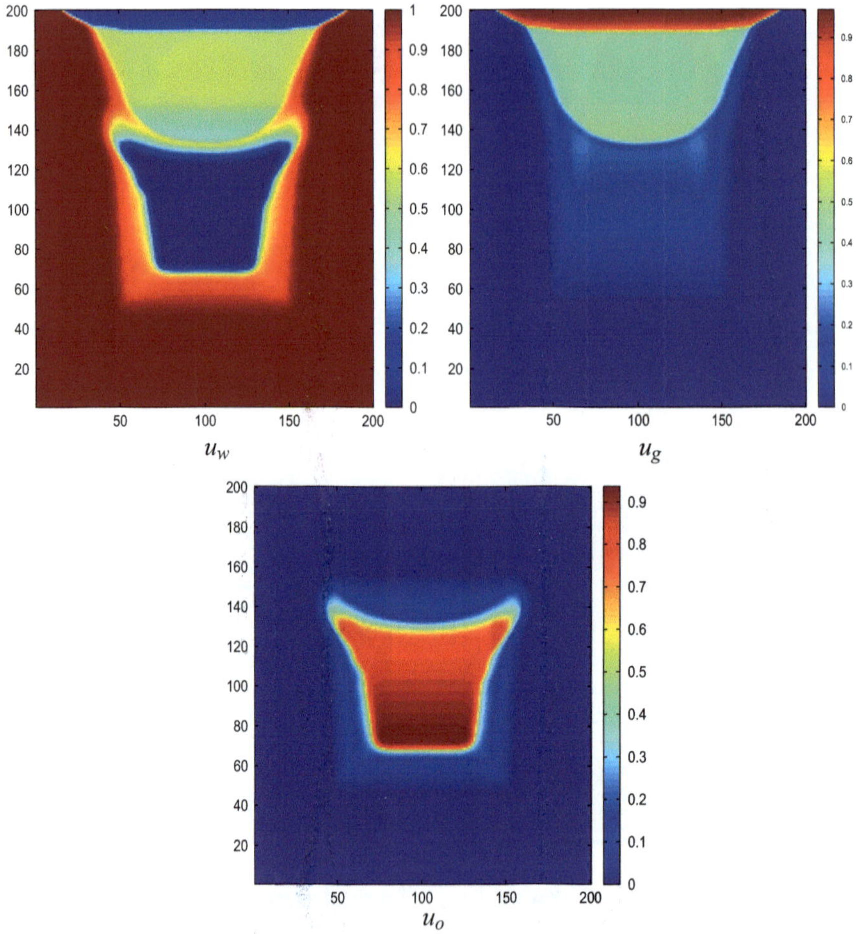

Fig. 4 Water, gas and oil saturations numerical solution for the test problem 2 at time $T = 1.0$, $m = 200$, $k/h = 0.5$

5.2.2 Test Problem 2

This experiment represents a simplified version of the evolution over time of a bubble of gas and oil trapped in the middle of an aquifer with pure water. The initial conditions are:

$$u = (u_w, u_g, u_o)(x, y) = \begin{cases} (0, 0.5, 0.5) & (x, y) \in [0.25, 0.75] \times [0.25, 0.75] \\ (1, 0, 0) & (x, y) \notin [0.25, 0.75] \times [0.25, 0.75]. \end{cases}$$

As in test problem 1, we are using WENO3 with $\alpha_x = \alpha_y = 0.8$.

The evolution over time is shown in Fig. 4 for $T = 1$. The main behavior observed is that of the gas. We observe how the gas leaves the bubble very soon reaching the top of the domain and accumulating there. On the other hand, oil seems to concentrate in the intermediate region and water occupies the regions left by gas and oil.

6 Conclusions

In this paper we have presented a numerical scheme that uses finite-difference WENO schemes for the simulation of multi-dimensional multi-phase flow problems in vertical equilibrium in a homogeneous porous medium. We have presented the derivation of the scheme in the one-dimensional setting using WENO-based numerical fluxes that yields divergence-free numerical fluxes not only on the continuous setting but also on the discrete setting. We have showed some one-dimensional numerical results to show that the behavior of the numerical solutions obtained with the WENO-based fluxes is physically correct and very close to the solution obtained with a scheme using an analytical expression of the pressure gradient.

We have presented a two-dimensional extension of the scheme showing that the numerical results obtained seem to be as good as the ones obtained in the one-dimensional setting, physically speaking, and preserving the divergence-free character of the continuous model equations.

However, this is a work in progress and there are some aspects that need to be improved: the computational time needed to obtain the numerical solutions needs to be diminished to test the scheme with finer meshes and we have to determine appropriate values for the numerical viscosities, α_x and α_y in the Lax-Friedrichs splitting, for both f^x and f^y fluxes.

Finally, the extension of the scheme to deal with non-zero capillary pressures, applying IMEX schemes, is currently under investigation.

Acknowledgements This research was partially supported by Ministerio de Economía y Competitividad under grant MTM2011-22741 and MTM2014-54388-P with the participation of FEDER. M.C. Martí and R. Bürger acknowledge support by CONICYT Postdoctoral 2015 Fondecyt project 3150140. R. Bürger acknowledges support by Fondecyt project 1130154; Conicyt project Anillo ACT1118 (ANANUM); Red Doctoral REDOC.CTA, MINEDUC project UCO1202 at Universidad de Concepción; BASAL project CMM, Universidad de Chile and Centro de Investigación en Ingeniería Matemática (CI^{2}MA), Universidad de Concepción; and Centro CRHIAM Proyecto Conicyt Fondap 15130015.

References

1. Aarnes, J., Kippe, V., Lie, K.A., Rustad, A.B.: Modelling of multiscale structures in flow simulations for petroleum reservoirs. In: Hasle, G., Lie, K.A. (eds.) Geometric Modeling, Numerical Simulation and Optimization: Applied Mathematics at SINTEF. Springer, Berlin (2007)

2. Aziz, K., Settari, A.: Petroleum Reservoir Simulations. Applied Science Publishers, London (1979)
3. Baeza, A., Mulet, P., Zorío, D.: High order boundary extrapolation technique for finite difference methods on complex domains with cartesian meshes. J. Sci. Comput. **66**(2), 761–791 (2016). doi:10.1007/s10915-015-0043-2
4. Chen, Z., Huan, G., Ma, Y.: Computational Methods for Multiphase Flows in Porous Media. Society for Industrial and Applied Mathematics, Philadelphia (2006)
5. Christie, M.A.: Upscaling of reservoir simulation. J. Pet. Technol. **48**(11), 1004–1010 (1996)
6. Cunha, M.C.C., Santos, M.M., Bonet, J.E.: Buckley-Leverett mathematical and numerical models describing vertical equilibrium process in porous media. Int. J. Eng. Sci. **42**, 1289–1303 (2004)
7. Darcy, H.: Les fontaines publiques de la ville de Dijon. Dalmont, Paris (1856)
8. Donat, R., Guerrero, F., Mulet, P.: IMEX WENO schemes for two-phase flow vertical equilibrium processes in a homogeneous porous medium. Appl. Math. Inf. Sci. **7**(5), 1865–1878 (2013)
9. Donat, R., Guerrero, F., Mulet, P.: Implicit-Explicit methods for models for vertical equilibrium multiphase flow. Comput. Math. Appl. **68(3)**, 363–383 (2014)
10. Golub, G.H., van Loan, C.F.: Matrix Computations, 4th edn. Johns Hopkins University Press, Baltimore, MD (2013)
11. Guerrero, F., Donat, R., Mulet, P.: Solving a model for 1-D three phase flow vertical equilibrium processes in a homogeneous porous medium by means of a weighted essentially non oscillatory numerical scheme. Comput. Math. Appl. **66**, 1284–1298 (2013)
12. Jiang, G.-S., Shu, C.-W.: Efficient implementation of weighted ENO schemes. J. Comput. Phys. **126**(1), 202–228 (1996)
13. Juanes, R.: Displacement theory and multiscale numerical modeling of three-phase flow in porous media. Ph.D. Thesis, University of Berkeley (2003)
14. Levy, D., Puppo, G., Russo, G.: Central weno schemes for hyperbolic systems of conservation laws. Math. Model. Numer. Anal. **33**, 547–571 (1999)
15. Liu, X.D., Osher, S., Chan, T.: Weighted essentially non-oscillatory schemes. J. Comput. Phys. **115**(1), 200–212 (1994)
16. Peaceman, D.W.: Fundamentals of Numerical Reservoir Simulation. Elsevier, New York (1977)
17. Shu, C.-W., Osher, S.: Efficient implementation of essentially non-oscillatory shock-capturing schemes, ii. J. Comput. Phys. **83**, 32–78 (1989)

Time Dependent Scattering in an Acoustic Waveguide Via Convolution Quadrature and the Dirichlet-to-Neumann Map

Li Fan, Peter Monk, and Virginia Selgas

Abstract We propose to use finite elements and BDF2 time stepping to solve the problem of computing a solution to the time dependent wave equation with a variable sound speed in an infinite sound hard pipe (waveguide). By using the Laplace transform and an appropriate Dirichlet-to-Neumann (DtN) map for the problem, we can prove that this problem can be reduced to a variational problem on a bounded domain that has a unique solution. This solution can be discretized in space using finite elements (projecting into a Fourier space on the two artificial boundaries to allow the rapid calculation of the DtN map). We discretize in time using the Convolution Quadrature (CQ) approach and in particular BDF2 time-stepping. Thanks to CQ we obtain a stable and convergent discretization of the DtN map, and hence of the fully discrete BDF2-finite element scheme without a CFL condition. We illustrate the method with some numerical results.

1 Introduction

Simulating sound propagation in pipes (also called waveguides) requires to solve the wave equation in a sound hard acoustic waveguide. In this paper we consider the use of a finite element time domain approach to the problem. We suppose that the waveguide encloses a bounded perturbation assumed to be a region in which the sound speed differs from the background speed in the rest of the waveguide. We refer to the perturbation as the scatterer. A sound wave is incident on this perturbation and produces a scattered wave that needs to be computed. For simplicity we will work in two spatial dimensions, but the algorithm we develop can be used for a true three dimensional pipe with obvious modifications.

L. Fan • P. Monk
Department of Mathematical Sciences, University of Delaware, Newark, DE 19716, USA
e-mail: fanli0218@gmail.com; monk@udel.edu

V. Selgas (✉)
Departamento de Matemáticas, Universidad de Oviedo, EPIG, 33203 Gijón, Spain
e-mail: selgasvirginia@uniovi.es

F. Ortegón Gallego et al. (eds.), *Trends in Differential Equations and Applications*, SEMA SIMAI Springer Series 8, DOI 10.1007/978-3-319-32013-7_18

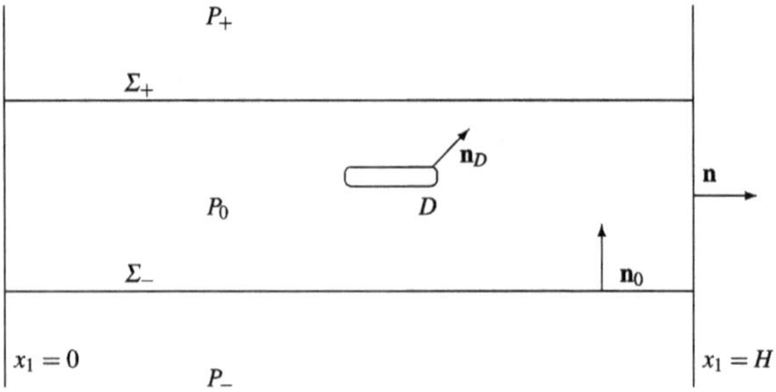

Fig. 1 Cartoon of the main geometric elements used in our analysis. Finite elements are used in the domain P_0 which includes the scatterer D. The artificial boundaries are Σ_- and Σ_+

Let us consider a waveguide $P = (0, H) \times \mathbb{R}$, containing an obstacle D which is assumed to be bounded and have a Lipschitz continuous boundary. Denote by $\mathbf{n}$ the unit outward normal on ∂P, i.e. $\mathbf{n} = (-1, 0)$ on $x_1 = 0$ and the opposite on $x_1 = H$. Similarly, we use the notation $\mathbf{n}_D$ for the unit outward normal on ∂D. Figure 1 shows a graphic of the computational domain.

The refractive index $\mathrm{n}(\mathbf{x})$ is assumed to be real and frequency independent, and such that $\mathrm{n}(\mathbf{x}) = 1$ if $\mathbf{x} \in P \setminus \overline{D}$ and $\mathrm{n}(\mathbf{x}) \neq 1$ if $\mathbf{x} \in D$. Later in the paper we will comment on impenetrable scatterers and frequency dependent coefficients. The speed of sound in the background waveguide outside D is a constant c_0.

We suppose that a given incident field u_{inc} hits the scatterer. The incident field is a bounded smooth solution of the background wave equation so that it satisfies the wave equation in the free waveguide:

$$\begin{aligned} \frac{1}{c_0^2}\partial_{tt}^2 u_{inc} &= \Delta u_{inc} \quad \text{in } \mathbb{R} \times P\,, \\ \partial_{\mathbf{n}} u_{inc} &= 0 \quad \text{on } \mathbb{R} \times \partial P\,, \end{aligned}$$

where ∂_{tt}^2 denotes the second time derivative, and $\partial_{\mathbf{n}}$ denotes the normal derivative. The boundary condition models a sound hard wall. In the sequel we assume that the incident field u_{inc} does not hit the scatterer D before $t = 0$, that is,

$$u_{inc} = \partial_t u_{inc} = \partial_{tt}^2 u_{inc} = 0 \quad \text{in } \overline{D}, \text{ for } t \leq 0\,. \tag{1}$$

In the time domain, the wave equation and boundary conditions for the total wave u and the scatterer field u_{sc} are

$$\begin{aligned} \frac{\mathrm{n}^2}{c_0^2}\partial_{tt}^2 u &= \Delta u \quad \text{in } P, \text{ for } t > 0\,, \\ u &= u_{inc} + u_{sc} \quad \text{in } P, \text{ for } t > 0\,, \\ \partial_{\mathbf{n}} u &= 0 \quad \text{on } \partial P, \text{ for } t > 0\,, \\ u &= 0 \quad \text{in } P, \text{ at } t = 0, \\ \partial_t u &= 0 \quad \text{in } P, \text{ at } t = 0. \end{aligned} \tag{2}$$

Here n is understood to be a function of positions $\mathbf{x}$. There is no need for a condition at infinity because u_{sc} propagates with finite velocity and so for any $t > 0$ there is a distance $M(t)$ such that $u_{sc}(t,\mathbf{x}) = 0$ for any $\mathbf{x} = (x_1, x_2)$ with $x_1 \in (0, H)$, $|x_2| > M(t)$. Our problem is to approximate u (or equivalently u_{sc}) and we shall use finite elements in space because they can approximate general boundaries of the scatterer easily.

To use finite elements we introduce a computational domain $P_0 := (0,H)\times(0,L)$ for $L > 0$ big enough to enclose the scatterer, that is, such that $\overline{D} \subset P_0$ (see Fig. 1). Then we can cover P_0 with finite elements (in our case using triangles). The only obstacle to a standard finite element approach in space is the need for a special artificial boundary condition at $x_2 = 0$ and $x_2 = L$ that takes care of the infinite waveguide on either side of P_0. This can be constructed using the Perfectly Matched Layer (PML) (see [9]); in fact, provided the PML is chosen to handle both traveling and evanescent modes in the solution this can be very successful. However the PML is difficult to analyze and requires an informed choice of the PML parameters so instead we propose to use a time domain Dirichlet-to-Neumann (DtN) map on the artificial interfaces $x_2 = 0$ and $x_2 = L$ following the approach of [6]. With this approach we need to store the solution on the artificial boundaries for all time steps, but, at least at low frequencies and in two dimensions this is not a crushing problem since only a few modes need to be stored for each time step.

We propose to use implicit time stepping to take care of possible refined meshes in some regions of the simulations, as well as to allow for changes in refractive index from place to place (this would change the CFL of an explicit scheme from place to place). In particular we shall use the Laplace transform to analyze the truncated problem (cf. [1, 7]) and convolution quadrature (cf. [10]) to prove that a family of time stepping schemes including Backward Differentiation Formula 2 (BDF2) give rise to stable and convergent time stepping method. An added bonus is that, at least before spatial discretization, the method shows how to construct perfect discrete DtN maps matched to the time stepping scheme. An alternative approach using more standard time stepping and integral equations on the interfaces might be constructed along the line of [3], but we do not pursue that here.

The paper proceeds as follows. In the next section we give details of how to reduce the problem to a family of Laplace domain equations posed on the computational domain P_0. Then in Sect. 3 we summarize the analysis of the Laplace domain problems, and then relate these back to a fully discrete time stepping scheme using convolution quadrature. The fully discrete scheme is shown to be optimally convergent. Then in Sect. 4 we provide a few numerical results from our method implemented using the multi-frequency approach of Banjai and Sauter proposed in [2].

2 Reduction to a Bounded Domain

It is convenient to perform the initial analysis using the scattered field $u_{sc} = u - u_{inc}$ which satisfies

$$\frac{\mathrm{n}^2}{c_0^2}\partial_{tt}^2 u_{sc} = \Delta u_{sc} + F \quad \text{in } P, \text{ for } t > 0\,,$$

$$\partial_{\mathbf{n}} u_{sc} = 0 \quad \text{on } \partial P, \text{ for } t > 0. \tag{3}$$

Above, we have set

$$F = \frac{1}{c_0^2}(1 - \mathrm{n}^2(\mathbf{x}))\partial_{tt}^2 u_{inc}\,.$$

Notice that $F = 0$ outside D, for any $t \in \mathbb{R}$, since $\mathrm{n}(\mathbf{x}) = 1$ there. Furthermore, $F = 0$ in the whole P_0 for any $t \leq 0$ according to (1). On the other hand, (1) also suggests that the scattered field is causal and, hence, we impose the initial conditions

$$u_{sc} = \partial_t u_{sc} = 0 \quad \text{in } P, \text{ at } t = 0.$$

In order to analyze (3), we transform it to the Laplace domain. More precisely, for any smooth and causal function $f(t)$, the Laplace transform we use is defined as

$$\mathscr{L}[f](s) = \int_0^\infty f(t)\exp(-st)\,dt \quad \text{for } s \in \mathbb{C}_\sigma\,,$$

where $\mathbb{C}_\sigma = \{s;\ s = \eta - i\omega \text{ with } \eta > \sigma\eta, \omega \in \mathbb{R}, \eta > \sigma,\ \omega \in \mathbb{R}\}$ for a fixed positive $\sigma \in \mathbb{R}$. Then, working formally with Eq. (3), we have that the Laplace transform of the scattered field, $\hat{u}_{sc} = \mathscr{L}[u_{sc}](s)$, solves

$$\frac{s^2\mathrm{n}^2}{c_0^2}\hat{u}_{sc} = \Delta\hat{u}_{sc} + \hat{F} \quad \text{in } P\,, \tag{4}$$

$$\partial_{\mathbf{n}}\hat{u}_{sc} = 0 \quad \text{on } \partial P\,. \tag{5}$$

Above, $\hat{F}$ stands for the Laplace transform of F; let us recall that $\hat{F} = 0$ in $P \setminus \overline{D}$.

As already mentioned, we make use of a bounded section of the pipe $P_0 = (0, H) \times (0, L)$ containing the scatterer D in its interior (see Fig. 1). Then, adopting the usual Galerkin strategy, the problem in P_0 consists in finding $\hat{u}_{sc} \in H^1(P_0)$ such that, for any $v \in H^1(P_0)$,

$$\int_{P_0}\frac{\mathrm{n}^2}{c_0^2}s^2\hat{u}_{sc}\,\overline{v} + \int_{P_0}\nabla\hat{u}_{sc}\cdot\nabla\overline{v} - \int_{\Sigma_+}\partial_{\mathbf{n}_0}\hat{u}_{sc}\,\overline{v} + \int_{\Sigma_-}\partial_{\mathbf{n}_0}\hat{u}_{sc}\,\overline{v} = \int_{P_0}\hat{F}\,\overline{v}\,, \tag{6}$$

where the unit vector $\mathbf{n}_0 = (0, 1)$ is normal to the artificial boundaries $\Sigma_- = (0, H) \times \{0\}$ and $\Sigma_+ = (0, H) \times \{L\}$.

In order to deal with the integrals on $\Sigma_\pm$, we next define the DtN maps $\hat{T}^s_\pm(\hat{u}_{sc}) = \pm\partial_{\mathbf{n}_0}\hat{u}_{sc}$ on $\Sigma_\pm$. Working in the remaining parts of the pipe $P \setminus \overline{P_0}$, where we have a homogeneous wave equation, we may obtain explicit expressions of the DtN maps.

More precisely, let us start by considering $P_- = (0, H) \times (-\infty, 0)$. Then $\hat{u}_{sc} \in H^1(P_-)$ satisfies

$$\frac{s^2}{c_0^2}\hat{u}_{sc} = \Delta\hat{u}_{sc} \quad \text{in } P_- , \tag{7}$$

$$\partial_{\mathbf{n}}\hat{u}_{sc} = 0 \quad \text{on } \partial P_- \setminus \Sigma_- . \tag{8}$$

Taking advantage of Eq. (8), we write the scattered field in P_- as

$$\hat{u}_{sc}(x_1, x_2) = \sum_{m=0}^{\infty} u_m(x_2) \cos\left(\frac{m\pi x_1}{H}\right) \quad \text{in } P_- , \tag{9}$$

where each $u_m(x_2)$ is bounded for $x_2 \to -\infty$. Then, Eq. (7) means that

$$-(u_m)'' + \frac{m^2\pi^2}{H^2}u_m + \frac{s^2}{c_0^2}u_m = 0 \quad \text{for } x_2 < 0 . \tag{10}$$

Also notice that $u_m(0) = u_{m,0}$, where $\{u_{m,0}\}_{m=0}^{\infty}$ are the complex Fourier expansion coefficients of $\hat{u}_{sc}$ on Σ_-:

$$\hat{u}_{sc} = \sum_{m=0}^{\infty} u_{m,0} \cos\left(\frac{m\pi x_1}{H}\right) \quad \text{on } \Sigma_- .$$

In consequence, denoting

$$\kappa_m \equiv \kappa_m(s) = \frac{s}{c_0}\sqrt{1 + \frac{m^2\pi^2}{H^2}\frac{c_0^2}{s^2}} , \tag{11}$$

and choosing $\Re(\kappa_m) > 0$ we have

$$\hat{u}_{sc}(x_1, x_2) = \sum_{m=0}^{\infty} u_{m,0} \cos\left(\frac{m\pi x_1}{H}\right) \exp(\kappa_m x_2) \quad \text{in } P_- . \tag{12}$$

In particular, it follows that

$$\partial_{\mathbf{n}_0}\hat{u}_{sc} = \partial_{x_2}\hat{u}_{sc} = \sum_{m=0}^{\infty} \kappa_m u_{m,0} \cos\left(\frac{m\pi x_1}{H}\right) \quad \text{on } \Sigma_- . \tag{13}$$

Summing up, we have the following explicit expression of the DtN map on Σ_-:

$$\hat{T}^s_- \xi = -\sum_{m=0}^{\infty} \kappa_m(s)\, \xi_m \cos\big(\frac{m\pi x_1}{H}\big) \quad \text{on } \Sigma_-\,, \tag{14}$$

for any ξ whose Fourier expansion on Σ_- is

$$\xi = \sum_{m=0}^{\infty} \xi_m \cos\big(\frac{m\pi x_1}{H}\big)\,, \tag{15}$$

where ξ_m $(m = 0, \ldots, \infty)$ are the complex expansion coefficients.

Similarly, we can work in P_+ to obtain an explicit expression of $\hat{T}^s_+$ on Σ_+. We now make use of the DtN maps to rewrite the variational formulation (6) of the model problem in the Laplace domain as follows: Find $\hat{u}_{sc} \in H^1(P_0)$ such that, for any $v \in H^1(P_0)$,

$$\int_{P_0} \frac{\mathrm{n}^2}{c_0^2} s^2 \hat{u}_{sc}\, \overline{v} + \int_{P_0} \nabla \hat{u}_{sc} \cdot \nabla \overline{v} - \int_{\Sigma_+} \hat{T}^s_+ \hat{u}_{sc}\, \overline{v} - \int_{\Sigma_-} \hat{T}^s_- \hat{u}_{sc}\, \overline{v} = \int_{P_0} \hat{F}\, \overline{v}\,. \tag{16}$$

3 Convergence Analysis

The analysis of existence, uniqueness and finite element convergence for the variational problem in the Laplace domain (16) follows the general steps of the analysis of the periodic grating problem in [6]. According to this, we only make an outline of the most important results of such analysis. To this end, we start introducing the following s-dependent norm on $H^1(P_0)$:

$$\|v\|_{s,H^1(P_0)} = \Big(\int_{P_0} \big(\frac{|s|^2}{c_0^2}|v|^2 + |\nabla v|^2\big)\, d\mathbf{x}\Big)^{1/2} \quad \text{for } v \in H^1(P_0)\,.$$

For each $r \in [0, 1]$, we define the following s-dependent norm on $H^r(\Sigma_\pm)$:

$$\|\xi\|_{s,H^r(\Sigma_\pm)} = \Big(\sum_{m=0}^{+\infty} \big(\frac{|s|^2}{c_0^2} + \frac{m^2\pi^2}{H^2}\big)^r |\xi_m|^2\Big)^{1/2},$$

for any $\xi \in H^r(\Sigma_\pm)$ written by means of its Fourier expansion (15). We also define the associated s-dependent norm on $H^{-r}(\Sigma_\pm)$ by duality.

Notice that the s-dependent $H^1(P_0)$ norm corresponds to a weighted energy for the field after inverse Laplace transforming back to the time domain. Besides, the s-dependent boundary norm on $H^r(\Sigma_\pm)$ is chosen so that both the trace of functions in $H^1(P_0)$ and the DtN maps can be estimated by appropriate norm bounds with explicit s-independence, as we detail in the following subsection.

3.1 Well-Posedness of the Variational Problem in the Laplace Domain

Following directly the argument in [6, Lemma 2.1] we can show the following bound of the trace operator $\gamma_{\Sigma_\pm} : H^1(P_0) \to H^{1/2}(\Sigma_\pm)$ in terms of weighted norms:

$$\|\gamma_{\Sigma_\pm} v\|_{s,H^{1/2}(\Sigma_\pm)} \le C_1 \|v\|_{s,H^1(P_0)} \quad \text{for } v \in H^1(P_0)\,,$$

where $C_1 = 2\sqrt{\frac{2c_0}{LH\sigma}}$.

Moreover, using the Fourier definition of the DtN operator (14) and reasoning as in the proof of [6, Lemma 2.2], we deduce the following bound of the DtN operators $\hat{T}^s_\pm : H^{1/2}(\Sigma_\pm) \to H^{-1/2}(\Sigma_\pm)$ in terms of weighted norms:

$$\|\hat{T}^s_\pm \xi\|_{s,H^{-1/2}(\Sigma_\pm)} \le C\|\xi\|_{s,H^{1/2}(\Sigma_\pm)} \quad \text{for } \xi \in H^{1/2}(\Sigma_\pm)\,,$$

where C is independent of s.

We can now analyze the variational formulation of the Laplace domain problem on P_0 applying the Lax-Milgram Lemma. To this end, we define the s-dependent sesquilinear form $a^s : H^1(P_0) \times H^1(P_0) \to \mathbb{C}$ associated to the variational formulation (16), given by

$$a^s(w,v) := \int_{P_0} \frac{\mathrm{n}^2}{c_0^2} s^2\, w\,\overline{v} + \int_{P_0} \nabla w \cdot \nabla \overline{v} - \int_{\Sigma_+} \hat{T}^s_+ w\,\overline{v} - \int_{\Sigma_-} \hat{T}^s_- w\,\overline{v}\,.$$

Let us emphasize the following properties of the sesquilinear form $a^s(\cdot,\cdot)$ in terms of $s \in \mathbb{C}_\sigma$:

- By using the definition of the s-dependent $H^1(P_0)$ norm, and the bounds on the trace operator and the DtN maps, we have the following continuity bound:

$$|a^s(w,v)| \le C_2 \|w\|_{s,H^1(P_0)} \|v\|_{s,H^1(P_0)}\,, \tag{17}$$

  where $C_2 = \max\{1, \|\mathrm{n}^2\|_{L^\infty(P_0)}\} + 8\frac{c_0}{LH\sigma}$.
- Using Bamberger and HaDuong's technique [1] as in the proof of [6, Lemma 3.1], we have the following coercivity bound in terms of s-dependent norms

$$\Re(a^s(v,sv)) \ge \sigma \inf_{\mathbf{x}\in P_0} \mathrm{n}^2(\mathbf{x})\, \|v\|^2_{s,H^1(P_0)}\,. \tag{18}$$

Notice that the estimates (17) and (18) make clear their dependence on both $s \in \mathbb{C}_\sigma$ and $w, v \in H^1(P_0)$. In particular, we can apply the Lax–Milgram theorem to guarantee the well-posedness of problem (16); moreover, we have the following bound on its unique solution:

$$\|\hat{u}_{sc}\|_{s,H^1(P_0)} \leq \frac{C}{\sigma} \|\hat{F}\|_{L^2(P_0)},$$

where C is independent of s, $\hat{u}_{sc}$ and $\hat{F}$.

3.2 Spatial Discretization of the Problem in the Laplace Domain

Discretization of $H^1(P_0)$ is by standard finite elements. More precisely, we consider a regular mesh family $\mathscr{T}_h$, $h > 0$, of P_0 consisting of triangles K of maximum diameter h and which can be mapped from the reference triangle element $\hat{K}$ using an affine mapping $m_K : \hat{K} \to K$. Then we define the finite element space $\mathbb{S}_h$ of continuous finite elements on $\mathscr{T}_h$. In particular

$$\mathbb{S}_h := \{f \in \mathscr{C}^0(P_0); f|_K = \hat{f} \circ m_K \text{ for some } \hat{f} \in \mathbb{P}_q \ \forall K \in \mathscr{T}_h\},$$

where $\mathbb{P}_q$ denotes the set of complex valued polynomials of total degree at most q.

The only remaining difficulty in discretizing the variational problem by means of the approximation space $\mathbb{S}_h$ is that we need to apply the DtN operators to traces of finite element functions. This could be done using an integral equation on $\Sigma_\pm$ as in [3], but for the simple geometry here we can truncate the Fourier expansions involved in the explicit expression of the DtN maps. This may be done efficiently by means of a trigonometric basis of $H^{1/2}(\Sigma_\pm)$, which is a common strategy in the frequency domain. More precisely, let us introduce the finite-dimensional space

$$\mathscr{P}_N := \text{span}\left\{\cos\left(\frac{m\pi x_1}{H}\right); m = 0, 1, \ldots, N\right\},$$

as well as the $L^2(\Sigma_\pm)$ orthogonal projections $p_{N,\pm} : L^2(\Sigma_\pm) \to \mathscr{P}_N$. We then approximate the operators $\hat{T}^s_\pm$ by means of $\hat{T}^s_{N,\pm} = \hat{T}^s_\pm \circ p_{N,\pm}$, and the s-dependent sesquilinear forms $a^s : H^1(P_0) \times H^1(P_0) \to \mathbb{C}$ by

$$a^s_{h,N}(w, v) := \int_{P_0} \frac{\mathrm{n}^2}{c_0^2} s^2 w\overline{v} + \int_{P_0} \nabla w \cdot \nabla \overline{v} - \int_{\Sigma_+} \left(\hat{T}^s_{N,+} w\right) p_{N,+}\overline{v}$$
$$- \int_{\Sigma_-} \left(\hat{T}^s_{N,-} w\right) p_{N,-}\overline{v}.$$

With this approach, the discrete counterpart of problem (16) consists of finding $\hat{u}_{sc,h,N} \in \mathbb{S}_h$ such that, for any $v \in \mathbb{S}_h$,

$$a^s_{h,N}(\hat{u}_{sc,h,N}, v) = \int_{P_0} \hat{F}\,\overline{v}\,. \tag{19}$$

Reasoning as at continuous level, we can see that the discrete sesquilinear form $a^s_{h,N}(\cdot,\cdot)$ is bounded and coercive in terms of s-dependent norms, here again with the same dependence on $s \in \mathbb{C}_\sigma$ and σ as in the continuous case. Indeed, properties (17) and (18) remain valid if we replace the sesquilinear form $a^s : H^1(P_0)\times H^1(P_0) \to \mathbb{C}$ by its discrete counterpart $a^s_{h,N} : H^1(P_0) \times H^1(P_0) \to \mathbb{C}$. In particular, this allows us to reason just as we did before to guarantee the existence of a unique solution of the discrete problem (19), $\hat{u}_{sc,h,N} \in \mathbb{S}_h$, and deduce the following bound:

$$\|\hat{u}_{sc,h,N}\|_{s,H^1(P_0)} \leq \frac{C}{\sigma}\,\|\hat{F}\|_{L^2(P_0)}\,.$$

This result is analogous to [6, Theorem 3.4].

We can then prove an error estimate based on Strang's second lemma in which we keep track of the dependence on the parameter $s \in \mathbb{C}_\sigma$. The analysis is similar to [6, Theorem 3.5]:

$$\begin{aligned}\|\hat{u}_{sc} - \hat{u}_{sc,h,N}\|_{s,H^1(P_0)} \leq \frac{|s|}{\sigma}\Bigg(&\Big(\frac{C_2}{\inf_{\mathbf{x}\in P_0} \mathrm{n}^2(\mathbf{x})} + 1\Big)\,\|\hat{u}_{sc} - \hat{w}_h\|_{s,H^1(P_0)}\\ &+C_1\,\|\gamma_{\Sigma_+}\hat{u}_{sc} - p_{N,+}\gamma_{\Sigma_+}\hat{u}_{sc}\|_{s,H^{1/2}(\Sigma_+)}\\ &\quad+C_1\,\|\gamma_{\Sigma_-}\hat{u}_{sc} - p_{N,-}\gamma_{\Sigma_-}\hat{u}_{sc}\|_{s,H^{1/2}(\Sigma_-)}\Bigg)\,,\end{aligned}$$

where $\hat{w}_h \in \mathbb{S}_h$ and C_1 and C_2 are the s-independent constants previously introduced (see (17) and (18)).

By taking the inverse Laplace transform of the above estimate, we can then derive an error estimate for the semi-discrete approximation $u_{sc,h,N}$ of u_{sc} (i.e. only discretizing in space). More precisely, following the approach of [10], let $T > 0$ denote the final time for the solution and set

$$H^r_0((0,T);X) = \{u \in H^r((-\infty,T);X)\ ;\ \ u(t,\cdot) = 0 \text{ for } t < 0\}\,, \tag{20}$$

where X stands for any Hilbert space. We then have the following theorem.

Theorem 1 *Assume that $\hat{F} \in L^2(\Omega)$, $s = \sigma - i\omega$ with $\sigma > \sigma_0$ and $n^2 > \delta$, for some constants $\sigma_0, \delta > 0$. Then there exists a unique solution $\hat{u}^s_{sc,h,N} \in \mathbb{S}_h$ to (19) and furthermore there is a constant C such that, for any $t \in (0,T)$ and*

$v_h \in H_0^2((0,T);\mathbb{S}_h)$,

$$\|u_{sc,h,N}(t) - u_{sc}(t)\|_{H^1(P_0)} \le C\Big(\|u_{sc} - v_h\|_{H_0^2((0,T);H^1(P_0))} + \|p_{N,+}u_{sc} - u_{sc}\|_{H_0^2((0,T);H^{1/2}(\Sigma_+))} + \|p_{N,-}u_{sc} - u_{sc}\|_{H_0^2((0,T);H^{1/2}(\Sigma_-))}\Big). \quad (21)$$

Here C depends on T but is independent of u_{sc} and t, and of the discretization parameters h and N.

3.3 Discretization in Time

Following the Convolution Quadrature (CQ) approach proposed in [10], to discretize in space and time we can use the discrete Laplace transform. To do this we need to choose a suitable time discretization. Let Δt denote the time step $\Delta t = T/N_t$ where N_t is the number of time steps, and let $t_n = n\Delta t$. As usual for CQ, a good choice of multistep method is BDF2 which approximates the solution $y(t)$ of $y' = f(t,y)$ using the difference equation

$$\frac{3}{2}y_{n+2} - 2y_{n+1} + \frac{1}{2}y_n = \Delta t f(t_{n+2}, y_{n+2}) \quad \text{for } n = -1, 0, 1, \ldots,$$

where $y_n = 0$ for $n \le 0$. The generating polynomial for this method is

$$\gamma(\zeta) = \frac{3}{2} - 2\zeta + \frac{1}{2}\zeta^2 \quad \text{for } \zeta \in \mathbb{C}.$$

The discrete time Laplace transform of the solution we wish to find is denoted $\hat{u}_{sc,h,N}^{\Delta t} \in \mathbb{S}_h$ and satisfies the Laplace domain variational problem with s replaced by $\gamma(\zeta)/\Delta t$:

$$a^{\gamma(\zeta)/\Delta t}(\hat{u}_{sc,h,N}^{\Delta t}, v_h) = \int_{P_0} \hat{F}|_{s=\gamma(\zeta)/\Delta t}\,\overline{v_h} \quad \text{for all } v_h \in \mathbb{S}_h, \quad (22)$$

where this equation holds for all $\zeta \in \mathbb{C}$ with $|\zeta| < 1$.

Taking the inverse discrete Laplace transform we obtain a fully discrete time stepping problem that determines $u_{sc,h,N}^{\Delta t,n} \in \mathbb{S}_h$ for $n = 0, 1, \ldots$. In particular, as in [6] we introduce a new variable

$$\hat{z}_{h,N}^{\Delta t} = \frac{\gamma(\zeta)}{\Delta t}\,\hat{u}_{sc,h,N}^{\Delta t}, \tag{23}$$

so that (22) can be rewritten as finding $\hat{u}_{sc,h,N}^{\Delta t} \in \mathbb{S}_h$ such that

$$\int_{P_0} \frac{\mathrm{n}^2}{c_0^2}\frac{\gamma(\zeta)}{\Delta t}\,\hat{z}_{h,N}^{\Delta t}\,\overline{v_h} + \int_{P_0} \nabla \hat{u}_{sc,h,N}^{\Delta t}\cdot\nabla\overline{v_h} - \int_{\Sigma_+} \hat{T}_+^{\gamma(\zeta)/\Delta t}(p_N \hat{u}_{sc,h,N}^{\Delta t})\,\overline{v_h}$$
$$- \int_{\Sigma_-} \hat{T}_-^{\gamma(\zeta)/\Delta t}(p_N \hat{u}_{sc,h,N}^{\Delta t})\,\overline{v_h} = \int_{P_0} \hat{F}|_{s=\gamma(\zeta)/\Delta t}\overline{v_h} \quad \text{for all } v_h \in \mathbb{S}_h\,. \tag{24}$$

Introducing the z-transform of the discrete time solution as

$$\hat{u}_{sc,h,N}^{\Delta t} = \sum_{m=0}^{\infty} u_{sc,h,N}^{\Delta t,m}\,\zeta^m, \qquad \hat{z}_{h,N}^{\Delta t} = \sum_{m=0}^{\infty} z_{h,N}^{\Delta t,m}\,\zeta^m\,,$$

and equating terms in ζ in (23) shows that the standard BDF2 equation is satisfied

$$\frac{1}{\Delta t}\left(\frac{3}{2}u_{sc,h,N}^{\Delta t,m} - 2u_{sc,h,N}^{\Delta t,m-1} + \frac{1}{2}u_{sc,h,N}^{\Delta t,m-2}\right) = z_{sc,h,N}^{\Delta t,m}$$

for each $m \geq 0$ where $u_{sc,h,N}^{\Delta t,p} = 0$ if $p \leq 0$.

To analyze (24) suppose that we have a finite Fourier series $w = \sum_{m=0}^{N} w_m \cos(m\pi x_1/H)$. Then from (14) we see that

$$\hat{T}^{\gamma(\zeta)/\Delta t} w = -\sum_{m=0}^{N} \kappa_m\Big(\frac{\gamma(\zeta)}{\Delta t}\Big)\, w_m\, \cos\Big(\frac{m\pi x_1}{H}\Big)\,,$$

where $\kappa_m(s)$ is given by (11). The same expansion holds for $\hat{T}_+^{\gamma(\zeta)/\Delta t}$. Expanding $\kappa_m(\gamma(\zeta)/\Delta t)$ in terms of ζ gives

$$\kappa_m\Big(\frac{\gamma(\zeta)}{\Delta t}\Big) = \sum_{j=0}^{\infty} \kappa_{m,j}^{\Delta t}\zeta^j,$$

for some coefficients $\kappa_{m,j}^{\Delta t}$ when $|\zeta| < 1$. These coefficients can be computed exactly for small values of j and in general computed numerically using a discrete approximation to the Cauchy integral formula as in [4, 6]. For example

$$\kappa_{m,0}^{\Delta t} = \frac{\sqrt{4\pi^2 c_0^2 (\Delta t)^2 m^2 + 9H^2}}{2\Delta t\, c_0 H},$$

$$\kappa_{m,1}^{\Delta t} = -6\,\frac{H}{\Delta t\, c_0 \sqrt{4\pi^2 c_0^2 (\Delta t)^2 m^2 + 9H^2}},$$

$$\kappa_{m,2}^{\Delta t} = \frac{\left(44\pi^2 c_0^2 (\Delta t)^2 m^2 + 27H^2\right) H}{2\Delta t\, c \left(4\pi^2 c_0^2 (\Delta t)^2 m^2 + 9H^2\right)^{3/2}},$$

and so on. Now define

$$\tilde{T}_{\pm}^{(j)} w = -\sum_{m=0}^{N} \kappa_{m,j}^{\Delta t}\, w_m \cos\!\left(\frac{m\pi x_1}{H}\right).$$

Equating powers of ζ in (24) gives

$$\int_{P_0} \frac{\mathrm{n}^2}{c_0^2 \Delta t}\left(\frac{3}{2} z_{sc,h,N}^{\Delta t,m} - 2 z_{sc,h,N}^{\Delta t,m-1} + \frac{1}{2} z_{sc,h,N}^{\Delta t,m-2}\right)\overline{v_h} + \int_{P_0} \nabla u_{sc,h,N}^{\Delta t,m}\cdot\nabla\overline{v_h}$$
$$-\sum_{j=0}^{m}\int_{\Sigma_+} \tilde{T}_{+}^{(j)}(p_N u_{sc,h,N}^{\Delta t,m-j})\,\overline{v_h} - \sum_{j=0}^{m}\int_{\Sigma_-} \tilde{T}_{-}^{(j)}(p_N u_{sc,h,N}^{\Delta t,m-j})\,\overline{v_h} = \int_{P_0} \hat{F}^{\Delta t,m}\overline{v}_h, \tag{25}$$

for all $v_h \in \mathbb{S}_h$.

We see that inside P_0 the method corresponds to using BDF2 and finite elements for the wave equation. On the artificial boundaries $\Sigma_\pm$ the method provides a discrete approximation to the DtN map that uses a discrete convolution at each time step. In particular, at time step m this requires access to the $N+1$ Fourier coefficients of $u_{sc,h,N}^{\Delta t,j}$ for $j = 0, \ldots, m$ on the two artificial boundaries. Thus storage requirements grow with time, but for pipes at low frequency there are few propagating modes and so N is not large (besides the propagating modes, some evanescent modes also need to be stored depending how far the artificial boundary is away from the scatterer).

At the expense of more notation, we can now eliminate $z_{sc,h,N}^{\Delta t,m}$ from the difference equation to obtain the discretization of a second order in time problem for u_{sc} alone.

Following Lubich's strategy [10] as in [6] we can prove the following fully discrete error estimate where $W_0^4((0,T);L^1(P_0))$ is defined analogously to $H_0^r((0,T);X)$ in (20):

Theorem 2 *Suppose we use BDF2 to discretize in time, and regular finite elements to discretize in space. In addition, suppose* $F \in W_0^4((0,T);L^1(P_0))$. *Then the time* *discrete finite element solution $u_{sc,h,N}^{\Delta t,n}$ is well defined for each time step $n = 0, 1, \ldots$ and satisfies the error estimate*

$$\|u_{sc,h,N}^{\Delta t,n} - u_{sc}(t_n)\|_{H^1(P_0)} \leq C\Bigg((\Delta t)^2 \int_0^T \int_{P_0} |\partial_t^4 F| + \|u_{sc} - v_h\|_{H_0^2((0,T);H^1(P_0))}$$
$$+\|p_{N,-}u_{sc} - u_{sc}\|_{H_0^2((0,T);H^{1/2}(\Sigma_-))} + \|p_{N,+}u_{sc} - u_{sc}\|_{H_0^2((0,T);H^{1/2}(\Sigma_+))}\Bigg)$$

for any $v_h \in H_0^2((0,T);\mathbb{S}_h)$. Here the constant C depends on T and Σ_+, but is independent of u_{sc} and v_h, and the discretization parameters h, N and Δt.

The theory we have outlined extends to impenetrable (sound hard or sound soft scatterers with little change). For frequency dependent refractive indices, the Laplace domain results can be proved under suitable conditions on the behavior of the refractive index in the Laplace domain (see for example [5]).

4 Numerical Results

Although the analysis of problem (16) and its discretization are written in terms of the Laplace transform of the scattered field, in practice we approximate the total field $u = u_{sc} + u_{inc}$. This avoids performing area integrals for F. Assuming the source of the incident wave is in the section of the pipe P_-, in the Laplace domain and after discretization in space, we seek $\hat{u}_{h,N} \in \mathbb{S}_h$ which is the unique solution of

$$a_{h,N}^s(\hat{u}_{h,N}, v_h) = \int_{\Sigma_-} \partial_{\mathbf{n}_0} \hat{u}_{inc}\,\overline{v_h} - \int_{\Sigma_-} \hat{T}_-^s \hat{u}_{inc}\,\overline{v_h}\,, \tag{26}$$

for any $v_h \in \mathbb{S}_h$. Notice that there is no need for a boundary condition on Σ_+ since the total field is outgoing there.

To deal with problem (26), in practice we approximate $\hat{T}_-^s \hat{u}_{inc}$ by $\hat{T}_{N,-}^s \hat{u}_{inc}$. Above we have shown how the Laplace domain problem can be converted into a time stepping problem using CQ as in [10]. Here, to demonstrate the method, we instead use the discrete Laplace transform approach from [4]. Suppose the final time of integration is T and we wish to take N_t timesteps. In Banjai and Sauter's approach (26) is solved for N_t choices of s chosen depending on the time-stepping method used (in fact fewer problems need to be solved in practice). An inverse discrete transform then gives the time dependent solution. We use the parameter choices from [4] even though the theory in that paper is for an integral equation based approach.

4.1 Convergence Rate

To obtain a simple exact solution we can consider an empty pipe. In this case the total field is simply given by the incident field, and the code must propagate the incident field through the finite element domain. We choose the computational domain to be $P_0 = (0, 0.6) \times (0, 1)$ with Σ_- at $x_2 = 0$ and Σ_+ at $x_2 = 1$, and the width of the pipe $H = 0.6$. The final time is $T = 6$ by which time the wave has almost left the computational domain. The incident field is a plane wave $u_{inc} = f(t - x_1/c)$ where $c = 1$ and

$$f(\tau) = \cos(2\pi(\tau - H/c))\exp(-1/(2\sigma^2)(\tau - L/c - t_p)^2)$$

where $t_p = 3$, $\sigma = 6/(2\pi\ b_w)$, and $b_w = 1.71$ denotes the bandwidth of the incident field; notice that the center frequency is 1. The parameters are chosen so that f is approximately zero in P_0 at $t = 0$. We choose a fixed spatial mesh shown in Fig. 2 (left panel) where the mesh size is $h \approx 0.016$, and use piecewise linear finite elements in space (using the FreeFem++ to implement the algorithm [8]) and BDF2 in time. Although only one mode is needed for the DtN maps in this case, we choose $N = 7$ for the Fourier spaces on Σ_- and Σ_+.

For simplicity we report the discrete maximum norm error at the nodes in the mesh as a function of N_t in the right hand panel of Fig. 2. The convergence rate is consistent with $O(N_t^{-2})$ convergence for at least part of the convergence history. We have no explanation for the increased rate at $N_t = 1024$. In any case the numerical results show that we can obtain accurate and convergent solutions over a wide range of time step sizes. Indeed, the coarsest time step is $\Delta t \approx 0.09$ and the finest time step is $\Delta t \approx 0.006$, and stability is seen across this range of time steps.

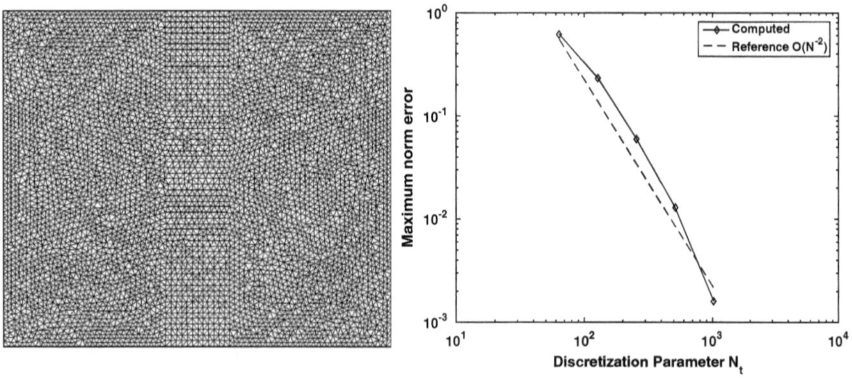

Fig. 2 Using the fixed spatial mesh shown in the *left hand* panel, we show the discrete maximum norm error at the spatial nodes as a function of the number of time steps in the *right hand panel*. We have predicted N_t^{-2} error in the $H^1(P_0)$ norm and see somewhat better than this rate at finer temporal discretization

4.2 *Scattering from a Penetrable and Impenetrable Obstacles*

Our next examples illustrate the flexibility of this approach since the finite element method can handle different boundary conditions and possible inhomogeneity of the scatterer. We start with a penetrable scatterer as analyzed in this paper. We choose $n(\mathbf{x}) = 1$ in the pipe, and $n(\mathbf{x}) = 2$ inside a disk of radius 0.3 centered at $(0, 0.6)$. In order to keep the ratio of mesh size to wavelength roughly constant, the mesh inside the scatterer is refined according to the local refractive index. In Fig. 3 we show the spatial mesh and three snapshots of the same incident field as used in the previous section choosing the number of time steps $N_t = 512$ (from the previous

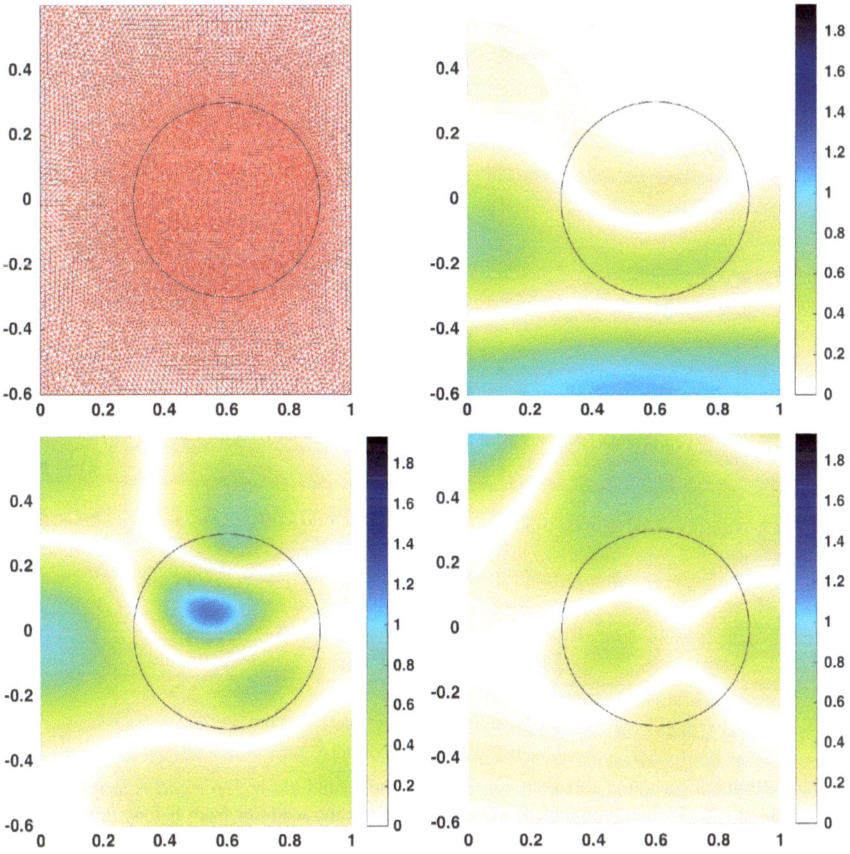

Fig. 3 Results for a penetrable scatterer. *Top left*: the spatial mesh, refined inside the scatterer. *Top right*: A snapshot of the total field at $t \approx 3$ when the incident wave is arriving at the scatterer from below. *Bottom left*: A snapshot of the total field at $t \approx 4$ when the maximum of the incident wave is at the scatterer. *Curved wave* fronts in the scatterer show that the wave has slowed there. *Bottom right*: A snapshot of the total field at $t \approx 5$ when the incident wave starts to pass the scatterer. A focal point is visible on the upper boundary of the scatterer

section we know the method propagates the incident wave with roughly 1% error when the obstacle is not present). Clearly, as expected, the waves slow down in the scatterer and are transmitted through the scatterer with a focal point on one side of the circle. No instability is evident.

In our second example we consider scattering from a sound soft obstacle. This corresponds to enforcing the Dirichlet boundary condition $u = 0$ on the boundary of the same disk as used in the previous example. Results are shown in Fig. 4.

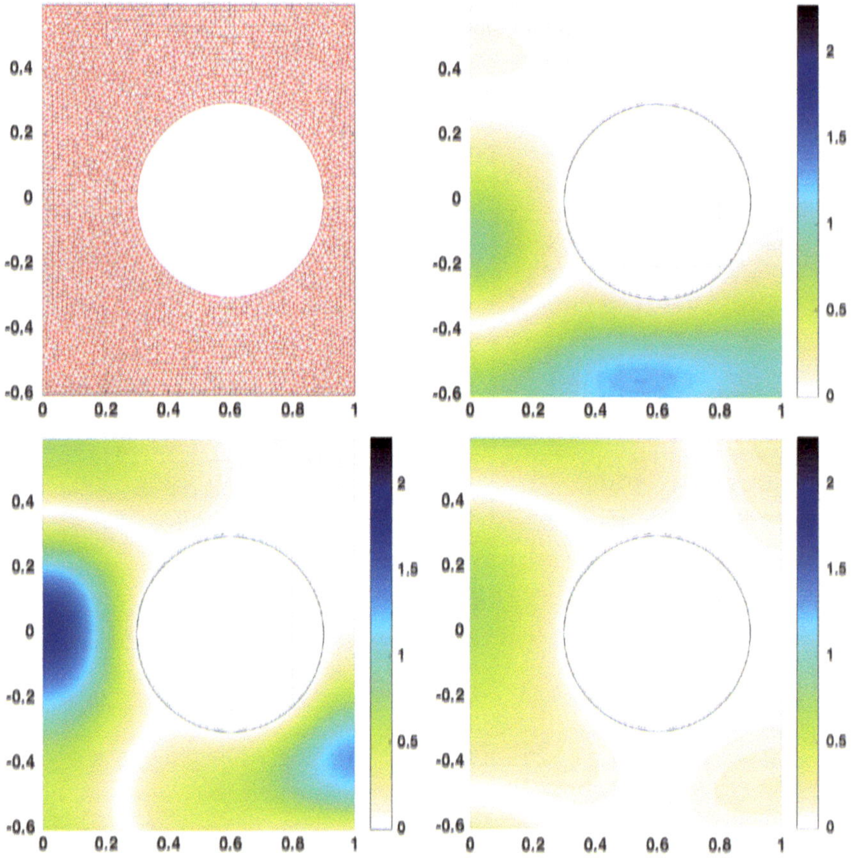

Fig. 4 Results for a sound soft scatterer. *Top left*: the spatial mesh. *Top right*: A snapshot of the total field at $t \approx 3$ when the incident wave is arriving at the scatterer from below. *Bottom left*: A snapshot of the total field at $t \approx 4$ when the maximum of the incident wave is at the scatterer. The incident wave is strongly reflected by the scatterer. *Bottom right*: A snapshot of the total field at $t \approx 5$ when the incident wave starts to pass the scatterer. Above the scatterer the wave is decreased in magnitude compared to Fig. 3 as is to be expected

5 Conclusions

In this paper we have shown how to derive and analyze a fully discrete time stepping method for the wave equation in an infinite pipe or waveguide. Using the DtN map to truncate the domain we obtain a coupled finite element and discrete DtN map for the discrete solution at each time step. Limited numerical results suggest the method is stable and accurate.

The main drawback of the method is that the solution needs to be recorded on the artificial boundaries to allow the convolution needed at each time step to be computed. However if there are only a few propagating modes in the solution this is not a crushing overhead unless very long solution times are required.

Acknowledgements The research of L. Fan and P. Monk was partially supported by NSF grant number DMS-1114889 and DMS-1125590. The research of V. Selgas by MTM2013-43671-P.

References

1. Bamberger, A., Duong, T.H.: Formulation variationnelle espace-temps pour le calcul par potentiel retarde de la diffraction d'une onde acoustique (I). Math. Meth. Appl. Sci. **8**, 405–435 (1986)
2. Banjai, L.: Time-domain Dirichlet-to-Neumann map and its discretization. IMA J. Numer. Anal. **34**, 1136–1155 (2014)
3. Banjai, L., Lubich, C., Sayas, F.: Stable numerical coupling of exterior and interior problems for the wave equation. Numer. Math. **129**(4), 611–646 (2015)
4. Banjai, L., Sauter, S.: Rapid solution of the wave equation in unbounded domains. SIAM J. Numer. Anal. **47**, 227–49 (2008)
5. Fan, L., Monk, P.: Time dependent scattering from a grating. J. Comput. Phys. **302**, 97–113 (2015)
6. Fan, L., Monk, P.: Time dependent scattering from a grating using convolution quadrature and the Dirichlet-to-Neumann map. (2015, submitted)
7. Ha-Duong, T.: On retarded potential boundary integral equations and their discretizations. In: Ainsworth, M., Davies, P., Duncan, D., Rynne, B., Martin, P. (eds.) Topics in Computational Wave Propagation: Direct and Inverse Problems, pp. 301–36. Springer, Berlin (2003)
8. Hecht, F.: New development in FreeFem++. J. Numer. Math. **20**, 251–265 (2012)
9. Lu, Y., Zhu, J.: Perfectly matched layer for acoustic waveguide modeling—benchmark calculations and perturbation analysis. Comput. Model. Eng. Sci. **22**, 235–248 (2007)
10. Lubich, C.: On the multistep time discretization of linear initial-boundary value problems and their boundary integral equations. Numer. Math. **67**, 365–89 (1994)

Location of Emergency Facilities with Uncertainty in the Demands

Luisa I. Martínez-Merino, Maria Albareda-Sambola, and Antonio M. Rodríguez-Chía

Abstract This work deals with the p-center problem, where the aim is to minimize the maximum distance between any customer with demand and his center, taking into account that each customer only has demand with a specific probability. We consider an integer programming formulation for the problem and extensive computational tests are reported, showing its potentials and limits on several types of instances. Finally, some improvements on the formulation have been developed obtaining in some cases much better resolution times.

1 Introduction

Discrete facility location models have been extensively studied in the literature. Different kinds of facilities have been modeled, such as routers or servers in communication networks, warehouses or distribution centers in supply chains, hubs or transhipment nodes in passenger transport networks, and hospital or emergency facilities in public service systems, among others. In general, the goal of these types of problems is to locate the facilities among a set of candidate sites and assign customers to the facilities optimizing some effectivity measure, that usually depends on the distances between the facilities and the customers, see for instance [7, 9, 18] and the references therein. The p Center Problem (pCP) is a well-known discrete optimization location problem which consists of locating p centers out of n sites and assigning (allocating) the remaining $n-p$ sites, to the centers, so as to minimize the maximum distance (cost) between a site and the corresponding center, see Chap. 4 of [18] and [1, 16]. It was shown in [16] that the pCP is NP-hard.

A straight application of the pCP is the location of emergency services like ambulances, hospitals or fire stations, since the whole population should be inside a

L.I. Martínez-Merino (✉) • A.M. Rodríguez-Chía
Departamento de Estadística e Investigación Operativa, Universidad de Cádiz, Cádiz, Spain
e-mail: luisa.martinez@uca.es; antonio.rodriguezchia@uca.es

M. Albareda-Sambola
Departament d'Estadística i Investigació Operativa, Universitat Politècnica de Catalunya, BarcelonaTech, Barcelona, Spain
e-mail: maria.albareda@upc.edu

F. Ortegón Gallego et al. (eds.), *Trends in Differential Equations and Applications*, SEMA SIMAI Springer Series 8, DOI 10.1007/978-3-319-32013-7_19

small radius around some emergency center. The *pCP* has been extensively studied, and both exact and heuristic algorithms have been proposed. Recent articles on the matter are [8, 10, 12–14, 21]. We also refer the reader to Chap. 5 of [7]. A recent survey on location of emergency services can be consulted in [3].

The uncertainties can be generally classified into three categories: provider-side uncertainty, receiver-side uncertainty and in-between uncertainty, depending on whether the uncertainties affect data concerning the facilities (capacities, availability, etc.—see, for instance, [4, 6, 23, 24, 26, 30]), the customers (demands, locations, number), or concerning the distribution network (transportation costs or times—see [25, 27] and references therein), respectively. In addition, two major categories of approaches have been adopted in the literature to deal with uncertain data in facility location models. Namely, stochastic programming (SP) and robust optimization (RO). The former has been used typically to deal with decision-making for facility locations in risk situations, in which the values of uncertain parameters are governed by discrete or continuous probability distributions that are known to a decision-maker. On the other hand, the RO approach attempts to optimize the worst-case system performance in uncertain situations that lack any information about the probability distributions of uncertain coefficients (e.g., [17]). Hence, the RO approach generally describes uncertain data using pre-specified intervals or ranges. Typical robustness measures include mini-max objective value and mini-max regret in an objective value.

In this paper we focus on the receiver-side uncertainty and in the SP approach. The SP approach has been widely applied to emergency logistics for short-notice disasters (e.g., hurricanes, flooding, and wild fires) by assuming that possible impacts of these disasters can be estimated based on historical and meteorological data. The usual goal of these stochastic location models is to optimize the expected value of a given objective function. A classical example of applying SP to disaster relief is the scenario-based, two-stage stochastic model proposed by Mete and Zabinsky [20], for medical supply pre-positioning and distribution in emergency management. Other examples can be found in, for instance, [2, 5, 28].

The paper is organized as follows. First we introduce the problem in Sect. 2, where some results related with the problem are presented. In Sect. 3 a mathematical formulation is given and it is afterwards strengthened by using variable fixing and adding valid inequalities. Then, in Sect. 4, we focus on a variant of the problem where only the K largest distances are considered. Section 5 is devoted to the computational results that both formulations provide and finally, in Sect. 6 we draw some conclusions of the work.

2 The Problem

Let $N = \{1, \ldots, n\}$ be the given set of sites or customers. Throughout the paper we assume, without loss of generality, that the set of candidate sites for centers is identical to N. Let $p \geq 2$ be the number of centers to be located. For each pair (i,j),

$i, j \in N$, let d_{ij} be the distance (cost, travel time) from i to j. We assume $d_{ii} = 0$ $\forall i \in N$ and $d_{ij} > 0\ \forall i, j \in N : i \neq j$. We do not assume other special properties like satisfaction of triangle inequality, that is to say, strictly speaking d is not necessarily a distance. But we need to do an additional assumption to deal with the case of ties among several distances from the same site. If this is the case, in order to break ties we suppose that there are preferences on the centers in such a way that sites undoubtedly will choose one of the centers before the others. In practice, ties can be broken by slightly perturbing the tied distances. Summarizing, we will also assume $d_{ia} \neq d_{ib}\ \forall i, a, b \in N : a \neq b$. Associated with each customer $i \in N$ is the probability of having demand $0 \leq q_i < 1$. The events of demand occurrence are assumed to be independent.

To describe a solution to the *PpCP* we will need to identify the set of p sites where facilities are open, and the assignment to one of those facilities of each of the potential customers, since at the moment of making the decision we do not know which customers will place a demand and which will not. In what follows, we will distinguish between the *assignment* cost of a customer and its *service cost*. The assignment cost corresponds to the distance between the customer and the facility it is assigned to a priori, whilst the service cost takes this same value but only in the scenarios where the customer does have demand.

In case of tie between a client and several plants, this will be assigned to the plant with the largest index. In case of ties between two clients and their plants we consider as the largest distance the one assigned to the client with the greatest index.

The goal of the *PpCP* is to identify the solution with the smallest expected value (among all possible scenarios) of the maximum service cost (among all customers with demand in that scenario). For any set of probabilities $(q_1, \ldots, q_n)$ with $0 \leq q_i \leq 1, i \in N$, any feasible solution of the *PpCP* is associated with a matrix $(\pi_{ij})_{i,j \in N}$, such that, π_{ij} represents the probability that there is not demand at the sites whose assignment cost (for this solution) is bigger than d_{ij} if site i is covered by plant j and 0 otherwise.

Lemma 1 *The matrix* $(\pi_{ij})_{i,j \in N}$ *satisfies:*

1. $\sharp\{j \in N : \pi_{ij} \neq 0\} \leq 1 \quad \forall i \in N.$
2. *If* $d_{(1)} \leqslant \cdots \leqslant d_{(n)}$ *is a non-decreasing sequence of distances between each customer and its assigned plant, and* $(1), \ldots, (n)$ *is the corresponding sequence of customers,*

$$\sum_{j=1}^{n} \pi_{(i)j} = \prod_{t=i+1}^{n} (1 - q_{(t)}).$$

3. *We have that*

$$\sum_{i=1}^{n} \sum_{j=1}^{n} \pi_{ij} q_i = 1 - \prod_{i=1}^{n} (1 - q_i) \leqslant 1. \tag{1}$$

This allows to compute the expected maximum service cost as

$$\sum_{i=1}^{n}\sum_{j=1}^{n} \pi_{ij} d_{ij} q_i.$$

Proof Each site is assumed to be served by just one plant. Then, for a given site $i \in N$, π_{ij} will be 0 for any $j \in N$ such that $j \neq j_i$, being j_i the plant covering i. Observe that the cardinality of the set $\{i \in N : \pi_{ij} \neq 0\}$ will be different from 0 whenever no site i' (covered by $j_{i'}$) exists satisfying that $d_{i'j'} > d_{ij}$ and $q_{i'} = 1$. The summation $\sum_{j=1}^{n} \pi_{(i)j}$ represents the probability that none of the sites (j) with $j > i$ has demand. Then, this probability is given by,

$$\prod_{j=i+1}^{n} (1 - q_{(j)}).$$

Finally, if π values are defined as above, for each of the distinct assignment cost d (distances between each customer and its assigned plant),

$$\sum\nolimits_{i:d_{(i)}=d} \sum\nolimits_{j \in N} \pi_{(i)j} q_i$$

gives the probability that the maximum service distance is d. Therefore, the sum of all π values gives the probability that some service is provided (there is some service cost). Equality (1) follows from the fact that the complement of this event is the scenario where no customer has demand. Clearly, this quantity will never exceed 1.

The following result shows that every customer is covered by its closest service facility.

Theorem 1 *The optimal value of the objective function above is achieved in a solution where every site is covered by its closest plant.*

Proof Assume that two plants are located at places j and j', and site i satisfies that $d_{ij'} \leq d_{ij}$. We will prove that the objective value when site i is covered by j is greater than or equal to the case where i is covered by j'. Assume that $d_{ij'}$ is in the s-th position of the non-increasing ordered sequence of distances between each site and its allocated plant, and d_{ij} is in the t-th position when site i is covered by j instead of j'. Let $F_{j'}$ and F_j be the objective value when i is allocated to plant j' and to j respectively. Also, let $d_{(1)}, \ldots, d_{(n)}$ and $d'_{(1)}, \ldots, d'_{(n)}$ the nondecreasing sequence of distances between each site and its allocated plant when i is allocated to plant j and to plant j', respectively.

$$d_{(1)} \leq \ldots \leq d_{(n)}$$
$$d'_{(1)} \leq \ldots \leq d'_{(n)}.$$

Assume that $d'_{(s)} = d_{ij'}$, $d_{(t)} = d_{ij}$, $q'_{(s)} = q_{(t)} = q_i$ and $d'_{(u)} = d_{(u-1)}$ and $q'_{(u)} = q_{(u-1)}$ for all $s+1 \leq u \leq t$.

$$
\begin{aligned}
F_{j'} - F_j &= \sum_{u=s}^{t} \prod_{v=u+1}^{n} (1-q'_{(v)})q'_{(u)}d'_{(u)} - \sum_{u=s}^{t} \prod_{v=u+1}^{n} (1-q_{(v)})q_{(u)}d_{(u)} \\
&= \prod_{v=t+1}^{n} (1-q_{(v)}) \left[\sum_{u=s}^{t-1} q'_{(u)}d'_{(u)} \prod_{v=u+1}^{t} (1-q'_{(v)}) + q'_{(t)}d'_{(t)} \right. \\
&\quad \left. - \sum_{u=s}^{t-1} q_{(u)}d_{(u)} \prod_{v=u+1}^{t} (1-q_{(v)}) - q_{(t)}d_{(t)} \right].
\end{aligned}
$$

For the sake of readability, let $F_{j'j} := \frac{F_{j'} - F_j}{\prod_{v=t+1}^{n}(1-q_{(v)})}$. Hence,

$$
\begin{aligned}
F_{j'j} &= q'_{(s)}d'_{(s)} \prod_{v=s+1}^{t} (1-q'_{(v)}) + \sum_{u=s+1}^{t-1} q'_{(u)}d'_{(u)} \prod_{v=u+1}^{t} (1-q'_{(v)}) + q'_{(t)}d'_{(t)} \\
&\quad - \sum_{u=s}^{t-1} q_{(u)}d_{(u)} \prod_{v=u+1}^{t} (1-q_{(v)}) - q_{(t)}d_{(t)} \\
&= q_{(t)}d'_{(s)} \prod_{v=s}^{t-1}(1-q_{(v)}) - q_{(t)}d_{(t)} + q_{(t)} \left[\sum_{u=s}^{t-2} q_{(u)}d_{(u)} \prod_{v=u+1}^{t-1} (1-q_{(v)}) + q_{(t-1)}d_{(t-1)} \right] \\
&= q_{(t)} \left[d'_{(s)} \prod_{v=s}^{t-1}(1-q_{(v)}) + \sum_{u=s}^{t-2} q_{(u)}d_{(u)} \prod_{v=u+1}^{t-1} (1-q_{(v)}) + q_{(t-1)}d_{(t-1)} - d_{(t)} \right] \\
&\leq q_{(t)}d_{(t)} \left[\prod_{v=s}^{t-1}(1-q_{(v)}) + \sum_{u=s}^{t-2} q_{(u)} \prod_{v=u+1}^{t-1} (1-q_{(v)}) + q_{(t-1)} - 1 \right] \\
&\leq 0.
\end{aligned}
$$

The last inequality is based on equality (1).

Corollary 1 *The closest assignment constraints (C.A.C.) can be used as valid inequalities for any formulation of the considered problem even if they are not needed to formulate it.*

Observe that, in fact, the service costs associated with customers that are very close to a facility will seldom be the ones yielding the largest service cost in a scenario, since many other customers (those with larger assignment costs) should have no demand for this to happen. Therefore, the probability that a small assignment cost becomes the actual largest service cost can be extremely low. For this

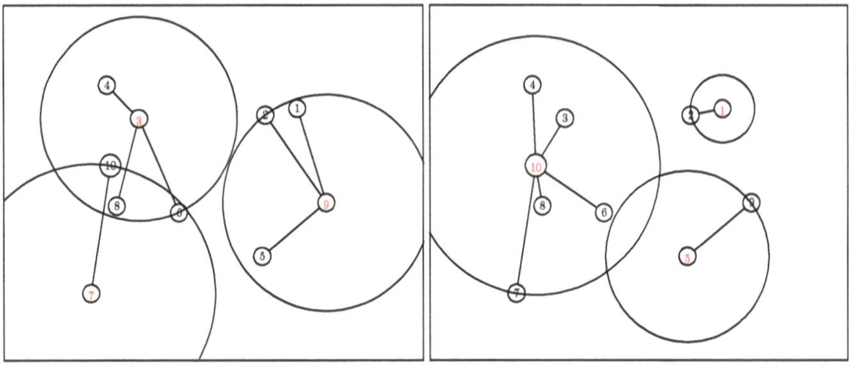

Fig. 1 Solutions of the problem without C.A.C. (*left*) and with C.A.C. (*right*)

reason, the approximation of the *PpCP* that consists in taking into account only the $K \leqslant p$ largest assignment costs can be very tight, even for moderate values of K (specially if demand probabilities q_i are large). From now on, we will refer to this approximation as *K-PpCP*.

Lemma 2 *In the K-PpCP, closest assignment constraints must be explicitly included in the formulation.*

In this case, closest assignment constraints must be added to the formulation. Otherwise we can't assure that the sites are assigned to their closest located facility. We show an example.

Example 1 Consider $n = 10$, $p = K = 3$ and the set of sites $a_1 = (81, 65)$, $a_2 = (71, 63)$, $a_3 = (32, 62)$, $a_4 = (22, 72)$, $a_5 = (70, 21)$, $a_6 = (44, 34)$, $a_7 = (17, 10)$, $a_8 = (25, 36)$, $a_9 = (90, 37)$ and $a_{10} = (23, 48)$. We also suppose that demand probabilities are $q_1 = 0.97$, $q_2 = 0.12$, $q_3 = 0.63$, $q_4 = 0.27$, $q_5 = 0.9$, $q_6 = 0.15$, $q_7 = 0.24$, $q_8 = 0.26$, $q_9 = 0.33$ and $q_{10} = 0.17$.

If C.A.C. are not included in the formulation, we obtain, as solution of the 3-*P3CP*, an objective value of 13.08 and the plants are located at 3, 7 and 9. This solution allocates sites 4, 6 and 8 to plant 3, site 10 to plant 7 and sites 1, 2 and 5 to plant 9. However, in this case the distance between site 10 and plant located at 3 ($d_{10,3} = 16.64$) is smaller than the distance between 10 and 7 ($d_{10,7} = 38.47$). The left side of Fig. 1 shows the solution for the problem if C.A.C. are not used. If we include the C.A.C., the objective value is 17.58 and the plants are located at sites 1, 5 and 10. We can see the solution of the problem using C.A.C. in the right side of Fig. 1.

Somehow, the *PpCP* can be seen as a tradeoff between other classical discrete location models, such as the p-center, the p-median and the k-centrum. Indeed, when all the demand probabilities coincide, the *PpCP* fits in the structure of the more general ordered median problem ([16, 20, 24]). We next show an example where the *K-PpCP* yields the solution of either of the above problems, depending on the values of the probabilities of demand.

In the next example, the solution of *K-PpCP* coincides with the solution of *p*-median problem for small values of *q*, with the solution of the *K*-centrum problem for values of *q* close to 0.5 and with the *p*-center problem for values of *q* close to 1. This illustrates the trend of behaviour of *PpCP* in comparison with classical criteria in location theory for different values of *q*. Indeed, when the probability of having demand in each site is small and they are almost identical, the probabilities of each assignment cost being the largest service cost become very similar and, therefore, the *PpCP* behaves very similarly to the *p*-median problem. As opposite, if these probabilities are close to 1, the probability that the largest assignment cost yields the largest service cost is close to 1 and, therefore, the weights of all the other assignment costs in the objective function are very small, leading thus to solutions that will be very close to those of the *p*-center problem.

Example 2 We have the following sites: $a_1 = (14, 70), a_2 = (40, 94), a_3 = (87, 5)$, $a_4 = (70, 70)$, $a_5 = (21, 48)$, $a_6 = (53, 16)$, $a_7 = (0, 47)$, $a_8 = (11, 11)$, $a_9 = (66, 75)$ and $a_{10} = (7, 68)$. We consider the 4P3CP and different probability vectors:

- $q_1 = (0.07, 0.15, 0.03, 0.13, 0.01, 0.09, 0.03, 0.12, 0.13, 0.14)$.
- $q_2 = (0.56, 0.41, 0.53, 0.53, 0.59, 0.6, 0.6, 0.42, 0.54, 0.47)$ and
- $q_3 = (0.88, 0.85, 0.95, 0.81, 0.98, 0.8, 0.99, 0.96, 0.86, 0.8)$.

As we can observe in Table 1, if we take demand probabilities equal to q_1, thc optimal set of plants for the 4-P3CP coincides with the optimal solution of the 3-median problem. Similarly, the solution obtained for the 4-3 centrum problem is optimal for the 4-P3CP if we take demand probabilities equal to q_2. Finally, Table 1 reports that the solution for the 4-*P*3*CP* and for the 3-center problem are the same if we consider q_3 as the vector of dcmand probabilities. Figure 2 shows the open plants and assignments that are obtained as solution of the 4-*P*3*CP* for the different probability vectors considered.

Table 1 Solutions corresponding to 4-*P*3*CP*, 3-median, 4-3centrum and 3-center respectively

	K-PpCP		*p*-median		K-pcentrum		*p*-center	
	Plants	O.V.	Plants	O.V.	Plants	O.V.	Plants	O.V.
q_1	{6,9,10}	10.32	{6,9,10}	170.74	{5,6,9}	130.68	{3,7,9}	37.64
q_2	{5,6,9}	32.53						
q_3	{3,7,9}	37.52						

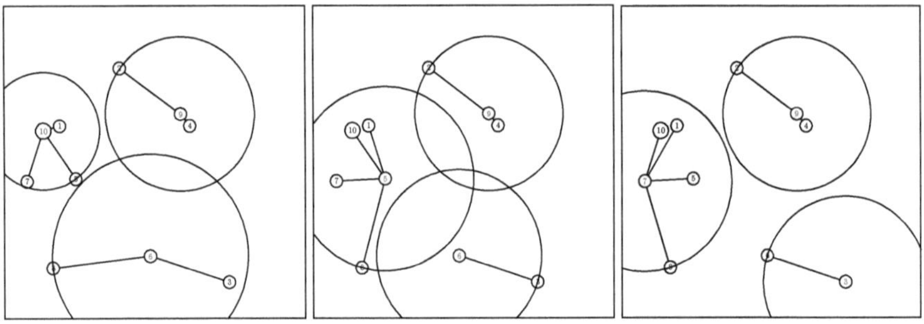

Fig. 2 Solutions corresponding to 3-median, 4-3 centrum and 3-center respectively

3 MIP Formulation for the *PpCP*

We next present a mixed integer programming formulation for the *PpCP*. As mentioned above, the assignment costs involved in a solution will contribute to the corresponding objective function value weighted with a probability that depends on the customers with larger assignment costs. Therefore, together with the classical binary location variables, other variables, used to sort the assignment costs and to compute these probabilities are required. In particular, we define the following two families of binary variables:

$$y_j = \begin{cases} 1, \text{ if a plant is opened at site } j, \\ 0, \text{ otherwise,} \end{cases} \qquad \text{for } j \in N,$$

$$x_{ijk\ell} = \begin{cases} 1, \text{ if } i \text{ is allocated to } j, k \text{ to } \ell \text{ and } d_{ij} \text{ is the first} \\ \quad \text{assignment cost larger than } d_{k\ell}, \\ 0, \text{ otherwise,} \end{cases}$$

for all $i, j, k, \ell \in N$ such that $d_{ij} > d_{k\ell}$ or if $d_{ij} = d_{k\ell}$ for all $i > k$. We also define a family of continuous variables

π_{ij} = probability that no site with allocation distance greater than d_{ij} has demand if i is allocated to j, and 0 otherwise.

Using these variables, the *PpCP* can be formulated as follows.

$$\text{(F1)} \min \sum_{i=1}^{n} \sum_{j=1}^{n} \pi_{ij} q_i d_{ij}$$

$$\text{s.t.} \sum_{j=1}^{n} y_j = p, \tag{2}$$

$$\sum_{k=1}^{n}\sum_{\ell=1}^{n} x_{ijk\ell} \leqslant y_j, \qquad \forall i,j \in N \tag{3}$$

$$\sum_{k=1}^{n}\sum_{\ell=1}^{n} x_{k\ell ij} \leqslant y_j, \qquad \forall i,j \in N \tag{4}$$

$$\sum_{j=1}^{n}\sum_{k=1}^{n}\sum_{\ell=1}^{n} x_{ijk\ell} \leqslant 1, \qquad \forall i \in N \tag{5}$$

$$\sum_{j=1}^{n}\sum_{k=1}^{n}\sum_{\ell=1}^{n} x_{k\ell ij} \leqslant 1, \qquad \forall i \in N \tag{6}$$

$$\sum_{i=1}^{n}\sum_{j=1}^{n}\sum_{k=1}^{n}\sum_{\ell=1}^{n} x_{ijk\ell} = n-1, \tag{7}$$

$$\sum_{k=1}^{n}\sum_{\ell=1}^{n}\sum_{\substack{j'=1\\ j'\neq j}}^{n} x_{ij'k\ell} + \sum_{k=1}^{n}\sum_{\ell=1}^{n} x_{k\ell ij} \leqslant 1, \qquad \forall i,j \in N \tag{8}$$

$$\pi_{k\ell} \geqslant (1-q_i)\pi_{ij} - 1 + x_{ijk\ell}, \qquad \forall i,j,k,\ell \in N \tag{9}$$

$$\pi_{ij} \geqslant \sum_{k=1}^{n}\sum_{\ell=1}^{n} x_{ijk\ell} - \sum_{k=1}^{n}\sum_{\ell=1}^{n} x_{k\ell ij}, \qquad \forall i,j \in N \tag{10}$$

$$y_j, x_{ijk\ell} \in \{0,1\}, \qquad \forall i,j,k,\ell \in N \tag{11}$$

$$\pi_{ij} \in [0,1], \qquad \forall i,j \in N. \tag{12}$$

In the objective function, $\pi_{ij}q_i$ gives the probability that d_{ij} is the largest service distance. Thus, the objective function accounts for the expected largest distance from a customer with demand to its plant. Constraint (2) ensures that p facilities are opened, and constraints (3) and (4) force that all assignments of customers are made to open facilities. The sorting of the used assignment distances is made through constraints (5)–(8), taking advantage of the variable definition (recall that $x_{ijk\ell}$ is not defined, or is fixed to zero, if $d_{ij} \lneqq d_{k\ell}$). In particular, constraints (5) and (6) ensure that the distance to cover site i is at most one immediately greater/smaller than another distance from a site and its plant. Constraints (8) together with (5) and (6) ensure that any site i is covered by at most one plant and constraints (7) guarantee that it is exactly one. Constraints (9)–(10) are used to guarantee that π and x variables take consistent values. Finally, the last families of constraints set the domains of the variables.

3.1 Valid Inequalities

1. As shown in Corollary 1, C.A.C. are valid. Several alternative sets of CAC have been proposed in the literature (see [11]). In this work, we have adapted the set presented in [29]:

$$\sum_{k=1}^{n}\sum_{\ell=1}^{n}\sum_{a=1;d_{ia}>d_{ij}}^{n} x_{iak\ell} + y_j \leqslant 1, \quad \forall i,j \in N, \tag{13}$$

$$\sum_{i=1}^{n}\sum_{j=1}^{n}\sum_{a=1;d_{ka}>d_{k\ell}}^{n} x_{ijka} + y_\ell \leqslant 1, \quad \forall k,\ell \in N. \tag{14}$$

   These constraints have proven very useful in preliminary computational results.
2. Constraint (7) can, in fact, be decomposed into the three constraints:

$$\sum_{i=1}^{n}\sum_{j\neq i}\sum_{k=1}^{n}\sum_{\ell\neq k} x_{ijk\ell} = n - p - 1, \tag{15}$$

$$\sum_{i=1}^{n}\sum_{j\neq i}\sum_{k=1}^{n} x_{ijkk} = 1, \tag{16}$$

$$\sum_{i=1}^{n}\sum_{k=1}^{n} x_{iikk} = p - 1. \tag{17}$$

   However, despite being theoretically better than (7), the use of this family of constraints does not reduce the computational effort required to solve formulation (F1).
3. A valid inequality for this formulation is

$$\sum_{j=1}^{n}\sum_{k=1}^{n}\sum_{\ell=1}^{n} x_{ijk\ell} + x_{kji\ell} \geqslant 1, \ \ \forall i \in N. \tag{18}$$

   Note that constraints (5) and (6) cannot be stated as equalities since, in both cases, the expression in the left hand side will take value 0 in exactly one customer, for any feasible solution (the customer with the smallest and the largest assignment cost, respectively), but it has to take value 1 in all the others. Since these two customers will undoubtedly be different, for any customer at least one of the two expressions will take value 1, as stated in valid inequality (18).

   In fact, combined with sets (5)–(7), it is not necessary to have both (8) and (18) together. Indeed, constraints (8) plus (5) and (6) guarantee that any site i cannot be assigned to more than one plant and constraints (18) ensure that any site is using (7), i.e. since there are $n-1$ variables different to 0, these two conditions are equivalent.

4. Other valid inequalities are

$$\pi_{k\ell} \leqslant (1-q_i)\pi_{ij} + 1 - x_{ijk\ell}, \quad \forall i,j,k,\ell \in N, \tag{19}$$

$$\pi_{ij} \leqslant \sum_{k=1}^{n}\sum_{\ell=1}^{n} x_{ijk\ell} + \sum_{k=1}^{n}\sum_{\ell=1}^{n} x_{k\ell ij}, \quad \forall i,j \in N. \tag{20}$$

If all distances are nonnegative, since we are minimizing and there are no capacity constraints, constraints (19) and (20) are not necessary.

5. The sum of π values is known (Lemma 1):

$$\sum_{i=1}^{n}\sum_{j=1}^{n} \pi_{ij} q_i = 1 - \prod_{i \in N}(1-q_i). \tag{21}$$

So it can also be added as a valid inequality.

3.2 *Fixing Variables*

In this section we describe a series of criteria to fix some of the variables. which can be useful to solve the problem. First, binary variables x_{ijkl} are defined for all $i,j,k,l \in N$ such that $d_{ij} > d_{kl}$ or if $d_{ij} = d_{kl}$ for all $i > k$. For this reason we can fix:

$$x_{ijk\ell} = 0, \quad \forall i,j,k,\ell \in N, \text{ such that, } d_{ij} < d_{k\ell} \text{ or } d_{ij} = d_{k\ell} \text{ and } i \leqslant k. \tag{22}$$

Note that the smallest p assignment distances are all equal to zero. As mentioned above, this tie could produce very inconvenient symmetries in the solutions. We will break those ties arbitrarily, beforehand, using (22), to avoid those awkward symmetries.

By the definition of these variables we also have that $x_{ijkl} = 0$ if i is not allocated to j or k is not allocated to l. Due to the formulation of the problem, a customer (i) can only be allocated to one facility and it is served by its closest plant. As a consequence,

$$\begin{aligned} x_{ijik} &= 0, \quad \forall i,j,k \in N. \\ x_{ijk\ell} &= 0, \quad \forall i,j,k,\ell \in N \text{ such that } d_{kj} < d_{k\ell}. \\ x_{ijk\ell} &= 0, \quad \forall i,j,k,\ell \in N \text{ such that } d_{i\ell} < d_{ij}. \end{aligned}$$

Since p facilities are opened, and each customer will be assigned to its closest open facility, and π variables are only non-null for pairs (i,j) being j the server assigned to customer i, we can set $\pi_{ij} = 0$ if $|\{j' \neq j : d_{ij'} \geqslant d_{ij}\}| < p$.

Now we see other fixing variables possibilities. Suppose that $d_{i(1)} \leqslant d_{i(2)} \leqslant \ldots \leqslant d_{i(n)}$ are the ordered distances of customer i to each site. Then, the variables can be fixed in the following way:

$$x_{ijk\ell} = 0 \text{ for all } i, j, k, \ell \in N \text{ such that } d_{ij} > d_{i(n-p+1)}.$$
$$x_{k\ell ij} = 0 \text{ for all } i, j, k, \ell \in N \text{ such that } d_{ij} > d_{i(n-p+1)}.$$

4 MIP Formulation for the *K-PpCP*

In this section we consider the variant of the previous model where only the K-largest distances are considered in the objective function. In order to do that, we need to use an additional family of variables:

$$z_{k\ell} = \begin{cases} 1, \text{ if } k \text{ is allocated to } \ell \text{ and the distance } d_{k\ell} \text{ is} \\ \quad \text{among the } n-K \text{ smallest distances,} \\ 0, \text{ otherwise.} \end{cases}$$

Taking only into account the K largest distances, we obtain the following formulation,

$$(\text{F1}^K) \ \min \sum_{i=1}^{n} \sum_{j=1}^{n} \pi_{ij} q_i d_{ij}$$
$$\text{s.t. constraints (2)–(7), (10)–(14), (18)}$$
$$\pi_{k\ell} \geqslant (1 - q_i)\pi_{ij} - 1 + x_{ijk\ell} - z_{k\ell}, \ \forall i, j, k, \ell \in N, \tag{23}$$
$$\sum_{i=1}^{n} \sum_{j=1}^{n} \sum_{\substack{k'=1,\ell'=1 \\ d_{k'\ell'} \geqslant d_{k\ell}}}^{n} x_{ijk'\ell'} \geq K z_{k\ell}, \ \forall k, \ell \in N. \tag{24}$$
$$z_{k\ell} \in \{0, 1\}, \ \forall k, \ell \in N. \tag{25}$$

Constraints (24) are used to guarantee that the z variables take consistent values. In addition, (24) is similar to (9) where term $-z_{kl}$ has been included to distinguish whether the distance d_{kl} is among K-largest distances.

We can add as valid inequalities for this formulation,

$$z_{k\ell} \leqslant \sum_{i=1}^{n} \sum_{j=1}^{n} x_{ijk\ell}, \ \forall k, \ell \in N, \tag{26}$$

$$z_{k\ell} \leqslant \sum_{i=1}^{n} \sum_{j=1}^{n} x_{k\ell ij}, \ \forall k, \ell \in N. \tag{27}$$

5 Computational Results

In this section we provide computational results for the formulations presented above. In order to know which valid inequalities improve the performance of this first formulation, Tables 2 and 3 show the results obtained when different inequality combinations are used. Both formulations were implemented using Mosel programming language and compiled by Xpress 7.7. Instances were run on a Intel(R) Core(TM) i7-4790K CPU 32 GB RAM. All the results reported in this work have been run in this same computer. We also established a time limit of 2 h to solve the problem.

In our experiments, we generated *PpCP* instances from the ORLIB p-median data electronically available at http://people.brunel.ac.uk/~mastjjb/jeb/orlib. We extracted several distance submatrices with different values of n from instances pmed1, pmed2, pmed3, pmed4 and pmed5. The demand probabilities were generated randomly in $[0, 1]$ and different values of p ranging from 2 to 10 were used.

Tables 2 and 3 report three columns of results for each variant of the formulation. The first column shows average gap (in percentage) between optimal solution and LP solution, the second column shows the average number of nodes that were used in the B&B tree and, finally, average running time (in seconds) appears in the third column. For each instance set the smallest average CPU times are marked in bold face. Superindices in the running time columns report the number of instances that were not solved in the limit time fixed to 7200 s.

It can be observed that running times considerably improve when C.A.C. are added to the formulation. While F1 finds optimal solutions until $n = 10$ and $p = 5$, the same formulation with C.A.C. can solve the *PpCP* for $n = 15$. We can observe this in the second group of columns of Table 2. In the third group of three columns we also use decomposed form of constraint (7) obtaining similar results.

We can also see in Table 2 the results of formulation $F1'$, where $F1' = F1 + [(13) - (17)] + (18) - (7) - (8)$. In this variant of the formulation C.A.C. are added, decomposed form of (7) is used and constraint (18) is used instead of (8). As we can see, in terms of CPU time, this variant provides the best results.

Since best time results are obtained with F1$'$ we continue the computational study using this formulation. In Table 3 one can see the results of formulation F1$'$ plus different combinations of inequalities (19)–(21). We can observe that $F1'$, $F1' + (20)$ and $F1' + (21)$ provide the best results.

In the case of the *K-PpCP*, results are reported in Table 4. As before, columns in Table 4 show average gaps at the root node, number of nodes and running times for each group of five instances. Again, bold face is used to mark the smallest times.

The first three columns correspond to formulation F1^K, the second group of columns corresponds to the same formulation plus constraints (26) and (27). In order to improve these results we use decomposed form of constraint (7).

Note that, surprisingly, the *K-PpCP* turned out to be more difficult to solve than the *PpCP*. One might expect that, in the case of the *PpCP*, part of the complexity arises from the fact that π variables have to take a value that is the product of many probabilities, which can cause numerical errors. However, as it happens in the p-

Table 2 Computational results for different possibilities of the *PpCP* formulation

		F1			F1+(13)+(14)			F1+[(13)–(17)]−(7)			F1′		
n	p	Gap	Nodes	Time	Gap	Nodes	Time	Gap	Nodes	Time	Gap	Nodes	Time
6	2	95.23	552	0.83	95.23	187	0.73	89.76	138	**0.60**	89.76	182	0.62
10	3	96.55	61507	576.20	96.55	419	12.52	93.88	509	15.37	93.88	516	**11.56**
10	5	96.19	660561	4601.11	96.06	469	**10.90**	93.05	445	10.96	93.05	696	11.74
13	3	98.18	105309	7200$^{(5)}$	98.15	1107	216.69	96.74	998	197.00	96.74	1243	**175.88**
13	5				97.08	2200	**351.09**	95.20	2859	395.11	95.20	4144	358.95
13	8				96.43	1511	180.21	93.51	1407	145.24	93.51	1586	**137.06**
15	3				99.16	1887	982.08	97.99	2038	841.92	97.99	2409	**742.73**
15	7				98.71	6420	2043.42	97.01	6162	1921.91	97.01	6665	**1718.50**
15	10				97.98	2217	554.01	94.15	1859	490.85	94.15	2078	**446.51**
20	3				99.48	1889	7206$^{(5)}$	98.81	2216	**7197**$^{(5)}$	98.81	3023	7204$^{(5)}$

Table 3 Computational results for further alternatives of formulation $F1'$

		F1′+(19)			F1′+(20)			F1′+(19)+(20)			F1′+(21)		
n	p	Gap	Nodes	Time	Gap	Nodes	Time	Gap	Nodes	Time	Gap	Nodes	Time
6	2	89.76	205	0.81	89.76	182	0.66	89.76	195	0.80	89.76	159	**0.62**
10	3	93.88	677	16.78	93.88	618	14.56	93.88	555	16.66	93.88	589	**12.05**
10	5	93.05	714	15.07	93.05	658	12.55	93.05	633	14.60	93.05	616	**10.29**
13	3	96.74	1372	219.82	96.74	1159	**173.54**	96.74	1258	215.33	96.74	1308	181.74
13	5	95.20	2967	356.31	95.20	2612	**277.97**	95.20	3164	368.94	95.20	3478	308.60
13	8	93.51	1539	150.55	93.51	1489	**124.42**	93.51	1517	136.96	93.51	1632	128.78
15	3	97.99	1762	**697.67**	97.99	2266	760.57	97.99	2211	801.92	97.99	2494	765.62
15	7	97.01	7022	2074.30	97.01	7033	1648.86	97.01	6882	1954.71	97.01	6392	**1566.36**
15	10	94.15	2081	526.15	94.15	2134	**450.59**	94.15	2133	563.44	94.15	2179	460.74
20	3	98.80	2950	7209(5)	98.79	3255	**7200**(5)	98.80	2330	7207(5)	98.77	2719	7205(5)

Table 4 Computational results of *K-PpCP* formulation

			$F1^K$			$F1^K$+(26)+(27)			$F1^K$+(26)+(27) + [(15)–(17)]−(7)		
n	p	K	Gap	Nodes	Time	Gap	Nodes	Time	Gap	Nodes	Time
6	2	2	92.71	144	**0.72**	92.71	175	0.85	84.43	136	0.77
10	3	3	95.35	541	57.65	95.35	417	**51.16**	91.72	488	53.71
10	5	3	95.80	878	64.57	95.80	845	**56.81**	92.59	849	65.51
13	3	4	97.79	1359	1554.02	97.79	1743	1568.11	96.10	1364	**1304.23**
13	5	4	96.93	3373	2382.23	96.93	3179	**2282.92**	94.90	3427	2309.56
13	8	4	96.32	1988	1344.41	96.32	1932	**1174.36**	93.24	1878	1165.70
15	3	4	99.05	1348	7220$^{(5)}$	98.99	1455	**7105**$^{(4)}$	97.84	1572	7172$^{(4)}$
15	7	4	98.51	2134	7215$^{(5)}$	98.54	1938	**7214**$^{(5)}$	96.82	2357	7223$^{(5)}$
15	10	4	97.89	2270	6299$^{(3)}$	97.86	2390	6210$^{(2)}$	93.75	2673	**5945**$^{(3)}$

center problem, when only one or a few assignment distances are taken into account in the objective function, many solutions exist with the same or very similar costs. We attribute the difficulty of solving the *K-PpCP* to this fact.

6 Conclusions

In this paper we have addressed a location problem where there is an uncertainty in the demand, i.e., the clients can have demand or not depending of a probability distribution. In spite of the large number of real situations that fit to this model, this has not been studied in the literature. We provide a first formulation for the problem and some interesting properties of the problem are also provided.

Acknowledgements This research has been partially supported by the Spanish Ministerio de Economía y Competitividad, grants numbers MTM2012-36163-C06-05 and MTM2013-46962-C2-2-P and Junta de Andalucía, grant number FQM-5849 and by ERDF funds. This support is gratefully acknowledged.

References

1. Albareda-Sambola, M., Díaz, J.A., Fernández, E.: Lagrangian duals and exact solution to the capacitated p-center problem. Eur. J. Oper. Res. **201**, 71–81 (2010)
2. Balcik, B., Beamon, B.M.: Facility location in humanitarian relief. Int. J. Logist. Res. Appl. **11**, 101–121 (2008)
3. Basar, A., Aatay, B., Unluyurt, T.: A taxonomy for emergency service station location problem. Optim. Lett. **6**, 1147–1160 (2012)
4. Berman, O., Krass, D., Menezes, M.B.C: Facility reliability issues in network p-median problems: strategic centralization and co-location effects. Oper. Res. **55**, 332–350 (2007)
5. Chang, M.-S., Tseng, Y.-L., Chen, J.-W.: A scenario planning approach for the flood emergency logistics preparation problem under uncertainty. Transp. Res. Part E **43**, 737–754 (2007)
6. Cui, T., Ouyang, Y., Shen Z.M.: Reliable facility location design under the risk of disruptions. Oper. Res. **58**, 996–1011 (2010)
7. Daskin, M.S.: Network and Discrete Location: Models, Algorithms, and Applications. Wiley, New York (1995)
8. Daskin, M.: A new approach to solving the vertex p-center problem to optimality: algorithm and computational results. Commun. Oper. Res. Soc. Jpn. **45**, 428–436 (2000)
9. Drezner, Z., Hamacher, H.: Facility Location: Applications and Theory. Springer, New York (2002)
10. Elloumi, S., Labbé, M., Pochet, Y.: New formulation and resolution method for the p-center problem. INFORMS J. Comput. **16**, 84–94 (2004)
11. Espejo, I., Marín, A., Rodríguez-Chía, A.: Closest assignment constraints in discrete location problems. Eur. J. Oper. Res. **219**, 49–58 (2012)
12. Huang, R., Kim, S., Menezes, M.B.C.: Facility location for large-scale emergencies. Ann. Oper. Res. **181**, 271–286 (2010)
13. Jia, H., Ordoñez, F., Dessouky, M.M.: A modeling framework for facility location of medical services for large-scale emergencies. IIE Trans. **39**, 41–55 (2007)

14. Jia, H., Ordoñez, F., Dessouky, M.M.: Solution approaches for facility location of medical supplies for large-scale emergencies. Comput. Ind. Eng. **52**, 257–276 (2007)
15. Kalcsics, J., Nickel, S., Puerto, J., Rodríguez-Chía, A.M.: Distribution systems design with role dependent objectives. Eur. J. Oper. Res. **202**, 491–501 (2010)
16. Kariv, O., Hakimi, S.L.: An algorithmic approach to network location problems I: the *p*-Centers. SIAM J. Appl. Math. **37**(3), 513–538 (1979)
17. Kouvelis, P., Yu, G.: Robust Discrete Optimization and Its Applications. Nonconvex Optimization and Its Applications, vol. 14, xvi, 356 p. Kluwer Academic Publishers, Dordrecht (1997)
18. Laporte, G., Nickel, S., Saldanha de Gama, F.: Location Science. Springer, Heidelberg (2015)
19. Marín, A., Nickel, S., Puerto, J., Velten, S.: A flexible model and efficient solution strategies for discrete location problems. Discret. Appl. Math. **157**, 1128–1145 (2009)
20. Mete, H.O., Zabinsky, Z.B.: Stochastic optimization of medical supply location and distribution in disaster management. Int. J. Prod. Econ. **126**, 76–84 (2010)
21. Mladenović, N., Labbé, M., Hansen, P.: Solving the p-center problem with tabu search and variable neighborhood search. Networks **42**, 48–64 (2003)
22. Nickel, S., Puerto, J.: Facility Location: A Unified Approach. Springer, Berlin (2005)
23. O'Hanley, J., Scaparra, M.P., García, S.: Probability chains: a general linearization technique for modeling reliability in facility location and related problems. Eur. J. Oper. Res. **230**, 63–75 (2013)
24. Shen, Z.J.M., Zhan, R.L., Zhang, J.: The reliable facility location problem: formulations, heuristics, and approximation algorithms. INFORMS J. Comput. **23**, 470–482 (2011)
25. Snyder L.V.: Facility location under uncertainty: a review. IIE Trans. **38**, 537–554 (2006)
26. Snyder, L.V., Daskin, M.S.: Reliability models for facility location: the expected failure cost case. Transp. Sci. **39**, 400–416 (2005)
27. Swamy, C., Shmoys, D.B.: Approximation algorithms for 2-stage stochastic optimization problems. In: Foundations of Software Technology and Theoretical Computer Science. Lecture Notes in Computer Science, vol. 4337, pp. 5–19. Springer, Berlin (2006)
28. Verma, A., Gaukler, G.M.: Pre-positioning disaster response facilities at safe locations: an evaluation of deterministic and stochastic modeling approaches. Comput. Oper. Res. **62**, 197–209 (2015)
29. Wagner, J.L., Falkson, L.M.: The optimal nodal location of public facilities with price-sensitive demand. Geogr. Anal. **7**, 69–83 (1975)
30. Zhan, R.L., Shen, Z.-J.M., Daskin, M.S.: System Reliability with Location-Specific Failure Probabilities. University of California, Berkeley (2007)

Regularized Inversion of Multi-Frequency EM Data in Geophysical Applications

Patricia Díaz de Alba and Giuseppe Rodriguez

Abstract The purpose of this work is to detect or infer, by non destructive investigation of soil properties, inhomogeneities in the ground or the presence of particular conductive substances such as metals, minerals and other geological structures. A nonlinear model is used to describe the interaction between an electromagnetic field and the soil. Starting from electromagnetic data collected by a ground conductivity meter, we reconstruct the electrical conductivity of the soil with respect to depth by a regularized Gauss–Newton method. We propose an inversion method, based on the low-rank approximation of the Jacobian of the nonlinear model, which depends both on a relaxation parameter and a regularization parameter, chosen by automatic procedures. Our numerical experiments on synthetic data sets show that the algorithm gives satisfactory results when the magnetic permeability in the subsoil takes small values, even when the noise level is compatible with real applications. The inversion problem becomes much harder to solve if the value of the permeability increases substantially, that is in the presence of ferromagnetic materials.

1 Introduction

Electromagnetic induction (EMI) is a non-invasive technique used to characterize the spatial variability of soil properties since the late 1970s. This technique has had widespread use in archaeological, hydrological, and geotechnical applications. In all cases, the soil property being investigated must influence its electrical conductivity either directly or indirectly, for EMI techniques to be effective. EM induction is becoming increasingly popular because it allows to collect large amount of data rapidly and inexpensively, and because in some situations it provides a better characterization of the spatial variations in soil properties than traditional techniques.

P. Díaz de Alba (✉) • G. Rodriguez
Department of Mathematics and Computer Science, University of Cagliari, Viale Merello 92, 09123 Cagliari, Italy
e-mail: patricia.diazdealba@gmail.com; rodriguez@unica.it

F. Ortegón Gallego et al. (eds.), *Trends in Differential Equations and Applications*, SEMA SIMAI Springer Series 8, DOI 10.1007/978-3-319-32013-7_20

A ground conductivity meter (GCM) is a device often used in applied geophysics. Its principle of operation is based on an alternating electrical current which flows through a small electric wire coils (the transmitter). A second coil (the receiver) is positioned at a fixed distance from the first one, and the two coil axes may be aligned either vertically or horizontally with respect to the surface of the soil. The transmitting coil generates an electromagnetic field above the ground, a portion of which propagates into it. This EM field, called the primary field H_P, induces eddy currents in the ground, in turn generating a secondary EM field H_S which propagates back to the surface and the air above. The second wire coil acts as a receiver, measuring the amplitude and phase components of both the primary and secondary EM fields.

The measurements obtained by a GCM depend on some instrument settings, like the orientation of the dipoles, the frequency of the alternating current, the inter-coil distance, and the height of the instrument above the ground.

Assuming a linear dependence between the GCM response and the subsurface electrical conductivity, a method was presented in [8] to estimate conductivities for a simple multilayered earth model, which is applicable for low induction numbers. The induction number, also called the "response parameter", combines many of the most significant parameters affecting the EM response into one single figure. It is defined as

$$B = r\sqrt{\frac{\mu_0 \omega \sigma}{2}},$$

where σ is the uniform electrical conductivity. The constant r is the inter-coil distance, $\mu_0 = 4\pi 10^{-7}$ H/m is the magnetic permeability of free space, and $\omega = 2\pi f$, being f the operating frequency of the device in Hz.

Adopting this linear model, Borchers et al. [3] implemented a Tikhonov inverse procedure to reconstruct conductivity profiles from measurements taken using a GCM at various heights above the ground. Then, to account for high values of the induction number, Hendrickx et al. [7] fitted the technique of Borchers et al. [3] to a nonlinear model described in Ward and Hohmann [11].

In this work, we extend a regularized inversion procedure based on the damped Gauss–Newton method, introduced by Deidda et al. [4]. The algorithm described therein takes into consideration the quadrature part of the measured signal, which is proportional to the "apparent" conductivity of the propagation medium. In order to investigate the possibility of getting more information from the available data, we consider either the in-phase or the quadrature component of the signal. Moreover, we introduce the possibility to process data collected at different operating frequencies, while in the mentioned paper the data to be inverted were obtained placing the instrument at various heights above the ground. This new approach is motivated by the availability of devices which use multiple frequencies simultaneously for each measurement.

The plan of the paper is the following. In Sect. 2 we describe the nonlinear forward problem which models the experimental setting. Section 3 describes the regularized inversion algorithm, and Sect. 4 reports the results of our numerical experiments. Section 5 contains concluding remarks and discusses possible future developments.

2 The Nonlinear Problem

The model here described is derived from Maxwell's equations, keeping into account the cylindrical symmetry of the problem. The input quantities are the distribution of the electrical conductivity and the magnetic permeability in the subsurface; the output is the instrument reading at height h.

The soil is assumed to possess a layered structure with n layers, each of thickness d_k, $k = 1, \dots, n$; see Fig. 1. As a consequence, the electromagnetic variables are piecewise constant. The thickness d_n of the bottom layer is assumed to be infinite. Let σ_k and μ_k be the electrical conductivity and the magnetic permeability in the k-th layer, respectively, and let $u_k(\lambda) = \sqrt{\lambda^2 + i\sigma_k\mu_k\omega}$, where λ is a variable of integration which has no particular physical meaning.

The characteristic admittance of the k-th layer is given by

$$N_k(\lambda) = \frac{u_k(\lambda)}{i\mu_k\omega}, \qquad k = 1, \dots, n.$$

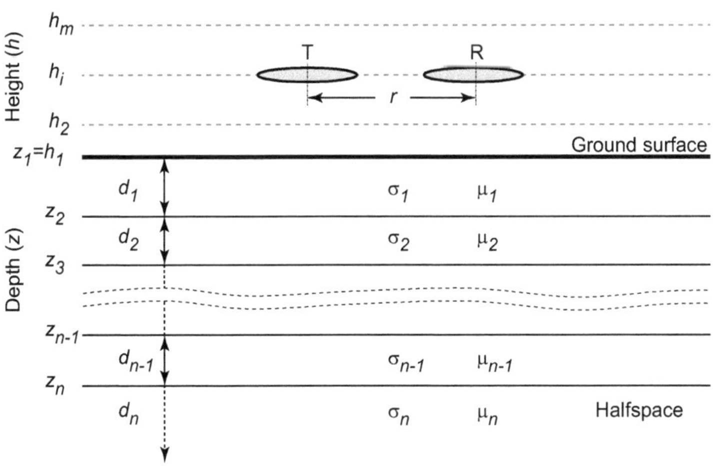

Fig. 1 Discretization and representation of the subsoil

The surface admittance at the top of the k-th layer is denoted by $Y_k(\lambda)$ and verifies the following recursion

$$Y_k(\lambda) = N_k(\lambda)\frac{Y_{k+1}(\lambda) + N_k(\lambda)\tanh(d_k u_k(\lambda))}{N_k(\lambda) + Y_{k+1}(\lambda)\tanh(d_k u_k(\lambda))},$$

for $k = n-1,\dots,1$. This recursion is initialized by setting $Y_n(\lambda) = N_n(\lambda)$ at the lowest layer.

Now, let the reflection factor be

$$R_0(\lambda) = \frac{N_0(\lambda) - Y_1(\lambda)}{N_0(\lambda) + Y_1(\lambda)},$$

where $N_0(\lambda) = \lambda/(\mathrm{i}\mu_0\omega)$. Then, assuming that the instrument coils are oriented vertically,

$$\frac{H_S}{H_P} = -r^3\int_0^\infty \lambda^2 e^{-2h\lambda} R_0(\lambda) J_0(r\lambda)\, d\lambda, \tag{1}$$

where H_P and H_S denote the primary and secondary magnetic field. A similar formula holds for the horizontal orientation of the coils [11]. We remark that (1) defines a complex valued function which can be evaluated by the Hankel transform.

In many previous works, only the quadrature component of (1) has been considered. This is justified by the fact that the imaginary part of H_S/H_P, scaled by the constant $4/(\mu_0\omega r^2)$, can be interpreted as an electrical conductivity, and is generally referred to as the *apparent conductivity*. In this work we consider either the in-phase or the quadrature component of the fields ratio, since they are both measured by a GCM.

To underline the role of the parameters which influence the measurements, we let $m(\boldsymbol{\sigma},\boldsymbol{\mu};\omega,h) := H_S/H_P$. The entries of the vectors $\boldsymbol{\sigma},\boldsymbol{\mu} \in \mathbb{R}^n$ are the conductivities and permeabilities of the ground layers, ω is the angular frequency of the instrument, and h is its height above the ground.

3 Solution of the Inverse Problem

In this paper, we assume the magnetic permeability to be known in each of the n layers. So the fields ratio (1) can be considered as a function of the values σ_k, $k = 1,\dots,n$, of the conductivity in the subsoil layers.

Multiple measurements are needed to recover the distribution of conductivity with respect to depth, so we assume that each measurement $b_{ij} \in \mathbb{C}$ is recorded at frequency ω_i, $i = 1,\dots,m_\omega$, and height h_j, $j = 1,\dots,m_h$. This amounts to $m = m_\omega m_h$ data points.

Let us consider the error in the model prediction, that is,

$$b_{ij} - m(\boldsymbol{\sigma}, \hat{\boldsymbol{\mu}}; \omega_i, h_j),$$

where $\hat{\boldsymbol{\mu}} = (\hat{\mu}_1, \dots, \hat{\mu}_n)^T$ is the known permeability distribution.

If the values b_{ij} and $m(\boldsymbol{\sigma}, \hat{\boldsymbol{\mu}}; \omega_i, h_j)$, with $i = 1, \dots, m_\omega$ and $j = 1, \dots, m_h$, are stacked in lexicographical order in the vectors $\mathbf{b}, \mathbf{m}(\boldsymbol{\sigma}) \in \mathbb{C}^m$ ($m = m_\omega m_h$), the residual vector can be written as

$$\mathbf{r}(\boldsymbol{\sigma}) = \mathscr{F}(\mathbf{b} - \mathbf{m}(\boldsymbol{\sigma})), \tag{2}$$

where $\mathscr{F}(\mathbf{z})$ denotes either the real or the imaginary part of the vector $\mathbf{z} \in \mathbb{C}^m$.

The problem of data inversion consists of computing the conductivity vector $\boldsymbol{\sigma} = (\sigma_1, \dots, \sigma_n)^T$ which determines a given data set $\mathbf{b} \in \mathbb{C}^m$. As it is customary, we use a least squares approach by solving the nonlinear problem

$$\min_{\boldsymbol{\sigma} \in \mathbb{R}^n} f(\boldsymbol{\sigma}), \quad f(\boldsymbol{\sigma}) = \frac{1}{2} \|\mathbf{r}(\boldsymbol{\sigma})\|^2, \tag{3}$$

where $\|\cdot\|$ represents the Euclidean norm.

The obvious choice of a solution algorithm for (3) is Newton's method. According to it, the step $\mathbf{s}_k$ in the iteration

$$\boldsymbol{\sigma}_{k+1} = \boldsymbol{\sigma}_k + \mathbf{s}_k$$

is chosen by solving the $n \times n$ linear system

$$\mathbf{f}''(\boldsymbol{\sigma}_k)\mathbf{s}_k = -\mathbf{f}'(\boldsymbol{\sigma}_k),$$

where $\mathbf{f}'(\boldsymbol{\sigma})$ is the gradient vector of $f(\boldsymbol{\sigma})$ and $\mathbf{f}''(\boldsymbol{\sigma}_k)$ is its Hessian matrix.

The analytical expression of $\mathbf{f}''(\boldsymbol{\sigma})$ is not always available, and its approximation often implies a large computational effort. To overcome this difficulty, we resort to the Gauss–Newton method, which minimizes at each step the norm of a linear approximation of the residual $\mathbf{r}(\boldsymbol{\sigma}_k + \mathbf{s}_k)$; see (2).

Let $\mathbf{r}(\boldsymbol{\sigma})$ be Fréchet differentiable and let $\boldsymbol{\sigma}_k$ denote the current approximation, then we can write

$$\mathbf{r}(\boldsymbol{\sigma}_{k+1}) \simeq \mathbf{r}(\boldsymbol{\sigma}_k) + J(\boldsymbol{\sigma}_k)\mathbf{s}_k,$$

where $J(\boldsymbol{\sigma})$ is the Jacobian of $\mathbf{r}(\boldsymbol{\sigma}) = (r_1(\boldsymbol{\sigma}), \dots, r_m(\boldsymbol{\sigma}))^T$, defined by

$$[J(\boldsymbol{\sigma})]_{ij} = \frac{\partial r_i(\boldsymbol{\sigma})}{\partial \sigma_j}, \qquad i = 1, \dots, m, \; j = 1, \dots, n.$$

The exact expression of the Jacobian matrix is given in [4, Theorem 3.2].

At each iteration k, the step length $\mathbf{s}_k$ is the solution of the linear least squares problem

$$\min_{\mathbf{s}\in\mathbb{R}^n} \|\mathbf{r}(\boldsymbol{\sigma}_k) + J_k\mathbf{s}\|, \tag{4}$$

with $J_k = J(\boldsymbol{\sigma}_k)$ or some approximation, leading to the following iterative method

$$\boldsymbol{\sigma}_{k+1} = \boldsymbol{\sigma}_k + \mathbf{s}_k = \boldsymbol{\sigma}_k - J_k^{\dagger}\,\mathbf{r}(\boldsymbol{\sigma}_k). \tag{5}$$

The symbol $J_k^{\dagger}$ denotes the Moore–Penrose pseudoinverse of the matrix J_k [1].

When the residuals $r_i(\boldsymbol{\sigma}_k)$ are small or mildly nonlinear in a neighborhood of the solution, the Gauss–Newton method is expected to behave similarly to Newton's method [1]. We remark that, while the physical problem is obviously consistent, this is not necessarily true in our case, where the conductivity $\sigma(z)$ is approximated by a piecewise constant function. Furthermore, in the presence of noise in the data the problem will certainly be inconsistent.

To ensure convergence, the damped Gauss–Newton method replaces the approximation (5) by

$$\boldsymbol{\sigma}_{k+1} = \boldsymbol{\sigma}_k + \alpha_k\mathbf{s}_k, \tag{6}$$

where α_k is a relaxation parameter to be determined. To choose it, we use the Armijo–Goldstein principle [9], which selects the step length α_k as the largest number in the sequence 2^{-i}, $i = 0, 1, \ldots,$ for which the following inequality holds

$$\|\mathbf{r}(\boldsymbol{\sigma}_k)\|^2 - \|\mathbf{r}(\boldsymbol{\sigma}_k + \alpha_k\mathbf{s}_k)\|^2 \geq \frac{1}{2}\alpha_k\|J_k\mathbf{s}_k\|^2. \tag{7}$$

This choice of α_k ensures convergence of the method, provided that $\boldsymbol{\sigma}_k$ is not a critical point [1], while the unrelaxed iteration may not converge at all.

The damped method allows us to include an important physical constraint in the inversion algorithm, i.e., the positivity of the solution. In our implementation, α_k is the largest step size which both satisfies the Armijo–Goldstein principle (7) and ensures that all the solution components are positive.

It is well known that problem (3) is extremely ill-conditioned. In particular, it has been observed in [4] that the Jacobian matrix $J(\boldsymbol{\sigma})$ has a large condition number virtually for each value of $\boldsymbol{\sigma}$ in the solution domain. A common remedy to overcome this difficulty consists of replacing the least-squares problem (4) by a nearby problem, whose solution is less sensitive to the error present in the data. This replacement is known as regularization.

A regularization method which particularly suits our problem, given the size of the matrices involved, is the truncated singular value decomposition (TSVD). The best rank ℓ approximation ($\ell \leq p = \text{rank}(J_k)$) to the Jacobian, according to the Euclidean norm, can be obtained by the SVD decomposition $J_k = U\Gamma V^T$, where

$\Gamma = \text{diag}(\gamma_1, \ldots, \gamma_p)$ and U, V are matrices with orthonormal columns $\mathbf{u}_i$, $\mathbf{v}_i$, respectively [1]. This factorization allows us to replace the ill-conditioned Jacobian matrix J_k by a well-conditioned low-rank matrix A_ℓ, such that

$$\|J_k - A_\ell\| = \min_{\text{rank}(A)=\ell} \|J_k - A\|.$$

Then, the regularized solution to (4) can be expressed as

$$\mathbf{s}^{(\ell)} = -A_\ell^\dagger \mathbf{r} = -\sum_{i=1}^{\ell} \frac{\mathbf{u}_i^T \mathbf{r}}{\gamma_i} \mathbf{v}_i,$$

where $\mathbf{r} = \mathbf{r}(\boldsymbol{\sigma}_k)$ and $\ell = 1, \ldots, p$ is the regularization parameter.

When some kind of a priori information on the problem is available, e.g., the solution is a smooth function, it is sometimes useful to introduce a regularization matrix $L \in \mathbb{R}^{t \times n}$ $(t \leq n)$, whose kernel approximately contains the sought solution. In this case, problem (4) is replaced by

$$\min_{\mathbf{s} \in \mathscr{S}} \|L\mathbf{s}\|, \qquad \mathscr{S} = \{\mathbf{s} \in \mathbb{R}^n \,:\, J_k^T J_k \mathbf{s} = -J_k^T \mathbf{r}(\boldsymbol{\sigma}_k)\},$$

under the assumption $\mathscr{N}(J_k) \cap \mathscr{N}(L) = \{0\}$. Very common choices for L are the discretization of the first derivative or the second derivative operators, which we will denote by D_1 and D_2, respectively.

The generalized singular value decomposition (GSVD) of the matrix pair (J_k, L) is the factorization

$$J_k = \tilde{U} \Sigma_J Z^{-1}, \qquad L = \tilde{V} \Sigma_L Z^{-1}, \tag{8}$$

where $\tilde{U}$ and $\tilde{V}$ are orthogonal matrices and Z is nonsingular. By the simultaneous factorization (8) it is possible to define a truncated GSVD (TGSVD) solution $\mathbf{s}_\ell$; see [4, 5] for details.

Our algorithm for the regularized solution of (3) applies either TSVD or TGSVD to each step of the damped Gauss–Newton method (6). For a fixed value of the regularization parameter ℓ, we substitute $\mathbf{s}_k$ in (6) by the truncated SVD or GSVD solution $\mathbf{s}^{(\ell)}$, obtaining the following iterative method

$$\boldsymbol{\sigma}_{k+1}^{(\ell)} = \boldsymbol{\sigma}_k^{(\ell)} + \alpha_k \mathbf{s}_k^{(\ell)}. \tag{9}$$

We denote by $\boldsymbol{\sigma}^{(\ell)}$ the solution at convergence.

The choice of the regularization parameter ℓ is crucial in order to obtain a good approximation $\boldsymbol{\sigma}^{(\ell)}$ of $\boldsymbol{\sigma}$. In real applications, experimental data are affected by noise, so the data vector in the residual function (2) must be expressed as $\mathbf{b} = \widehat{\mathbf{b}} + \mathbf{e}$, where $\widehat{\mathbf{b}}$ contains the exact data and $\mathbf{e}$ is the noise vector. If the noise is Gaussian and an accurate estimate of $\|\mathbf{e}\|$ is available, the discrepancy principle [5] determines ℓ

as the smallest index such that

$$\|\mathbf{b} - \mathbf{m}(\sigma^{(\ell)})\| \leq \kappa \|\mathbf{e}\|,$$

where $\kappa > 1$ is a user-specified constant independent of $\|\mathbf{e}\|$.

In the absence of a trustful estimate of the noise level, many *heuristic* methods have been introduced to approximate a regularization parameter. The L-curve, introduced by Hansen [5], is the curve which connects the points with coordinates

$$\left(\log \|\mathbf{r}(\sigma^{(\ell)})\|, \log \|L\sigma^{(\ell)}\|\right), \quad \ell = 1, \ldots, p.$$

In many discrete ill-posed problems this curve exhibits a typical "L" shape. The L-curve criterion selects the index ℓ corresponding to the vertex of the "L". This choice often produces a smooth solution with a sufficiently small residual. In our experiments, the corner is identified by means of the L-corner algorithm [6], which in this particular situation proved to be the most effective technique; see [10] for a review of methods.

4 Numerical Experiments

To assess the performance of our algorithm and to understand, at the same time, which experimental setting is the most effective for the investigation of the soil properties, we performed a set of numerical experiments on synthetic data sets. The computations were executed in double precision using MATLAB R2015a on an Intel Core i7 computer with 8 GB RAM, under the Linux operating system.

To model the conductivity of the subsoil with respect to depth, expressed in meters, we chose the test function $f(z)$ depicted in Fig. 2. For fixed n and $h = (3.5\,\text{m})/n$, we let $\sigma_q = f(qh)$ and $\widehat{\mu}_q = \mu_0$ for $q = 1, \ldots, n$. Then, we apply the forward model described in Sect. 2 to generate the instrument readings

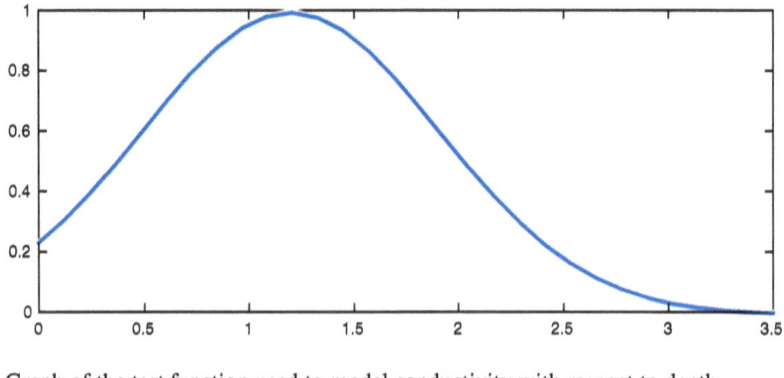

Fig. 2 Graph of the test function used to model conductivity with respect to depth

$$\hat{b}_{ij} = m(\boldsymbol{\sigma}, \widehat{\boldsymbol{\mu}}; \omega_i, h_j),$$

with $i = 1, \ldots, m_\omega$ and $j = 1, \ldots, m_h$, corresponding to frequency $\omega_i = 2\pi f_i$ and height h_j. Finally, we add Gaussian noise to the synthetic data by the formula

$$\mathbf{b} = \hat{\mathbf{b}} + \frac{\tau \|\hat{\mathbf{b}}\|}{\sqrt{m}} \mathbf{w},$$

where $\mathbf{w}$ is a vector with normally distributed entries with zero mean and unitary variance, $m = m_\omega m_h$, and τ is the noise level.

In order to simulate the use of a particular multi-frequency device, the Geophex GEM-2 conductivity meter, we consider the coils to be in the vertical orientation at a fixed distance $r = 1.66$ m. The measurement height h is either 1 m ($m_h = 1$) or 0.5 m and 1 m ($m_h = 2$). Each data set is recorded simultaneously at the operating frequencies $f_i = 775, 1175, 3925, 9825, 21725, 47025$ (all expressed in Hertz), that is, $m_\omega = 6$. This instrument setting is currently being used to process data collected in the Venice lagoon [2].

In the first experiment we investigate how to choose some of the parameters of the methods, namely, the regularization matrix L, the heights number m_h, the number of layers n, and whether the function $\mathscr{F}(z)$ in the residual (2) should be either the real or the imaginary part of z. For each choice of the parameters, we apply the above procedure to compute a synthetic data set, we add noise at level $\tau = 10^{-3}, 10^{-2}$, and we generate 20 realizations of the random noise vector $\mathbf{w}$, to produce 40 test problems. For each test, we measure the relative error

$$E_{\ell_{\text{opt}}} = \frac{\|\boldsymbol{\sigma} - \boldsymbol{\sigma}^{(\ell_{\text{opt}})}\|}{\|\boldsymbol{\sigma}\|},$$

where the regularization parameter ℓ_{opt} has been chosen in order to minimize the value of E_ℓ, $\ell = 1, \ldots, p$, so that the accuracy attained by the method is maximal.

For each combination of the selected parameters, we report in Table 1 the average of the values of $E_{\ell_{\text{opt}}}$ across the available 40 test problems. The table confirms that the choice of the regularization matrix $L = I$ produces the least accurate results, as observed in [4], while D_1 and D_2 are more or less equivalent. The method is not very sensitive upon the number of layers n and the accuracy does not improve substantially when $m_h = 2$, with respect to $m_h = 1$. Since increasing m_h implies a larger data acquisition time, in our next experiments we will set $L = D_2$, $m_h = 1$, and $n = 30$. Regarding the choice of the function $\mathscr{F}$, both the real and imaginary part of the signal seem to contain the same amount of information about the solution, with the quadrature component reaching a slightly better accuracy. This suggests that both components should be considered in the solution of the least squares problem (3). We plan to extend the algorithm in this sense in our future work.

Table 1 Best accuracy attainable by the method for selected choices of the parameters

	L	m_h	$n = 20$	$n = 30$	$n = 40$
$\mathscr{F} = \mathscr{R}$	I	1	3.6e − 01	3.7e − 01	3.7e − 01
		2	4.5e − 01	4.5e − 01	4.4e − 01
	D_1	1	3.0e − 01	3.3e − 01	2.9e − 01
		2	2.4e − 01	2.4e − 01	2.3e − 01
	D_2	1	2.4e − 01	2.1e − 01	2.8e − 01
		2	2.5e − 01	2.5e − 01	2.3e − 01
$\mathscr{F} = \mathscr{I}$	I	1	3.2e − 01	3.9e − 01	4.0e − 01
		2	3.0e − 01	3.4e − 01	3.4e − 01
	D_1	1	1.9e − 01	2.3e − 01	1.9e − 01
		2	2.1e − 01	1.9e − 01	1.9e − 01
	D_2	1	2.3e − 01	2.0e − 01	2.4e − 01
		2	2.1e − 01	2.2e − 01	2.1e − 01

Each entry of the table is the average of $E_{\ell_{\text{opt}}}$ across 40 experiments, with two noise levels and 20 noise realizations

Table 2 Results for different values of the relative magnetic permeability μ_r

	μ_0	$\mu_r = 10$	$\mu_r = 10^2$	$\mu_r = 10^3$
Optimal-$\mathscr{R}$	2.3e − 01	4.3e − 01	5.3e − 01	5.5e − 01
	0	13	9	19
Optimal-$\mathscr{I}$	2.4e − 01	5.3e − 01	4.5e − 01	7.1e − 01
	0	6	4	12
L-curve-$\mathscr{R}$	2.6e − 01	6.3e − 01	4.7e − 01	5.4e − 01
	0	20	18	27
L-curve-$\mathscr{I}$	2.6e − 01	4.2e − 01	5.5e − 01	7.4e − 01
	0	23	10	16

Each row displays the average error and the number of *failures* across 40 experiments; see text. The upper block concerns the optimal choice of ℓ, the bottom block the choice by the L-curve

In our next experiment, we consider the presence of electromagnetic materials in the subsoil ($\mu > \mu_0$ in some layer) and analyze the effectiveness of the L-curve as a method to choose the regularization parameter ℓ. Table 2 is divided into two main blocks: the first two rows concern the optimal choice $\ell = \ell_{\text{opt}}$, the last two rows the choice $\ell = \ell_{\text{L-curve}}$, produced by the L-curve. The integer number on the bottom of each row represents the number of *failures*, that is, how many of the 40 experiments produced a relative error larger than 1.5. We verified that when the error is below this limit it is still possible to recover from the solution significant information, e.g., the localization in depth of the maximal conductivity. The real number on the top of each row represents the average of $E_{\ell_{\text{opt}}}$ (first two rows) and $E_{\ell_{\text{L-curve}}}$ (last two rows) across the acceptable errors. The first column contains the result corresponding to $\hat{\mu}_q = \mu_0$, $q = 1, \ldots, n$, as for the previous experiment. In the second to fourth column, the magnetic permeability of each layer is set to

$$\hat{\mu}_q = \mu_r \mu_0 f(qh) + \mu_0, \quad q = 1, \ldots, n,$$

where $\mu_r = 10, 10^2, 10^3$, and $f(z)$ is the function of Fig. 2. The largest value of μ_r roughly correspond to the magnetic permeability of iron.

From Table 2, it is immediately evident that the inversion problem is much harder to solve when $\mu_r > 1$. The considerable number of experiments whose relative error is larger than 1.5 (the *failures*) suggests that the algorithm, originally conceived for constant permeability μ_0, should be modified in order to deal with the general situation. Nevertheless, when the algorithm does not fail the error for $\mu_r > 1$ is only slightly larger than for $\mu = \mu_0$. Preliminary results on field data (see [2]) suggest that the solutions produced by the method are still accurate for moderate values of μ_r.

When the regularization parameter ℓ is chosen by the L-curve, rather than optimally, the performance of the method gets worse, in terms of number of failures, but the error is still acceptable. This experiment confirms that both the real and imaginary part of the signal contain substantial information about the solution.

To illustrate the effect of regularization on the computed solutions we depict in Fig. 3 the first four regularized solutions $\sigma^{(\ell)}$, that is, the limit solutions of the iterative scheme (9) when $\ell = 1, 2, 3, 4$. This experiment is characterized by constant permeability μ_0 and noise level $\tau = 10^{-3}$; the solution (thick line) is computed by minimizing the real part of the signal. The exact solution is displayed in each graph by a thin line. The graphs show that when the parameter is smaller than the optimal value, the solution is over-regularized and it is just a sketch of the correct conductivity profile. On the contrary, when ℓ is too large there are no constraints on the error propagation, and the under-regularized solution exhibits abnormal oscillations.

In Fig. 4 we compare the solution obtained by minimizing the real part of the data (left column) to the one corresponding to the imaginary part (right column). The thick line is the exact solution, the thin line is the optimal solution, the dashed line represents the L-curve solution. The graphs in the top row correspond to $\mu_r = 10$ and $\tau = 10^{-3}$. When $\mathscr{F} = \mathscr{R}$, the L curve selects the optimal parameter $\ell = 2$, with an error $E_{\ell_{\text{opt}}} = E_{\ell_{\text{L-curve}}} = 0.37$; when $\mathscr{F} = \mathscr{I}$, the algorithm fails.

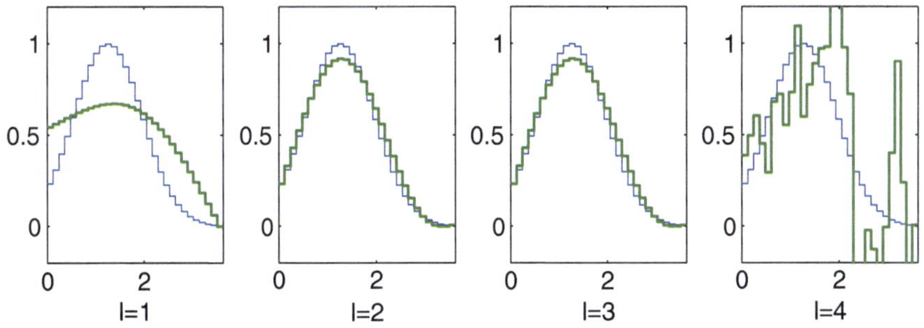

Fig. 3 Plot of the first four regularized solutions, computed by minimizing the real part of the signal, compared to the exact solution. The magnetic permeability $\mu = \mu_0$ is constant, the noise level is $\tau = 10^{-3}$

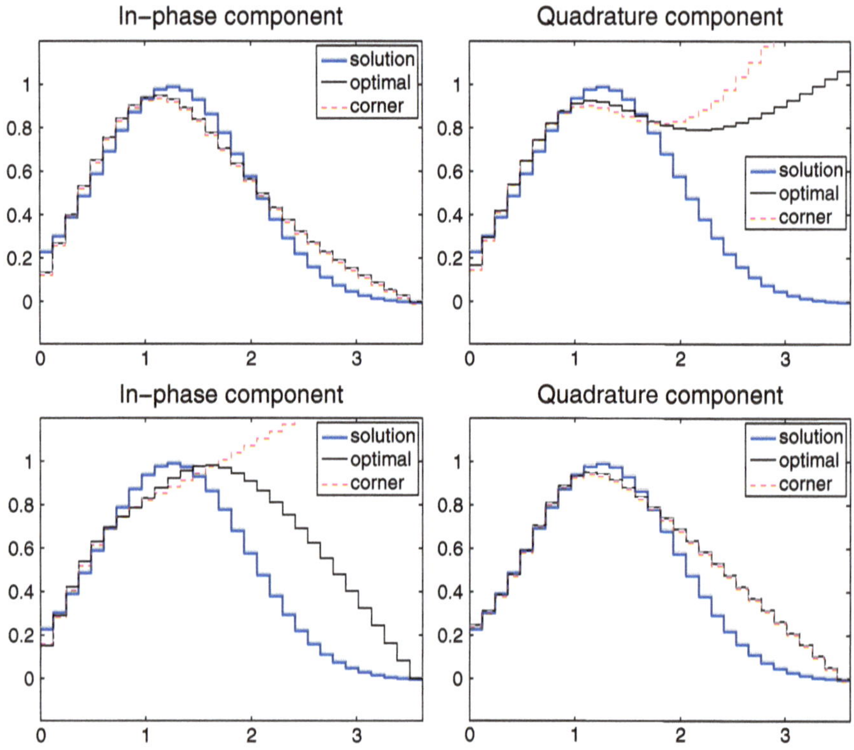

Fig. 4 Solution obtained by minimizing the real part of the data (*left*) or the imaginary part (*right*); $\tau = 10^{-3}$, $\mu_r = 10$ in the *top row*, $\mu_r = 10^2$ in the *bottom row*. The value of ℓ is chosen either optimally or by the L-curve

The bottom row of Fig. 4 displays a similar experiment, with $\mu_r = 10^2$. In this case, the L-curve correctly identifies the regularization parameter for the quadrature part minimization ($\ell = 3$), with an error $E_{\ell_{\text{opt}}} = E_{\ell_{\text{L-curve}}} = 0.70$. On the contrary, while the optimal error for the real part is $E_{\ell_{\text{opt}}} = 1.36$ ($\ell = 3$), the L-curve chooses the parameter $\ell = 2$, producing an incorrect solution.

5 Conclusions

In this paper we presented an extension of the algorithm proposed in [4] for the inversion of EMI data in a geophysical setting. The algorithm has been generalized in order to process data collected by multi-frequency devices, and to fit either the real or the imaginary part of the signal. Moreover, we took into consideration the presence of ferromagnetic materials in the subsoil.

While the results corresponding to low magnetic permeability ($\mu = \mu_0$) are satisfactory, considering larger permeabilities makes the algorithm less robust. In the near future, we plan to modify the inversion procedure in order to deal with the complex signal as a whole, and with large values of μ. Another problem we are facing is the determination of both the conductivity and the permeability starting from the data.

The new algorithm has been applied to EM data collected in the Venice lagoon, using a multi-frequency device, to which the algorithm from [4] could not have been applied. The results obtained seem to have correctly identified the structure of the subsoil in the surveyed area; see [2].

Acknowledgements The authors would like to thank the referees for carefully reading the manuscript and for comments that lead to improvements of the presentation. The work of the authors was partially supported by INdAM-GNCS.

References

1. Björck, A.A.: Numerical Methods for Least Squares Problems. SIAM, Philadelphia (1996)
2. Boaga, J., Ghinassi, M., D'Alpaos, A., Deidda, G.P., Rodriguez, G., Cassiani, G.: Unravelling the vestiges of ancient meandering channels in tidal landscapes via multi-frequency inversion of Electro-Magnetic data (2016, submitted)
3. Borchers, B., Uram, T., Hendrickx, J. M.: Tikhonov regularization of electrical conductivity depth profiles in field soils. Soil Sci. Soc. Am. J. **61**, 1004–1009 (1997)
4. Deidda, G.P., Fenu, C., Rodriguez, G.: Regularized solution of a nonlinear problem in electromagnetic sounding. Inverse Prob. **30**, 125014 (2014)
5. Hansen, P.C.: Rank-Deficient and Discrete Ill-Posed Problems. SIAM, Philadelphia (1998)
6. Hansen, P.C., Jensen, T.K., Rodriguez, G.: An adaptive pruning algorithm for the discrete L-curve criterion. J. Comput. Appl. Math. **198**, 483–492 (2007)
7. Hendrickx, J.M.H., Borchers, B., Corwin, D.L., Lesch, S.M., Hilgendorf, A.C., Schlue, J.: Inversion of soil conductivity profiles from electromagnetic induction measurements. Soil Sci. Soc. Am. J. **66**, 673–685 (2002)
8. McNeill, J.D.: Electromagnetic terrain conductivity measurement at low induction numbers. Technical Report TN-6 Geonics Limited (1980)
9. Ortega, J.M., Rheinboldt, W.C.: Iterative Solution of Nonlinear Equations in Several Variables. Academic Press, New York (1970)
10. Reichel, L., Rodriguez, G.: Old and new parameter choice rules for discrete ill-posed problems. Numer. Algorithm. **63**, 65–87 (2013)
11. Ward, S.H., Hohmann, G.W.: Electromagnetic theory for geophysical applications. In: Nabighian, M.N. (ed.) Electromagnetic Methods in Applied Geophysics, Volume 1. Theory, Volume 3 of Investigation in Geophysics, pp. 131–311. Society of Exploration Geophysicists, Tulsa, OK (1987)

Total Positivity: A New Inequality and Related Classes of Matrices

A. Barreras and J.M. Peña

Abstract In this paper we present the extension of some results to classes of matrices related to total positivity. First, we survey some properties and results for matrices with Signed Bidiagonal Decomposition (SBD matrices), a class of matrices that contains Totally Positive (TP) matrices and their inverses. We also extend the affirmative answer of an inequality conjectured for the Frobenius norm of the inverse of matrices whose entries belong to [0, 1] to the class of nonsingular totally positive matrices.

1 Introduction

There are some classes of structured matrices very important in applications and that also present many advantages under a mathematical and a computational point of view. In this last aspect, we can mention that recent research in Numerical Linear Algebra has shown that certain classes of matrices allow us to perform many computations to high relative accuracy, independently of the size of the condition number (cf. [14]). For instance, the computation of their singular values, eigenvalues or inverses. These classes of matrices are defined by special sign or other structure and require to know some natural parameters to high relative accuracy, and they are related to some subclasses of P-matrices. Let us recall that a square matrix is called a P-matrix if all its principal minors are positive. Subclasses of P-matrices with many applications are the nonsingular totally nonnegative matrices and the nonsingular M-matrices (a nonsingular matrix A with nonpositive off-diagonal entries is an M-matrix if A^{-1} has nonnegative entries). Usually, accurate spectral computation (eigenvalues, singular values) or accurate inversion is assured when an accurate matrix factorization with a suitable pivoting is provided. For instance, the bidiagonal decomposition in the case of totally positive matrices (see [24]) or an

A. Barreras (✉)
Centro Universitario de la Defensa/IUMA, Universidad de Zaragoza, Zaragoza, Spain
e-mail: albarrer@unizar.es

J.M. Peña
Dep. Applied Mathematics/IUMA, Universidad de Zaragoza, Zaragoza, Spain
e-mail: jmpena@unizar.es

F. Ortegón Gallego et al. (eds.), *Trends in Differential Equations and Applications*,
SEMA SIMAI Springer Series 8, DOI 10.1007/978-3-319-32013-7_21

LDU factorization after a symmetric pivoting in the case of diagonally dominant matrices (cf. [12, 29]).

Let us recall that a matrix is *totally positive* (TP) if all its minors are nonnegative. Let us remind that TP matrices are also called in the literature totally nonnegative matrices (cf. [17]). The class of totally positive matrices is representative of the properties mentioned in the first sentence of this introduction. It has applications to many fields, presents interesting theoretical properties and has very nice stability and computational properties (see [2, 17–19, 22, 24, 28, 31]).

In this paper, we survey the extension of some properties valid for TP matrices to a more general class of matrices, called SBD matrices (matrices with signed bidiagonal decomposition). We also extend the affirmative answer of an inequality conjectured for the Frobenius norm of the inverse of matrices whose entries belong to $[0, 1]$ to the class of nonsingular TP matrices. In Sect. 2 we present the class of SBD matrices and several results extending properties of TP matrices to the new class (see also [7]). In Sect. 3 we introduce the basic concepts related to the conjecture. We start with the case of tridiagonal TP matrices and provide a proof using elementary arguments for the sake of self completeness. In Sect. 4 we prove the extension to any class of P-matrices closed under Schur complements and satisfying the Fisher inequality. As a consequence, we derive the result for nonsingular TP matrices.

2 A Class of Matrices Related to Total Positivity

Let $Q_{k,n}$ be the set of sequences of k $(\leq n)$ positive integers less than or equal to n. Given $\alpha \in Q_{k,n}$ and $\beta \in Q_{l,n}$ we denote by $A[\alpha|\beta]$ the $k \times l$ submatrix of A containing rows numbered by α and columns numbered by β. If $\alpha = \beta$ then we denote the principal submatrix by $A[\alpha] := A[\alpha|\alpha]$. We use α^c to denote the increasingly rearranged complement of α, $\alpha^c = \{1, \ldots, n\} \setminus \alpha$. Besides we denote $A(\alpha|\beta] := A[\alpha^c|\beta]$, $A[\alpha|\beta) := A[\alpha|\beta^c]$ and $A(\alpha) := A[\alpha^c]$.

Some classes of matrices, including TP matrices, can be decomposed as a product of bidiagonal matrices. These decompositions use matrices of the form

$$L^{(k)} = \begin{pmatrix} 1 & & & & & & \\ 0 & 1 & & & & & \\ & \ddots & \ddots & & & & \\ & & 0 & 1 & & & \\ & & & l_{n-k}^{(k)} & 1 & & \\ & & & & \ddots & \ddots & \\ & & & & & l_{n-1}^{(k)} & 1 \end{pmatrix},$$

$$U^{(k)} = \begin{pmatrix} 1 & 0 & & & & & \\ & \ddots & \ddots & & & & \\ & & 1 & 0 & & & \\ & & & 1 & u_{n-k}^{(k)} & & \\ & & & & \ddots & \ddots & \\ & & & & & 1 & u_{n-1}^{(k)} \\ & & & & & & 1 \end{pmatrix},$$

where $k = 1, \ldots, n-1$.

Let A be a nonsingular $n \times n$ matrix. Suppose that we can write A as a product of bidiagonal matrices

$$A = L^{(1)} \cdots L^{(n-1)} D U^{(n-1)} \cdots U^{(1)}, \tag{1}$$

where $D = \operatorname{diag}(d_1, \ldots, d_n)$, and, for $k = 1, \ldots, n-1$, $L^{(k)}$ and $U^{(k)}$ are matrices as mentioned above satisfying:

1. $d_i \neq 0$ for all i,
2. $l_i^{(k)} = u_i^{(k)} = 0$ for $i < n-k$,
3. $l_i^{(k)} = 0 \Rightarrow l_{i+s}^{(k-s)} = 0$ for $s = 1, \ldots, k-1$ and
   $u_i^{(k)} = 0 \Rightarrow u_{i+s}^{(k-s)} = 0$ for $s = 1, \ldots, k-1$.

Then we denote (1) by $\mathscr{BD}(A)$ a bidiagonal decomposition of A. It was proved, in [4, Proposition 2.2], that this decomposition is unique.

Let us recall an important result that characterizes nonsingular TP matrices in terms of it bidiagonal decomposition (it is a consequence of Theorem 4.2 of [20]).

Theorem 1 *A nonsingular $n \times n$ matrix A is TP if and only if there exists a (unique) $\mathscr{BD}(A)$ such that:*

1. $d_i > 0$ *for all i.*
2. $l_i^{(k)} \geq 0$, $u_i^{(k)} \geq 0$ *for* $1 \leq k \leq n-1$ *and* $n-k \leq i \leq n-1$.

Let us now define a class of matrices with bidiagonal decomposition with special sign conditions on the entries of bidiagonal matrices. This class will generalize the class of nonsingular TP matrices. Recall that a vector $\varepsilon = (\varepsilon_1, \ldots, \varepsilon_m)$ with $\varepsilon_j \in \{-1, 1\}$ for $j = 1, \ldots, m$ is called *signature*.

Definition 1 Given a signature $\varepsilon = (\varepsilon_1, \ldots, \varepsilon_{n-1})$ and a nonsingular $n \times n$ matrix A, we say that A is SBD with signature ε if there exists a $\mathscr{BD}(A)$ such that:

1. $d_i > 0$ for all i,
2. $l_i^{(k)} \varepsilon_i \geq 0$, $u_i^{(k)} \varepsilon_i \geq 0$ for $1 \leq k \leq n-1$ and $n-k \leq i \leq n-1$.

We say that A is SBD, if it is SBD with some signature ε. Observe that these matrices are nonsingular.

Given a signature $\varepsilon = (\varepsilon_1, \ldots, \varepsilon_{n-1})$, let us define a diagonal matrix $K = \mathrm{diag}(k_1, \ldots, k_n)$ with k_i satisfying

$$k_i \in \{-1, 1\} \ \forall\, i = 1, \ldots, n, \quad k_i k_{i+1} = \varepsilon_i \ \forall\, i = 1, \ldots, n-1. \tag{2}$$

Now, we present three results that provide characterizations of SBD matrices. Some of these characterizations are given in terms of important matrices decompositions, such as *LDU* decomposition or *UL* decomposition.

The following theorem appeared in [4, Theorem 3.1] and provides several characterizations of SBD matrices.

Theorem 2 *Let $A = (a_{ij})_{1\le i,j\le n}$ be a nonsingular matrix and let $\varepsilon = (\varepsilon_1, \ldots, \varepsilon_{n-1})$ be a signature. Then the following properties are equivalent:*

(i) *A is SBD with signature ε.*
(ii) *$KAK = |A|$ is TP, where K is any diagonal matrix satisfying (2).*
(iii) *A^{-1} is SBD with signature $-\varepsilon = (-\varepsilon_1, \ldots, -\varepsilon_{n-1})$.*
(iv) *$|A|$ is TP and, for all $1 \le i, j \le n$,*

$$sign(a_{ij}) = \begin{cases} \varepsilon_j \cdots \varepsilon_{i-1}, & \text{if } i > j \\ 1 & , \text{if } i = j \\ \varepsilon_i \cdots \varepsilon_{j-1}, & \text{if } i < j. \end{cases}$$

Observe that an SBD matrix with signature $(1, \ldots, 1)$ is a nonsingular TP matrix. As a corollary of Theorem 2 we have the following result, which corresponds with [4, Corollary 3.3]:

Corollary 1 *Let A be a nonsingular matrix. Then the following properties are equivalent:*

(i) *A is SBD with signature $(1, \ldots, 1)$.*
(ii) *A is TP.*
(iii) *A^{-1} is SBD with signature $(-1, \ldots, -1)$.*

Let us now present a characterization of SBD matrices in terms of their *LDU* decomposition (cf. [4, Proposition 3.5]).

Proposition 1 *An $n \times n$ matrix A is SBD with signature $\varepsilon = (\varepsilon_1, \ldots, \varepsilon_{n-1})$ if and only if $A = LDU$, where L (resp., U) is a lower (resp., an upper) triangular matrix with unit diagonal and SBD with signature ε and D is a diagonal matrix whose diagonal entries are positive.*

Let us recall that, given a matrix A, a factorization $A = BC$ is called an *UL decomposition* if B is upper triangular and C is lower triangular. In order to characterize SBD matrices in terms of their *UL* decomposition, we need to introduce a new class of matrices (presented in [5]). We say that a matrix A is *signature similar to TP with signature ε*, denoted by SSTP with signature ε, if $A = KBK$, where B is TP and K satisfies (2). The following proposition (which corresponds with [5, Proposition 3.10]) gives a characterization of SBD matrices by their *UL* decomposition.

Proposition 2 *A matrix A is SBD with signature* ε *if and only if there exists a lower and an upper triangular SSTP matrices with signature* ε, A_L *and* A_U, *such that* $A = A_U A_L$.

An algorithm can be performed with *high relative accuracy* if it does not include subtractions (except of the initial data), that is, if it only includes products, divisions, sums of numbers of the same sign and subtractions of the initial data (cf. [14]). Up to now, we only have algorithms with high relative accuracy for a reduced number of classes of matrices, related with total positivity (cf. [1, 10, 11, 24, 26]) or diagonal dominance (cf. [12, 29]). In the problem of finding algorithms with high relative accuracy, the choice of adequate parameters is crucial to avoid subtractions during the algorithm. For nonsingular TP matrices, if we know with high relative accuracy the entries of (1), then algorithms with high relative accuracy can be applied (cf. [23, 24]). We recall that, with the same parameters, these algorithms can be used to compute with high relative accuracy the singular values, eigenvalues, inverses or the *LDU* decomposition of SBD matrices (cf. [4]).

Given an SBD matrix A, let us observe that, from (1) and taking into account that $K^2 = I$, we have

$$\begin{aligned} KAK = {} & (KL^{(1)}K) \cdots (KL^{(n-1)}K)(KDK) \\ & (KU^{(n-1)}K) \cdots (KU^{(1)}K), \end{aligned} \tag{3}$$

which is the $\mathscr{BD}(KAK)$. Besides, taking into account (2), it can be checked that all factors of $\mathscr{BD}(KAK)$ are nonnegative.

As shown in recent references [13, 15, 23–26], the diagonal entries of the diagonal matrix D of the $\mathscr{BD}(A)$ (see Eq. (1)) and the off-diagonal entries of the remaining factors of (1) can be considered natural parameters associated with A. In the computation of these parameters, Neville elimination (see [26]) has been frequently a useful tool. Let us see that if we assume that we know these parameters with high relative accuracy for SBD matrices, then we can find algorithms with high relative accuracy to compute their singular values, their eigenvalues, their inverses or to solve certain linear systems $Ax = b$ (those with Kb with a chessboard pattern of signs).

For all the mentioned computations we can follow a procedure that were presented in [4] and it can be summarized by the following steps:

1. From $\mathscr{BD}(A)$, we obtain $\mathscr{BD}(|A|)$, given by (3).
2. We can apply known algorithms with high relative accuracy for TP matrices to $\mathscr{BD}(|A|)$. Recall that, by Theorem 2, $|A|$ is TP if A is SBD.
3. From the information obtained for $|A|$, we can get the corresponding result for A.

Let us now explain how to perform each of the previous steps.

As for Step 1, let us assume that we know the $\mathcal{BD}(A)$ (see Eq. (1)) with high relative accuracy for a given SBD matrix. Then $|A| = KAK$ for a diagonal matrix K satisfying (2) and so we can deduce from (3) that

$$|A| = |L^{(1)}| \cdots |L^{(n-1)}||D||U^{(n-1)}| \cdots |U^{(1)}| \tag{4}$$

is the $\mathcal{BD}(|A|)$. Since all factors of $\mathcal{BD}(KAK)$ are nonnegative, we have that $|L^{(j)}| = KL^{(j)}K$, $|U^{(j)}| = KU^{(j)}K$ for all $j = 1, \ldots, n-1$. Thus, (4) follows from (3).

As for Step 2, we apply the corresponding algorithm for TP matrices with high relative accuracy, using $\mathcal{BD}(|A|)$ (given by (4)). In particular, we consider the following accurate computations with TP matrices:

A. The eigenvalues of $|A|$ can be obtained by the method of [23, Sect. 5].
B. The singular values of $|A|$ can be obtained by the method of [23, Sect. 6].
C. The inverse of $|A|$ can be obtained by the method of [24, p. 736].
D. Observe that $Ax = b$ is equivalent to solving $(KAK)(Kx) = Kb$, that is, $|A|(Kx) = Kb$. Then, $|A|^{-1}$ can be calculated accurately by the procedure of the previous case. By Ando [2, Theorem 3.3], $|A|^{-1}$ has a chessboard pattern of signs and so, since Kb has also a chessboard pattern of signs, $Kx = |A|^{-1}(Kb)$ can be calculated without subtractions and therefore with high relative accuracy.

As for Step 3, we have the following cases corresponding to each of the cases of Step 2:

A. We have that $|A| = KAK = K^{-1}AK$ and so they are similar matrices and have the same eigenvalues.
B. The singular values of A and $|A|$ coincide because $|A| = KAK$, that is, $|A|$ and A coincide up to unitary matrices.
C. We have that $|A|^{-1} = (KAK)^{-1} = KA^{-1}K$ and so $A^{-1} = K|A|^{-1}K$.
D. If we know Kx, then $x = K(Kx)$.

In addition, let us show that if we have the $\mathcal{BD}(A)$ (see Eq. (1)) with high relative accuracy, then we can also calculate the LDU decomposition of A with high relative accuracy, and even obtain the matrix A with high relative accuracy. In fact, by the uniqueness of the LDU decomposition of a matrix, it can be checked that

$$L = L^{(1)} \cdots L^{(n-1)}, \qquad U = U^{(n-1)} \cdots U^{(1)}. \tag{5}$$

Since the bidiagonal matrices $L^{(k)}$, $U^{(k)}$ satisfy sign properties of Definition 1, then we have that matrices L and U can be calculated without subtractions and so with high relative accuracy. Then we can also compute $A = LDU$ with high relative accuracy.

Several properties of SBD matrices have been studied in [3, 4]. We summarize some of them in the following result.

Proposition 3 *Let A, B be two $n \times n$ SBD matrices with the same signature ε. Then*

(i) A^T is also SBD with signature ε.

(*ii*) *Any principal submatrix of A is SBD.*
(*iii*) *AB is also SBD with signature ε.*
(*iv*) *A is a P-matrix, that is, it has all its principal minors positive.*

Given two matrices $A = (a_{ij})_{1\leq i,j\leq n}$ and $B = (b_{ij})_{1\leq i,j\leq n}$, we define the *Hadamard product*, or entrywise product, of A and B as the matrix $A \circ B := (a_{ij}b_{ij})_{1\leq i,j\leq n}$. The *Hadamard core* (cf. [5, 9]) of the $n \times n$ TP matrices is given by

$$CTP := \{A : B \text{ is TP } \Rightarrow A \circ B \text{ is TP}\}. \tag{6}$$

It is known, by Fallat and Johnson [17, Theorem 8.2.5], that tridiagonal TP matrices are in the CTP. Then, by Fallat and Johnson [17, Corollary 8.3.2], it can be deduced the following result (cf. [5, Proposition 3.1]).

Proposition 4 *Let A be an $n \times n$ tridiagonal TP matrix and B an $n \times n$ TP matrix. Then* $\det(A \circ B) \geq \det A \det B$.

Given an $n \times n$ TP matrix A, then we say that A is *oscillatory* if a certain power of A, A^k, becomes strictly totally positive; that is, all the minors of A^k are strictly positive (see [2]). Recall that, by Ando [2, Theorem 4.2], a nonsingular TP matrix $A = (a_{ij})_{1\leq i,j\leq n}$ is oscillatory if and only if $a_{i,i+1} > 0$ and $a_{i+1,i} > 0$ for all $i = 1, \ldots, n-1$. Moreover, observe that since an oscillatory matrix is TP and nonsingular, we have that $a_{ii} > 0$ for all $i = 1, \ldots, n$ (cf. [2, Corollary 3.8]). Thus, a tridiagonal oscillatory matrix $A = (a_{ij})_{1\leq i,j\leq n}$ satisfies $a_{ij} \neq 0$ for $|i-j| \leq 1$.

It is known (cf. [31, Proposition 4.12], [9, Corollary 2.7] or [17, Corollary 8.2.6]) that the Hadamard product of two $n\times n$ tridiagonal TP matrices is again a tridiagonal TP matrix. Taking into account that the nonsingularity of tridiagonal TP matrices is also preserved by the Hadamard product (see Proposition 4), we can extend the previous fact to nonsingular tridiagonal TP matrices. Thus, in [5, Proposition 3.2], a generalization of [27, Theorem 1] from the class of tridiagonal oscillatory matrices to the class of nonsingular tridiagonal TP matrices was given.

Proposition 5 *Let A, B be two nonsingular $n \times n$ tridiagonal TP matrices. Then $A \circ B$ is a nonsingular tridiagonal TP matrix.*

We shall now extend Proposition 4 to the class of SBD matrices. Analogously to (6), we define the Hadamard core of the $n \times n$ SBD matrices by

$$CSBD := \{A : B \text{ is SBD } \Rightarrow A \circ B \text{ is SBD}\}. \tag{7}$$

The following result (cf. [5, Proposition 3.4]) shows that tridiagonal SBD matrices belong to CSBD.

Proposition 6 *Let A be an $n\times n$ tridiagonal SBD matrix and B an $n\times n$ SBD matrix. Then $A \circ B$ is SBD.*

The set of tridiagonal SBD matrices form a semigroup with respect to the Hadamard product as the following corollary [5, Corollary 3.5] shows. It extends Proposition 5 to tridiagonal SBD matrices and it is a direct consequence of Proposition 6.

Corollary 2 *Let A, B be two $n \times n$ tridiagonal SBD matrices. Then the matrix $A \circ B$ is a tridiagonal SBD matrix.*

Besides, Corollary 2 cannot be extended to SBD matrices that are not tridiagonal, as the following example shows. Observe that the following matrix is a nonsingular TP matrix

$$A = \begin{pmatrix} 1.1 & 1 & 1 \\ 1 & 1 & 1 \\ 0 & 1 & 1.1 \end{pmatrix},$$

so, by Theorem 2, A is SBD. However, the matrix

$$A \circ A^T = \begin{pmatrix} 1.1^2 & 1 & 0 \\ 1 & 1 & 1 \\ 0 & 1 & 1.1^2 \end{pmatrix}$$

satisfies that $\det(A \circ A^T) = -0.9559 < 0$. So $A \circ A^T$ is not a P-matrix. Recall that, by Proposition 3, SBD matrices are P-matrices, and then, we conclude that $A \circ A^T$ it is not SBD.

Let us recall that given A an $n \times n$ matrix, if $A[\alpha \mid \beta]$ is invertible for some $\alpha, \beta \in Q_{k,n}$, $1 \le k \le n$, then the *Schur complement* of $A[\alpha \mid \beta]$ in A, denoted by $A/A[\alpha \mid \beta]$, is defined as

$$A/A[\alpha \mid \beta] := A(\alpha \mid \beta) - A(\alpha \mid \beta]A[\alpha \mid \beta]^{-1}A[\alpha \mid \beta). \tag{8}$$

If $\alpha = \beta$, we denote $A/A[\alpha \mid \alpha]$ by $A/A[\alpha]$.

If A is invertible, we can use formula (1.29) of [2] to derive the following formula for Schur complement of principal submatrices:

$$(A/A[\alpha])^{-1} = A^{-1}(\alpha) = A^{-1}[\alpha^c]. \tag{9}$$

In [17, Proposition 1.5.1], it is shown that the Schur complement of principal submatrices using contiguous index sets, $A/A[\alpha]$ with $\alpha = (i, i+1, \ldots, i+k-1)$, of a nonsingular TP matrix, is TP. However, this result is not valid for general Schur complements of TP matrices. For SBD matrices, in [5, Theorem 3.6] it was proved that general Schur complements of principal submatrices of SBD matrices are again SBD.

Theorem 3 *Let A be an SBD matrix. Then $A/A[\alpha]$, the Schur complement of $A[\alpha]$ in A, is SBD for all $\alpha \in Q_{k,n}$, $1 \le k \le n$.*

Finally, let us present a lower bound for a minimal eigenvalue of an SBD matrix. Let us recall that the well-known Gerschgorin's Circles Theorem provides a lower bound for an eigenvalue with minimal absolute value, λ_*, of a matrix $A = (a_{ij})_{1\leq i,j\leq n}$:

$$|\lambda_*| \geq \min_i \left\{ |a_{ii}| - \sum_{j\neq i} |a_{ij}| \right\}.$$

The following result improves this bound for SBD matrices. The next index subset is used in following result: given $i \in \{1,\ldots,n\}$ let

$$J_i := \{j| \, |j-i| \text{ is odd}\}. \tag{10}$$

Observe that, by Theorem 2, we know that $K|A|K$ is SBD, where $|A|$ is a TP matrix; that is, SBD matrices are similar to TP matrices. So A and $|A|$ have the same eigenvalues. Recall that an eigenvalue with minimal absolute value of a nonsingular TP matrix is positive (cf. [2, Corollary 6.6]). Thus, we know that SBD matrices also satisfy this property. The following result extends to SBD matrices the bound obtained in [30, Theorem 4.4] for nonsingular TP matrices and corresponds with [5, Corollary 2.7].

Proposition 7 *Let A be an $n \times n$ SBD matrix and let λ_* be an eigenvalue of A with minimal absolute value. For each $i \in \{1,\ldots,n\}$, let J_i be the index subset defined by (10). Then*

$$\lambda_* \geq \min_i \left\{ a_{ii} - \sum_{j\in J_i} |a_{ij}| \right\}. \tag{11}$$

The following example (which is included in [5, Example 2.8]) shows that the bound given by Proposition 7 cannot be improved.

Example 1 Let us consider the SBD matrix

$$A = \begin{pmatrix} 12 & -7 & -1 \\ 0 & 6 & 1 \\ 0 & 3 & 8 \end{pmatrix}.$$

The eigenvalues of A are 12, 9 and 5, which coincides with the eigenvalues of the TP matrix $|A|$. Observe that the bound given by (11), $\lambda_* \geq 5$, cannot be improved, because this bound is achieved by the smallest eigenvalue. Observe also that the lower bound given by the Gerschgorin's Circles Theorem is $\lambda_* \geq \min\{4,5,5\} = 4$, which is worse than the previous one.

3 Inequality for Tridiagonal TP Matrices

We present in this section another property of tridiagonal TP matrices. In particular, a lower bound for the norm of the inverse of a tridiagonal TP matrix with entries in [0, 1] is presented.

Recall that, given a nonsingular $n \times n$ matrix A, the procedure of Gaussian elimination without pivoting provides as result a sequence of $n-1$ matrices:

$$A = A^{(1)} \longrightarrow A^{(2)} \longrightarrow \cdots \longrightarrow A^{(n)}, \tag{12}$$

where $A^{(t)}$ has zeros below its main diagonal in the first $t-1$ columns:

$$A^{(t)} = \begin{pmatrix} a_{11}^{(t)} & a_{12}^{(t)} & \cdots & \cdots & \cdots & a_{1n}^{(t)} \\ 0 & a_{22}^{(t)} & \cdots & \cdots & \cdots & a_{2n}^{(t)} \\ \vdots & 0 & \ddots & & & \vdots \\ \vdots & \vdots & & \ddots & & \vdots \\ \vdots & \vdots & & & a_{tt}^{(t)} \ldots & a_{tn}^{(t)} \\ \vdots & \vdots & & & \vdots & \vdots \\ 0 & 0 & \cdots & \cdots & a_{nt}^{(t)} \ldots & a_{nn}^{(t)} \end{pmatrix}.$$

Given a real matrix $A = (a_{ij})_{1 \le i,j \le n}$ its Frobenius norm is defined as

$$\|A\|_F := \left(\sum_{i=1}^{n} \sum_{j=1}^{n} a_{ij}^2 \right)^{1/2}.$$

A Hadamard matrix of order n is a matrix $A = (a_{ij})_{1 \le i,j \le n}$ such that $a_{ij} \in \{-1, 1\}$ whose rows and columns are mutually orthogonal; that is, $AA^T = nI_n$, where I_n is the identity matrix of order n. An S-matrix of order n is a matrix $A = (a_{ij})_{1 \le i,j \le n}$ such that $a_{ij} \in \{0, 1\}$, formed by considering a Hadamard matrix of order $n+1$ in which the entries in the first row and column are 1, changing 1's to 0's and -1's to 1's, and deleting the first row and the first column.

Let $\mathcal{D}_n$ denote the set of all $n \times n$ matrices A whose entries are in the interval [0, 1]. Sloane and Harwit proposed (see [32]) the following conjecture concerning matrices in $\mathcal{D}_n$.

Conjecture 1 If $A \in \mathcal{D}_n$ is a nonsingular matrix, then

$$\|A^{-1}\|_F \ge \frac{2n}{n+1},$$

where the equality holds if and only if A is an S-matrix.

This conjecture appeared from a problem in the field of spectroscopy (cf. [21]). The conjecture were proved for matrices of odd order by Cheng in 1987 (see [8, 16]).

We now present a result that proves the conjecture for nonsingular tridiagonal TP matrices that lie in the set $\mathscr{D}_n$.

Proposition 8 *Let $A \in \mathscr{D}_n$ be a nonsingular tridiagonal TP matrix. Then*

$$\|A^{-1}\|_F \geq \frac{2n}{n+1}.$$

Proof We proceed by induction on n. If $n = 2$ it is known (see [16]) that $\|A^{-1}\|_F \geq \sqrt{2} > 4/3$.

Suppose that the result holds for matrices of order $n-1$; that is, given $\tilde{A} \in \mathscr{D}_{n-1}$ nonsingular tridiagonal TP, then $\|\tilde{A}^{-1}\|_F \geq \frac{2n-2}{n}$. Consider now $A \in \mathscr{D}_n$ a nonsingular tridiagonal TP matrix. Since, by Ando [2, Corollary 3.8], a nonsingular TP matrix have positive principal minors, we have that $a_{11} > 0$. Then, after the first step in Gaussian elimination, we have $A^{(2)} = (a_{ij}^{(2)})_{1\leq i,j\leq n}$ (see Eq. (12))

$$A^{(2)} = \left(\begin{array}{c|c} a_{11} & a_{12} \\ \hline 0 & \\ \vdots & A^{(2)}[2,\ldots,n] \\ 0 & \end{array}\right)$$

where $a_{i1}^{(2)} = 0$ for all $i \in \{2,\ldots,n\}$ and $a_{11}^{(2)} = a_{11} \neq 0$. Observe that we can express $A = L_1^{-1}A^{(2)}$ $(=: (a_{ij})_{1\leq i,j\leq n})$, where

$$L_1 = \begin{pmatrix} 1 & & & \\ \frac{-a_{21}}{a_{11}} & 1 & & \\ 0 & & 1 & \\ \vdots & & & \ddots & \\ 0 & & & & 1 \end{pmatrix}.$$

Thus, we have that

$$A^{-1} = (L_1^{-1}A^{(2)})^{-1} = (A^{(2)})^{-1}L_1$$

$$= \left(\begin{array}{c|ccc} \frac{1}{a_{11}} & \beta_2 & \cdots & \beta_n \\ \hline 0 & & & \\ \vdots & & (A^{(2)}[2,\ldots,n])^{-1} & \\ 0 & & & \end{array}\right) \begin{pmatrix} 1 & & \\ \frac{-a_{21}}{a_{11}} & 1 & \\ & & \ddots & \\ & & & 1 \end{pmatrix}$$

$$= \left(\begin{array}{c|ccc} (A^{-1})_{11} & \beta_2 & \cdots & \beta_n \\ \hline \gamma_2 & & & \\ \vdots & & (A^{(2)}[2,\ldots,n])^{-1} & \\ \gamma_n & & & \end{array} \right) \tag{13}$$

where β_i, γ_i are real numbers for all $i \in \{2,\ldots,n\}$ and $(A^{-1})_{11}$ denotes the (1,1) entry of A^{-1}. Since $A^{(2)}$ is nonsingular we have that $B := A^{(2)}[2,\ldots,n]$ is also nonsingular. Taking into account that $A \in \mathcal{D}_n$ is tridiagonal TP and considering Gaussian elimination, we have that

$$a_{ij}^{(2)} = a_{ij} \in [0,1]$$

for all $i,j \in \{2,\ldots,n\}$, $(i,j) \neq (2,2)$ and we deduce that

$$a_{22}^{(2)} = a_{22} - \frac{a_{21}}{a_{11}} a_{12} \leq a_{22} \leq 1$$

and

$$a_{22}^{(2)} = a_{22} - \frac{a_{21}}{a_{11}} a_{12} = \frac{\det A[1,2|1,2]}{a_{11}} > 0.$$

Thus $B \in \mathcal{D}_{n-1}$. Furthermore, observe that B is also tridiagonal and it can be expressed as the Schur complement $B = A/A[1]$ and this Schur complement in a TP matrix is TP (see [2, Theorem 3.3]). Thus, by the induction hypothesis, we have that

$$\|B^{-1}\|_F \geq \frac{2n-2}{n}. \tag{14}$$

Observe that, since $A \in \mathcal{D}_n$ is TP and considering the (1,1) cofactor of A and formula (2) of [6] for the determinant of a tridiagonal matrix ($\det A = a_{11} \det A[2,\ldots,n] - a_{21}a_{12} \det A[3,\ldots,n]$), we have

$$\begin{aligned} (A^{-1})_{11} &= \frac{\det A[2,\ldots,n]}{\det A} = \frac{\det A[2,\ldots,n]}{a_{11} \det A[2,\ldots,n] - a_{12}a_{21} \det A[3,\ldots,n]} \\ &\geq \frac{\det A[2,\ldots,n]}{a_{11} \det A[2,\ldots,n]} \geq 1. \end{aligned} \tag{15}$$

Thus, by (13)–(15) we can derive

$$\|A^{-1}\|_F^2 = \|B^{-1}\|_F^2 + \left((A^{-1})_{11}\right)^2 + \sum_{i=2}^{n} \beta_i^2 + \sum_{i=2}^{n} \gamma_i^2 \geq \|B^{-1}\|_F^2 + C, \tag{16}$$

for any $0 \leq C \leq 1$. Let us consider $\hat{C} = \frac{8n^2-4}{n^2(n+1)^2}$, observe that then $0 \leq \hat{C} \leq 1$ for all $n > 2$ and thus, by (14) and (16), we have

$$\|A^{-1}\|_F^2 \geq \|B^{-1}\|_F^2 + \hat{C} \geq \left(\frac{2n-2}{n}\right)^2 + \frac{8n^2-4}{n^2(n+1)^2} = \left(\frac{2n}{n+1}\right)^2$$

and the results holds.

4 Inequality for a General Class of Matrices

In this section, we shall prove that the inequality of the conjecture recalled in the previous section also holds for more general classes of matrices and, in particular, for nonsingular TP matrices in $\mathscr{D}_n$.

Our classes of matrices will be closed under Schur complements and will be formed by P-matrices (all its principal minors are positive) satisfying, in addition, a classical inequality called the Fisher inequality:

$$\det A \leq \det A[\alpha] \det A(\alpha)$$

for any $\alpha \in Q_{k,n}$ and $1 \leq k < n$.

Theorem 4 *Let $A \in \mathscr{C}_n \cap \mathscr{D}_n$, where $\mathscr{C}_n$ is any class of $n \times n$ P-matrices closed under Schur complements and satisfying the Fisher inequality. Then*

$$\|A^{-1}\|_F \geq \frac{2n}{n+1}.$$

Proof We proceed by induction on n. If $n = 2$ it is known (see [16]) that $\|A^{-1}\|_F \geq \sqrt{2} > 4/3$.

Suppose that the result holds for matrices of order $n-1$; that is, given $\tilde{A} \in \mathscr{C}_{n-1} \cap \mathscr{D}_{n-1}$, then $\|\tilde{A}^{-1}\|_F \geq \frac{2n-2}{n}$. Consider now $A \in \mathscr{C}_n \cap \mathscr{D}_n$. Since A is a P-matrix, we have that $a_{11} > 0$. Then, after the first step in Gaussian elimination, we have $A^{(2)} = (a_{ij}^{(2)})_{1 \leq i,j \leq n}$ (see Eq. (12))

$$A^{(2)} = \left(\begin{array}{c|c} a_{11} & a_{12} \\ \hline 0 & \\ \vdots & A^{(2)}[2, \ldots, n] \\ 0 & \end{array}\right)$$

where $a_{i1}^{(2)} = 0$ for all $i \in \{2, \dots, n\}$ and $a_{11}^{(2)} = a_{11} \neq 0$. Observe that we can express $A = L_1^{-1}A^{(2)}$ $(=: (a_{ij})_{1\le i,j\le n})$, where

$$L_1 = \begin{pmatrix} 1 & & & \\ \frac{-a_{21}}{a_{11}} & 1 & & \\ \vdots & & \ddots & \\ \frac{-a_{n1}}{a_{11}} & & & 1 \end{pmatrix}.$$

Thus, we have that (13) holds. Since $A^{(2)}$ is nonsingular we have that $B := A^{(2)}[2, \dots, n]$ is also nonsingular. Taking into account that $A \in \mathcal{D}_n$ is a P-matrix and considering Gaussian elimination, we have that either

$$a_{ij}^{(2)} = a_{ij} \in [0, 1]$$

or

$$a_{ij}^{(2)} = a_{ij} - \frac{a_{i1}}{a_{11}} a_{1j} \le a_{ij} \le 1$$

and

$$a_{ij}^{(2)} = a_{ij} - \frac{a_{i1}}{a_{11}} a_{1j} = \frac{\det A[1, i|1, j]}{a_{11}} > 0.$$

Thus $B \in \mathcal{D}_{n-1}$. Furthermore, observe that B can be expressed as the Schur complement $B = A/A[1]$ and so $B \in \mathcal{C}_{n-1}$. In conclusion, $B \in \mathcal{C}_{n-1} \cap \mathcal{D}_{n-1}$. Thus, by the induction hypothesis, we have that (14) holds. Since $A \in \mathcal{C}_n$, Fisher's inequality implies that $\det A \le a_{11} \det A[2, \dots, n]$ and, taking also into account that $A \in \mathcal{D}_n$, we have

$$(A^{-1})_{11} = \frac{\det A[2, \dots, n]}{\det A} \ge \frac{\det A[2, \dots, n]}{a_{11} \det A[2, \dots, n]} \ge 1. \tag{17}$$

Thus, by (13), (14) and (17) we can derive (16) for any $0 \le C \le 1$. Let us consider $\hat{C} = \frac{8n^2-4}{n^2(n+1)^2}$, observe that then $0 \le \hat{C} \le 1$ for all $n > 2$ and thus, by (14) and (16), we have

$$\|A^{-1}\|_F^2 \ge \|B^{-1}\|_F^2 + \hat{C} \ge \left(\frac{2n-2}{n}\right)^2 + \frac{8n^2-4}{n^2(n+1)^2} = \left(\frac{2n}{n+1}\right)^2$$

and the results holds.

As a consequence of the previous result, we can extend the result of the previous section to all nonsingular TP matrices.

Corollary 3 *Let $A \in \mathscr{D}_n$ be a nonsingular TP matrix. Then*

$$\|A^{-1}\|_F \geq \frac{2n}{n+1}.$$

Proof By Theorem 4, it is sufficient to see that the class of nonsingular TP matrices is a class of P-matrices closed under Schur complements and satisfying the Fisher inequality. By Ando [2, Corollary 3.8], nonsingular TP matrices are P-matrices. It is well known that they are closed under Schur complements (cf. [2, Theorem 3.3]). Finally, it is also well known that they satisfy the Fisher inequality (cf. [17]).

If we consider a nonsingular (tridiagonal) TP matrix A, observe that the lower bound provided in Sects. 3 and 4 for the norm of A^{-1} could imply an ill conditioning of A. However we have presented, in Sect. 2, accurate computations for these classes of matrices that do not depend on the conditioning of the initial matrix.

Acknowledgements This work has been partially supported by the Spanish Research Grant MTM2015-65433, by Gobierno the Aragón and Fondo Social Europeo.

References

1. Alonso, P., Delgado, J., Gallego, R., Peña, J.M.: Conditioning and accurate computations with Pascal matrices. J. Comput. Appl. Math. **252**, 21–26 (2013)
2. Ando, T.: Totally positive matrices. Linear Algebra Appl. **90**, 165–219 (1987)
3. Barreras, A., Peña, J.M.: Bidiagonal decompositions, minors and applications. Electron. J. Linear Algebra **25**, 60–71 (2012)
4. Barreras, A., Peña, J.M.: Accurate computations of matrices with bidiagonal decomposition using methods for totally positive matrices. Numer. Linear Algebra Appl. **20**, 413–424 (2013)
5. Barreras, A., Peña, J.M.: On the extension of some total positivity inequalities. Linear Algebra Appl. **448**, 153–167 (2014)
6. Barreras, A., Peña, J.M.: On tridiagonal sign regular matrices and generalizations. Advances in Differential Equations and Applications. SEMA SIMAI Springer Series, vol. 4, pp. 239–247. Springer, Cham (2014)
7. Barreras, A., Peña, J.M.: Classes of structured matrices related with total positivity. In: Díaz, J.M., Díaz, J.C., García, C., Medina, J., Ortegóm, F., Pérez, C., Redondo, M.V., Rodríguez, J.R. (eds.) Proceedings of the XXIV Congress of Differential Equations and Applications/XIV Congress on Applied Mathematics, pp. 745–750 (2015). ISBN: 978-84-9828-527-7
8. Cheng, C.S.: An application of the Kiefer-Wolfowitz equivalence theorem to a problem in Hadamard transform optics. Ann. Stat. **15**, 1593–1603 (1987)
9. Crans, A.S., Fallat, S.M., Johnson, C.R.: The Hadamard core of the totally nonnegative matrices. Linear Algebra Appl. **328**, 203–222 (2001)
10. Delgado, J., Peña, J.M.: Accurate computations with collocation matrices of rational bases. Appl. Math. Comput. **219**, 4354–4364 (2013)
11. Delgado, J., Peña, J.M.: Fast and accurate algorithms for Jacobi-Stirling matrices. Appl. Math. Comput. **236**, 253–259 (2014)
12. Demmel, J., Koev, P.: Accurate SVDs of weakly diagonally dominant M-matrices. Numer. Math. **98**, 99–104 (2004)

13. Demmel, J., Koev, P.: The accurate and efficient solution of a totally positive generalized Vandermonde linear system. SIAM J. Matrix Anal. Appl. **27**, 142–152 (2005)
14. Demmel, J., Gu, M., Eisenstat, S., Slapnicar, I., Veselic, K., Drmac, Z.: Computing the singular value decomposition with high relative accuracy. Linear Algebra Appl. **299**, 21–80 (1999)
15. Dopico, F.M., Koev, P.: Accurate symmetric rank revealing and eigen decompositions of symmetric structured matrices. SIAM J. Matrix Anal. Appl. **28**, 1126–1156 (2006)
16. Drnovšek, R.: On the S-matrix conjecture. Linear Algebra Appl. **439**, 3555–3560 (2013)
17. Fallat, S.M., Johnson, C.R.: Totally Nonnegative Matrices. Princeton University Press, Princeton/Oxford (2011)
18. Gantmacher, F.P., Krein, M.G.: Oscillation Matrices and Kernels and Small Vibrations of Mechanical Systems (revised edn.). AMS Chelsea, Providence, RI (2002)
19. Gasca, M., Micchelli, C.A. (eds.): Total Positivity and Its Applications. Mathematics and Its Applications, vol. 359. Kluwer Academic Publisher, Dordrecht (1996)
20. Gasca, M., Peña, J.M.: On factorizations of totally positive matrices. In: Gasca, M., Micchelli, C.A. (eds.) Total Positivity and Its Applications. Mathematics and Its Applications, vol. 359, pp. 109–130. Kluwer Academic Publishers, Dordrecht (1996)
21. Harwit, M., Sloane, N.J.A.: Hadamard Transform Optics. Academic Press. New York (1979)
22. Karlin, S.: Total Positivity, vol. I. Stanford University Press, Stanford (1968)
23. Koev, P.: Accurate eigenvalues and SVDs of totally nonnegative matrices. SIAM J. Matrix Anal. Appl. **27**, 1–23 (2005)
24. Koev, P.: Accurate computations with totally nonnegative matrices. SIAM J. Matrix Anal. Appl. **29**, 731–751 (2007)
25. Marco, A., Martínez, J.J.: A fast and accurate algorithm for solving Bernstein-Vandermonde linear systems. Linear Algebra Appl. **422**, 616–628 (2007)
26. Marco, A., Martínez, J.J.: Accurate computations with Said-Ball-Vandermonde matrices. Linear Algebra Appl. **432**, 2894–2908 (2010)
27. Markham, T.L.: A semigroup of totally nonnegative matrices. Linear Algebra Appl. **3**, 157–164 (1970)
28. Peña, J.M. (ed.): Shape Preserving Representations in Computer Aided Geometric Design. Nova Science Publishers, Commack, NY (1999)
29. Peña, J.M.: LDU decompositions with L and U well conditioned. Electron. Trans. Numer. Anal. **18**, 198–208 (2004)
30. Peña, J.M.: Eigenvalue bounds for some classes of P-matrices. Numer. Linear Algebra Appl. **16**, 871–882 (2009)
31. Pinkus, A.: Totally Positive Matrices. Cambridge Tracts in Mathematics, Num. 181. Cambridge University Press, Cambridge (2010)
32. Sloane, N.J.A, Harwit, M.: Masks for Hadamard transform optics, and weighing designs. Appl. Opt. **15** 107–114 (1976)

Applications of $\mathscr{C}^\infty$-Symmetries in the Construction of Solvable Structures

Adrián Ruiz and Concepción Muriel

Abstract A complete set of first integrals for a third order ordinary differential equation (ODE) that admits the non-solvable symmetry algebra $\mathfrak{sl}(2,\mathbb{R})$ can be found by quadratures. These first integrals arise from a solvable structure that can be constructed in terms of two first integrals associated to $\mathscr{C}^\infty$-symmetries of a reduced second order ODE. The general procedure is illustrated by an explicit example where three independent first integrals of the third order equation are provided in terms of a complete set of solutions to a second order linear ODE.

1 Introduction

If an nth order ordinary differential equation (ODE) admits a k-dimensional Lie symmetry algebra, $\mathscr{G}$, then its general solution can be obtained by means of the general solution of an $(n-k)$th order reduced equation and the solution of a kth order auxiliary equation. If $\mathscr{G}$ is solvable, then the general solution of the corresponding auxiliary equation can be obtained by k successive quadratures [8, 9, 14, 16]. However, if $\mathscr{G}$ is non-solvable, this step by step method of reduction is no longer available. The reason is that, at a certain stage of the reduction process, at least one of the generators of $\mathscr{G}$ cannot be used to proceed with the order reduction. In this case we say, roughly speaking, that the corresponding symmetries have been lost for the reduced equation.

Lost symmetries have been widely studied in the literature [1–5]. These lost symmetries are called type I hidden symmetries and they are difficult to study because there are no general methods for determining them. These type I hidden symmetries can be found in the case of an ODE which admits the non-solvable

A. Ruiz (✉) • C. Muriel
Department of Mathematics, University of Cádiz, Campus Universitario de Puerto Real, 11510 Cádiz, Spain
e-mail: adrian.ruizservan@alum.uca.es; concepcion.muriel@uca.es

F. Ortegón Gallego et al. (eds.), *Trends in Differential Equations and Applications*, SEMA SIMAI Springer Series 8, DOI 10.1007/978-3-319-32013-7_22

symmetry algebra $\mathfrak{sl}(2,\mathbb{R})$. A basis of generators $\{\mathbf{v}_1,\mathbf{v}_2,\mathbf{v}_3\}$ of $\mathfrak{sl}(2,\mathbb{R})$ can be chosen verifying the following commutation relations:

$$[\mathbf{v}_1,\mathbf{v}_3]=\mathbf{v}_1,\quad [\mathbf{v}_1,\mathbf{v}_2]=2\mathbf{v}_3,\quad [\mathbf{v}_3,\mathbf{v}_2]=\mathbf{v}_2. \tag{1}$$

If we use the vector field $\mathbf{v}_1$ (resp. $\mathbf{v}_2$) to reduce the order of the equation, the Lie symmetry $\mathbf{v}_3$ can be recovered as Lie symmetry of the reduced equation, but $\mathbf{v}_2$ (resp. $\mathbf{v}_1$) is lost as Lie symmetry. On the other hand, if we use $\mathbf{v}_3$ at the first step, then both vector fields $\mathbf{v}_1$ and $\mathbf{v}_2$ are lost as Lie point symmetries of the reduced equation.

In 2001, Muriel and Romero [10] showed that many of the known reduction processes can be explained by the invariance of the equation under a class of vector fields called $\mathcal{C}^\infty$-symmetries or λ-symmetries. In particular, in [11], the case of the non-solvable algebra $\mathfrak{sl}(2,\mathbb{R})$ was studied and it has been proved that the lost symmetries can be recovered as λ-symmetries of the reduced equation. Besides, the general solution of the original equation can be recovered by quadratures from the solution of the reduced equation (see page 489 in [11] for details).

On the other hand, in 1991, Basarab-Horwath [7] introduced the concept of solvable structure for involutive systems of vector fields. In [7] it is proved the equivalence between the existence of a solvable structure and the integrability by quadratures of an involutive system of vector fields. A solvable symmetry algebra of an ODE is a particular case of solvable structure for the (trivially involutive) system formed by the vector field associated to the ODE.

Our goal in this paper is to study some connections between solvable structures and λ-symmetries for third order ordinary differential equations admitting the non-solvable symmetry algebra $\mathfrak{sl}(2,\mathbb{R})$. The main result states that if a third order ordinary differential equation admits $\mathfrak{sl}(2,\mathbb{R})$ as symmetry algebra, then a solvable structure with respect to the vector field associated to the equation can be explicitly computed from such symmetry algebra by using λ-symmetries. Once the solvable structure is known, the equation can be solved by quadratures, as in the case of solvable Lie algebras.

The paper is organized as follows. In Sect. 2 we include the concepts, notation and previous results that will be used throughout the paper. In Sect. 3 we compute explicitly a solvable structure for any third order ordinary differential equation which admits $\mathfrak{sl}(2,\mathbb{R})$ as symmetry algebra. The adopted approach uses a reduced ODE of second order, which inherits two non-equivalent $\mathcal{C}^\infty$-symmetries from two of the generators of $\mathfrak{sl}(2,\mathbb{R})$. According to the results obtained in the preprint [13], a solvable structure for this reduced equation can be explicitly constructed from the inherited $\mathcal{C}^\infty$-symmetries. Once this solvable structure is known, two independent first integrals of the reduced equation associated to the inherited $\mathcal{C}^\infty$-symmetries can be found by quadratures. Such first integrals are used in Theorem 2 to construct a solvable structure with respect to the original third order equation.

In Sect. 4 we use that solvable structure to integrate by quadratures the equation, using the techniques given in [6, 7, 15]. Finally, in Sect. 5 we include an example of a third order ordinary differential equation which admits the non-solvable Lie algebra $\mathfrak{sl}(2,\mathbb{R})$ to illustrate how to construct a solvable structure from that algebra. This solvable structure is used to give a complete set of first integrals of the equation in terms of two independent solutions of a second order linear ODE.

2 Preliminaries

2.1 Generalized $\mathscr{C}^\infty$-Symmetries, Equivalence of Generalized $\mathscr{C}^\infty$-Symmetries and Common First Integrals

Throughout this paper M will denote an open subset of the space of the independent and dependent variables (x, u) of a given ordinary differential equation (ODE):

$$u_n = \phi(x, u, u_1, \cdots, u_{n-1}). \tag{2}$$

Let $(x, u^{(n)})$ denote the coordinates on the open set $M^{(n)}$ of the corresponding nth order jet space, where $u_j = \frac{d^j u}{dx^j}$, for $1 \le j \le n$. By following [14, p. 288], we consider smooth functions $P[u]$, where the bracket notation means that P depends on x, u and derivatives of u with respect to x, up to some finite, but unspecified order. Let

$$\mathbf{v} = \xi[u]\partial_x + \eta[u]\partial_u, \tag{3}$$

be a generalized vector field (in the sense of the Definition 5.1 in [14]) and consider a smooth function $\lambda = \lambda[u]$. We define the nth order λ-prolongation of $\mathbf{v}$ as the vector field on $M^{(n)}$

$$\mathbf{v}^{[\lambda,(n)]} = \mathbf{v} + \sum_{i=1}^{n} \eta^{[\lambda,(i)]}[u]\partial_{u_i}, \quad n \ge 1, \tag{4}$$

whose coefficients are determined by the recursive formula

$$\eta^{[\lambda,(i)]}[u] = (D_x + \lambda)^i(Q[u]) + \xi[u]u_{i+1}, \quad 1 \le i \le n, \tag{5}$$

where $Q[u] = \eta[u] - \xi[u]u_1$ is the characteristic of the vector field $\mathbf{v}$.

Definition 1 Let $\mathbf{v}$ be a generalized vector field of the form (3) and let λ be a differential function. We will say that the pair $(\mathbf{v}, \lambda)$ is a generalized $\mathscr{C}^\infty$-symmetry of Eq. (2) if

$$\mathbf{v}^{[\lambda,(n)]}(u_n - \phi) = 0 \quad \text{when} \quad u_n = \phi(x, u^{(n-1)}). \tag{6}$$

Remark 1 If the infinitesimals ξ, η or the function λ of a generalized $\mathscr{C}^\infty$-symmetry $(\mathbf{v}, \lambda)$ of (2) depend on derivatives of u of order $j \geq n$, we substitute them according to Eq. (2) and its differential consequences. In this way in what follows we can consider, without loss of generality, only generalized $\mathscr{C}^\infty$-symmetries such that $\xi, \eta, \lambda \in \mathscr{C}^\infty(M^{(n-1)})$.

The vector field associated to Eq. (2) will be denoted by

$$\mathbf{A}_{(x,u)} = \partial_x + u_1\partial_u + \cdots + \phi\partial_{u_{n-1}}$$

where the subscript (x, u) is used to make clear the coordinates that we are using in the equation. The following characterization of generalized $\mathscr{C}^\infty$-symmetries can be proved as in [10, Theorem 2.1]: the pair $(\mathbf{v}, \lambda)$ is a generalized $\mathscr{C}^\infty$-symmetry of (2) if and only if

$$[\mathbf{v}^{[\lambda,(n-1)]}, \mathbf{A}_{(x,u)}] = \lambda\mathbf{v}^{[\lambda,(n-1)]} - (\mathbf{A}_{(x,u)} + \lambda)(\xi)\mathbf{A}_{(x,u)}. \tag{7}$$

We recall now the concept of $\mathbf{A}_{(x,u)}$-equivalent generalized $\mathscr{C}^\infty$-symmetries [13]:

Definition 2 We will say that two generalized $\mathscr{C}^\infty$-symmetries $(\mathbf{v}_1, \lambda_1)$ and $(\mathbf{v}_2, \lambda_2)$ of Eq. (2) are $\mathbf{A}_{(x,u)}$-equivalent (or simply equivalent) if the corresponding vector fields

$$\left\{\mathbf{A}_{(x,u)}, \mathbf{v}_1^{[\lambda_1,(n-1)]}, \mathbf{v}_2^{[\lambda_2,(n-1)]}\right\}$$

are dependent over $\mathscr{C}^\infty(M^{(n-1)})$. In this case we will write

$$(\mathbf{v}_1, \lambda_1) \overset{\mathbf{A}_{(x,u)}}{\sim} (\mathbf{v}_2, \lambda_2).$$

It can be checked that $\overset{\mathbf{A}_{(x,u)}}{\sim}$ is an equivalence relation in set of generalized $\mathscr{C}^\infty$-symmetries of (2). In the equivalence class of a given generalized $\mathscr{C}^\infty$-symmetry $(\mathbf{v}, \lambda)$ there are two distinguished elements:

$$(Q[u]\partial_u, \lambda) \tag{8}$$

and

$$(\partial_u, \lambda_Q), \quad \text{where } \lambda_Q = \lambda + \frac{\mathbf{A}_{(x,u)}(Q)}{Q}, \tag{9}$$

that will be called the evolutionary and the canonical representative of the class, respectively.

2.2 Some Results on Solvable Structures

We recall the notion of solvable structure and some of its properties [6, 7, 15]. In this section $\mathscr{M}_n$ denotes an nth-dimensional manifold, $\boldsymbol{\Omega} = dx_1 \wedge \cdots \wedge dx_n$ denotes the volume form in a local system of coordinates on some open set of $\mathscr{M}_n$ and $\lrcorner$ denotes the interior product.

Definition 3 Let $\mathscr{A} = \{\mathbf{A}_1, \cdots, \mathbf{A}_{n-p}\}$ be an involutive system of independent smooth vector fields defined on $\mathscr{M}_n$, where $p \in \mathbb{N}$ and $p \leq n-1$.

1. A system $\{\mathbf{X}_1, \cdots, \mathbf{X}_m\}$ of independent vector fields, where $m \leq p$, is a system of symmetries of $\mathscr{A}$ if $\{\mathbf{A}_1, \cdots, \mathbf{A}_{n-p}, \mathbf{X}_1, \cdots, \mathbf{X}_m\}$ are independent and there exist some functions $c_{i,j}^k \in \mathscr{C}^\infty(\mathscr{M}_n)$, for $1 \leq i \leq m$ and $1 \leq j, k \leq n-p$, such that
$$[\mathbf{X}_i, \mathbf{A}_j] = \sum_{k=1}^{n-p} c_{i,j}^k \mathbf{A}_k.$$
2. Let $\mathscr{S} = \langle \mathbf{X}_1, \cdots, \mathbf{X}_p \rangle$ be an ordered set of independent vector fields on $\mathscr{M}_n$. We will say that the ordered system $\mathscr{A} \cup \mathscr{S} = \langle \mathbf{A}_1, \cdots, \mathbf{A}_{n-1}, \mathbf{X}_1, \cdots, \mathbf{X}_p \rangle$ is a solvable structure with respect to $\mathscr{A}$ if $\mathscr{S}_j = \{\mathbf{A}_1, \cdots, \mathbf{A}_{n-1}, \mathbf{X}_1, \cdots, \mathbf{X}_j\}$ is in involution, $\mathbf{X}_1$ is a symmetry of $\mathscr{A}$ and $\mathbf{X}_{j+1}$ is a symmetry of $\mathscr{S}_j$ for $j = 1, \cdots, p-1$.

When a solvable structure with respect to an involutive system $\mathscr{A}$ is known, then a complete set of first integrals for $\mathscr{A}$ can be constructed by quadratures as follows [6, 7, 15]:

Theorem 1 *If $\langle A_1, \cdots, A_{n-p}, X_1, \cdots, X_p \rangle$ is a solvable structure with respect to $\mathscr{A} = \{A_1, \cdots, A_{n-p}\}$ then, locally,*

$$\omega_1 = \frac{X_{p-1} \lrcorner \cdots \lrcorner X_1 \lrcorner A_{n-p} \lrcorner \cdots \lrcorner A_1 \lrcorner \boldsymbol{\Omega}}{X_p \lrcorner \cdots \lrcorner X_1 \lrcorner A_{n-p} \lrcorner \cdots \lrcorner A_1 \lrcorner \boldsymbol{\Omega}} \quad \textit{is exact},$$
$$\omega_2 = \frac{X_p \lrcorner X_{p-2} \lrcorner \cdots \lrcorner X_1 \lrcorner A_{n-p} \lrcorner \cdots \lrcorner A_1 \lrcorner \boldsymbol{\Omega}}{X_p \lrcorner \cdots \lrcorner X_1 \lrcorner A_{n-p} \lrcorner \cdots \lrcorner A_1 \lrcorner \boldsymbol{\Omega}} \quad \textit{is exact module } \omega_1,$$
$$\vdots$$
$$\omega_p = \frac{X_p \lrcorner \cdots \lrcorner X_2 \lrcorner A_{n-p} \lrcorner \cdots \lrcorner A_1 \lrcorner \boldsymbol{\Omega}}{X_p \lrcorner \cdots \lrcorner X_1 \lrcorner A_{n-p} \lrcorner \cdots \lrcorner A_1 \lrcorner \boldsymbol{\Omega}} \quad \textit{is exact module } \omega_1, \cdots, \omega_{p-1},$$

and the corresponding primitives are first integrals of the system $\{A_1, \cdots, A_{n-p}\}$.

3 Solvable Structures and the Non-solvable Symmetry Algebra $\mathfrak{sl}(2,\mathbb{R})$ for Third Order Ordinary Differential Equations

Let us consider a third order ODE

$$u_3 = \phi(x, u, u_1, u_2), \tag{10}$$

defined for $(x, u) \in M$. Let us suppose that (10) admits the non-solvable Lie algebra $\mathfrak{sl}(2,\mathbb{R})$ as a symmetry algebra. A base of generators $\{\mathbf{v}_1, \mathbf{v}_2, \mathbf{v}_3\}$ of $\mathfrak{sl}(2,\mathbb{R})$ can be chosen verifying the following commutation relations:

$$[\mathbf{v}_1, \mathbf{v}_3] = \mathbf{v}_1, \quad [\mathbf{v}_1, \mathbf{v}_2] = 2\mathbf{v}_3, \quad [\mathbf{v}_3, \mathbf{v}_2] = \mathbf{v}_2. \tag{11}$$

If we introduce canonical coordinates for the vector field $\mathbf{v}_3 = \xi_3(x,u)\partial_x + \eta_3(x,u)\partial_u$, then the order of (10) can be reduced by one: let

$$\varphi(x, u) = (y(x, u), \alpha(x, u)) \tag{12}$$

be a local change of variables such that $\varphi_*(\mathbf{v}_3) = \partial_\alpha$. We denote $w = \alpha_1 = \frac{d\alpha}{dy}$ and for $i \geq 1$, $w_i = \alpha_{i+1} = \frac{d^i w}{dy^i}$. Locally, Eq. (10) can be written in terms of the invariants $\{y, w, w_1, w_2\}$ of $\mathbf{v}_3$ as the reduced equation:

$$w_2 = \tilde{\phi}(y, w, w_1), \tag{13}$$

defined for $(y, w) \in M_1$, for some open set M_1. We denote by $\mathbf{A}_{(y,w)}$ the vector field associated to (13). We define the projection

$$\begin{aligned} \pi^{(1)}_{\mathbf{v}_3} : \varphi^{(1)}(M^{(1)}) &\to M_1 \\ (y, \alpha, w) &\to (y, w). \end{aligned}$$

A vector field $\mathbf{V}$ on $M^{(1)}$ is called $\pi^{(1)}_{\mathbf{v}_3}$-projectable if $[\mathbf{v}_3^{(1)}, \mathbf{V}] = f\mathbf{v}_3^{(1)}$, for some $f \in \mathscr{C}^\infty(M^{(1)})$. The corresponding projection is denoted by $(\pi^{(1)}_{\mathbf{v}_3})_*(\mathbf{V})$.

In what follows, the functions and the vector fields defined on $M^{(n)}$ will be denoted by the same symbol in coordinates $\{x, u, u_1, \cdots, u_n\}$ and in coordinates $\{y, \alpha, w, w_1, \cdots, w_{n-1}\}$, with the omission of the change of variables $\varphi^{(n)}$.

According to [10, Theorem 3] the vector fields $\mathbf{v}_1$ and $\mathbf{v}_2$ can be used to obtain two independent $\mathscr{C}^\infty$-symmetries of Eq. (13) by using the following procedure:

1. Let $\varsigma_1, \varsigma_2 \in \mathscr{C}^\infty(M)$ be such that: $\mathbf{v}_3(\varsigma_1) = \varsigma_1$ and $\mathbf{v}_3(\varsigma_2) = -\varsigma_2$.
2. The vector fields $\varsigma_1\mathbf{v}_1^{(1)}$ and $\varsigma_2\mathbf{v}_2^{(1)}$ are $\mathbf{v}_3^{(1)}$-projectable. If, for $i = 1, 2$ we denote

$$\mathbf{Y}_i = (\pi^{(1)}_{\mathbf{v}_3})_*(\varsigma_i\mathbf{v}_i^{(1)}), \tag{14}$$

and

$$\lambda_i = -\frac{\mathbf{A}_{(x,u)}(\varsigma_i)}{\varsigma_i}, \tag{15}$$

then the pairs

$$(\mathbf{Y}_1, \lambda_1) \quad \text{and} \quad (\mathbf{Y}_2, \lambda_2) \tag{16}$$

are $\mathcal{C}^\infty$-symmetries of Eq. (13).

At this stage we have the second order ODE (13) which admits two non-equivalent $\mathcal{C}^\infty$-symmetries. According to [12, Sect. 5] there exist two functions $I_1 = I_1(y, w, w_1)$ and $I_2 = I_2(y, w, w_1)$ which are functionally independent first integrals of $\mathbf{A}_{(y,w)}$ and such that

$$\begin{array}{ll} \mathbf{Y}_1^{[\lambda_1,(1)]}(I_1) = \mathbf{Y}_2^{[\lambda_2,(1)]}(I_2) = 0, & \\ \mathbf{Y}_1^{[\lambda_1,(1)]}(I_2) \neq 0, & \mathbf{Y}_2^{[\lambda_2,(1)]}(I_1) \neq 0. \end{array} \tag{17}$$

An algorithm that can be followed to ease the determination of $I_1 = I_1(y, w, w_1)$ and $I_2 = I_2(y, w, w_1)$ by using solvable structures and integration by quadratures can be seen in [13].

Next we show that these two first integrals, I_1 and I_2, written in terms of the original variables $\{x, u, u_1, u_2\}$ are also first integrals of the original third order equation (10):

Proposition 1 *Let $I_1 = I_1(y, w, w_1)$ and $I_2 = I_2(y, w, w_1)$ be two functionally independent first integrals associated to the vector field $\mathbf{A}_{(y,w)}$. Then the corresponding functions*

$$I_i = I_i(y(x, u), w(x, u, u_1), w_1(x, u, u_1, u_2)), \quad (i = 1, 2) \tag{18}$$

are two functionally independent first integrals of the vector field $\mathbf{A}_{(x,u)}$ associated to the original Eq. (10).

Proof Let φ be the local change of variables defined in (12). The vector field $\mathbf{A}_{(x,u)}$ written in the new coordinates is:

$$\varphi_*^{(2)}(\mathbf{A}_{(x,u)}) = \frac{1}{D_x(y) \circ \varphi^{-1}}(\mathbf{A}_{(y,w)} + w\partial_\alpha).$$

Since $\mathbf{A}_{(y,w)}(I_i) = 0$ and $\dfrac{\partial I_i}{\partial \alpha} = 0$, we obtain that $\varphi_*^{(2)}(\mathbf{A}_{(x,u)})(I_i) = 0$, for $i = 1, 2$. The result follows by writing these relations in terms of the original variables $\{x, u, u_1, u_2\}$. ■

So far we have a third order ordinary differential equation and two functionally independent first integrals I_1 and I_2 constructed from two first integrals associated to the $\mathscr{C}^\infty$-symmetries (16). We need another independent first integral in order to complete the integration of the equation. Our next goal is the construction of a solvable structure with respect to $\mathbf{A}_{(x,u)}$, by using I_1 and I_2, in order to compute a remaining first integral.

Proposition 2 *Let* $I_1 = I_1(y, w, w_1)$ *and* $I_2 = I_2(y, w, w_1)$ *be two functionally independent first integrals associated to the vector field* $\mathbf{A}_{(y,w)}$. *Then:*

1. $\dfrac{\partial(I_1, I_2)}{\partial(w, w_1)} \neq 0$ *on some open set* $V_1 \subset M_1^{(2)}$.
2. $\dfrac{\partial(w, w_1)}{\partial(u_1, u_2)} \neq 0$ *on some open set* $V_2 \subset M^{(2)}$.
3. *The function defined as* $\Phi = \Phi(x, u, u_1, u_2) = (x, u, I_1, I_2)$ *is a local change of variables, where*

$$I_i = I_i(y(x, u), w(x, u, u_1), w_1(x, u, u_1, u_2)), \quad (i = 1, 2).$$

Proof

1. If $\dfrac{\partial(I_1, I_2)}{\partial(w, w_1)} = 0$, then there exists a function $h = h(y, w, w_1)$ such that $\dfrac{\partial I_1}{\partial w} = h\dfrac{\partial I_2}{\partial w}$ and $\dfrac{\partial I_1}{\partial w_1} = h\dfrac{\partial I_2}{\partial w_1}$. Let $(\partial_w, \lambda_{Q_2})$ be the canonical representative of $(\mathbf{Y}_2, \lambda_2)$ and denote

$$\mathbf{X}_2 = (\partial_w)^{[\lambda_{Q_2},(1)]} = \frac{\partial}{\partial w} + \lambda_{Q_2}\frac{\partial}{\partial w_1}.$$

Since $\mathbf{X}_2(I_2) = 0$, we can write,

$$\mathbf{X}_2(I_1) = \frac{\partial I_1}{\partial w} + \lambda_{Q_2}\frac{\partial I_1}{\partial w_1} = h\frac{\partial I_2}{\partial w} + \lambda_{Q_2} h\frac{\partial I_2}{\partial w_1} = h\mathbf{X}_2(I_2) = 0. \tag{19}$$

Since $(\partial_w, \lambda_{Q_2}) \overset{\mathbf{A}_{(y,w)}}{\sim} (\mathbf{Y}_2, \lambda_2)$, (19) implies that $\mathbf{Y}_2^{[\lambda_2,(1)]}(I_1) = 0$, which cannot happen because of (17). Consequently $\dfrac{\partial(I_1, I_2)}{\partial(w, w_1)} \neq 0$ on some open set $V_1 \subset M_1^{(2)}$.

2. Let φ be local change of variables defined in (12). We have that, locally, $J\varphi^{(2)}(x, u, u_1, u_2) \neq 0$. On the other hand:

$$J\varphi^{(2)}(x, u, u_1, u_2) = \frac{\partial(y, \alpha)}{\partial(x, u)} \cdot \frac{\partial(w, w_1)}{\partial(u_1, u_2)},$$

and $\dfrac{\partial(y,\alpha)}{\partial(x,u)} \neq 0$; therefore $\dfrac{\partial(w,w_1)}{\partial(u_1,u_2)} \neq 0$ on some open set $V_2 \subset M^{(2)}$.

3. It is clear that $J\Phi(x,u,u_1,u_2) = \dfrac{\partial(I_1,I_2)}{\partial(u_1,u_2)}$. By the chain rule we obtain that:

$$\frac{\partial(I_1,I_2)}{\partial(u_1,u_2)} = \frac{\partial(I_1,I_2)}{\partial(w,w_1)} \cdot \frac{\partial(w,w_1)}{\partial(u_1,u_2)}.$$

By applying items 1 and 2 of this proposition we conclude that $J\Phi(x,u,u_1,u_2) \neq 0$. ■

Now we are ready to explicitly provide a solvable structure with respect to the vector field associated to the original third order equation.

Theorem 2 *Let $\mathbf{v}_1, \mathbf{v}_2, \mathbf{v}_3$ be generators of the symmetry algebra $\mathfrak{sl}(2,\mathbb{R})$ of Eq. (10) satisfying the conditions given in (11). For $i = 1,2$, let I_i be the function defined in (18) and define $\mathbf{Z}_i = \Phi_*^{-1}(\partial_{I_i})$, where $\Phi(x,u,u_1,u_2) = (x,u,I_1,I_2)$. Then the set*

$$\langle \mathbf{A}_{(x,u)}, \mathbf{v}_3^{(2)}, \mathbf{Z}_1, \mathbf{Z}_2 \rangle \tag{20}$$

is a solvable structure with respect to $\langle \mathbf{A}_{(x,u)} \rangle$.

Proof Let us consider the local change of variables $\Phi = \Phi(x,u,u_1,u_2) = (x,u,I_1,I_2)$ defined in Proposition 2. By applying the Proposition 1 we can write:

$$\Phi_*(\mathbf{A}_{(x,u)}) = \partial_x + \tau(x,u,I_1,I_2)\partial_u, \tag{21}$$

where the function τ is u_1 written in the coordinates $\{x,u,I_1,I_2\}$. Since $\mathbf{v}_3^{(2)}(I_1) = \mathbf{v}_3^{(2)}(I_2) = 0$, then

$$\Phi_*(\mathbf{v}_3^{(2)}) = \xi_3(x,u)\partial_x + \eta_3(x,u)\partial_u.$$

Since $\mathbf{v}_3^{(2)}$ is a Lie symmetry of Eq. (10), we have that:

$$[\Phi_*\mathbf{A}_{(x,u)}, \Phi_*\mathbf{v}_3^{(2)}] = (\mathbf{A}_{(x,u)}(\xi_3) \circ \Phi^{-1}) \cdot \Phi_*\mathbf{A}_{(x,u)} \tag{22}$$

and it is clear that:

$$[\partial_{I_1}, \Phi_*(\mathbf{v}_3^{(2)})] = 0, \quad [\partial_{I_1}, \partial_{I_2}] = 0, \quad [\partial_{I_2}, \Phi_*\mathbf{v}_3^{(2)}] = 0. \tag{23}$$

Finally we obtain that, for $i = 1,2$:

$$[\Phi_*(\mathbf{A}_{(x,u)}), \partial_{I_i}] = \frac{-\tau_{I_i}}{\eta_3 - \tau\xi_3}\left(\Phi_*(\mathbf{v}_3^{(2)}) - \xi_3\Phi_*(\mathbf{A}_{(x,u)})\right). \tag{24}$$

By using (22)–(24) we conclude that

$$\langle \Phi_*(\mathbf{A}_{(x,u)}), \Phi_*(\mathbf{v}_3^{(2)}), \partial_{I_1}, \partial_{I_2} \rangle$$

is a solvable structure with respect to $\langle \Phi_*(\mathbf{A}_{(x,u)}) \rangle$. Coming back to the original coordinates and defining

$$\mathbf{Z}_i = \Phi_*^{-1}(\partial_{I_i}), \quad \text{for} \quad i = 1, 2, \tag{25}$$

we conclude that

$$\langle \mathbf{A}_{(x,u)}, \mathbf{v}_3^{(2)}, \mathbf{Z}_1, \mathbf{Z}_2 \rangle$$

is a solvable structure with respect to $\langle \mathbf{A}_{(x,u)} \rangle$. ■

4 Complete System of First Integrals of $\mathbf{A}_{(x,u)}$

In the previous discussion we have shown how to construct a solvable structure $\langle \mathbf{A}_{(x,u)}, \mathbf{v}_3^{(2)}, \mathbf{Z}_1, \mathbf{Z}_2 \rangle$ with respect to $\langle \mathbf{A}_{(x,u)} \rangle$ from the non-solvable symmetry algebra $\mathfrak{sl}(2, \mathbb{R})$. In this section, we apply the theory of solvable structures in order to compute a complete system of first integrals of $\mathbf{A}_{(x,u)}$. We denote the volume form $\boldsymbol{\Omega} = dx \wedge du \wedge du_1 \wedge du_2$ and consider the following differential 1-forms, as in Theorem 1:

$$\begin{aligned}
\boldsymbol{\omega}_1 &= \frac{\mathbf{Z}_1 \lrcorner \mathbf{v}_3^{(2)} \lrcorner \mathbf{A}_{(x,u)} \lrcorner \boldsymbol{\Omega}}{\mathbf{Z}_2 \lrcorner \mathbf{Z}_1 \lrcorner \mathbf{v}_3^{(2)} \lrcorner \mathbf{A}_{(x,u)} \lrcorner \boldsymbol{\Omega}}, \\
\boldsymbol{\omega}_2 &= \frac{\mathbf{Z}_2 \lrcorner \mathbf{v}_3^{(2)} \lrcorner \mathbf{A}_{(x,u)} \lrcorner \boldsymbol{\Omega}}{\mathbf{Z}_2 \lrcorner \mathbf{Z}_1 \lrcorner \mathbf{v}_3^{(2)} \lrcorner \mathbf{A}_{(x,u)} \lrcorner \boldsymbol{\Omega}}, \\
\boldsymbol{\omega}_3 &= \frac{\mathbf{Z}_2 \lrcorner \mathbf{Z}_1 \lrcorner \mathbf{A}_{(x,u)} \lrcorner \boldsymbol{\Omega}}{\mathbf{Z}_2 \lrcorner \mathbf{Z}_1 \lrcorner \mathbf{v}_3^{(2)} \lrcorner \mathbf{A}_{(x,u)} \lrcorner \boldsymbol{\Omega}}.
\end{aligned}$$

According to Theorem 1, the differential 1-form $\boldsymbol{\omega}_1$ is exact, and a function Θ_1 such that

$$d\Theta_1 = \boldsymbol{\omega}_1 \tag{26}$$

is a common first integral of the system of vector fields $\{\mathbf{A}_{(x,u)}, \mathbf{v}_3^{(2)}, \mathbf{Z}_1\}$ (see [15] for details).

Secondly, we also know that $\boldsymbol{\omega}_2$ is exact module $\boldsymbol{\omega}_1$. Nevertheless, since by relations (22)–(24) the ordered set $\langle \mathbf{A}_{(x,u)}, \mathbf{v}_3^{(2)}, \mathbf{Z}_2, \mathbf{Z}_1 \rangle$ is a solvable structure with respect to $\langle \mathbf{A}_{(x,u)} \rangle$, then the differential 1-form $\boldsymbol{\omega}_2$ is closed and hence locally exact.

A function Θ_2 such that

$$d\Theta_2 = \boldsymbol{\omega}_2 \tag{27}$$

is a common first integral of the system $\{\mathbf{A}_{(x,u)}, \mathbf{v}_3^{(2)}, \mathbf{Z}_2\}$.

Finally we have that $\boldsymbol{\omega}_3$ is exact module $\boldsymbol{\omega}_1$ and $\boldsymbol{\omega}_2$, and a function Θ_3 such that

$$d\Theta_3 = \boldsymbol{\omega}_3, \quad \text{mod } \boldsymbol{\omega}_1, \boldsymbol{\omega}_2 \tag{28}$$

completes the set of independent first integrals of the vector field $\mathbf{A}_{(x,u)}$.

The functions I_1 and I_2 given in (18) can be used to simplify the above-described procedure as follows: these functions I_1 and I_2 are two independent first integrals of $\mathbf{A}_{(x,u)}$, that are also first integrals of $\mathbf{v}_3^{(2)}$. In fact, by (25), I_1 is a common first integral of the set $\{\mathbf{A}_{(x,u)}, \mathbf{v}_3^{(2)}, \mathbf{Z}_2\}$ and I_2 is a common first integral of the set $\{\mathbf{A}_{(x,u)}, \mathbf{v}_3^{(2)}, \mathbf{Z}_1\}$. On the other hand, by definition of $\boldsymbol{\omega}_1$, I_1 is a common first integral of the set $\{\mathbf{A}_{(x,u)}, \mathbf{v}_3^{(2)}, \mathbf{Z}_1\}$, which implies that Θ_1 and I_1 must be functionally dependent. Similarly Θ_2 and I_2 must be functionally dependent. Therefore, a function Θ_3 satisfying (28) can be found as a primitive of the restriction of ω_3 to the submanifold of $M^{(2)}$ where I_1 and I_2 are constant.

In the next section we present an example where this simplified method is illustrated.

5 Example

In this section we apply the results obtained in this paper to a particular third order equation that admits the non-solvable symmetry algebra $\mathfrak{sl}(2,\mathbb{R})$. The presented method can be used to construct a solvable structure and derive three functionally independent first integrals for any other equation admitting $\mathfrak{sl}(2,\mathbb{R})$.

Let us consider the following apparently simple but nontrivial third order ordinary differential equation

$$u^2 u_3 + 1 = 0. \tag{29}$$

The vector field associated to (29) is given by

$$\mathbf{A}_{(x,u)} = \partial_x + u_1 \partial_u + u_2 \partial_{u_1} - \frac{1}{u^2}\partial_{u_2}.$$

The classical symmetry algebra (point and contact symmetries) of Eq. (29) is 3-dimensional and generated by

$$\mathbf{v}_1 = \partial_x, \quad \mathbf{v}_2 = x^2\partial_x + 2\,u\,x\,\partial_u, \quad \mathbf{v}_3 = x\partial_x + u\partial_u.$$

These vector fields satisfy the relations (11).

We can use the vector field $\mathbf{v}_3$ to get an order reduction for Eq. (29). By introducing the canonical coordinates

$$y = \frac{u}{x}, \qquad \alpha = \ln(x) \tag{30}$$

for $\mathbf{v}_3$, we obtain the following reduced equation:

$$w_2 = -\frac{y^2w^4 - w^5 - 3\,y^2w_1^2}{wy^2}. \tag{31}$$

Let $\mathbf{A}_{(y,w)}$ denote the vector field associated to this reduced equation. The Lie symmetries $\mathbf{v}_1$ and $\mathbf{v}_2$ can be recovered as non-equivalent $\mathcal{C}^\infty$-symmetries of the reduced equation by following the procedure described in Sect. 3. Two functions ζ_1 and ζ_2 satisfying $\mathbf{v}_3(\zeta_1) = \zeta_1$ and $\mathbf{v}_3(\zeta_2) = -\zeta_2$ can be easily calculated in variables (30): by choosing $\zeta_1 = \mathrm{e}^{\alpha}y^{-1}$ and $\zeta_2 = \mathrm{e}^{-\alpha}y^{-1}$, the vector fields $\varsigma_1\mathbf{v}_1^{(1)}$ and $\varsigma_2\mathbf{v}_2^{(1)}$ are $\pi_{\mathbf{v}_3}^{(1)}$-projectable, and the expressions of the respective projections, in coordinates (y, w), are:

$$\begin{aligned} \mathbf{Y}_1 &= (\pi_{\mathbf{v}_3}^{(1)})_*(\varsigma_1\mathbf{v}_1^{(1)}) = -\partial_y - w^2\partial_w, \\ \mathbf{Y}_2 &= (\pi_{\mathbf{v}_3}^{(1)})_*(\varsigma_2\mathbf{v}_2^{(1)}) = \partial_y - w^2\partial_w. \end{aligned} \tag{32}$$

The pairs $(\mathbf{Y}_1, \lambda_1)$ and $(\mathbf{Y}_2, \lambda_2)$ are two non-equivalent $\mathcal{C}^\infty$-symmetries of Eq. (31) for the respective functions

$$\lambda_1 = -\frac{\mathbf{A}_{(y,\alpha)}(\varsigma_1)}{\varsigma_1} = -w + y^{-1} \quad \text{and} \quad \lambda_2 = -\frac{\mathbf{A}_{(y,\alpha)}(\varsigma_2)}{\varsigma_2} = w + y^{-1}. \tag{33}$$

In order to determine two functions $I_1 = I_1(y, w, w_1)$ and $I_2 = I_2(y, w, w_1)$ such that I_i is a first integral of $\{\mathbf{A}_{(y,w)}, \mathbf{Y}_i^{[\lambda_i,(1)]}\}$, for $i = 1, 2$, we consider the local system of coordinates $\{s, m_1, m_2\}$ where

$$s = -\frac{y^2w^3 + 2\,w_1y + w}{2\,w^3}, \qquad m_1 = -\frac{wy - 1}{2w}, \qquad m_2 = \frac{wy + 1}{2w}, \tag{34}$$

in which $\mathbf{Y}_1^{[\lambda_1,(1)]}$ and $\mathbf{Y}_2^{[\lambda_2,(1)]}$ are simultaneously straightened, i.e., $\mathbf{Y}_1^{[\lambda_1,(1)]} = \partial_{m_1}$ and $\mathbf{Y}_2^{[\lambda_2,(1)]} = \partial_{m_2}$. In these coordinates $\mathbf{A}_{(y,w)}$ becomes

$$\mathbf{A}_{(y,w)} = \frac{1}{{m_1}^2 - {m_2}^2}\left(\partial_s - (m_1^2 + \frac{s}{2})\partial_{m_1} - (m_2^2 + \frac{s}{2})\partial_{m_2}\right). \tag{35}$$

In consequence, for $i = 1, 2$, the equation

$$m_i'(s) + m_i(s)^2 + \frac{s}{2} = 0 \tag{36}$$

is the reduced equation of (31) associated to the $\mathscr{C}^\infty$-symmetry $(\mathbf{Y}_i, \lambda_i)$. Equation (36) is a Riccati-type equation that can be converted into the second order linear ODE

$$\psi''(s) + \frac{s}{2}\psi(s) = 0 \tag{37}$$

through the standard transformation $m_i(s) = \frac{\psi'(s)}{\psi(s)}$. It can be checked that two first integrals $I_1 = I_1(s, m_1, m_2)$ and $I_2 = I_2(s, m_1, m_2)$ of (35) can be expressed in terms of two independent solutions $\psi_1 = \psi_1(s)$ and $\psi_2 = \psi_2(s)$ of the Airy-type equation (37) in the form

$$I_i(s, m_i) = \frac{m_i\psi_1(s) - \psi_1'(s)}{m_i\psi_2(s) - \psi_2'(s)}, \quad \text{for} \quad i = 1, 2. \tag{38}$$

These functions written in terms of the variables $\{y, w, w_1\}$ by using (34) are two independent first integrals for the reduced Eq. (31).

According to Proposition 1, by writing these functions in terms of the original variables by using the expressions of (34) in variables $\{x, u, u_1, u_2\}$

$$s = u\,u_2 - \frac{1}{2}u_1^2, \quad m_1 = \frac{xu_1 - 2u}{2x}, \quad m_2 = \frac{u_1}{2}, \tag{39}$$

we obtain the following first integrals of $\mathbf{A}_{(x,u)}$:

$$I_1 = \frac{(xu_1 - 2\,u)\psi_1\,(s) - 2\,x\,\psi_1'\,(s)}{(xu_1 - 2\,u)\psi_2\,(s) - 2\,x\,\psi_2'\,(s)}, \qquad I_2 = \frac{u_1\psi_1\,(s) - 2\,\psi_1'\,(s)}{u_1\psi_2\,(s) - 2\,\psi_2'\,(s)}, \tag{40}$$

where s is given in (39).

Next we use these two first integrals to construct a solvable structure with respect to $\langle \mathbf{A}_{(x,u)} \rangle$ by following the procedure of Theorem 2. In this example we use a slight modification of the change of variables that appears in the proof of this theorem in order to simplify the calculations. We define $\widetilde{\Phi}(x, u, u_1, u_2) = (x, s, I_1, I_2)$, where s is given in (39) and I_1, I_2 are given by (40). It can be checked that the Jacobian of this transformation becomes

$$J\widetilde{\Phi}(x, u, u_1, u_2) = \frac{-8\,u_1\,x\,(\psi_1(s)\psi_2'(s) - \psi_2(s)\psi_1'(s))^2}{(((xu_1 - 2\,u)\psi_2\,(s) - 2\,x\,\psi_2'\,(s))(u_1\psi_2\,(s) - 2\,\psi_2'\,(s)))^2}.$$

Since the Wronskian of ψ_1 and ψ_2

$$W(\psi_1, \psi_2)(s) = \psi_1(s)\psi_2'(s) - \psi_2(s)\psi_1'(s) \tag{41}$$

is not identically zero, we conclude that $\widetilde{\Phi}$ defines a local change of variables on $M^{(2)}$.

We define, as in Theorem 2, $\widetilde{\mathbf{Z}}_i = \widetilde{\Phi}_*^{-1}(\partial_{I_i})$, for $i = 1, 2$. Next we check that

$$\langle \mathbf{A}_{(x,u)}, \mathbf{v}_3^{(2)}, \widetilde{\mathbf{Z}}_1, \widetilde{\mathbf{Z}}_2 \rangle$$

is a solvable structure with respect to $\langle \mathbf{A}_{(x,u)} \rangle$. For simplicity, we compute the necessary Lie brackets in coordinates $\{x, s, I_1, I_2\}$. The respective expressions of $\mathbf{v}_3^{(2)}$ and $\mathbf{A}_{(x,u)}$ in coordinates $\{x, s, I_1, I_2\}$ become

$$\begin{aligned} \widetilde{\Phi}_*(\mathbf{v}_3^{(2)}) &= x\partial_x, \\ \widetilde{\Phi}_*(\mathbf{A}_{(x,u)}) &= \partial_x + \frac{(I_2\psi_2(s) - \psi_1(s))\,(I_1\psi_2(s) - \psi_1(s))}{x\,(I_2 - I_1)\,W(\psi_1, \psi_2)}\partial_s. \end{aligned} \tag{42}$$

Therefore,

$$\begin{aligned} &[\widetilde{\Phi}_*(\mathbf{A}_{(x,u)}), \widetilde{\Phi}_*(\mathbf{v}_3^{(2)})] = \widetilde{\Phi}_*(\mathbf{A}_{(x,u)}), \\ &\left[\widetilde{\Phi}_*(\mathbf{A}_{(x,u)}), \partial_{I_1}\right] = \rho_1(\widetilde{\Phi}_*(\mathbf{v}_3^{(2)}) - x\widetilde{\Phi}_*(\mathbf{A}_{(x,u)})), \\ &\left[\widetilde{\Phi}_*(\mathbf{v}_3^{(2)}), \partial_{I_1}\right] = 0, \\ &\left[\widetilde{\Phi}_*(\mathbf{A}_{(x,u)}), \partial_{I_2}\right] = \rho_2(\widetilde{\Phi}_*(\mathbf{v}_3^{(2)}) - x\widetilde{\Phi}_*(\mathbf{A}_{(x,u)})), \\ &\left[\widetilde{\Phi}_*(\mathbf{v}_3^{(2)}), \partial_{I_2}\right] = [\partial_{I_1}, \partial_{I_2}] = 0, \end{aligned} \tag{43}$$

where

$$\rho_1 = \frac{I_2\,\psi_2(s) - \psi_1(s)}{(I_2 - I_1)(I_1\psi_2(s) - \psi_1(s))x}, \quad \rho_2 = -\frac{I_1\,\psi_2(s) - \psi_1(s)}{(I_2 - I_1)(I_2\psi_2(s) - \psi_1(s))x}.$$

Relations (43) prove that $\langle \mathbf{A}_{(x,u)}, \mathbf{v}_3^{(2)}, \widetilde{\mathbf{Z}}_1, \widetilde{\mathbf{Z}}_2 \rangle$ is a solvable structure with respect to $\langle \mathbf{A}_{(x,u)} \rangle$. In this way, we have constructed a solvable structure from the non-solvable algebra $\mathfrak{sl}(2, \mathbb{R})$.

Now we use this solvable structure to find a first integral of $\mathbf{A}_{(x,u)}$ functionally independent with I_1 and I_2. It can be checked that the corresponding differential 1-form

$$\omega_3 = \frac{\widetilde{\mathbf{Z}}_2 \lrcorner \widetilde{\mathbf{Z}}_1 \lrcorner \mathbf{A}_{(x,u)} \lrcorner \boldsymbol{\Omega}}{\widetilde{\mathbf{Z}}_2 \lrcorner \widetilde{\mathbf{Z}}_1 \lrcorner \mathbf{v}_3^{(2)} \lrcorner \mathbf{A}_{(x,u)} \lrcorner \boldsymbol{\Omega}}$$

in coordinates $\{x, s, I_1, I_2\}$ becomes

$$\widetilde{\Phi}^* \boldsymbol{\omega}_3 = \frac{1}{x}dx - \frac{W(\psi_1, \psi_2)\,(I_2 - I_1)}{(I_2\,\psi_2(s) - \psi_1(s))\,(I_1\,\psi_2(s) - \psi_1(s))}ds. \tag{44}$$

A primitive of $\widetilde{\Phi}^*\omega_3$, restricted to the submanifold defined by $I_1 = C_1$ and $I_2 = C_2$, for $C_1, C_2 \in \mathbb{R}$, is given by

$$\ln(x) - \ln\left(\frac{C_2\psi_2(s) - \psi_1(s)}{C_1\psi_2(s) - \psi_1(s)}\right).$$

Therefore, the function $I_3 = x\,\dfrac{I_2\psi_2(s) - \psi_1(s)}{I_1\psi_2(s) - \psi_1(s)}$, where s, I_1, I_2 are given in (39) and (40), respectively, is a first integral of $\mathbf{A}_{(x,u)}$ functionally independent with I_1 and I_2.

In the original variables this first integral becomes

$$I_3(x, u, u_1, u_2) = \frac{(xu_1 - 2u)\psi_2(s) - 2x\psi_2'(s)}{u_1\psi_2(s) - 2\psi_2'(s)}.$$

The general solution of Eq. (29) can be expressed, in implicit form, as:

$$\begin{cases} \dfrac{(xu_1 - 2\,u)\psi_1\,(s) - 2\,x\,\psi_1'\,(s)}{(xu_1 - 2\,u)\psi_2\,(s) - 2\,x\,\psi_2'\,(s)} = C_1, \\[2ex] \dfrac{u_1\psi_1\,(s) - 2\,\psi_1'\,(s)}{u_1\psi_2\,(s) - 2\,\psi_2'\,(s)} = C_2, \\[2ex] \dfrac{(xu_1 - 2u)\psi_2(s) - 2x\psi_2'(s)}{u_1\psi_2(s) - 2\psi_2'(s)} = C_3 \end{cases}$$

where $C_i \in \mathbb{R}$, for $i = 1, 2, 3$ and $s = u\,u_2 - \frac{1}{2}u_1^2$.

6 Conclusions

The process of integration by quadratures for ODEs that admit solvable symmetry algebras is well known. For ODEs admitting non-solvable symmetry algebras, as the Lie algebra $\mathfrak{sl}(2, \mathbb{R})$, this procedure is no longer available. Nevertheless, the determination of a solvable structure for these ODEs lets to find by quadratures a complete set of first integrals. Effective methods to construct solvable structures are therefore very important in solving these ODEs.

In this paper we show how a solvable structure for the vector field of a third order ODE admitting the non-solvable symmetry algebra $\mathfrak{sl}(2, \mathbb{R})$ can be constructed by using two functionally independent first integrals associated to two non-equivalent $\mathscr{C}^\infty$-symmetries of the reduced Eq. (13). Once the solvable structure has been determined, a complete set of first integrals for the equation can be found by quadratures, although the symmetry algebra admitted by the equation is non-solvable.

Acknowledgements This research was partially supported by the University of Cádiz and Junta de Andalucía research group FQM 377. A. Ruiz gratefully acknowledges financial support from the Doctoral School of the University of Cádiz for the registration fee of the XXIV Congress on Differential Equations and Applications and XIV Congress on Applied Mathematics. The authors also thank Prof. Romero for his assistance, patience and always valuable comments and suggestions.

References

1. Abraham-Shrauner, B.: Hidden symmetries and nonlocal group generators for ordinary differential equations. IMA J. Appl. Math. **56**(3), 235–252 (1996). doi:10.1093/imamat/56.3.235
2. Abraham-Shrauner, B.: Hidden symmetries, first integrals and reduction of order of nonlinear ordinary differential equations. J. Nonlinear Math. Phys. **9**(suppl. 2), 1–9 (2002). Special issue in honor of P. G. L. Leach on the occasion of his 60th birthday. doi:10.2991/jnmp.2002.9.s2.1
3. Abraham-Shrauner, B., Guo, A.: Hidden and nonlocal symmetries of nonlinear differential equations. In: Modern Group Analysis: Advanced Analytical and Computational Methods in Mathematical Physics (Acireale, 1992), pp. 1–5. Kluwer Academic Publisher, Dordrecht (1993). doi:10.1007/978-94-011-2050-0
4. Abraham-Shrauner, B., Leach, P.G.L., Govinder, K.S., Ratcliff, G.: Hidden symmetries and contact symmetries of ordinary differential equations. J. Phys. A: Math. Gen. **29**(23), 6707–6716 (1995). doi:10.1088/0305-4470/28/23/020
5. Abraham-Shrauner, B., Govinder, K.S., Leach, P.G.L.: Integration of second order ordinary differential equations not possessing Lie point symmetries. Phys. Lett. A. **203**(4), 169–174 (1995). doi:10.1016/0375-9601(95)00426-4
6. Barco, M.A., Prince, G.E.: Solvable symmetry structures in differential form applications. Acta Appl. Math. **66**(1), 89–121 (2001). doi:10.1023/A:1010609817442
7. Basarab-Horwath, P.: Integrability by quadratures for systems of involutive vector fields. Ukr. Math. Zh. **43**(10), 1236–1242 (1991). doi:10.1007/BF01061807
8. Hydon, P.E.: Symmetry Methods for Differential Equations: A Beginner's Guide. Cambridge Texts in Applied Mathematics. Cambridge University Press, Cambridge (2000)
9. Ibragimov, N.H.: A Practical Course in Differential Equation and Mathematical Modelling: Classical and New Methods. Nonlinear Mathematical Models, Symmetry and Invariance Principles. ALGA Publications, Karlskrona (2004)
10. Muriel, C., Romero, J.L.: New methods of reduction for ordinary differential equations. IMA J. Appl. Math. **66**(2), 111–125 (2001). doi:10.1093/imamat/66.2.111
11. Muriel, C., Romero, J.L.: $\mathscr{C}^{\infty}$-symmetries and non-solvable symmetry algebras. IMA J. Appl. Math. **66**, 477–498 (2001). doi:10.1093/imamat/66.5.477
12. Muriel, C., Romero, J.L.: First integrals, integrating factors and λ-symmetries of second-order differential equations. J. Phys. A: Math. Theor. **42**(36), 365207 (2009). doi:10.1088/1751-8113/42/36/365207

13. Muriel, C., Romero, J.L., Ruiz, A.: $\lambda-$Symmetries and integrability by quadratures. Preprint (2015)
14. Olver, P.J.: Applications of Lie Groups to Differential Equations. Graduate Texts in Mathematics. Springer, New York (2000)
15. Prince, G., Sherring, J.: Geometric aspects of reduction of order. Trans. Am. Math. Soc. **334**, 433–453 (1992). doi:10.1090/S0002-9947-1992-1149125-6
16. Stephani, H., MacCallum, M.: Differential Equations: Their Solutions Using Symmetries. Cambridge University Press, Cambridge (1989)

Travelling Wave Solutions of a Generalized Variable-Coefficient Gardner Equation

R. de la Rosa and M.S. Bruzón

Abstract In this paper, a simple way to construct exact solutions by using equivalence transformations is shown. We consider a generalized variable-coefficient Gardner equation from the point of view of Lie symmetries in partial differential equations. We obtain the continuous equivalence transformations of the equation in order to reduce the number of arbitrary functions and give a clearer formulation of the results. Furthermore, we calculate Lie symmetries of the reduced equation. Then, we determine the similarity variables and the similarity solutions which allow us to reduce our equation into an ordinary differential equation. Finally, we obtain some exact travelling wave solutions of the equation by using the simplest equation method.

1 Introduction

Over the last years, nonlinear equations with variable coefficients have become increasingly important due to these describe many nonlinear phenomena more realistically than equations with constant coefficients. The variable coefficient Gardner equation, the variable coefficient reaction-diffusion equation or the nonlinear Schrödinger equation are just some examples. Gardner equation is widely used in different fields of physics, for instance, fluid dynamics, plasma physics and quantum field theory. Moreover, it also appears as a useful model to describe wave phenomena in plasma and solid state.

In [6], it was considered a generalized variable-coefficient Gardner equation given by:

$$u_t + A(t)u^n\, u_x + C(t)\, u^{2n} u_x + B(t)\, u_{xxx} + Q(t)\, u = 0, \tag{1}$$

where n is an arbitrary positive integer, $A(t)$, $B(t) \neq 0$, $C(t) \neq 0$ and $Q(t)$ are arbitrary smooth functions of t.

R. de la Rosa (✉) • M.S. Bruzón
Departamento de Matemáticas, Universidad de Cádiz, Campus de Puerto Real, 11510 Cádiz, Spain
e-mail: rafael.delarosa@uca.es; m.bruzon@uca.es

F. Ortegón Gallego et al. (eds.), *Trends in Differential Equations and Applications*, SEMA SIMAI Springer Series 8, DOI 10.1007/978-3-319-32013-7_23

The problem lies in the fact that the analysis of equations involving arbitrary functions seems rather difficult. Equivalence transformations arise for determining an exhaustive solution of the problem. Furthermore, equivalence transformations allow us to consider complete equivalence classes instead of individual equations.

An equivalence transformation is a non-degenerate change of variables acting on dependent and independent variables so that it takes any equation of the form (1) into an equation of the same form, except maybe the form of the arbitrary functions $A(t)$, $B(t)$, $C(t)$ and $Q(t)$. Equivalence transformations play an important role in the study of partial differential equations involving arbitrary functions due to the fact that these allow an exhaustive study and a simple and clear formulation of the results [8, 9, 14, 15, 18, 19].

In [6] was proved that the equivalence group of Eq. (1) is given by the set of transformations

$$\tilde{t} = \alpha(t), \qquad \tilde{x} = (x + \epsilon_2)e^{\epsilon_1}, \qquad \tilde{u} = e^{\epsilon_1 - \epsilon_r r(t)}u, \tag{2}$$

where ϵ_1, ϵ_2, ϵ_r are arbitrary constants, and $\alpha = \alpha(t)$, $r = r(t)$ are arbitrary functions verifying $\alpha_t \neq 0$. The new arbitrary elements are related with the old ones by means of the transformations

$$\tilde{A} = \frac{e^{n\epsilon_r r + (1-n)\epsilon_1}}{\alpha_t}A, \qquad \tilde{B} = \frac{e^{3\epsilon_1}}{\alpha_t}B, \qquad \tilde{C} = \frac{e^{2n\epsilon_r r + (1-2n)\epsilon_1}}{\alpha_t}C,$$
$$\tilde{Q} = \frac{Q + \epsilon_r r_t}{\alpha_t}, \qquad \tilde{n} = n. \tag{3}$$

The point that two arbitrary elements $\alpha(t)$ and $r(t)$ appear in the group of transformations (3) enables us to establish two of the arbitrary functions. Thus, by using the equivalence transformation

$$\tilde{t} = e^{3\epsilon_1}\int B(t)dt, \quad \tilde{x} = (x + \epsilon_2)e^{\epsilon_1}, \quad \tilde{u} = e^{\frac{-\epsilon_1}{n}}\left(\frac{B(t)}{C(t)}\right)^{-\frac{1}{2n}}u, \tag{4}$$

Eq. (1) takes the form

$$\tilde{u}_{\tilde{t}} + \tilde{A}(\tilde{t})\tilde{u}^n\tilde{u}_{\tilde{x}} + \tilde{u}^{2n}\tilde{u}_{\tilde{x}} + \tilde{u}_{\tilde{x}\tilde{x}\tilde{x}} + \tilde{Q}(\tilde{t})\tilde{u} = 0, \tag{5}$$

where

$$\tilde{A}(\tilde{t}) = \frac{e^{-\epsilon_1}A(t)}{\sqrt{B(t)\,C(t)}},$$

and

$$\tilde{Q}(\tilde{t}) = e^{-3\epsilon_1}\left(\frac{Q(t)}{B(t)} + \frac{C(t)}{2nB(t)^2}\left(\frac{B(t)}{C(t)}\right)_t\right).$$

This enables us to consider without losing generality the class

$$u_t + A(t)u^n u_x + u^{2n} u_x + u_{xxx} + Q(t)u = 0, \tag{6}$$

due to the study of symmetries and exact solutions of class (6) can be extended to (1) undoing transformation (4).

In this paper we study Eq. (6) from the point of view of symmetry reductions in partial differential equations. We use the Lie classical symmetries of Eq. (6) obtained in [6] for functions $A(t)$ and $Q(t)$. From these symmetries, we obtain the similarity variables and the similarity solutions which allow us to reduce the equation to an ordinary differential equation. Then, we show some exact travelling wave solutions for an ordinary differential equation by using the simplest equation method given by Kudryashov.

2 Classical Symmetries of Class (6)

In the nineteenth century, Sophus Lie developed a method to study differential equations known today as Lie classical method. This method is based on the determination of the symmetry group of a differential equation by using a one-parameter group of transformations. In other words, the largest transformation group which acts in both dependent and independent variables of the equation so that solutions of the equation are transformed into other solutions.

Lie theory is one of the most important and powerful methods used to study differential equations. It is well known due to its many applications in mathematics and physics. Among them, it is noted that symmetry groups can be used to obtain exact solutions of partial differential equations, directly [10, 17, 20] or by obtaining the similarity variables and the similarity solutions [7, 12]; or determine conservation laws [2–4, 16].

A symmetry generator of Eq. (6) is a vector field

$$\mathbf{v} = \tau(t,x,u)\partial_t + \xi(t,x,u)\partial_x + \eta(t,x,u)\partial_u, \tag{7}$$

where τ, ξ and η are called infinitesimals, such that

$$pr^{(3)}\mathbf{v}(\Delta) = 0 \quad \text{when} \quad \Delta = 0, \tag{8}$$

where Δ represents Eq. (6) and

$$pr^{(3)}\mathbf{v} = \mathbf{v} + \zeta^t \partial u_t + \zeta^x \partial u_x + \zeta^{xxx} \partial u_{xxx}, \tag{9}$$

is the third prolongation of the vector field (7). The functions ζ^J are given by

$$\zeta^J(t,x,u^{(3)}) = D_J(\eta - \tau u_t - \xi u_x) + \tau u_{Jt} + \xi u_{Jx},$$

with $J = (j_1, \ldots, j_k)$, $1 \leq j_k \leq 2$, $1 \leq k \leq 3$, and $u^{(3)}$ denotes the set of partial derivatives up to third order [13].

Invariance criterion (8) yields a determining system for the infinitesimals. From this determining system if n, $A(t)$ and $Q(t)$ are arbitrary we obtain

$$\mathbf{v}_1 = \partial_x.$$

New symmetries were obtained for the case $n \neq 1$ in [6] and, for $n = 1$ in [5], which are shown below:

2.1 *Case 1: n ≠ 1*

If the functions $A(t)$ and $Q(t)$ are given by,

$$A(t) = c_1(k_1 t + k_2)^{-\frac{1}{3}}, \tag{10}$$

$$Q(t) = c_2(k_1 t + k_2)^{-1}, \tag{11}$$

where k_1, k_2, c_1 and c_2 are arbitrary constants, we get the following symmetry

$$\mathbf{v} = a_2\,(k_1 t + k_2)\,\partial_t + \left(a_2 k_1 \frac{x}{3} + a_1\right)\partial_x - a_2 k_1 \frac{u}{3n}\partial_u, \tag{12}$$

with a_1 and a_2 arbitrary constants, which is spanned by the generators

$$\mathbf{v}_1, \quad \mathbf{v}_2 = (k_1 t + k_2)\,\partial_t + k_1 \frac{x}{3}\partial_x - k_1 \frac{u}{3n}\partial_u. \tag{13}$$

Finally, in the case that $A(t) = Q(t) = 0$ the algebra is three dimensional with generators

$$\mathbf{v}_1, \quad \mathbf{v}_4 = \partial_t, \quad \mathbf{v}_5 = t\partial_t + \frac{x}{3}\partial_x - \frac{u}{3n}\partial_u. \tag{14}$$

2.2 *Case 2: n = 1*

In this case, we have the generator

$$\mathbf{v} = (k_1 t + k_2)\,\partial_t + \left(\beta + k_1 \frac{x}{3}\right)\partial_x + \left(\gamma - k_1 \frac{u}{3}\right)\partial_u, \tag{15}$$

where $A(t)$, $Q(t)$, $\beta(t)$, $\gamma(t)$, k_1 and k_2 must satisfy the following conditions:

$$(k_1 t + k_2)A_t + \frac{k_1}{3}A + 2\gamma = 0, \tag{16}$$

$$(k_1 t + k_2)Q_t + k_1 Q = 0, \tag{17}$$

$$\gamma A - \beta_t = 0, \tag{18}$$

$$\gamma Q + \gamma_t = 0. \tag{19}$$

3 Reductions and Exact Solutions

By using the Lie symmetries of Eq. (6), we can obtain the similarity variables and the similarity solutions. This allows us to transform Eq. (6) into an ordinary differential equation (ODE), solving the characteristic system

$$\frac{dt}{\tau} = \frac{dx}{\xi} = \frac{du}{\eta}. \tag{20}$$

3.1 *Reductions for Case 1*

In the following reductions we distinguish:

Reduction 1.1 Considering that $A(t)$ and $Q(t)$ are given by

$$A(t) = c_1 t^{-\frac{1}{3}}, \qquad Q(t) = c_2 t^{-1}, \tag{21}$$

we have Eq. (6) admits the generator

$$\mathbf{v} = t\partial_t + \frac{x}{3}\partial_x - \frac{u}{3n}\partial_u. \tag{22}$$

Solving the characteristic system we obtain the similarity variable and the similarity solution

$$z = \frac{x^3}{t}, \qquad u = \frac{h(z)}{x^{\frac{1}{n}}}. \tag{23}$$

Taking into account (21) and (23) into Eq. (6), this equation is transformed into the ODE

$$
\begin{aligned}
&27h'''n^3z^3 + 54h''n^3z^2 - h'n^3z^2 - 27h''n^2z^2 + 3h^{2n}h'n^3z \\
&+6h'n^3z + c_2hn^3z - 9h'n^2z + 9h'nz + 3c_1h^nh'n^3z^{\frac{2}{3}} \\
&-c_1h^{n+1}n^2z^{-\frac{1}{3}} - h^{2n+1}n^2 - 2hn^2 - 3hn - h = 0.
\end{aligned} \tag{24}
$$

Reduction 1.2 Now, if we assume that

$$A(t) = c_1, \qquad Q(t) = c_2, \tag{25}$$

from (13), Eq. (6) admits the following generator

$$\mathbf{v}_2 - \alpha\mathbf{v}_1 = \partial_t - \alpha\partial_x, \tag{26}$$

where $\alpha \neq 0$ is an arbitrary constant. Solving characteristic system (20) we obtain the similarity variable and the similarity solution

$$z = x + \alpha t, \qquad u = h(z). \tag{27}$$

Substituting (25) and (27) into Eq. (6) we get the following ODE

$$h''' + h^{2n}h' + c_1h^nh' + \alpha h' + c_2h = 0. \tag{28}$$

3.2 *Reductions for Case 2*

Analogously to the previous section, we differentiate:

Reduction 2.1 Assuming afresh that $A(t)$ and $Q(t)$ are given by (21), we get generator (22). Taking into account (20) we obtain

$$z = \frac{x^3}{t}, \qquad u = \frac{h(z)}{x}. \tag{29}$$

From (21) and (29), Eq. (6) can be transformed into the following ODE

$$
\begin{aligned}
&27h'''z^{\frac{10}{3}} + 27h''z^{\frac{7}{3}} - h'z^{\frac{7}{3}} + 3h^2h'z^{\frac{4}{3}} - 6hz^{\frac{1}{3}} \\
&+6h'z^{\frac{4}{3}} + c_2hz^{\frac{4}{3}} + 3c_1hh'z - h^3z^{\frac{1}{3}} - c_1h^2 = 0.
\end{aligned} \tag{30}
$$

Reduction 2.2 We consider that the functions $A(t)$ and $Q(t)$ are given by

$$A(t) = c_2 - 2c_1 t, \qquad Q(t) = 0. \tag{31}$$

Then, from Eqs. (16)–(19) we have

$$\beta(t) = c_3 + c_1 c_2 t - c_1^2 t^2, \qquad \gamma(t) = c_1, \tag{32}$$

where c_3 is an arbitrary constant. Thus, we get the generator

$$\mathbf{v} = \partial_t + \left(c_3 + c_1 c_2 t - c_1^2 t^2\right) \partial_x + c_1 \partial_u.$$

By using (20) we obtain

$$z = x + \frac{c_1^2}{3} t^3 - \frac{c_1 c_2}{2} t^2 - c_3 t, \qquad u = c_1 t + h(z). \tag{33}$$

By means of (31) and (33), Eq. (6) is transformed into

$$h''' + h^2 h' + c_2 h h' - c_3 h' + c_1 = 0. \tag{34}$$

Finally, from (32), if we consider that $\beta = -\alpha$ and $\gamma = 0$, with $\alpha \neq 0$ an arbitrary constant, we obtain the travelling wave generator (26) which leads us to the following ODE

$$h''' + h^2 h' + c_1 h h' + \alpha h' = 0. \tag{35}$$

3.3 *Travelling Waves Solutions*

Considering the case $A(t)$ and $Q(t)$ constants for which the equation admits translations in time and in the space (26) we obtained the reduced Eqs. (28) and (35). In (28) we suppose $c_2 = 0$, therefore Eq. (35) is a particular case of Eq. (28). The reduced Eq. (28) can be integrated with respect to z

$$h'' + \frac{1}{2n+1} h^{2n+1} + \frac{c_1}{n+1} h^{n+1} + \alpha h + c_3 = 0, \tag{36}$$

where c_3 is the constant of integration. Multiplying Eq. (36) by h' and integrating once with respect to z we obtain

$$\left(h'\right)^2 = -\frac{h^{2n+2}}{(n+1)\,(2n+1)} - \frac{2\,c_1\,h^{n+2}}{(n+1)\,(n+2)} - \alpha\,h^2 - 2\,c_3\,h. \tag{37}$$

Let us assume that Eq. (37) has a solution of the form

$$h(z) = aF^b(z), \tag{38}$$

where a and b are parameters to be determined later. By substituting (38) into (37) we obtain

$$(F')^2 = -\frac{1}{(2n^3+7n^2+7n+2)\,a\,b^2\,F^b}\Big[(n+2)\,a^{2n+1}\,F^{(2n+1)\,b+2} + (4\,c_I\,n+2\,c_I)\,a^{n+1}\,F^{(n+1)\,b+2} + \left(2\,\alpha\,n^3+7\,\alpha\,n^2+7\,\alpha\,n+2\,\alpha\right)a\,F^{b+2} + \left(4\,c_3\,n^3+14\,c_3\,n^2+14\,c_3\,n+4\,c_3\right)F^2\Big]. \tag{39}$$

In the following we will determine the exponents and coefficients of Eq. (39) so that Eq. (39) is solvable in terms of Jacobi elliptic function, i.e. Eq. (39) becomes

$$(F')^2 = r + mF^2 + qF^4, \tag{40}$$

where r, m and q are constants. We may choose them properly such that the corresponding solution F of the ODE (40) is one of the Jacobi elliptic, combined Jacobi elliptic functions. If $r = 1$, $m = -(1+k^2)$, $q = 2k^2$, then the solution is

$$h_1 = a\,[\mathrm{sn}(z|k)]^b, \tag{41}$$

or

$$h_2 = a\,[\mathrm{cd}(z|k)]^b \equiv a\left[\frac{\mathrm{cn}(z|k)}{\mathrm{dn}(z|k)}\right]^b,$$

where $0 \le k \le 1$, is called modulus of Jacobi elliptic functions, and $\mathrm{sn}(z|k)$ is the Jacobi elliptic sine function [1]. If $r = 1-k^2$, $m = 2k^2-1$, $q = -2k^2$, the solution is

$$h_3 = a\,[\mathrm{cn}(z|k)]^b,$$

where $\mathrm{cn}(z|k)$ is the Jacobi elliptic cosine function. If $r = k^2-1$, $m = 2-k^2$, $q = -2$, the solution is

$$h_4 = a\,[\mathrm{dn}(z|k)]^b,$$

where $\mathrm{dn}(z|k)$ is the third Jacobi elliptic function. By comparing the exponents and the coefficients of Eqs. (39) and (40) we can obtain exact solutions for Eq. (37) with

$n = 1$, $c_3 = 0$ and $\alpha = -1$ when $a = \sqrt{6}$, $c_1 = 0$ and $b = 1$,

$$h = \sqrt{6}\,\mathrm{sech}(z).$$

In this case Eq. (6) is the Korteweg-de Vries equation.

3.4 The Simplest Method

Suppose that the nonlinear partial differential equation for $u(x,t)$ is in the form

$$\Delta(u, u_t, u_x, u_{xx}, \ldots) = 0,$$

where Δ is polynomial in $u(x,t)$ and its partial derivatives, in which the highest order derivatives and nonlinear terms are involved. In order to obtain exact solitary wave solutions of the equation, we have to pursue the following fundamental steps [11]:

Step 1: We consider the travelling wave variable

$$u(t,x) = h(z) = h(x + \alpha t), \tag{42}$$

where α represents the speed of the travelling wave. The wave variable (42) carries Eq. (6) into the following ordinary differential equation

$$\Delta'(h, \alpha h', h', h'', \ldots) = 0. \tag{43}$$

Step 2: To seek the travelling wave solution of Eq. (43), we assume that (43) has a solution in the following form

$$h(z) = k_0 + k_1 Y + \cdots + k_N Y^N + \kappa_1 \left(\frac{Y'}{Y}\right) + \cdots + \kappa_N \left(\frac{Y'}{Y}\right)^N, \tag{44}$$

where k_n ($n = 0, 1, \ldots, N$) and κ_n ($n = 1, 2, \ldots, N$) are unknown constants to be calculated, and $Y(z)$ is the general solution of the Riccati equation:

$$Y'(z) + Y^2(z) - aY(z) - b = 0, \tag{45}$$

with a and b constants which must be determined.

Step 3: The positive integer N in (44) can be determined by taking into consideration the homogeneous balance between the highest order linear terms and the nonlinear terms of highest order occurring in (43).

Step 4: Inserting (44) and the derivatives $h', h'', \ldots$, into (43) we get a polynomial in $Y(z)$ and its derivatives. Requiring the vanishing of the different powers of the function $Y(z)$, we obtain an overdetermined system of equations which must be solved to find k_n, κ_n and α. This complete the determination of the solution of the ODE.

We will make use of this method to construct travelling wave solutions to the partial differential equation (6) by means of Reduction 2.2 when $k_3 = 1$, $\gamma = 0$ and $\beta = -\alpha$. Thus, Eq. (6) takes the form

$$u_t + c_1 u u_x + u^2 u_x + u_{xxx} = 0. \tag{46}$$

This equation is transformed into (35) which can be integrated with respect to z

$$h'' + \frac{h^3}{3} + \frac{c_1 h^2}{2} + \alpha h + k = 0, \tag{47}$$

where k is the constant of integration. We consider Eq. (35), taking the homogeneous balance between the highest order derivative h''' and the nonlinear term of highest order $h^2 h'$ we obtain $N = 2$. Therefore, the solution of (35) takes the following form

$$h = k_0 + k_1 Y + k_2 Y^2 + \kappa_1 \left(\frac{Y'}{Y}\right) + \kappa_2 \left(\frac{Y'}{Y}\right)^2, \tag{48}$$

where $k_0, k_1, k_2, \kappa_1, \kappa_2$ are constant to be determined later and $Y(z)$ satisfies Eq. (45). By Step 4 we obtain a system. By solving this system we obtain $b = 0$, $\kappa_2 = -k_2$ and $\alpha, a, k_0, k_1, k_2, \kappa_1$ must satisfy the following equations

$$\begin{aligned}
&-2a^6k_2^3 + 6a^5k_2^2\kappa_1 + 3a^4c_1k_2^2 - 6a^4k_2\kappa_1^2 + 6a^4k_0k_2^2 + 12a^4k_2 - 6a^3c_1k_2\kappa_1 \\
&-12a^3k_0k_2\kappa_1 - 12a^3k_2 + 2a^3\kappa_1^3 - 6a^3\kappa_1 - 6a^2c_1k_0k_2 + 3a^2c_1\kappa_1^2 - 6a^2k_2\alpha \\
&+6a^2k_0\kappa_1^2 - 6a^2k_0^2k_2 + 6a^2\kappa_1 + 6ac_1k_0\kappa_1 + 6ak_0^2\kappa_1 \\
&+6a\alpha\kappa_1 + 3c_1k_0^2 + 6k_0\alpha + 2k_0^3 + 6k = 0,
\end{aligned}$$

$$\begin{aligned}
&2a^5k_2^3 - 5a^4k_2^2\kappa_1 + a^4k_1k_2^2 - 2a^3c_1k_2^2 + 4a^3k_2\kappa_1^2 - 2a^3k_1k_2\kappa_1 - 4a^3k_0k_2^2 \\
&+6a^3k_2 + 3a^2c_1k_2\kappa_1 - a^2c_1k_1k_2 + a^2k_1\kappa_1^2 + 6a^2k_0k_2\kappa_1 + a^2k_1 - 2a^2k_0k_1k_2 \\
&+2a^2k_2 - a^2\kappa_1^3 - 4a^2\kappa_1 + ac_1k_1\kappa_1 + 2ac_1k_0k_2 - ac_1\kappa_1^2 + 2ak_2\alpha \\
&-2ak_0\kappa_1^2 + 2ak_0k_1\kappa_1 + 2ak_0^2k_2 - c_1k_0\kappa_1 + c_1k_0k_1 + k_1\alpha - k_0^2\kappa_1 + k_0^2k_1 - \alpha\kappa_1 = 0,
\end{aligned} \tag{49}$$

$$\begin{aligned}
&8a^4k_2^3 - 16a^3k_2^2\kappa_1 + 8a^3k_1k_2^2 - 4a^2c_1k_2^2 + 10a^2k_2\kappa_1^2 - 12a^2k_1k_2\kappa_1 - 8a^2k_0k_2^2 \\
&+2a^2k_1^2k_2 + 32a^2k_2 + 4ac_1k_2\kappa_1 - 4ac_1k_1k_2 + 4ak_1\kappa_1^2 - 2ak_1^2\kappa_1 \\
&+8ak_0k_2\kappa_1 + 4ak_1 - 8ak_0k_1k_2 - 2a\kappa_1^3 - 12a\kappa_1 + 2c_1k_1\kappa_1
\end{aligned}$$

$$-c_1k_1^2 - c_1\kappa_1^2 - 2k_0\kappa_1^2 + 4k_0k_1\kappa_1 - 2k_0k_1^2 = 0,$$

$$8a^3k_2^3 - 12a^2k_2^2\kappa_1 + 12a^2k_2^2k_1 - 12ak_2k_1\kappa_1 + 6ak_2\kappa_1^2 - 6\kappa_1 \\ +6ak_2k_1^2 + 24ak_2 - 3k_1^2\kappa_1 + 3k_1\kappa_1^2 + k_1^3 + 6k_1 - \kappa_1^3 = 0.$$

Equation (45) with $b = 0$ is the Bernoulli equation, so we can obtain the corresponding solution h of the ODE (43) in terms of this equation. As a result, the solution of the Bernoulli equation is

$$Y(z) = a\left(\frac{Y_1 + Y_2}{1 + Y_1 + Y_2}\right), \tag{50}$$

where $Y_1(z) = \sinh(a(z+c))$, $Y_2(z) = \cosh(a(z+c))$ and c is an arbitrary constant. Substituting (50) into (48) we obtain the following solution

$$h(z) = k_0 + \frac{a}{2}\left[k_1\left(1 + \tanh\left(\frac{a}{2}(z+c)\right)\right)\right. \\ \left. +2\,a\,k_2\tanh\left(\frac{a}{2}(z+c)\right) + \kappa_1\left(1 - \tanh\left(\frac{a}{2}(z+c)\right)\right)\right].$$

By using transformation (27) we can obtain a solution of Eq. (46). In Figs. 1 and 2 we show two exact solutions of Eq. (46).

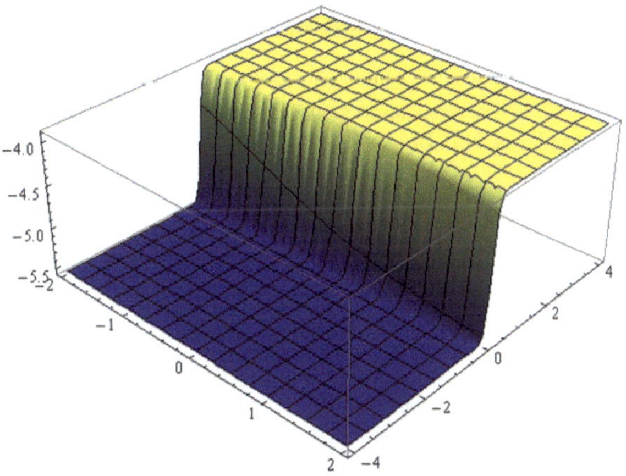

Fig. 1 Solution $u(x,t)$ of Eq. (46) for $a = c = k_1 = 1$, $k_0 \approx -7.71$, $k_2 \approx 1.18$, $\kappa_1 = 5$, $\alpha \approx -17.12$ and $k \approx -54.57$

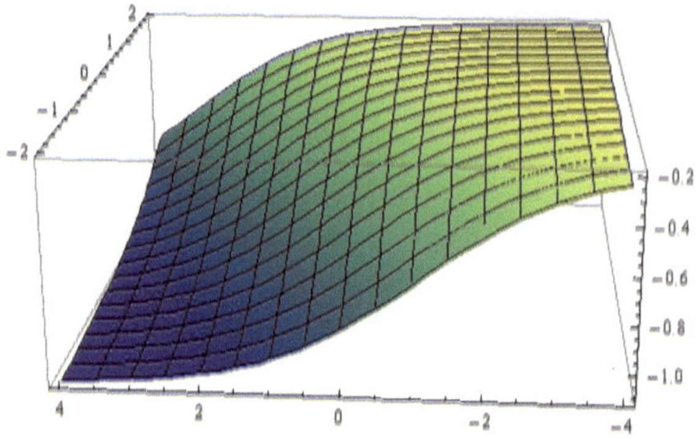

Fig. 2 Solution $u(x, t)$ of Eq. (46) for $a = c = k_1 = 1$, $k_0 \approx -0.66$, $k_2 \approx -0.53$, $\kappa_1 = -1$, $\alpha \approx -0.92$ and $k \approx -1.19$

4 Conclusions

In this paper, a generalized variable-coefficient Gardner equation has been considered. By using equivalence transformations we can restrict our study to a subclass (6) of Eq. (1) with fewer number of arbitrary functions. Symmetry analysis of Eq. (6) with respect the time dependent functions has been presented. From the symmetries of Eq. (6) we obtain the similarity reductions which transform Eq. (6) into an ODE. By means of the similarity reductions and the simplest equation method, we have obtained some exact travelling wave solutions. In view of the analysis, we see how equivalence transformations can be used to simplify the search of exact solutions of the equation and allow us to present these solutions in a simple and clear form.

Acknowledgements We warmly thank the referee his valuable comments and suggestions. The authors acknowledge the financial support from Junta de Andalucía group FQM–201, Universidad de Cádiz. The first author expresses his sincere gratitude to the Plan Propio de Investigación 2013 de la Universidad de Cádiz and the Comisión Académica del Programa de Doctorado en Matemáticas de la Universidad de Cádiz for their support. The second author also acknowledges the support of DGICYT project MTM2009-11875 with the participation of FEDER.

References

1. Abramowitz, M., Stegun, I.A.: Handbook of Mathematical Functions. Dover, New York (1972)
2. Bruzón, M.S., Gandarias, M.L., Ibragimov, N.H.: Self-adjoint sub-classes of generalized thin film equations. J. Math. Anal. Appl. **357**, 307–313 (2009)
3. Bruzón, M.S., Gandarias, M.L., de la Rosa, R.: Conservation laws of a Gardner equation with time-dependent coefficients. J. Appl. Nonlinear Dyn. **4**(2), 169–180 (2015)
4. de la Rosa, R., Gandarias, M.L., Bruzón, M.S.: A study for the microwave heating of some chemical reactions through Lie symmetries and conservation laws. J. Math. Chem. **53**, 949–957 (2015)
5. de la Rosa, R., Gandarias, M.L., Bruzón, M.S.: On symmetries and conservation laws of a Gardner equation involving arbitrary functions. Appl. Math. Comput. (preprint)

6. de la Rosa, R., Gandarias, M.L., Bruzón, M.S.: Equivalence transformations and conservation laws for a generalized variable-coefficient Gardner equation. Commun. Nonlinear Sci. Numer. Simul. **40**, 71–79 (2016)
7. Gandarias, M.L., Bruzón, M.S.: Some conservation laws for a forced KdV equation. Nonlinear Anal. Real World Appl. **13**, 2692–2700 (2012)
8. Gandarias, M.L., Ibragimov, N.H.: Equivalence group of a fourth-order evolution equation unifying various non-linear models. Commun. Nonlinear Sci. Numer. Simul. **13**, 259–268 (2008)
9. Gandarias, M.L., Torrisi, M., Tracinà, R.: On some differential invariants for a family of diffusion equations. J. Phys. A: Math. Gen. **40**(30), 8803–8813 (2007)
10. Hubert, M.B., Betchewe, G., Doka, S.Y., Crepin, K.T.: Soliton wave solutions for the nonlinear transmission line using the Kudryashov method and the (G'/G)-expansion method. Appl. Math. Comput. **239**, 299–309 (2014)
11. Kudryashov, N.A.: Simplest equation method to look for exact solutions of nonlinear differential equations. Chaos, Solitons Fractals. **24**, 1217–1231 (2005)
12. Liu, H., Li, J., Liu, L.: Painlevé analysis, Lie symmetries, and exact solutions for the time-dependent coefficients Gardner equations. Nonlinear Dyn. **59**, 497–502 (2010)
13. Olver, P.: Applications of Lie Groups to Differential Equations. Springer, New York (1993)
14. Ovsiannikov, L.V.: Group Analysis of Differential Equations. Academic Press, New York (1982)
15. Torrisi, M., Tracinà, R.: Equivalence transformations and symmetries for a heat conduction model. Int. J. Non-Linear Mech. **33**(3), 473–487 (1998)
16. Tracinà, R., Bruzón, M.S., Gandarias, M.L., Torrisi, M.: Nonlinear self-adjointness, conservation laws, exact solutions of a system of dispersive evolution equations. Commun. Nonlinear Sci. Numer. Simul. **19**, 3036–3043 (2014)
17. Triki, H., Wazwaz, A.M.: Traveling wave solutions for fifth-order KdV type equations with time-dependent coefficients. Commun. Nonlinear Sci. Numer. Simul. **19**, 404–408 (2014)
18. Tsaousi, C., Tracinà, R., Sophocleous, C.: Laplace type invariants for variable coefficient mKdV equations. J. Phys.: Conf. Ser. **621**, 012015 (2015)
19. Vaneeva, O., Kuriksha, O., Sophocleous, C.: Enhanced group classification of Gardner equations with time-dependent coefficients. Commun. Nonlinear Sci. Numer. Simul. **22**, 1243–1251 (2015)
20. Zhang, L.H., Dong, L.H., Yan, L.M.: Construction of non-travelling wave solutions for the generalized variable-coefficient Gardner equation. Appl. Math. Comput. **203**, 784–791 (2008)

A Second Order Local Projection Lagrange-Galerkin Method for Navier-Stokes Equations at High Reynolds Numbers

Rodolfo Bermejo and Laura Saavedra

Abstract We present a stabilized Backward Difference Formula of order 2-Lagrange Galerkin method for the incompressible Navier-Stokes equations at high Reynolds numbers. The stabilization of the conventional Lagrange-Galerkin method is done via a local projection technique for inf-sup stable finite elements. We have proven that for the Taylor-Hood finite element the a priori error estimate for velocity in the $l^{\infty}(L^2(\Omega)))$-norm is $O(h^2 + \Delta t^2)$ whereas the error for the pressure in the $l^2(L^2(\Omega)))$-norm is $O(h^2 + \Delta t^2)$, with error constants that are independent of the inverse of the Reynolds number. Numerical examples at high Reynolds numbers show the robustness of our method.

1 Introduction

In most of industrial problems, we have to deal with flows at high Reynolds numbers, namely, convection-dominated problems, appearing the hyperbolic nature of time-dependent Navier-Stokes (NS) equations. The classic Lagrange-Galerkin (LG) method consists of the discretization of the material derivative along the trajectory of the fluid particles, using a finite difference scheme. This is a natural way, from a physical point of view, to introduce upwinding and transforms the NS equations into a linear Stokes problem. Therefore, at each time step, one has to solve an algebraic linear system of equations which is more manageable than the algebraic nonlinear system of equations produced by conventional implicit time marching schemes. A priori, these advantages make LG methods look like efficient methods to integrate NS equations. However, they have drawbacks concerned with the calculation of some integrals which appear in the formulation of the numerical

R. Bermejo
Departamento de Matemática Aplicada, E.T.S.I. Industriales, Universidad Politécnica de Madrid, Madrid, Spain
e-mail: rbermejo@etsii.upm.es

L. Saavedra (✉)
Departamento de Matemática Aplicada a la Ingeniería Aeroespacial, E.T.S.I. Aeronáutica y del Espacio, Universidad Poliécnica de Madrid, Madrid, Spain
e-mail: laura.saavedra@upm.es

F. Ortegón Gallego et al. (eds.), *Trends in Differential Equations and Applications*, SEMA SIMAI Springer Series 8, DOI 10.1007/978-3-319-32013-7_24

solution and whose integrands are defined in two different meshes. These integrals have to be calculated very accurately to maintain the stability and the accuracy of the method, see [2, 9], requiring thus the use of high order quadrature rules. Since each quadrature point has an associated foot of characteristic curve, this means that many systems of differential equations have to be solved backward in time. Hence, LG methods may become less efficient than they look at first.

It is relatively easy to prove that LG methods are unconditionally stable if the aforementioned integrals are calculated exactly. However, when the time step, Δt, is small and the viscosity is not sufficiently high to kill the instabilities there are intervals of values of Δt in which the solution becomes either unstable or significantly less accurate. In order to fix this drawback, we present in this work the stabilization of LG methods in the spirit of the local projection stabilization approach of Braack and Burman [3] and Ganesan and Tobiska [4], just to cite a few. This stabilization technique is well suited to LG methods because is relatively easy to incorporate to any LG method code and maintains the symmetry of the linear system that has to be solved in every time step.

2 A Lagrange-Galerkin Method for Navier Stokes Equations

Let $\Omega \subset \mathbb{R}^d$ $(d = 2, 3)$, be a bounded domain with Lipschitz boundary $\Gamma = \partial\Omega$ and let $[0, T]$ denote a time interval. We consider the Navier-Stokes equations for a fluid of constant density $(\rho = 1)$ under the action of an external force field $f : \overline{\Omega} \times [0, T] \to \mathbb{R}^d$ and with a known initial condition $v(x, 0) = v^0(x)$,

$$\begin{aligned} \frac{\partial v}{\partial t} + (v \cdot \nabla)v - \nu\Delta v + \nabla p = f, \\ \operatorname{div} v = 0, \\ v_{|\Gamma} = 0, \end{aligned} \tag{1}$$

where $v : \Omega \times [0, T) \to \mathbb{R}^d$ is the flow velocity, $p : \Omega \times [0, T) \to \mathbb{R}$ is the pressure and ν is the kinematic viscosity coefficient, which is assumed to be constant.

Lagrange-Galerkin methods are based on the discretization of Navier-Stokes equations (1) along the characteristics of the operator $\frac{D}{Dt} = \frac{d}{dt} + v \cdot \nabla$, known as material derivative. Thus, we introduce the mapping $X(x, t; \cdot) : (0, T) \to \mathbb{R}^d$ solution of the initial value problem

$$\begin{cases} \dfrac{dX(x, t; s)}{ds} = v(X(x, t; s), s), \\ X(x, t; t) = x, \end{cases} \tag{2}$$

which is called the characteristic curve through point (x, t). The point $X(x, t; s)$ represents the position occupied at instant s by the point that is in x at instant t and moves with velocity v. If $v \in L^1(0, T; \mathbf{W}^{1,\infty}(\Omega))$ the problem (2) has a unique solution X defined in $[0, T]$ for each initial condition (x, t) as

$$X(x, t; s) = x - \int_t^s v(X(x, t; \tau), \tau) d\tau. \tag{3}$$

Along the characteristic curves the time derivative is equal to the material derivative,

$$\begin{aligned} \frac{d}{ds} v(X(x, t; s), s) &= v'(X(x, t; s), s) + \nabla v(X(x, t; s), s) v(X(x, t; s), s) \\ &= \frac{D}{Ds} v(X(x, t; s), s), \end{aligned} \tag{4}$$

therefore we can discretize the material derivative using a finite difference scheme.

In order to obtain our numerical schemes, we divide the interval $[0, T]$ into N subintervals of $\Delta t = T/N$ size and approximate the material derivative of velocity along the characteristic curves using a Backwards Differential Formula of second order (BDF2).

Let $\Omega_h = \bigcup_{j=1}^{N_e} T_j$ be a regular quasi-uniform triangulation of the region Ω, T_j be a simplex of dimension d and h be the maximum diameter of elements. We associate with Ω_h the H^1-conforming finite element spaces $\mathbf{V}_h \subset H^1(\Omega)$, $\mathbf{V}_{0h} = \mathbf{V}_h \cap \mathbf{H}_0^1(\Omega)$ and $M_h \subset L_0^2(\Omega)$.

In each time step we approximate the weak solution of problem (1) by two functions $(v_h^{n+1}, p_h^{n+1}) \in \mathbf{V}_{h0} \times M_h$. Then the characteristic curve $X(x, t_{n+1}; \cdot)$ is replaced by $X_h(x, t_{n+1}; \cdot)$, which is the numerical solution of the initial value problem (2) replacing v with v_h. Since $v_h(\cdot, t)$ may not exist if $t \notin \{t_0, \ldots, t_N\}$, it is usually calculated by some extrapolation formula using certain values in the set $\{v_h^m\}_{m=0}^n$.

Discrete problem BDF2-LG: Find $\{(v_h, \pi_h)\}_{n=2}^N \in (\mathbf{V}_{0h} \times M_h)^N$ such that

$$\begin{aligned} \frac{1}{\Delta t}\left(D_h^n v_h, w_h\right) + \nu\left(\nabla v_h^{n+1}, \nabla w_h\right) + \left(p_h^{n+1}, \text{div} w_h\right) & \\ = \left(f^{n+1}, w_h\right), \forall w_h \in \mathbf{V}_{0h}, & \end{aligned} \tag{5}$$

$$\left(\text{div} v_h^{n+1}, q_h\right) = 0, \forall q \in M_h \tag{6}$$

for $n \in \{1, \ldots, N-1\}$ with

$$D_h^n u := \frac{1}{2}\left(3u^{n+1} - 4u^n \circ X_h^{n,n+1} + u^{n-1} \circ X_h^{n-1,n+1}\right),$$

$v_h^0 = R_h v^0 \in \mathbf{V}_{0h}$, R_h the $\mathbf{L}^2$ elliptic projector onto $\mathbf{V}_{0h}$ and $X_h^{j,n+1}(x) = X_h(x, t_{n+1}, t_j)$. The approximate values $v_h^1 \in \mathbf{V}_{0h}$ and $p_h^1 \in M_h$ can be obtained by a single step scheme.

3 Local Projection Stabilized Lagrange-Galerkin Method

The local projection stabilized LG method (LPS) can be interpreted as a variational multiscale method. Such methods are based on the scale separation of turbulent flows. When a flow becomes turbulent very different time and space scales appear that make difficult or even impossible to predict the behaviour of the flow with precision. Three type of scales are considered:

1. Large scales: in the Kolmogorov cascade these are the scales containing energy. They are the scales of the mean flow, which are perfectly captured in the simulation.
2. Small resolved scales: the ones of the inertial range, known as subfilter scales. Kinetic energy is merely transferred to smaller scales, inertial effects are still much larger than viscous effects. These scales are supposed to be captured by the mesh.
3. Small unresolved scales: kinetic energy is dissipated by molecular viscosity at these scales. Their effect on the other scales has to be modeled.

The finite element spaces $\mathbf{V}_h$ and M_h are decomposed as

$$\mathbf{V}_h = \overline{\mathbf{V}}_h \oplus \mathbf{V}'_h, M_h = \overline{M}_h \oplus M'_h, \tag{7}$$

where $\overline{\mathbf{V}}_h, \overline{M}_h$ are the finite dimension spaces for the large scales and $\mathbf{V}'_h, M'_h$ are the finite element spaces for the small resolved scales. In projection-based variational multiscale methods the influence of the unresolved small scales on the large scales is assumed to be negligible. Furthermore, in most of these methods the action of unresolved small scales on the small resolved scales is modeled through a term of added viscous stresses of the form

$$c(u'_h, v'_h) = \sum_{K \in \Omega_h} (\tau_K (I - \Pi_h)\nabla u_h, (I - \Pi_h)\nabla v_h), \tag{8}$$

where τ_K is a mesh-dependent coefficient, $I : L^2(\Omega) \to L^2(\Omega)$ is the identity operator, and $\Pi_h : L^2(\Omega) \to \mathbf{G}_h$, is the projection defined by the relation $\Pi_h q = \Pi_K(q_{|K})$, being $\Pi_K : L^2(K) \to \mathbf{G}_h(K)$ a local projection operator in the finite dimensional space $\mathbf{G}_h(K) \subseteq \{\nabla v_{h|K} / v_h \in \overline{\mathbf{V}}_h\}$. This term appears on the equation of momentum for the scales resolved obtaining the final problem

$$\frac{1}{\Delta t}\left(D_h^n v_h, w_h\right) + \nu\left(\nabla v_h^{n+1}, \nabla w_h\right) + c(v_h^{n+1'}, w'_h) + \left(p_h^{n+1}, \mathrm{div} w_h\right)$$

$$= \left(f^{n+1}, w_h\right), \forall w_h \in \mathbf{V}_{0h}, \tag{9}$$

$$\left(\mathrm{div} v_h^{n+1}, q_h\right) = 0, \forall q \in M_h. \tag{10}$$

In the LPS method proposed in this work the spaces $\mathbf{V}_h$ and M_h consist of $\mathbf{P}_2$ and P_1 finite elements, respectively, and $\mathbf{G}_h(K)$ is formed by $\mathbf{P}_0$ finite elements. The added viscosity is taken as $\tau_K = c_{add} h_k^2$, with c_{add} a constant whose value has to be adjusted in every simulation.

The error estimate that we have obtained for the velocity, under some regularity assumptions, is

$$\|v - v_h\|_{l^\infty(\mathbf{L}^2(\Omega))} = O(h^2 + \Delta t^2) + \min\left(C\frac{\Delta t}{f(\nu)}, C\frac{\Delta t}{h}, 2\right)\frac{h^2}{\Delta t}, \tag{11}$$

with C a constant and $f(\nu)$ a function of the viscosity.

4 Numerical Results

In this section we test the behavior of the local projection LG method through two numerical examples. The first one is an academic example with known analytical solution, proposed in Notsu and Tabata [10]. The second example is the flow past an airfoil at zero angle of attack. This test was proposed in Guermond et al. [6] to assess the behavior of the subgrid method proposed in this work.

4.1 Two Dimensional Flow in a Square Domain

This problem was solved for the modified Lagrange-Galerking methods in Bermejo and Saavedra [1]. On a domain $\Omega = (0,1)^2$ we impose suitable initial, boundary conditions and an external force term such that the exact solution of incompressible Navier-Stokes equations, (v, p) is given by

$$v_1(x,t) = (1 + \sin(\pi t))\sin^2(\pi x_1)\sin(2\pi x_2), \tag{12}$$

$$v_2(x,t) = -(1 + \sin(\pi t))\sin^2(\pi x_2)\sin(2\pi x_1), \tag{13}$$

$$p(x,t) = (1 + \sin(\pi t))\cos(\pi x_1)\cos(\pi x_2). \tag{14}$$

The final time was set to $T = 1$. The results shown below have been obtained with dynamic viscosity values $\nu = 10^{-5}$ and $\nu = 10^{-7}$. We calculate the numerical solution in a family of structured meshes Ω_{h_j} formed by right triangles whose edge lengths are $h_j = 1/2^j$, $j \in \{3, 4, 5, 6, 7\}$. The feet of the characteristic curves are calculated by a Runge-Kutta method of order 4.

In our previous work [1] we have seen that due to the use of numerical integration to compute the terms

$$\int_{\Omega} v_h(X^{n,n+1}(x))w_h(x)dx, \tag{15}$$

instabilities appear when the diffusion is too small. If a BDF2 formula is used for the discretization of the material derivative, these instabilities are more significant and appear before when refining the time step. We have also demonstrated that the use of high order quadrature rules delays the appearance of the instability and this becomes weaker or even disappears completely. The following results show that if a subgrid viscosity technique is used, these types of instabilities are smoothed out. We have compared the influence on the stability of the increment of the order of the quadrature rule with the increment of the added viscosity. Furthermore, the accuracy achieved with the stabilized and standard LG methods is similar. Unless otherwise specified, a Gaussian quadrature rule of order ten (25 nodes) is used to approximate (15).

First, the results for $Re = 10^5$ are presented. The norm of the errors obtained for the velocity are plotted in Fig. 1 with $\Delta t = 0.001$. This figure shows that second order convergence is obtained, as was expected for the theoretical results. In Fig. 2 the time convergence curve is plotted. We notice that the errors of LG method, are nearly the same as those obtained with the stabilized method, until the solution becomes unstable. As can be seen, the stability interval of the LPS method is larger than the one of the convectional LG method. Figure 2 also shows that second-order convergence in time is achieved with the different methods for both the pressure and the velocity.

As in the previous case, the solution obtained with standard LG method for $\nu = 10^{-7}$) becomes unstable when the time step decreases, even more if a low-order quadrature rule is used. The same applies to stabilized methods although the solutions are more stable. If we use a high order quadrature rule, increasing the computational cost, we can see that the instabilities appear for smaller time steps. In Fig. 3 the $l^{\infty}(L^2)$-norm of the velocity error is plotted using two different quadrature

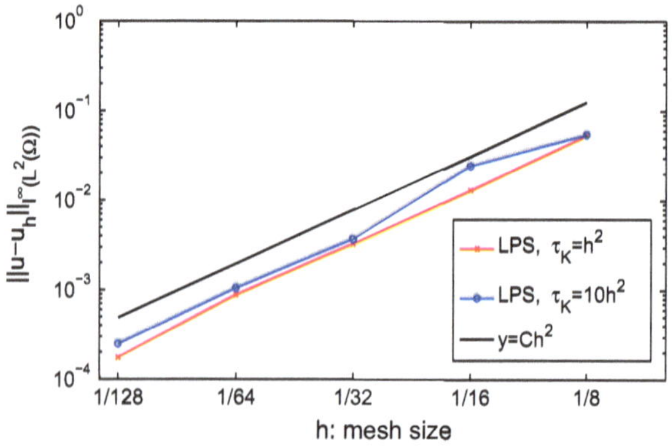

Fig. 1 Space convergence curve of the velocity, $Re = 10^5$, $\Delta t = 0.001$

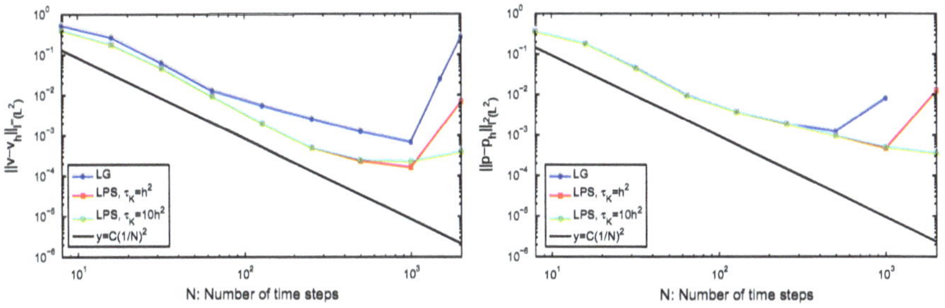

Fig. 2 Temporal error estimation for the velocity and the pressure obtained for $h = 1/128$ at $Re = 10^5$

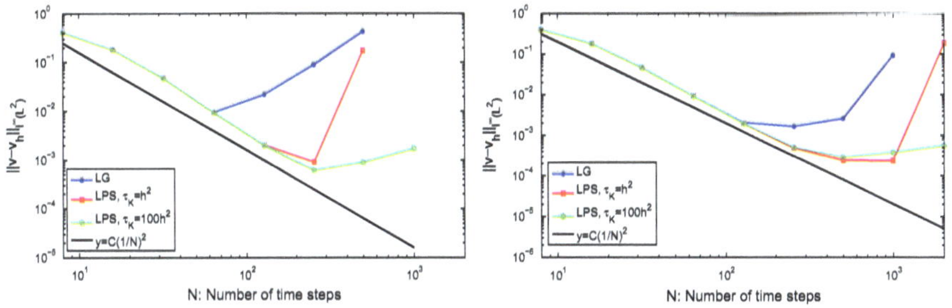

Fig. 3 Time convergence curve of the velocity obtained with sixth-order (*left panel*) and tenth-order (*right panel*) quadrature rules, $Re = 10^7$, $h = 1/128$

rules (order 6 and 10) with $\tau_K = h^2$ and $\tau_K = 100h^2$. We notice that the accuracy in the evaluation of the integrals (15) has a stronger effect on the stability than the subgrid viscosity.

For the sake of completeness we present the results obtained with the BDF formula of order one and LPS method with $\tau_K = h^2$ and a quadrature rule of order 10. The time convergence curve can be seen in Fig. 4 where no instabilities appear. The errors of the velocity and the pressure are computed with the norm $l^2(H^1)$ because with this norm the instabilities should be seen before.

4.2 Flow Past a NACA0012 Airfoil at Zero Angle of Attack

Now, we show the simulations of the flow past the NACA0012 airfoil at zero angle of attack at $Re = 10^5$ and $Re = 3 \times 10^6$. The first value of the Reynolds number corresponds to a laminar flow and the second one to a fully turbulent flow.

The definition of geometry of the airfoil NACA0012 is well-known and the formula to create an airfoil between $x = 0$ and $x = 1$ can be found in a wide number of references. To obtain some data for our simulations, as the types of meshes and the experimental data for the comparison, we follow the work of the Langley Research Center [11] where a validation NACA0012 airfoil case for turbulence models is given.

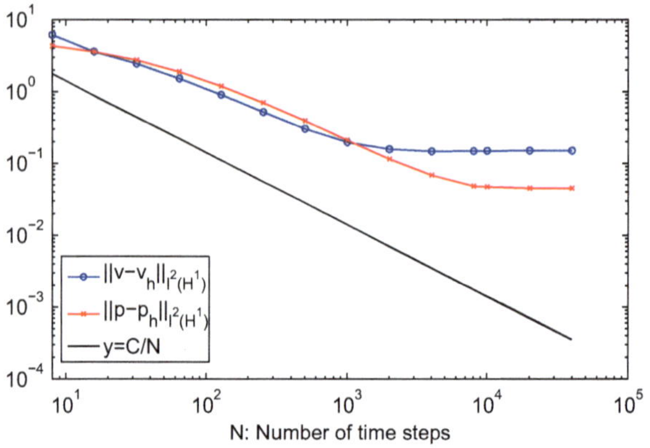

Fig. 4 Time convergence curve for LPS-BDF1 method, $Re = 10^7$, $h = 1/128$

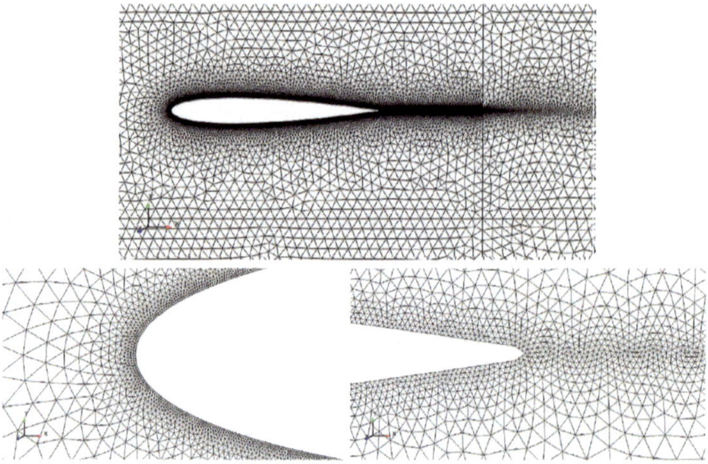

Fig. 5 Mesh around de airfoil (*top*) and near the leading and trailing edges (*bottom*)

The computational domain is $\Omega = [-5, 10] \times [-5, 5]$ and we simulate until $T = 5s$. The no-slip boundary condition is used on the airfoil, a unit horizontal velocity is imposed on the boundary $\{(x, y) \in \Omega/x = -5\} \cap \{(x, y) \in \Omega/y = -5, 5\}$ and the convectional do-nothing condition is set in $\{(x, y) \in \Omega/x = 10\}$.

The mesh consists of 71,785 elements, 182,893 velocity nodes and 37,036 pressure nodes, the mesh size inside the boundary layer is $h = 10^{-3}$. This spatial discretization is adequate to capture the viscous boundary layer for $Re = 10^5$ but in the case of $Re = 3 \times 10^6$ our simulations could not give accurate results. In fact, the mesh is too coarse to obtain a solution with classical Lagrange Galerkin methods without a turbulence model. We have obtained solutions that blow up in final times. Nevertheless, we have kept this mesh to check the behavior of stabilized methods, even without including turbulence models. We shown in Fig. 5 the mesh around the profile and details of the regions near the leading and trailing edges.

Taking into account the conclusions extracted from the academic test, these simulations are carried out with the first-order BDF formula and the quadrature rule of order 10. For the stabilized method the parameter is $\tau_K = c_{add} h_K^2$, where c_{add} is a constant value that will be specified in each case. We set the initial condition to zero. The time step employed for $Re = 10^5$ is $\Delta t = 10^{-3}$ and when $Re = 3 \times 10^6$ is $\Delta t = 5 \times 10^{-4}$.

4.2.1 Results for $Re = 10^5$

The purpose of this experiment is to show that, at high Reynolds numbers, the stabilized LG method gives a stable solution whereas the convectional LG method fails.

In Fig. 6 we can see the contours of the velocity and the pressure at $t = 5s$ obtained by the standard LG method and the stabilized method, with $C_{add} = 0.1$ and $C_{add} = 1$. The differences between the two methods are clearly observed in these images and also the influence of the added viscosity in smoothing the solution. In Fig. 7 the vorticity field at $t = 3s$ is plotted. As can be seen, the instabilities had already appeared at this time in the solution obtained with the standard LG method.

4.2.2 Results for $Re = 3 \times 10^6$

In this section we show the simulations at $Re = 3 \times 10^6$ with $c_{add} = 1$ and $c_{add} = 2$, to see the effect of the added viscosity on the simulation. In this case the boundary layers should be turbulent over more than half of the airfoil. We show in Fig. 8 the upper and lowed pressure coefficients $c_p = 2(p - p_\infty)/\rho U_\infty^2$ at $t = 5s$. We compare our results with the experimental data given in Gregory and O'Reilly [5]. At this point, we must say that, although the experiment is three dimensional, according to [11] the data given in [5] are appropriate for the validation of surface pressures in two-dimensional simulations.

In order to assess the influence of the subgrid viscosity on the flow, we plot in Fig. 9 the vorticity field and the contours of the modulus of the velocity at $t = 2.5s$, for $c_{add} = 1$ and $c_{add} = 2$. We can see that the vorticity field is very similar in the two simulations, however there are differences in the modulus of the velocity. The velocity is fully smooth for $c_{add} = 2$, whereas for $c_{add} = 1$ the contours are less smooth. Nevertheless, in Fig. 8, the pressure coefficients show more oscillations if the artificial viscosity is higher. We think that this small oscillatory character of both solutions is due to the fact that the boundary layer is not well resolved.

We compare the results achieved with the local stabilized projection method with those obtained using a RANS $k - \omega$ turbulence model using the commercial ANSYS-Fluent code. In Fig. 10 the upper surface pressure coefficients predicted at $t = 5s$ with both methods are plotted. We observe that our results are close to both the experimental and the RANS model results. As can be expected, the latter are

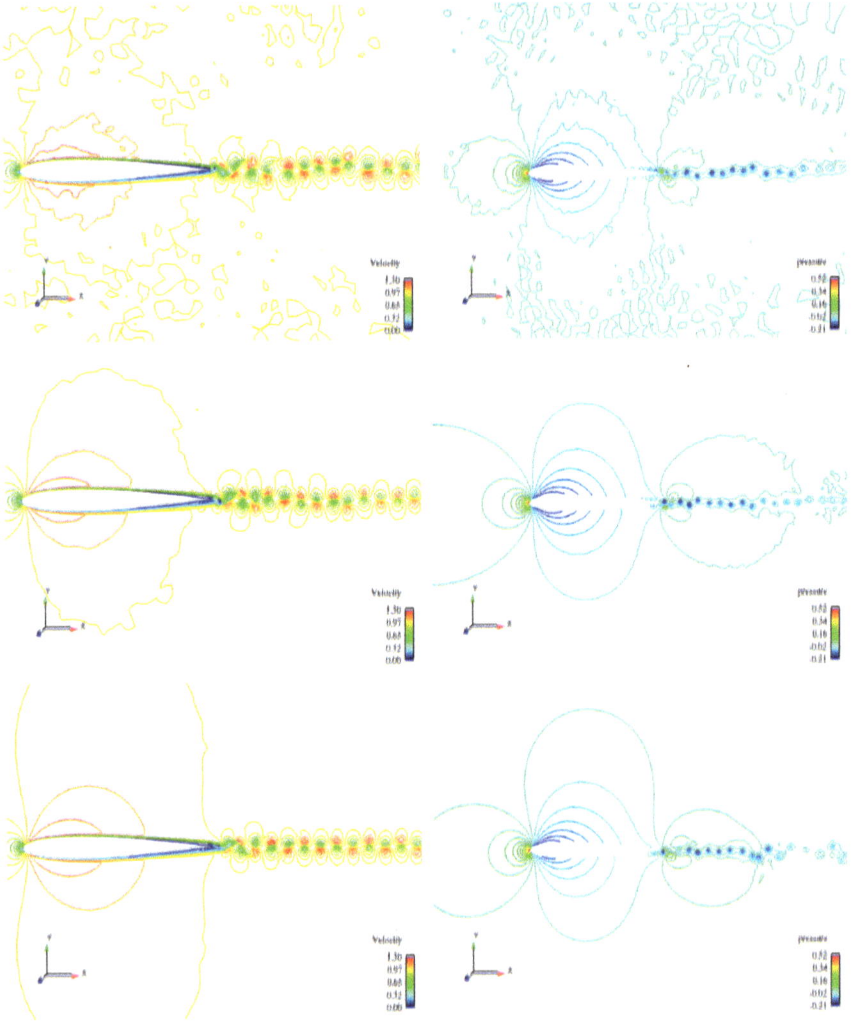

Fig. 6 Velocity and pressure contours at $t = 5s$ obtained with LG method (*top*) and with the LPS method: $c_{add} = 1$ (*middle*) and $c_{add} = 0.1$ (*bottom*)

also oscillatory but with a smaller amplitude than that obtained with the stabilized method.

We should notice the good results obtained with the coarse mesh used for this Reynolds number and without the coupling of any turbulence model. In spite of this, a more detailed study about the best choice of parameter c_{add} values should be done instead of a simple trial and error strategy. As the only conclusion of our last tests we can say that if we increase slightly the value of the added viscosity the solution seems to be a little smother but not necessarily more precise. Furthermore,

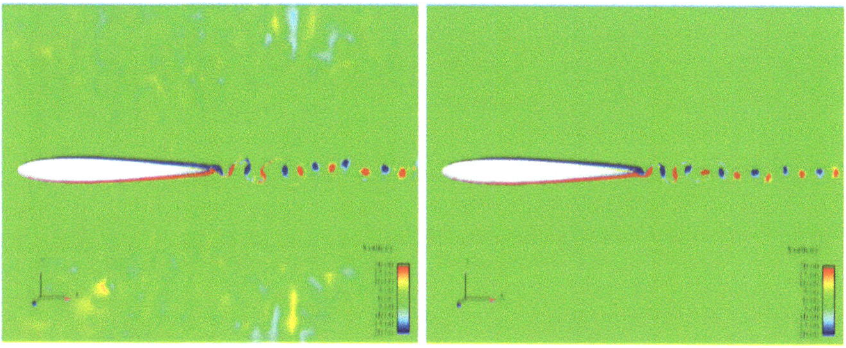

Fig. 7 Vorticity field at $t = 3s$ for LG method (*left*) and stabilized LG method with $c_{add} = 0.1$ (*right*)

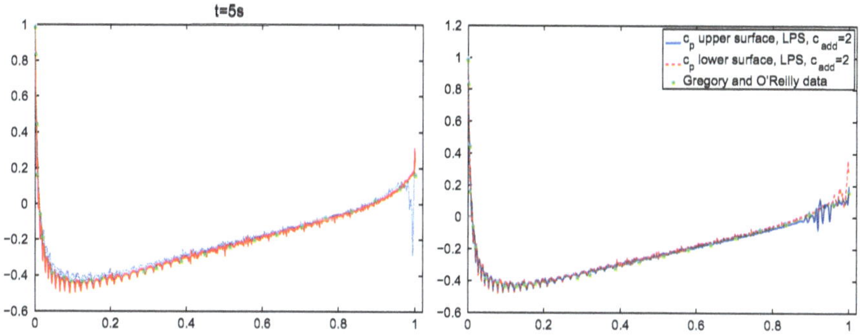

Fig. 8 Pressure coefficient on the upper and lower surfaces at $t = 5s$ with $c_{add} = 1$ (*left*) and $c_{add} = 2$ compared to the experimental results at $Re = 3 \times 10^6$

studies will be performed which will include a Smagorinsky-type turbulent viscosity ν_{add} following the works of John et al. [7, 8] (among others).

5 Conclusions

We have introduced a local projection BDF2-LG method for incompressible flows at high Reynolds numbers. We have shown that this method is easy to implement in a standard LG code. Numerical experiments show that, at very high Reynolds numbers, the method is more stable than the conventional LG method, and when both conventional and local projected stabilized LG methods are stable, the latter is more accurate. However, to maintain the stability of the method when Δt is small we still need to use high order quadrature rules together with the stabilization term. In future works, a more detailed study of the choice of the parameter c_{add} should be done, taking into account the different scales of the problem.

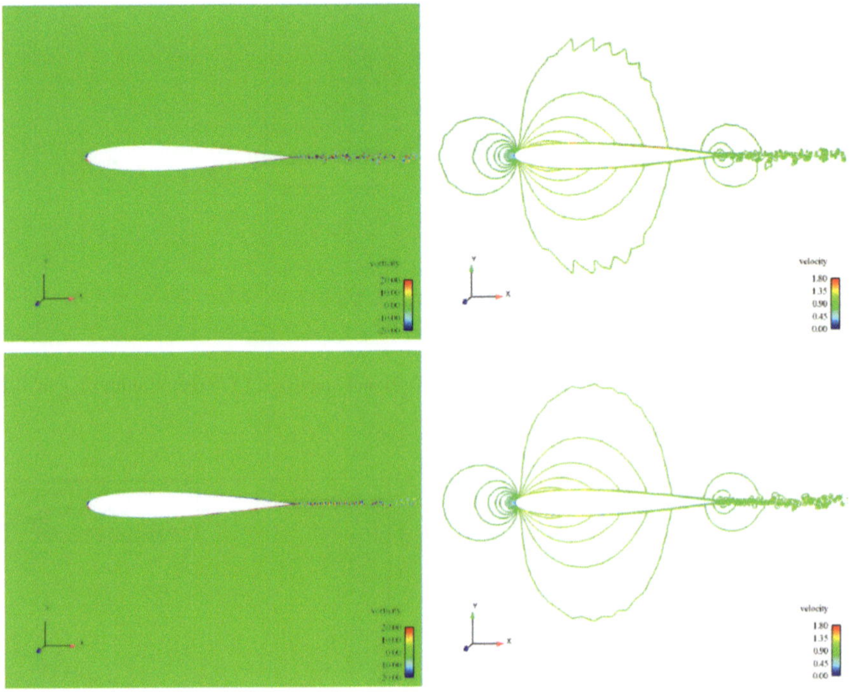

Fig. 9 Form *left to right*, vorticity field and contours of velocity at $t = 2.5s$ and from *top to bottom*: $c_{add} = 1$ and $c_{add} = 2$

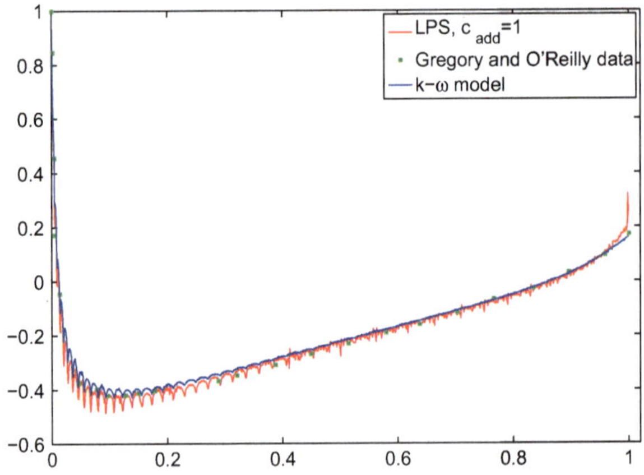

Fig. 10 c_p profiles at $t = 5s$ with the stabilized LG method and the $k - \omega$ model

References

1. Bermejo, R., Saavedra, L.: Modified Lagrange-Galerkin methods to integrate time dependent Navier-Stokes equations. SIAM J. Sci. Comput. (2014, submitted)
2. Bermejo, R., Galán del Sastre, P., Saavedra, L.: A second order in time modified Lagrange-Galerkin finite element method for the incompressible Navier-Stokes equations. SIAM J. Numer. Anal. **50**, 3084–3109 (2012)
3. Braack, M., Burman, E.: Local projection stabilization of the Oseen problem and its interpretation as a variational multiscale method. SIAM J. Numer. Anal. **43**, 2544–2566 (2006)
4. Ganesan, S., Tobiska, L.: Stabilization by local projection for convection–diffusion and incompressible flow problems. J. Sci. Comput. **43**, 326–342 (2010)
5. Gregory, N., O'Reilly, C.L.: Low-speed aerodynamic characteristics of NACA 0012 aerofoil sections including the effects of upper-surface roughness simulation hoar frost. NASA R&M 3726, Jan 1970
6. Guermond, J.-L., Marra, A., Quartapelle, L.: Subgrid stabilized projection method for 2D unsteady flows at high Reynolds numbers. Comput. Methods Appl. Mech. Eng. **195**, 5857–5876 (2006)
7. John, V., Roland, M.: Simulations of the turbulent channel flow at $Re_\tau = 180$ with projection-based finite element variational multiscale methods. Int. J. Numer. Meth. Fluids **55**, 407–429 (2007)
8. John, V., Kaya, S., Kindl, A.: Finite element error analysis for projection-based variational multiscale method with nonlinear eddy viscosity. J. Math. Anal. Appl. **344**, 627–641 (2008)
9. Morton, K.W., Priestley, A., Suli, E.: Stability of the Lagrange-Galerkin method with non-exact integration. M2AN Math. Model. Numer. Anal. **22**, 625–653 (1988)
10. Notsu, H., Tabata, M.: A single-step characteristic-curve finite element scheme of second order in time for the incompressible Navier-Stokes equations. J. Sci. Comput. **38**(1), 1–14 (2009)
11. Rumsey, C.: Langley Research Center. Turbulence Modeling Resource: http://turbmodels.larc.nasa.gov/index.html. Last updated: 09/2014

Finite Element Approximation of Hydrostatic Stokes Equations: Review and Tests

Francisco Guillén-González and J. Rafael Rodríguez-Galván

Abstract We present a review of a theory of stability and accuracy of Finite Element (FE) schemes for the Hydrostatic Stokes system which has been recently developed in Guillén-González and Rodríguez-Galván (Numer Math 130(2):225–256, 2015; SIAM J Numer Anal 53(4):1876–1896, 2015). Moreover, some new numerical results, not previously published, will be shown. This theory makes possible numerical simulations for classical FE (without the need of vertical integration required by most hydrostatic schemes in literature) and works even for anisotropic (not purely hydrostatic) models. The key is that stability of mixed approximation for Hydrostatic Stokes equations requires, besides the well-known Ladyzenskaja-Babuška-Brezzi (LBB) condition, an extra inf-sup condition. Some new numerical experiments are presented in this work. They suggest that for ($\mathscr{P}_1$ + bubble)–$\mathscr{P}_1$ one can reduce the number of degrees of freedom and also computational effort, without significantly worsening error orders. Some other unpublished numerical experiments are also presented here, in singular 2D domains and in realistic 3D domains (Gibraltar Strait).

1 Introduction

In this work we outline some formulations for the Hydrostatic Stokes equations (a linearized version of the Primitive Equations, were the Boussinesq approximation is taken into account and the Coriolis force is not considered) for which usual Stokes-stable Finite Elements (FE) are also Hydrostatic-stable. As result, Hydrostatic Stokes equations can be formulated as a mixed (Stokes-like) problem, which can be

F. Guillén-González
Dpto. Ecuaciones Diferenciales y Análisis Numérico and IMUS, Universidad de Sevilla, Aptdo. 1160, 41080 Sevilla, Spain
e-mail: guillen@us.es

J.R. Rodríguez-Galván (✉)
Dpto. de Matemáticas, Universidad de Cádiz,
Facultad de Ciencias, 11510, Puerto Real, Cádiz, Spain
e-mail: rafael.rodriguez@uca.es

F. Ortegón Gallego et al. (eds.), *Trends in Differential Equations and Applications*, SEMA SIMAI Springer Series 8, DOI 10.1007/978-3-319-32013-7_25

approximated by standard FE tools and for which vertical integrated formulations (commonly used in most schemes in Oceanography) are avoided.

The equations of geophysical fluid dynamics governing the motion of the ocean are derived from the conservation laws from physics. In the case of large scale ocean (see e.g. [6]), the resulting system is too complex and, from a practical point of view, numerous simplifications use to be introduced, including the "small layer" hypothesis:

$$\varepsilon = \frac{\text{vertical scale}}{\text{horizontal scale}} \quad \text{is very small,}$$

for example a few Kms over some thousand Kms, that is $\varepsilon \simeq 10^{-3}, 10^{-4}$.

Considering the Boussinesq approximation, we focus on the momentum laws, with yield to the Navier-Stokes (with homogeneous viscosity $\nu = \nu_x = \nu_y = \nu_z$ equations). On the other hand, applying the rigid lid hypothesis and a vertical scaling to the physical domain, it is transformed into the following isotropic or adimensional domain, for which the ratio (horizontal scales)/(vertical scales) is of the order of unity:

$$\Omega = \big\{(\mathbf{x}, z) \in \mathbb{R}^3 \;/\; \mathbf{x} = (x, y) \in S,\; -D(\mathbf{x}) < z < 0\big\}.$$

Note that vertical scaling of the domain implies a modification of the vertical momentum equation (see (2) below) due to which anisotropic viscosity ($\varepsilon^2\nu_z < \nu_x = \nu_y$) is introduced. For details, see e.g. [3, 4] and references therein.

We decompose the boundary of Ω into three parts: the surface, $\Gamma_s = \overline{S} \times \{0\}$, the bottom, $\Gamma_b = \{(\mathbf{x}, -D(\mathbf{x})) \;/\; \mathbf{x} = (x, y) \in S\}$, and the talus or lateral walls, $\Gamma_l = \{(\mathbf{x}, z) \;/\; \mathbf{x} \in \partial S, -D(\mathbf{x}) < z < 0\}$.

Finally, an ε-dependent scaling of vertical velocity is introduced (see [3]), leading to the following equations (called *Anisotropic* or *Quasi-Hydrostatic Navier-Stokes Equations* and, for the limit case $\epsilon = 0$, *Hydrostatic Navier-Stokes* or *Primitive Equations*) in the time-space domain $(0, T) \times \Omega$:

$$\partial_t \mathbf{u} + (\mathbf{u} \cdot \nabla_{\mathbf{x}})\mathbf{u} + v\partial_z \mathbf{u} - \Delta_\nu \mathbf{u} + \nabla_{\mathbf{x}} p = \mathbf{F}, \tag{1}$$

$$\varepsilon^2\big\{\partial_t v + (\mathbf{u} \cdot \nabla_{\mathbf{x}})v + v\partial_z v - \Delta_\nu v\big\} + \partial_z p = g, \tag{2}$$

$$\nabla_{\mathbf{x}} \cdot \mathbf{u} + \partial_z v = 0, \tag{3}$$

where $\nabla_{\mathbf{x}} = (\partial_x, \partial_y)^T$, $\nabla_{\mathbf{x}} \cdot \mathbf{u} = \partial_x u_1 + \partial_y u_2$, $\Delta_\nu = \nu_x\partial^2_{xx} + \nu_y\partial^2_{yy} + \nu_z\partial^2_{zz}$, being $\nu = (\nu_x, \nu_y, \nu_z)$ the (adimensional kinematic) viscosity. The unknowns are the 3D velocity field, $(\mathbf{u}, v) : \Omega \times (0, T) \to \mathbb{R}^3$ and the pressure, $p : \Omega \times (0, T) \to \mathbb{R}$. The term $\mathbf{F} = (f_1, f_2)^T$ models a given horizontal force while g involves vertical forces due to gravity. In this paper, we focus on the constant density case, therefore g can be written in potential form and incorporated into the pressure term, hence $g = 0$ can be assumed in (2). Anyway, it is important to note that, unlike most Primitive Equations schemes, we focus on the mixed problem without injecting vertical forces into the horizontal motion equation. Therefore variable density can be treated in a straightforward way. In a forthcoming paper, we deal into the general (transient, nonlinear, variable-density case). The effects due to Coriolis acceleration are not considered in this work because they are linear terms not affecting to the results presented below.

The system is endowed with initial values for the velocity field, $(\mathbf{u}, v)|_{t=0} = (\mathbf{u}_0, v_0)$ and adequate boundary conditions, for instance:

$$\nu_z \partial_z \mathbf{u}|_{\Gamma_s} = \mathbf{g}_s, \quad v|_{\Gamma_s} = 0, \tag{4}$$

$$\mathbf{u}|_{\Gamma_b \cup \Gamma_l} = 0, \quad v|_{\Gamma_b} = 0, \tag{5}$$

$$\varepsilon^2 \nabla_{\mathbf{x}} v \cdot \mathbf{n}_{\mathbf{x}}|_{\Gamma_l} = 0, \tag{6}$$

where $\mathbf{g}_s$ represents the wind stress and $\mathbf{n}_{\mathbf{x}}$ is the horizontal part of the normal vector.

The limit of the Hydrostatic equations (1)–(3) when $\varepsilon \to 0$ is studied on rigorous mathematical grounds in [4] (stationary case) and [3] (evolutive case). As far as we know, existence and regularity results for the *differential* problem (1)–(3), and also numerical schemes, are based on the introduction of an equivalent *integral-differential problem*, by doing a vertical integration. From the numerical point of view, this idea has advantages (it is only necessary to compute a $2D$ pressure) but also some drawbacks (for instance, standard FE in unstructured meshes, variable density and non-hydrostatic cases are difficult to handle).

In Sect. 2 we review the stability and accuracy results of the FE approximation for *differential* formulation (1)–(3), which have been recently obtained in [9]. The main difficulty lies on the strong anisotropy of these equations when ε is small, which affects their stability and invalidates its approximation by means of standard Stokes stable combinations of FE, such as Taylor-Hood $\mathscr{P}_2 - \mathscr{P}_1$, or the mini-element, $\mathscr{P}_{1,b} - \mathscr{P}_1$. The reason is that a new "hydrostatic" inf-sup condition, see $(IS)_h^V$ below, must be taken into account (in addition to the usual Stokes LBB inf-sup condition, see $(IS)_h^P$ below) which is not satisfied by standard Stokes FE like $\mathscr{P}_2 - \mathscr{P}_1$ and $\mathscr{P}_{1,b} - \mathscr{P}_1$. Then, we give some "non-standard" combinations of FE which are stable for (1)–(3).

In Sects. 3 and 4 we go into some *stabilized* reformulations of the discrete hydrostatic Stokes system related to (1)–(3) which are discussed in [10]. The first reformulation (in Sect. 3) avoids the restriction $(IS)_h^V$ and allows the use of standard Stokes FE combinations, like $\mathscr{P}_2 - \mathscr{P}_1$ (for which order $O(h^2)$ can be proved) or $\mathscr{P}_{1,b} - \mathscr{P}_1$ (of order $O(h)$). The second reformulation (Sect. 4) allows to control the pressure in a stronger norm (adding $\partial_z p_h \in L^2(\Omega)$). Order $O(h)$ is obtained for $\mathscr{P}_{1,b} - \mathscr{P}_1$ (including $\partial_z p$), although order $O(h^2)$ is not clear for $\mathscr{P}_2 - \mathscr{P}_1$.

Finally, in Sect. 5 we introduce some innovative numerical experiments. First and second ones show the power of stabilized reformulations in "critical" domains, with discontinuous depth or without sidewall talus. Third test compares error order for the most significant FE and formulations. Fourth one develops a new interesting idea starting from the non-standard FE $(\mathscr{P}_{1,b}, \mathscr{P}_1) - \mathscr{P}_1$ (presented in Sect. 2) based on eliminating bubbles from one component of the velocity. Then computation time is reduced while error orders are not significantly different. Last numerical tests exploits the facilities of our schemes for *3D* tests in domains which have been derived from real data.

2 Stability of Finite Elements Approximations for the Hydrostatic Equations

Let us consider the hydrostatic linear steady variational equations related to (1)–(3) for the less favorable case $\varepsilon = 0$: find $(\mathbf{u}, v, p) \in \mathbf{U} \times V \times P$ such that

$$\nu(\nabla \mathbf{u}, \nabla \overline{\mathbf{u}}) - (p, \nabla_{\mathbf{x}} \cdot \overline{\mathbf{u}}) = \langle \mathbf{f}, \overline{\mathbf{u}} \rangle \quad \forall \overline{\mathbf{u}} \in \mathbf{U}, \tag{7}$$

$$(p, \partial_z \overline{v}) = 0 \quad \forall \overline{v} \in V, \tag{8}$$

$$(\nabla \cdot (\mathbf{u}, v), \overline{p}) = 0 \quad \forall \overline{p} \in P, \tag{9}$$

where $(\cdot,\cdot)$ is the $L^2(\Omega)$ scalar product and $\langle \cdot,\cdot \rangle$ denotes duality in $\mathbf{U}'$, where

$$\mathbf{U} = \mathbf{H}^1_{b,l}(\Omega) = \Big\{ \mathbf{u} \in H^1(\Omega)^2 / \mathbf{u}|_{\Gamma_b \cup \Gamma_l} = 0 \Big\},$$

$$V = H^1_{z,0}(\Omega) = \Big\{ v \in L^2(\Omega) \;/\; \partial_z v \in L^2(\Omega),\; v|_{\Gamma_s \cup \Gamma_b} = 0 \Big\},$$

$$P = L^2_0(\Omega) = \Big\{ p \in L^2(\Omega) \;/\; \textstyle\int_\Omega p = 0 \Big\}.$$

$\mathbf{U}$ is equipped with the norm $\|\nabla \mathbf{u}\|$ (hereafter $\|\cdot\|$ denotes the $L^2(\Omega)$-norm) while in V we consider $\|\partial_z v\|$, which is a norm owing to homogeneous Dirichlet condition on $\Gamma_s \cup \Gamma_b$ and vertical Poincaré inequality. We take in P the usual $L^2(\Omega)$-norm. The function $\mathbf{f}$ in (7) results from gathering the horizontal force $\mathbf{F}$ and the Neumann boundary condition (4).

Considering $\mathbf{u}$ as the only "coercive variable" in (7)–(9) and v, p as Lagrange multipliers, the following inf-sup conditions are introduced:

$$\sup_{0 \neq (\mathbf{u},v) \in \mathbf{U} \times V} \frac{(\nabla \cdot (\mathbf{u}, v), p)}{\|(\nabla \mathbf{u},\ \partial_z v)\|} \geq \beta_p \|p\| \qquad \forall p \in P, \tag{IS^P

$$\sup_{0 \neq p \in P} \frac{(\partial_z v, p)}{\|p\|} \geq \beta_v \|\partial_z v\| \quad \forall v \in V, \tag{IS^V

where $\|(\nabla \mathbf{u},\ \partial_z v)\|$ is the usual norm of $\mathbf{U} \times V$.

It is not difficult to show that $(IS)^P$ and $(IS)^V$ hold if β_p is the Stokes LBB constant and $\beta_v = 1$ (specifically, the first condition follows from Stokes LBB condition while for the second one is enough to take $\widetilde{p} = \partial_z v$, for each $v \in V$).

Then we can achieve well-posedness of (7)–(9) by the following result:

Theorem 1 *The following statements are equivalent*

1. *$\mathbf{U}$, V and P satisfy $(IS)^P$ and $(IS)^V$.*
2. *Problem (7)–(9) is well-posed in $\mathbf{U} \times V \times P$.*

In this case, there exists a unique weak solution $(\mathbf{u}, v, p) \in \mathbf{U} \times V \times P$ of (7)–(9) and the following estimates hold:

$$\|\nabla \mathbf{u}\| \le \frac{1}{\nu}\|\mathbf{f}\|_{\mathbf{U}'}, \quad \|\partial_z v\| \le \frac{1}{\beta_v}\|\mathbf{f}\|_{\mathbf{U}'}, \quad \|p\| \le \frac{2}{\beta_p}\|\mathbf{f}\|_{\mathbf{U}'}, \tag{10}$$

where $\|\mathbf{f}\|_{\mathbf{U}'}$ denotes dual norm.

Theorem above has been proved in [9] (and previously in [2], using a similar approach). In [9], a different proof is also presented (it is based on the saddle point theory for mixed problems and is sharper in the sense of illustrating the role of $(IS)^V$).

In the discrete case, let $\mathcal{T}_h$ be a regular family of meshes in $\overline{\Omega}$ satisfying the usual regularity condition: there exists $\sigma > 1$ such that $h_T \le \sigma \rho_T$ for every $T \in \mathcal{T}_h$, where h_T is the diameter of the triangle T and ρ_T is the maximum diameter of all circles contained in T. Note that no kind of structure is assumed in $\mathcal{T}_h$ (in particular, vertical integration is not necessary).

Let $\mathbf{U}_h \subset \mathbf{U}$, $V_h \subset V$ and $P_h \subset P$ be conforming FE spaces and let us consider the following FE approximation of (7)–(9): find $(\mathbf{u}_h, v_h, p_h) \in \mathbf{U}_h \times V_h \times P_h$ such that

$$\nu(\nabla \mathbf{u}_h, \nabla \overline{\mathbf{u}}_h) - (p_h, \nabla_{\mathbf{x}} \cdot \overline{\mathbf{u}}_h) = (\mathbf{f}, \overline{\mathbf{u}}_h) \quad \forall \overline{\mathbf{u}}_h \in \mathbf{U}_h, \tag{11}$$

$$(p_h, \partial_z \overline{v}_h) = 0 \qquad \forall \overline{v}_h \in V_h, \tag{12}$$

$$(\nabla \cdot (\mathbf{u}_h, v_h), \overline{p}_h) = 0 \qquad \forall \overline{p}_h \in P_h. \tag{13}$$

Let us introduce the discrete inf-sup conditions

$$\sup_{0 \ne (\mathbf{u}_h, v_h) \in \mathbf{U}_h \times V_h} \frac{(\nabla \cdot (\mathbf{u}_h, v_h), p_h)}{\|\nabla \mathbf{u}_h, \partial_z v_h\|} \ge \gamma_p \|p_h\| \qquad \forall p_h \in P_h, \tag{$(IS)_h^P$}$$

$$\sup_{0 \ne p_h \in P_h} \frac{(p_h, \partial_z v_h)}{\|p_h\|} \ge \gamma_v \|\partial_z v_h\| \quad \forall v_h \in V_h, \tag{$(IS)_h^V$}$$

where $\gamma_p, \gamma_v > 0$. Note that $(IS)_h^P$ condition is similar to the Stokes LBB inf-sup condition and in fact, one can see that every Stokes stable FE combination satisfy $(IS)_h^P$. In [1, 2, 9], it is shown the following result:

Theorem 2 *The following statements are equivalent:*

1. $\mathbf{U}_h$, V_h *and* P_h *satisfy conditions* $(IS)_h^P$ *and* $(IS)_h^V$.
2. *Scheme (11)–(13) is well-posed.*

In this case, there exists a unique solution $(\mathbf{u}_h, v_h, p_h) \in \mathbf{U}_h \times V_h \times P_h$ *of (11)–(13) and the following estimates hold:*

$$\|\nabla \mathbf{u}_h\| \leq \frac{1}{\nu}\|\mathbf{f}\|_{\mathbf{U}'}, \quad \|\partial_z v_h\| \leq \frac{1}{\nu\gamma_v}\|\mathbf{f}\|_{\mathbf{U}'}, \quad \|p_h\| \leq \frac{2}{\gamma_p}\|\mathbf{f}\|_{\mathbf{U}'}. \tag{14}$$

The proof is a mere translation to the discrete case of the proof of Theorem 1. Then we can apply the Cea's lemma in the Banach-Necas-Babuska framework (see [9] and references therein) obtaining:

Corollary 1 (Error Estimates) *Let* $(\mathbf{u}, v, p)$ *and* $(\mathbf{u}_h, v_h, p_h)$ *be the solutions of (7)–(9) and (11)–(13) respectively. There is a constant* $C > 0$ *(depending on constants* γ_p *and* γ_v *of* $(IS)_h^P$ *and* $(IS)_h^V$*) such that:*

$$\|(\nabla(\mathbf{u}-\mathbf{u}_h), \partial_z(v-v_h), p-p_h)\|$$
$$\leq C\left(\inf_{\overline{\mathbf{u}}_h\in\mathbf{U}_h}\|\nabla(\mathbf{u}-\overline{\mathbf{u}}_h)\| + \inf_{\overline{v}_h\in V_h}\|\partial_z(v-\overline{v}_h)\| + \inf_{\overline{p}_h\in P_h}\|p-\overline{p}_h\|\right).$$

Finally, the following necessary conditions (which can be shown from an inspection of the FE linear system (11)–(13)) will be useful to analyze stability of specific FEs:

Lemma 1 *Let* $N_u = \dim \mathbf{U}_h$, $N_v = \dim V_h$ *and* $N_p = \dim P_h$.

1. *If* $(IS)_h^P$ *holds then* $N_p \leq N_u + N_v$.
2. *If* $(IS)_h^V$ *holds then* $N_v \leq N_p$.

Let $(\mathcal{P}_k, \mathcal{P}_l) - \mathcal{P}_m$ denote a FE approximation with $\mathcal{P}_k$, $\mathcal{P}_l$ and $\mathcal{P}_m$ elements for $\mathbf{U}_h$, V_h and P_h, respectively. Note that usual Stokes conforming Lagrange FE $\mathcal{P}_k - \mathcal{P}_m$ correspond to the case $k = l$, for instance Taylor-Hood $\mathcal{P}_2 - \mathcal{P}_1$ is also denoted as $(\mathcal{P}_2, \mathcal{P}_2) - \mathcal{P}_1$. We say that $(\mathcal{P}_k, \mathcal{P}_l) - \mathcal{P}_m$ is *hydrostatic stable* if the related FE spaces satisfy $(IS)_h^P$ and $(IS)_h^V$. Using the theory developed in this section we can conclude:

Corollary 2

1. $\mathcal{P}_{1,b} - \mathcal{P}_1$ *is not hydrostatic stable (it does not satisfy* $(IS)_h^V$*).*
2. $\mathcal{P}_2 - \mathcal{P}_1$ *is not hydrostatic stable.*
3. $(\mathcal{P}_2, \mathcal{P}_1) - \mathcal{P}_0$ *is hydrostatic stable.*

Proof A simple count of the degrees of freedom of $\mathcal{P}_{1,b} - \mathcal{P}_1$ and $\mathcal{P}_2 - \mathcal{P}_1$ in a simple structured mesh shows that N_v (number of dofs. for v) is greater that N_p, then (according to Lemma 1) they do not satisfy $(IS)_h^V$.

On the other hand, $(\mathscr{P}_2, \mathscr{P}_1) - \mathscr{P}_0$ satisfies Stokes LBB (see [11] and references therein), therefore $(IS)_h^P$ holds. And, for each $v_h \in V_h \sim \mathscr{P}_1$ it is easy to verify $(IS)_h^V$ (taking $\widetilde{p_h} = \partial_z v_h \in P_h \sim \mathscr{P}_0$ for each $v_h \in V_h$).

Remark 1 Two interesting FE are $(\mathscr{P}_{1,b}, \mathscr{P}_1) - \mathscr{P}_1$ and $(\mathscr{P}_2, \mathscr{P}_1) - \mathscr{P}_1$, which result eliminating degrees of freedom on V_h from the mini-element and the Taylor-Hood element, respectively.

In [11] is shown that these FE are Stokes-stable in uniformly unstructured meshes. Therefore $(IS)_h^P$ holds in those meshes. On the other hand, they satisfy the necessary conditions expressed in Lemma 1. Numerical tests in [9] suggest that $(IS)_h^V$ or a similar condition is satisfied in convex domains, but some new experiments given in Sect. 5.1 suggest that $(IS)_h^V$ does not hold in non-convex domains with discontinuous bottom function.

3 Stabilization of Vertical Velocity

Now we will try to recover, in the Hydrostatic case, the stability of classical Stokes FEs. Let us consider the following reformulation of (7)–(9): find $(\mathbf{u}, v, p) \in \mathbf{U}\times V\times P$ such that

$$\nu(\nabla\mathbf{u}, \nabla\overline{\mathbf{u}}) - (p, \nabla_{\mathbf{x}} \cdot \overline{\mathbf{u}}) = \langle \mathbf{f}, \overline{\mathbf{u}} \rangle \quad \forall \overline{\mathbf{u}} \in \mathbf{U}, \tag{15}$$

$$\nu(\nabla \cdot (\mathbf{u}, v), \partial_z \overline{v}) - (p, \partial_z \overline{v}) = 0 \quad \forall \overline{v} \in V, \tag{16}$$

$$(\nabla \cdot (\mathbf{u}, v), \overline{p}) = 0 \quad \forall \overline{p} \in P. \tag{17}$$

This new system is obtained by adding to (8) the *consistent* term $\nu(\nabla\cdot(\mathbf{u}, v), \partial_z \overline{v})$ (which vanishes in the continuous problem). Indeed, (9) or (17) imply $\nabla \cdot (\mathbf{u}, v) = 0$ almost everywhere in Ω, hence system (15)–(17) coincides with (7)–(9) and therefore, (15)–(17) is well-posed.

In the discrete case, let $\mathbf{U}_h \subset \mathbf{U}$, $V_h \subset V$ and $P_h \subset P$ be three conforming FE spaces. Let us consider the problem: find FE functions $\mathbf{u}_h \in \mathbf{U}_h$, $v_h \in V_h$ and $p_h \in P_h$ such that

$$\nu(\nabla\mathbf{u}_h, \nabla\overline{\mathbf{u}}_h) - (p_h, \nabla_{\mathbf{x}} \cdot \overline{\mathbf{u}}_h) = \langle \mathbf{f}, \overline{\mathbf{u}}_h \rangle, \tag{18}$$

$$\nu(\nabla \cdot (\mathbf{u}_h, v_h), \partial_z \overline{v}_h) - (p_h, \partial_z \overline{v}_h) = 0, \tag{19}$$

$$(\nabla \cdot (\mathbf{u}_h, v_h), \overline{p}_h) = 0, \tag{20}$$

for all $(\overline{\mathbf{u}}_h, \overline{v}_h, \overline{p}_h) \in \mathbf{U}_h\times V_h\times P_h$. This system can be seen as the result of introducing the stabilizing term $\nu(\nabla \cdot (\mathbf{u}_h, v_h), \partial_z \overline{v}_h)$ in (11)–(13). But the equivalence of both discrete systems cannot be guaranteed because, in this case, (20) cannot be applied to deduce that $\nu(\nabla \cdot (\mathbf{u}_h, v_h), \partial_z \overline{v}_h) = 0$, due to $\partial_z V_h \not\subset P_h$ in general.

Anyway, as it is going to be outlined in this section (more details are presented in [10]), $(IS)_h^P$ is a sufficient condition for the well-posedness of (18)–(20), i.e. the discrete hydrostatic stability constraint $(IS)_h^V$ is not necessary when (8) is reformulated as (19) (in consequence, any standard LBB-stable FE is stable for (18)–(20)). And both stability and error estimates can be provided for this scheme.

Towards this end, we make use of the theory for mixed FE (see e.g. [5]). Problem (15)–(17) can be written as: find $\mathbf{w} \in \mathbf{W}$ such that

$$a(\mathbf{w}, \overline{\mathbf{w}}) + b(p, \overline{\mathbf{w}}) = \langle \mathbf{f}, \overline{\mathbf{w}} \rangle \quad \forall \overline{\mathbf{w}} \in \mathbf{W}, \tag{21}$$

$$b(\overline{p}, \mathbf{w}) = 0 \qquad \forall \overline{p} \in P, \tag{22}$$

for the following bilinear and linear forms:

$$\begin{aligned} a(\mathbf{w}, \overline{\mathbf{w}}) &:= \nu(\nabla \mathbf{u}, \nabla \overline{\mathbf{u}}) + \nu(\partial_z v, \partial_z \overline{v}) + \nu(\nabla_{\mathbf{x}} \cdot \mathbf{u}, \partial_z \overline{v}), \\ b(p, \overline{\mathbf{w}}) &:= -(p, \nabla_{\mathbf{x}} \cdot \overline{\mathbf{u}}) - (p, \partial_z \overline{v}), \\ \langle \mathbf{f}, \overline{\mathbf{w}} \rangle &:= \langle (\mathbf{f}, 0), \overline{\mathbf{w}} \rangle_{\mathbf{W}', \mathbf{W}} = \langle \mathbf{f}, \overline{\mathbf{u}} \rangle_{\mathbf{U}', \mathbf{U}}, \end{aligned}$$

where we denote $\mathbf{W} = \mathbf{U} \times V$, $\mathbf{w} = (\mathbf{u}, v)$ and $\overline{\mathbf{w}} = (\overline{\mathbf{u}}, \overline{v}) \in \mathbf{W}$. Similarly, problem (18)–(20) can be written as:

$$a(\mathbf{w}_h, \overline{\mathbf{w}}_h) + b(p_h, \overline{\mathbf{w}}_h) = \langle \mathbf{f}, \overline{\mathbf{w}}_h \rangle \quad \forall \overline{\mathbf{w}}_h \in \mathbf{W}_h, \tag{23}$$

$$b(\overline{p}_h, \mathbf{w}_h) = 0 \qquad \forall \overline{p}_h \in P_h. \tag{24}$$

Let us denote by $B_h : \mathbf{W}_h \to P_h'$ and $B_h^t : P_h \to \mathbf{W}_h'$ the linear forms defined as $\langle B_h \mathbf{w}_h, p_h \rangle = b(\mathbf{w}_h, p_h) = \langle \mathbf{w}_h, B_h^t p_h \rangle$, for all $\mathbf{w}_h \in \mathbf{W}_h$, $p_h \in P_h$. From mixed methods theory, one has well posedness (existence and uniqueness of solution and stability estimates) of (23)–(24) if:

1. $a(\cdot, \cdot)$ is coercive on $\ker B_h$, i.e. exists $\alpha > 0$ such that

$$a(\mathbf{w}_h, \mathbf{w}_h) \geq \alpha \|\mathbf{w}_h\|_{\mathbf{W}}^2 \quad \forall \mathbf{w}_h \in \ker B_h. \tag{25}$$

2. And $b(\cdot, \cdot)$ satisfies an inf-sup condition, i.e. there exists $\beta > 0$ such that

$$\sup_{\mathbf{w}_h \in \mathbf{W}_h} \frac{b(\mathbf{w}_h, p_h)}{\|\mathbf{w}_h\|_{\mathbf{W}}} \geq \beta \|p_h\|_{P/\ker B_h^t} \quad \forall p_h \in P_h. \tag{26}$$

Using a technical result for controlling $\|\nabla_{\mathbf{x}} \cdot \mathbf{u}\|$ by means of $\|\nabla_{\mathbf{x}} \mathbf{u}\|$, one can achieve (25), while (26) is easy to show provided $(IS)_h^P$ holds. This way we can prove the following result (see[10] for details):

Theorem 3 (Stability) *Let $\mathbf{U}_h \subset \mathbf{U}$, $V_h \subset V$ and $P_h \subset P$ be families of FE in a regular partition $\mathcal{T}_h$ of Ω satisfying the inf-sup condition $(IS)_h^P$. Then scheme (18)–(20) has a unique solution $(\mathbf{u}_h, v_h, p_h) \in \mathbf{U}_h \times V_h \times P_h$, which satisfies the following* *stability estimates:*

$$\|\mathbf{u}_h\|^2 + \|\partial_z v_h\|^2 \le \frac{4}{\nu^2}\|\mathbf{f}\|^2_{\mathbf{U}'}, \quad \|p_h\| \le \frac{5}{\gamma_p}\|\mathbf{f}\|_{\mathbf{U}'}, \tag{27}$$

where γ_p is the constant in $(IS)_h^P$.

Error estimates from mixed FE theory (see [5, 10]) conduce also to the following result:

Theorem 4 (Error Estimates) *Under conditions of Theorem 3, let $(\mathbf{w}, p) = (\mathbf{u}, v, p)$ be the solution of problem (7)–(9) (or (15)–(17)) and let $(\mathbf{w}_h, p_h) = (\mathbf{u}_h, v_h, p_h)$ be the solution of scheme (18)–(20). Assume that there exists a positive constant $\gamma_p > 0$ satisfying $(IS)_h^P$. Then*

$$\|\mathbf{w} - \mathbf{w}_h\|_{\mathbf{W}} \le c_1 \inf_{\overline{\mathbf{w}}_h \in \mathbf{W}_h} \|\mathbf{w} - \overline{\mathbf{w}}_h\|_{\mathbf{W}} + c_2 \inf_{\overline{p}_h \in P_h} \|p - \overline{p}_h\|_P \tag{28}$$

$$\|p - p_h\|_P \le c_3 \inf_{\overline{\mathbf{w}}_h \in \mathbf{W}_h} \|\mathbf{w} - \overline{\mathbf{w}}_h\|_{\mathbf{W}} + c_4 \inf_{\overline{p}_h \in P_h} \|p - \overline{p}_h\|_P, \tag{29}$$

where $c_1, \ldots, c_4$ are constants depending on ν, γ_p.

Remark 2 The form $a(\cdot,\cdot)$ is not symmetric, duc to the stabilization term. A symmetric bilinear form can be defined introducing the consistent term $-\nu\nabla_{\mathbf{x}}(\nabla \cdot (\mathbf{u}, v))$ in the horizontal momentum equation (15), thus one has:

$$\hat{a}(\mathbf{w}, \overline{\mathbf{w}}) = \nu(\nabla\mathbf{u}, \nabla\overline{\mathbf{u}}) + \nu(\nabla_{\mathbf{x}} \cdot \mathbf{u}, \nabla_{\mathbf{x}} \cdot \overline{\mathbf{u}}) + \nu(\partial_z v, \nabla_{\mathbf{x}} \cdot \overline{\mathbf{u}}) + \nu(\partial_z v, \partial_z \overline{v}) + \nu(\nabla_{\mathbf{x}} \cdot \mathbf{u}, \partial_z \overline{v}).$$

It can be show that this bilinear form is coercive, hence we can get similar results to Theorems 3 and 4.

Remark 3 Following a standard reasoning in theory of finite elements (see [10] and references therein), error estimates (28), (29) conduce to convergence estimates. For instance, assuming that $\mathbf{u}_h$ and v_h are approximated in the same space $\mathcal{P}_r$ $(r \ge 2)$ and p_h is approximated in $\mathcal{P}_{r-1}$, one has:

$$\|(\mathbf{u} - \mathbf{u}_h)\| + \|\partial_z(v - v_h)\| + \|p - p_h\| \le Ch^r\Big(\|(\mathbf{u}, v)\|_{H^{r+1}(\Omega)^d} + \|p\|_{H^r(\Omega)}\Big).$$

In particular, order $O(h^2)$ is obtained in the $\mathcal{P}_2 - \mathcal{P}_1$ case if $(\mathbf{u}, v) \in H^3(\Omega)$ and $p \in H^2(\Omega)$. In a similar way, order $O(h)$ can also be obtained for $\mathcal{P}_{1,b} - \mathcal{P}_1$ if $(\mathbf{u}, v) \in H^2(\Omega)$ and $p \in H^1(\Omega)$. Our numerical results agree this statement, see Sect. 5.

Remark 4 The theory presented above can be extended to more general (not purely hydrostatic) problems, with $\varepsilon > 0$, so that more realistic problems from Oceanography can be modeled. Let us consider the following linear steady variational problem, where the stabilizing term $(\nabla \cdot (\mathbf{u}, v), \partial_z \overline{v})$ is introduced:

$$\nu(\nabla \mathbf{u}, \nabla \overline{\mathbf{u}}) - (p, \nabla_{\mathbf{x}} \cdot \overline{\mathbf{u}}) = \langle \mathbf{f}, \overline{\mathbf{u}} \rangle \quad \forall \overline{\mathbf{u}} \in \mathbf{U}, \tag{30}$$

$$\varepsilon^2(\nabla v, \nabla \overline{v}) + \nu(\nabla \cdot (\mathbf{u}, v), \partial_z \overline{v}) - (p, \partial_z \overline{v}) = 0 \quad \forall \overline{v} \in V, \tag{31}$$

$$(\nabla \cdot (\mathbf{u}, v), \overline{p}) = 0 \quad \forall \overline{p} \in P. \tag{32}$$

The discrete FE problem related to (30)–(32) can be set in the saddle point framework (21)–(22), by defining the bilinear form

$$a(\mathbf{w}, \overline{\mathbf{w}}) := \nu(\nabla \mathbf{u}, \nabla \overline{\mathbf{u}}) + \varepsilon^2(\nabla v, \nabla \overline{v}) + \nu(\partial_z v, \partial_z \overline{v}) + \nu(\nabla_{\mathbf{x}} \cdot \mathbf{u}, \partial_z \overline{v}).$$

Then, arguing like in previous paragraphs, Theorems 3 and 4 (and Remarks 2 and 3) can be extended to the non-hydrostatic problem (30)–(32).

4 Regularization of the Vertical Derivative of Pressure

The formulation (15)–(17) of the Hydrostatic Stokes equations (7)–(9) can be revised for obtaining accuracy rate also for the $L^2(\Omega)$-norm of $\partial_z p$. The idea is to introduce an additional consistent term to the stabilized problem (15)–(17).

Concretely, let us consider the pressure space

$$\widehat{P} = H_z^1(\Omega) \cap L_0^2(\Omega) = \{ p \in L_0^2(\Omega) \,/\, \partial_z p \in L^2(\Omega) \}, \tag{33}$$

endowed with the norm $\|p\|_{\widehat{P}}^2 = \|p\|^2 + \|\partial_z p\|^2$. Then, let us consider the following reformulation or (15)–(17): find $(\mathbf{u}, v) \in \mathbf{W} = \mathbf{U} \times V$ and $p \in \widehat{P}$ such that

$$\nu(\nabla \mathbf{u}, \nabla \overline{\mathbf{u}}) - (p, \nabla_{\mathbf{x}} \cdot \overline{\mathbf{u}}) = (\mathbf{f}, \overline{\mathbf{u}}), \quad \forall\, \overline{\mathbf{u}} \in \mathbf{U}, \tag{34}$$

$$\nu(\nabla \cdot (\mathbf{u}, v), \partial_z \overline{v}) - (p, \partial_z \overline{v}) = 0, \quad \forall\, \overline{v} \in V, \tag{35}$$

$$(\nabla \cdot (\mathbf{u}, v), \overline{p}) + (\partial_z p, \partial_z \overline{p}) = 0, \quad \forall\, \overline{p} \in \widehat{P}. \tag{36}$$

This system is obtained by adding to (17) the term $(\partial_z p, \partial_z \overline{p})$, which is consistent in the sense that it vanishes if p satisfies (8). Indeed, by a density argument, (8) imply $\partial_z p = 0$ almost everywhere in Ω, hence it is clear that the solution of (15)–(17) satisfies (34)–(36). Since we are going to prove uniqueness of solution of problem (34)–(36), both problems are equivalent. In particular the solution of (34)–(36) satisfies energy estimates. Anyway, a new estimate, involving also the $L^2(\Omega)$-norm of $\partial_z p$, is provided below.

In this case, the saddle-point approach used in Sect. 3 can not be reproduced, because it is not obvious how to obtain inf-sup conditions, similar to $(IS)_h^P$, involving the H_z^1-norm of the pressure space $\widehat{P}$ defined in (33). Therefore, a different approach is adopted, based on the Banach-Necas-Babuska Theorem (which can be interpreted as a generalized Lax-Milgram theorem, see for instance [7], Theorem 2.6): Let us define the following (non-symmetric) bilinear form on $\mathbf{X} \times \mathbf{X}$:

$$\mathscr{A}(\chi, \overline{\chi}) := \nu(\nabla \mathbf{u}, \nabla \overline{\mathbf{u}}) + \nu(\partial_z v, \partial_z \overline{v}) + (\partial_z p, \partial_z \overline{p}) \\ + \nu(\nabla_{\mathbf{x}} \cdot \mathbf{u}, \partial_z \overline{v}) - (p, \nabla \cdot (\overline{\mathbf{u}}, \overline{v})) + (\nabla \cdot (\mathbf{u}, v), \overline{p}),$$

where $\overline{\chi} = (\overline{\mathbf{w}}, \overline{p}) \in \mathbf{X}$, with $\overline{\mathbf{w}} = (\overline{\mathbf{u}}, \overline{v})$. Then problem (34)–(36) can be written as:

$$\text{Find } \chi \in \mathbf{X} \text{ such that } \quad \mathscr{A}(\chi, \overline{\chi}) = \langle F, \overline{\chi} \rangle_{\mathbf{X}', \mathbf{X}} \quad \forall \overline{\chi} \in \mathbf{X},$$

where $\langle \cdot, \cdot \rangle_{\mathbf{X}', \mathbf{X}}$, denotes the duality product and

$$\langle F, \overline{\chi} \rangle_{\mathbf{X}', \mathbf{X}} := \langle (\mathbf{f}, 0, 0), \overline{\chi} \rangle_{\mathbf{X}', \mathbf{X}} = \langle \mathbf{f}, \overline{\mathbf{u}} \rangle_{\mathbf{U}', \mathbf{U}}.$$

One can prove that $\mathscr{A}$ satisfies the inf-sup and coercivity hypothesis of the Banach-Necas-Babuska Theorem and therefore the following results can be stated (see[10] for details):

Theorem 5 *If $\mathbf{U}_h$, V_h and $\widehat{P}_h$ are FE spaces satisfying $(IS)_h^P$ with constant γ_p (independent of h), there exists a unique solution $(\mathbf{u}_h, v_h, p_h) \in \mathbf{U}_h \times V_h \times \widehat{P}_h$ of the discrete problem related to of (34)–(36), which satisfies the following a priori estimates:*

$$\|(\nabla \mathbf{u}_h,\ \partial_z v_h,\ p_h,\ \partial_z p_h)\| \leq \frac{1}{\tau} \|\mathbf{f}\|_{\mathbf{U}'},$$

where $\tau \in (0, 1/2]$ is a constant independent of h (in fact, τ only depends on ν and γ_p).

Theorem 6 *Assume that $(IS)_h^P$ holds with constant $\gamma_p > 0$ (independent of h). Then there is a constant $C > 0$ depending on γ_p (and independent of h) such that*

$$\|(\nabla(\mathbf{u} - \mathbf{u}_h),\ \partial_z(v - v_h),\ (p - p_h),\ \partial_z(p - p_h))\| \leq \\ C\Big(\inf_{\overline{\mathbf{u}}_h \in \mathbf{U}_h} \|\mathbf{u} - \overline{\mathbf{u}}_h\| + \inf_{\overline{v}_h \in V_h} \|\partial_z(v - \overline{v}_h)\| + \inf_{\overline{p}_h \in \widehat{P}_h} \|p - \overline{p}_h\|_{H_z^1} \Big).$$

Remark 5 For the $\partial_z p$ regularized scheme, order $O(h)$ is obtained, even for $\partial_z p$ in $L^2(\Omega)$, for the combination $\mathscr{P}_{1,b} - \mathscr{P}_1$ if $\mathbf{u}, v, p \in H^2(\Omega)$, what improves the results obtained in the non $\partial_z p$-regularized scheme. But order $O(h^2)$ cannot be reached for

$\mathscr{P}_2 - \mathscr{P}_1$ (because a best approximation, for instance $\mathscr{P}_2$, would be required for pressure).

Remark 6 In the anisotropic case (with $\varepsilon > 0$), a reformulation like (34)–(36) suffers an important modification, because Eq. (35) must be replaced by

$$(\nabla \cdot (\mathbf{u}, v), \overline{p}) + (\partial_z p - \varepsilon^2 \Delta v, \partial_z \overline{p}) = 0, \quad \forall \overline{p} \in \widehat{P},$$

whose treatment is not straightforward.

5 Numerical Simulations

Here we present numerical experiments with a triple objective: First, to exploit the flexibility of the schemes for approximations in domains with singularity in the bottom (Sect. 5.1) or vanishing sidewall (Sect. 5.2). Second, in Sect. 5.3 we introduce a set of experiments to compare the stabilized and unstabilized schemes presented in previous sections, reaching conclusions which, so far, had not been observed. Specifically, we show that stabilized $(\mathscr{P}_{1,b}, \mathscr{P}_1) - \mathscr{P}_1$ FE is faster than stabilized $\mathscr{P}_{1,b} - \mathscr{P}_1$, while error orders are comparable. Finally, in Sect. 5.5 we present a realistic $3D$ numerical simulation (in the Gibraltar Strait) which, had never been published in papers.

5.1 Experiments in a Domain with Discontinuous Bottom

We consider a non convex $2D$ domain which presents, as particularity, a bottom step producing a discontinuity point in the depth function $D(x)$. More specifically, given the surface interval $S = [0, 1]$ we define $D(x) = 0.5$ in $[0, 1/2)$ while $D(x) = 1$ in $(1/2, 0]$.

The standard FE software *FreeFem++* [8] was employed. The domain is discretized using a non-structured mesh which is defined, using FreeFem++ meshing capabilities, by 50 sub-intervals on the surface boundary, $\overline{S}$, and also 50 sub-intervals on the right, left and bottom boundaries. After computing the solution (u_h, v_h, p_h), the mesh is refined using FreeFem++ `adaptmesh` function. As as mesh adapting indicator, we pass to this function the following data: $u_h/\|u_h\|_\infty + v_h/\|v_h\|_\infty + p_h/\|p_h\|_\infty$. Then the solution is recomputed in the refined mesh.

In this test we compare the raw (with no stabilization) Hydrostatic Stokes scheme (11)–(13) with the v-stabilized (and non $\partial_z p$-regularized) formulation (18)–(20). Note that these theoretical formulations are focused on the less favorable case $\varepsilon = 0$ (i.e. vanishing vertical viscosity) while horizontal viscosity is $\nu = 1$. In a forthcoming paper, we deal into more realistic cases, from the point of view of selection of viscosity, variable density, and other parameters.

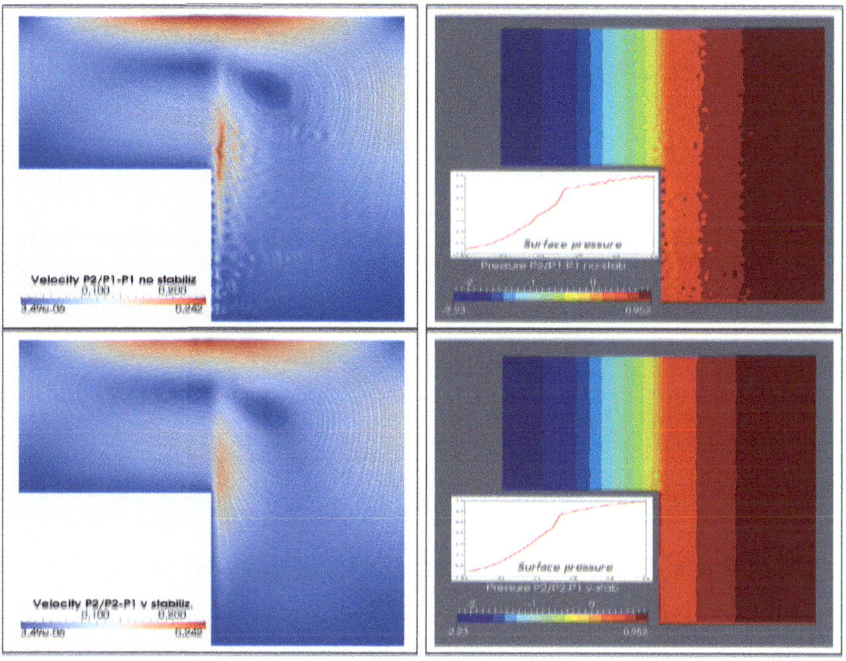

Fig. 1 Velocity streamlines *(left)* and pressure *(right)*. *Top*: non v-stabilized (and non p-regularized) scheme. *Bottom*: v-stabilized (and non p-regularized) scheme

For the first scheme, we used $(\mathscr{P}_2, \mathscr{P}_1) - \mathscr{P}_1$ FE, which satisfy $(IS)_h^P$ in most unstructured meshes. Previous experiments presented in [9] (but just in domains where D is continuous) suggested that $(IS)_h^V$ also holds in those domains. But now, velocity streamlines and pressure, presented in Fig. 1 (top), show some instabilities as consequence of the stress produced by the discontinuity in the bottom.

For the second scheme, v-stabilization makes possible the use of $\mathscr{P}_2 - \mathscr{P}_1$ FE and spurious oscillations are damped. Results are slightly better for the p-regularized formulation (34)–(35) although for sake of brevity they are not shown here.

5.2 *Cavity Test in a Convex Domain Without Sidewall Talus*

Now we consider a $2D$ convex domain with no sidewall talus, namely $\Gamma_l = \emptyset$. Note that this kind of tests are not easy to develop for usual integro-differential formulations of the primitive equations, where normally the imposition of a talus ($D(x) > D_{min} > 0$) is required.

Specifically, the domain is defined by the surface interval $\overline{S} = [0, 1]$ and the depth function $D(x) = (x - 1/2)^2 + 1/4$ (vanishing on $x = 0$ and $x = 1$). We fix $u = x(1 - x)$ and $v = 0$ on Γ_s, while u and v vanish on $\Gamma_l \cup \Gamma_b$. For mesh adaptivity we

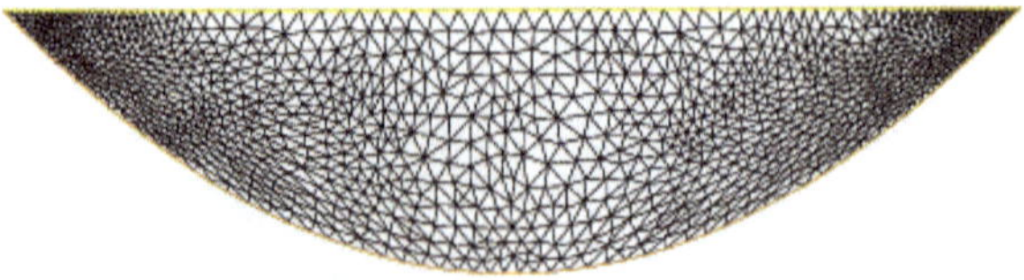

Fig. 2 Mesh in a domain with no sidewall talus

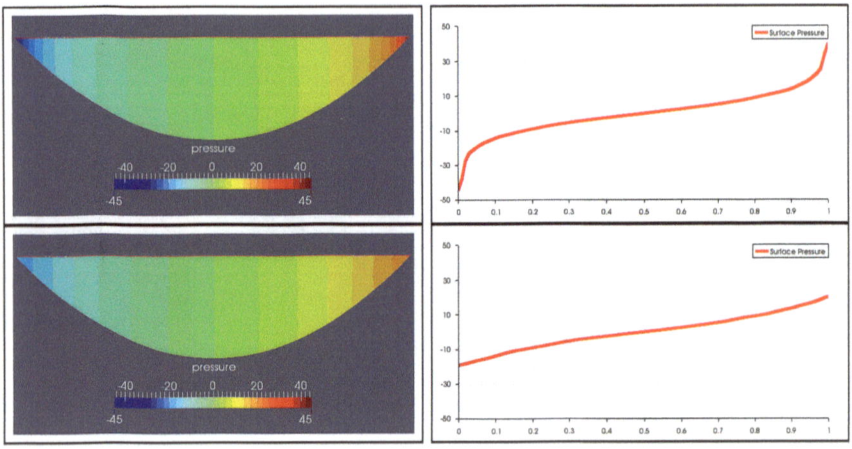

Fig. 3 Pressure in a convex domain without talus, $\mathscr{P}_2 - \mathscr{P}_1$ FE. *Top to bottom*: pressure for the v-stabilized scheme and pressure for the $\partial_z p$-regularized scheme. *Left*: pressure iso-values. *Right*: pressure on surface

used the indicator commented in Sect. 5.1. Figure 2 shows the resulting mesh, after one solving+adaptation step, starting from a triangulation with $n = 50$ subintervals on Γ_s and Γ_b.

Figure 3 shows pressure isolines for the v-stabilized scheme (18)–(20) and for the $\partial_z p$-regularized related to (34)–(36). Results show a quite different behaviour for both schemes: maximum absolute values are different (≈ 40 and ≈ 20, respectively), and pressure isolines are almost uniformly distributed in the $\partial_z p$-regularized case, while for the v-stabilized scheme they are accumulated in corners $x = 0$ and $x = 1$, suggesting a singularity where $D(x) = 0$.

Velocity streamlines are not shown for current test because they present a standard behaviour for both schemes (with an identical maximum velocity magnitude, equal to 0.2253936).

5.3 Error Orders

With the aim of comparing numerically error orders for unstabilized and stabilized schemes, we have considered in the unit square domain in $\mathbb{R}^2$ the following velocity and pressure functions, which constitute an exact solution of the Hydrostatic Problem (7)–(9):

$$\mathbf{u}(x,y) = \cos(2\pi x)\sin(2\pi y) - \sin(2\pi y), \quad v(x,y) = -\mathbf{u}(y,x),$$
$$p(x,y) = 2\pi\cos(2\pi x).$$

The solution for the unstabilized scheme (11)–(13) has been approximated for different mesh sizes and norms, using both $(\mathscr{P}_{1,b},\mathscr{P}_1)-\mathscr{P}_1$ and $(\mathscr{P}_2,\mathscr{P}_1)-\mathscr{P}_1$ (which satisfy $(IS)_h^P$ but $(IS)_h^V$ is not clear). Also the v-stabilized scheme (18)–(20) was used with $\mathscr{P}_{1,b}-\mathscr{P}_1$ and $\mathscr{P}_2-\mathscr{P}_1$

Table 1 compares the results obtained for $(\mathscr{P}_{1,b},\mathscr{P}_1)-\mathscr{P}_1$ and stabilized $\mathscr{P}_{1,b}-\mathscr{P}_1$. We have similar orders for $\mathbf{u}$ (optimal error order $O(h^2)$ in L^2 and $O(h)$ in H^1). Orders for v are also comparable in L^2-norm (over $O(h^{3/2})$) although better for $\mathscr{P}_{1,b}-\mathscr{P}_1$. Only in H_z^1-norm the order is clearly poorer for $(\mathscr{P}_{1,b},\mathscr{P}_1)-\mathscr{P}_1$ (about $O(h^{1/2})$ is suggested). Orders for pressure in L^2 are similar for the two cases (about order $O(h^{3/2})$).

Table 2 compares the results obtained for $(\mathscr{P}_2,\mathscr{P}_1)-\mathscr{P}_1$ and $\mathscr{P}_2-\mathscr{P}_1$. In this case, optimal orders for $\mathbf{u}$ ($O(h^3)$ in L^2 and $O(h^2)$ in H^1 norms) arise for stabilized $\mathscr{P}_2-\mathscr{P}_1$ but only $O(h^{3/2})$ and $O(h^{1/2})$ are reached for $(\mathscr{P}_2,\mathscr{P}_1)-\mathscr{P}_1$. Orders for v are better for stabilized $\mathscr{P}_2-\mathscr{P}_1$, specially in H_z^1-norm (where surprisingly $O(h^2)$ is obtained) and also they are better for pressure.

In brief, stabilized $\mathscr{P}_{1,b}-\mathscr{P}_1$ and $\mathscr{P}_2-\mathscr{P}_1$ elements reach (and even surpass, in some cases) optimal order and seem to be the best choice for solving the Hydrostatic Stokes equations. Although, nonstabilized $(\mathscr{P}_{1,b},\mathscr{P}_1)-\mathscr{P}_1$ FE does not seem a bad choice: it reaches also optimal order for horizontal velocity and pressure (but not for vertical velocity in energy norm).

Table 1 Error orders for $(\mathscr{P}_{1,b},\mathscr{P}_1)-\mathscr{P}_1$ and v-stabilized $\mathscr{P}_{1,b}-\mathscr{P}_1$

		$(\mathscr{P}_{1,b},\mathscr{P}_1)-\mathscr{P}_1$					$\mathscr{P}_{1,b}-\mathscr{P}_1$				
h		2^{-3}	2^{-4}	2^{-5}	2^{-6}	2^{-7}	2^{-3}	2^{-4}	2^{-5}	2^{-6}	2^{-7}
$\mathbf{u}$	L^2	2.442	1.998	2.187	1.945	2.044	1.616	1.908	1.982	1.999	2.002
	H_0^1	1.229	1.001	1.102	0.976	1.020	0.936	1.001	1.004	1.002	1.001
v	L^2	1.752	1.688	1.629	1.510	1.578	1.591	1.775	1.867	1.888	1.857
	$H_{z,0}^1$	0.463	1.117	0.490	0.327	0.508	0.830	0.947	0.992	1.000	1.001
p	L^2	1.702	1.423	1.767	1.903	1.643	1.625	1.812	1.733	1.641	1.579

Table 2 Error orders for $(\mathscr{P}_2, \mathscr{P}_1) - \mathscr{P}_1$ and v-stabilized $\mathscr{P}_2 - \mathscr{P}_1$

		$(\mathscr{P}_2, \mathscr{P}_1) - \mathscr{P}_1$					$\mathscr{P}_2 - \mathscr{P}_1$				
h		2^{-3}	2^{-4}	2^{-5}	2^{-6}	2^{-7}	2^{-3}	2^{-4}	2^{-5}	2^{-6}	2^{-7}
u	L^2	2.912	2.803	2.369	2.458	2.561	3.129	3.047	3.029	3.018	3.010
	H_0^1	2.086	1.872	1.605	1.676	1.697	1.967	1.987	1.999	2.001	2.001
v	L^2	1.706	2.222	0.687	1.686	1.617	1.572	1.887	1.972	1.993	1.998
	$H_{z,0}^1$	0.641	1.240	−0.062	0.754	0.593	1.846	1.899	1.964	1.989	1.997
p	L^2	2.370	1.790	1.480	1.702	1.687	2.532	2.267	2.109	2.042	2.017

5.4 *Comparison of Computation Times*

Although numerical test Sect. 5.3 confirms that error order for stabilized $\mathscr{P}_{1,b} - \mathscr{P}_1$ and $\mathscr{P}_2 - \mathscr{P}_1$ beat, in general, the nonstabilized $(\mathscr{P}_{1,b}, \mathscr{P}_1) - \mathscr{P}_1$ and $(\mathscr{P}_2, \mathscr{P}_1) - \mathscr{P}_1$ ones. We must not forget that this latter schemes conduce to systems with fewer number of unknowns and, maybe, to less computational effort (unless unstabilized Hydrostatic schemes lead to ill conditioned systems).

So here we intend to check if the reduction of unknowns makes $(\mathscr{P}_{1,b}, \mathscr{P}_1) - \mathscr{P}_1$ and $(\mathscr{P}_2, \mathscr{P}_1) - \mathscr{P}_1$ to be faster (from the point of view of reduction of the computation time) than $\mathscr{P}_{1,b} - \mathscr{P}_1$ and $\mathscr{P}_2 - \mathscr{P}_1$ FEs on stabilized schemes.

With this in mind, for each one of those FEs we measure the CPU times for standard cavity tests with decreasing mesh size (and then, increasing number of unknowns). More in detail, for each FE, we have run four cavity tests in the unity square, taking unstructured meshes defined by $n_1 = 30$, $n_2 = 60$, $n_3 = 90$ and $n_4 = 120$ subintervals on each one of its four edges.

Results are shown in Fig. 4, where we use the notation `KLM` to denote nonstabilized $(\mathscr{P}_K, \mathscr{P}_L) - \mathscr{P}_M$ and `KLMs` to denote *stabilized* $(\mathscr{P}_K, \mathscr{P}_L) - \mathscr{P}_M$ spaces. For instance, `221s` denotes *stabilized* $(\mathscr{P}_2, \mathscr{P}_2) - \mathscr{P}_1$ (i.e $\mathscr{P}_2 - \mathscr{P}_1$) FE.

Note that stabilized $\mathscr{P}_{1,b} - \mathscr{P}_1$ and $\mathscr{P}_2 - \mathscr{P}_1$ require less computational effort than nonstabilized $(\mathscr{P}_{1,b}, \mathscr{P}_1) - \mathscr{P}_1$ and $(\mathscr{P}_2, \mathscr{P}_1) - \mathscr{P}_1$ FEs, presumably because of better conditioned linear systems.

Note also that we introduced $(\mathscr{P}_{1,b}, \mathscr{P}_1) - \mathscr{P}_1$ and $(\mathscr{P}_2, \mathscr{P}_1) - \mathscr{P}_1$ in the stabilized scheme, getting interesting conclusions:

- Stabilized $(\mathscr{P}_2, \mathscr{P}_1) - \mathscr{P}_1$ FE is faster than $\mathscr{P}_2 - \mathscr{P}_1$. The drawback is that (according to test Sect. 5.3) $(\mathscr{P}_2, \mathscr{P}_1) - \mathscr{P}_1$ conduce to worse error orders and one cannot hope to improve them by stabilizing.
- Stabilized $(\mathscr{P}_{1,b}, \mathscr{P}_1) - \mathscr{P}_1$ FE is faster than $\mathscr{P}_{1,b} - \mathscr{P}_1$. And here we have really an *interesting result* because error orders for $(\mathscr{P}_{1,b}, \mathscr{P}_1) - \mathscr{P}_1$ are similar to $\mathscr{P}_{1,b} - \mathscr{P}_1$ orders, according to test Sect. 5.3.

In conclusion, from the point of view of computational time (and error orders), $(\mathscr{P}_{1,b}, \mathscr{P}_1) - \mathscr{P}_1$ and $(\mathscr{P}_2, \mathscr{P}_1) - \mathscr{P}_1$ are worse than stabilized $\mathscr{P}_{1,b} - \mathscr{P}_1$ and $\mathscr{P}_2 - \mathscr{P}_1$, but if we stabilize $(\mathscr{P}_{1,b}, \mathscr{P}_1) - \mathscr{P}_1$ we reach lower CPU time with orders which are similar to stabilized $\mathscr{P}_{1,b} - \mathscr{P}_1$.

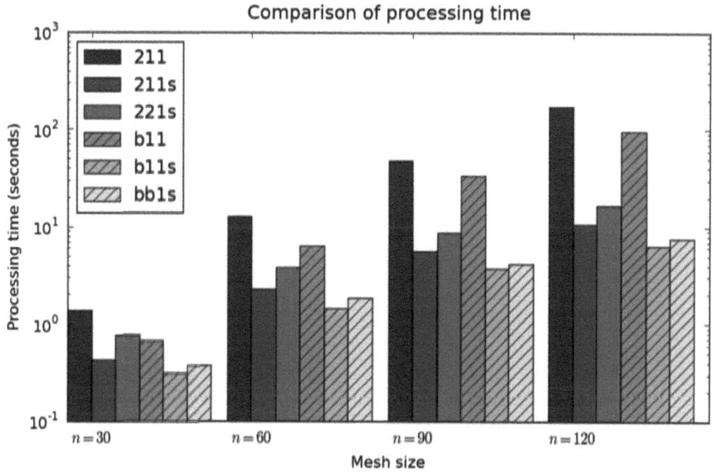

Fig. 4 Evolution of CPU times for different FEs when the number of dof. increases

5.5 *Realistic 3D Test in the Gibraltar Strait*

We have exploited the facilities of schemes presented above for $3D$ tests in unstructured meshes, automatically created in domains which have been defined by real data.

Specifically, a $3D$ mesh (using the *GMSH* format) of the Gibraltar strait has been constructed (using a specific Python script) from data which is available with free license: coast lines were obtained from a map dataset available in naturalearthdata.com and bathymetry data was downloaded from the U.S.A. National Geophysical Data Center (ETOPO2v2).

Then FreeFem++[8] has been employed for programming the 3D $\mathscr{P}_2 - \mathscr{P}_1$ v-stabilized Hydrostatic Stokes scheme (18)–(20), where (besides rigid lid $v = 0$ on surface) non-slip Dirichlet conditions have been imposed for $\mathbf{u}$ and v on the bottom, including both European and African coast boundaries. On the remaining boundaries, Neumann boundary conditions have been defined, specifically wind traction for $\mathbf{u}$ on Γ_s ($\nu \partial_z \mathbf{u} = \mathbf{g}_s$), where $\mathbf{g}_s = 1$, and null flux $\nabla(\mathbf{u}, v) \cdot \mathbf{n} = 0$ on the east (Mediterranean) and west (Atlantic) artificial boundaries. We chose, as in previous tests, horizontal viscosity $\nu = 1$ and, now, we selected $\epsilon = 10^{-4}$ (so that vertical viscosity is $\epsilon^2 = 10^{-8}$).

Figure 5 shows velocity streamlines and pressure. Hydrostatic restriction $\partial_z p = 0$ is satisfactory approximated (note that, due to computational efficiency, mesh size is not small). Extreme pressure values are concentrated in coast regions where depth is extremely small, what suggest the convenience of improving the meshing process in these regions.

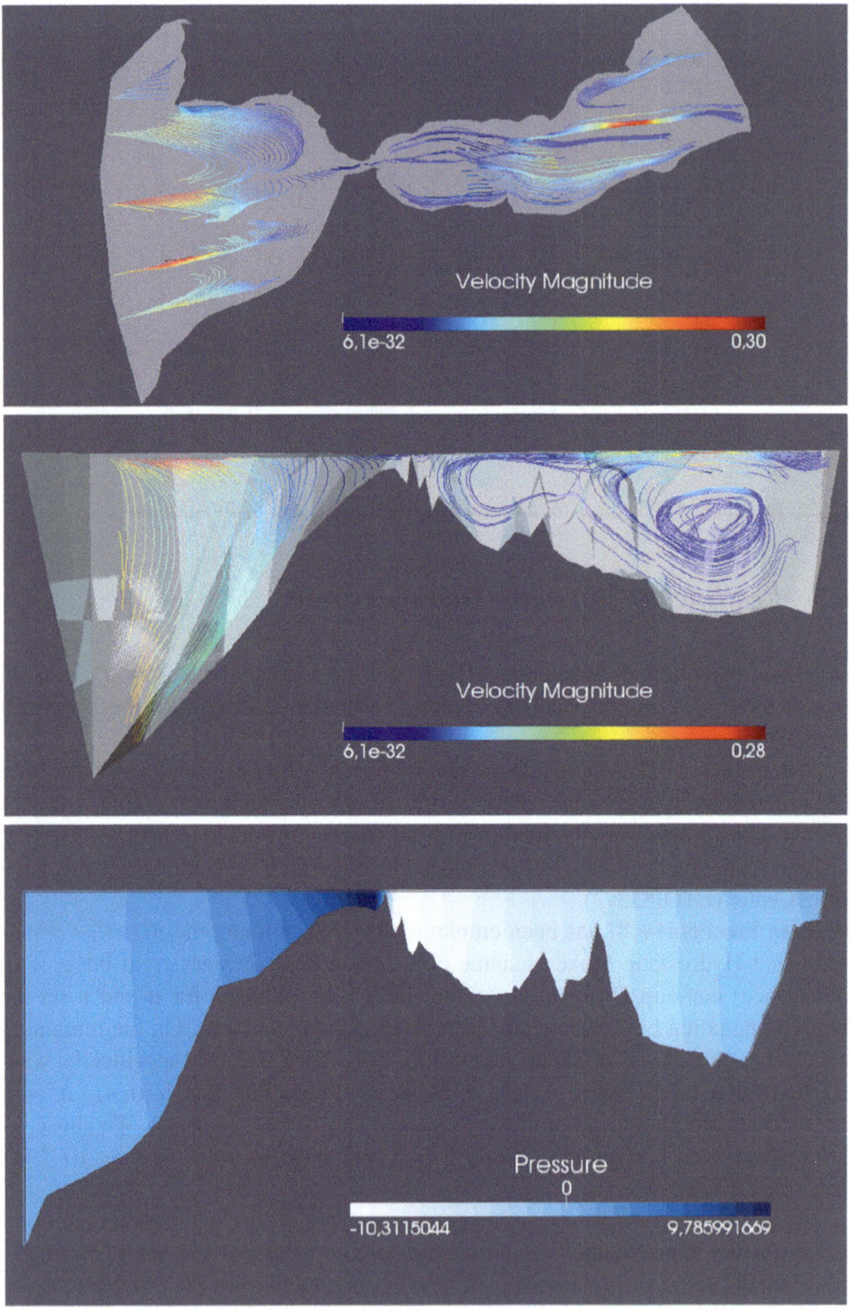

Fig. 5 3D Gibraltar strait test. *Top to bottom*: stream lines ((x, y) and (x, z) planes) and pressure ((x, z) plane)

Acknowledgements The work was partially supported by MINECO grant MTM2012-32325 with the participation of FEDER. We would like to thank the referees for providing useful comments which served to improve the paper.

References

1. Azérad, P.: Analyse et approximation du problème de Stokes dans un bassin peu profond. C. R. Acad. Sci. Paris Sér. I Math. **318**(1), 53–58 (1994)
2. Azérad, P.: Analyse des équations de Navier-Stokes en bassin peu profond et de l'équation de transport. Ph.D. thesis, Neuchâtel (1996)
3. Azérad, P., Guillén, F.: Mathematical justification of the hydrostatic approximation in the primitive equations of geophysical fluid dynamics. SIAM J. Math. Anal. **33**(4), 847–859 (2001)
4. Besson, O., Laydi, M.R.: Some estimates for the anisotropic Navier-Stokes equations and for the hydrostatic approximation. Math. Model. Numer. Anal. **26**(7), 855–865 (1992)
5. Brezzi, F., Fortin, M.: Mixed and Hybrid Finite Element Methods. Springer, New York (1991)
6. Cushman-Roisin, B., Beckers, J.M.: Introduction to Geophysical Fluid Dynamics - Physical and Numerical Aspects. Academic Press, San Diego (2009)
7. Ern, A., Guermond, J.-L.: Theory and Practice of Finite Elements. Springer, New York (2004)
8. Hecht, F.: New development in FreeFem++. J. Numer. Math. **20**(3–4), 251–265(2012)
9. Guillén-González, F., Rodríguez-Galván, J.R.: Analysis of the hydrostatic stokes problem and finite-element approximation in unstructured meshes. Numer. Math. **130**(2), 225–256 (2015)
10. Guillén-González, F., Rodríguez-Galván, J.R.: Stabilized schemes for the hydrostatic Stokes equations. SIAM J. Numer. Anal. **53**(4), 1876–1896 (2015)
11. Guillén-González, F., Rodríguez-Galván, J.R.: On the stability of approximations for the Stokes problem using different finite element spaces for each component of the velocity. Appl. Numer. Math. **99**, 51–76 (2016)

MIX
Papier aus verantwortungsvollen Quellen
Paper from responsible sources
FSC® C105338

If you have any concerns about our products,
you can contact us on
ProductSafety@springernature.com

In case Publisher is established outside the EU,
the EU authorized representative is:
Springer Nature Customer Service Center GmbH
Europaplatz 3, 69115 Heidelberg, Germany

Printed by Libri Plureos GmbH
in Hamburg, Germany